Lecture Notes in Computer Science 13174

More information about this subseries at https://link.springer.com/bookseries/7407

Petra Mutzel · Md. Saidur Rahman ·
Slamin (Eds.)

WALCOM: Algorithms and Computation

16th International Conference and Workshops, WALCOM 2022
Jember, Indonesia, March 24–26, 2022
Proceedings

 Springer

Editors
Petra Mutzel 🄳
University of Bonn
Bonn, Germany

Slamin 🄳
Universitas Jember
Jember, Indonesia

Md. Saidur Rahman 🄳
Bangladesh University of Engineering
and Technology (BUET)
Dhaka, Bangladesh

ISSN 0302-9743   ISSN 1611-3349 (electronic)
Lecture Notes in Computer Science
ISBN 978-3-030-96730-7   ISBN 978-3-030-96731-4 (eBook)
https://doi.org/10.1007/978-3-030-96731-4

LNCS Sublibrary: SL1 – Theoretical Computer Science and General Issues

This Springer imprint is published by the registered company Springer Nature Switzerland AG
The registered company address is: Gewerbestrasse 11, 6330 Cham, Switzerland

Preface

WALCOM 2022, the 16th International Conference and Workshops on Algorithms and Computation, was held during March 24-26, 2022 at the Universitas Jember, Jember, East Java, Indonesia. The workshop covered diverse areas of algorithms and computation, namely, approximation algorithms, computational complexity, computational geometry, graph algorithms, graph drawing, visualization, online algorithms, parameterized complexity and property testing. The quality of the workshop was ensured by a Program Committee comprising 32 researchers of international reputation from Australia, Bangladesh, Canada, China, Czech Republic, France, Germany, Hong Kong, India, Indonesia, Italy, Japan, the Netherlands, Norway, Singapore, Taiwan and USA.

This proceedings volume contains 30 contributed papers and three invited papers presented at WALCOM 2022. The Program Committee thoroughly reviewed each of the 89 submissions from 29 countries and accepted 30 of them for presentation at the conference after elaborate discussions on 286 review reports prepared by Program Committee members together with 103 external reviewers. Among the highest scored contributed papers, the Program Committee selected the paper "Parameterized Complexity of Immunization in the Threshold Model" as the best paper and the paper "Reverse Shortest Path Problem in Weighted Unit-Disk Graphs" as the best student paper. The image of the workshop was highly enhanced by the three invited talks of eminent and well-known researchers Prof. Hans L. Bodlaender of Utrecht University and Technical University Eindhoven, The Netherlands, Prof. Tiziana Calamoneri of University of Rome "Sapienza", Italy and Prof. Takehiro Ito of Tohoku University, Japan.

As editors of this proceedings, we would like to thank all the authors who submitted their papers to WALCOM 2022. We also thank the members of the Program Committee and external reviewers for their hard work in reviewing the manuscripts. Our sincere appreciation goes to the invited speakers for delivering wonderful talks from which researchers of this field have been benefited immensely. We acknowledge the continuous encouragements of the advisory board members Prof. M. Kaykobad, Prof. Takao Nishizeki and Prof. C. Pandu Rangan. The Steering Committee members of WALCOM always supported us with their valuable suggestions. We sincerely thank the honorable rector of Universitas Jember Dr. Ir. Iwan Taruna for his all around support for organizing WALCOM 2022. We sincerely thank the Organizing Committee led by Prof. Slamin, Prof. Antonius C. Prihandoko and Prof. Dafik for their excellent services that made the workshop a grand success.

We would like to thank Springer for publishing this proceedings in their prestigious LNCS series and also for supporting the best paper award and the best student paper award. Finally, we acknowledge the EasyChair conference management system for providing a beautiful platform for conference administration.

March 2022   Petra Mutzel
                                                                          Md. Saidur Rahman
                                                                                  Slamin

WALCOM Organization

WALCOM Steering Committee

Tamal Dey	Purdue University, USA
Seok-Hee Hong	University of Sydney, Australia
Costas S. Iliopoulos	King's College London, UK
Giuseppe Liotta	University of Perugia, Italy
Petra Mutzel	University of Bonn, Germany
Shin-ichi Nakano	Gunma University, Japan
Subhas Chandra Nandy	ISI, Kolkata, India
Md. Saidur Rahman	BUET, Bangladesh
Ryuhei Uehara	JAIST, Japan

WALCOM 2022 Organizer

WALCOM 2022 Committees

Program Committee

Aritra Banik	NISER, India
Arijit Bishnu	Indian Statistical Institute, Kolkata, India
Hans L. Bodlaender	Utrecht University, The Netherlands
Siu-Wing Cheng	Hong Kong University of Science and Technology, Hong Kong
Marek Chrobak	University of California, Riverside, USA
Dafik	Universitas Jember, Indonesia
Minati De	Indian Institute of Technology, Delhi, India
Mark de Berg	TU Eindhoven, The Netherlands
Henning Fernau	Trier University, Germany
Guillaume Fertin	University of Nantes, France
Fedor Fomin	University of Bergen, Norway
Neelima Gupta	University of Delhi, India
Xin Han	Dalian University of Technology, China
Seok-Hee Hong	University of Sydney, Australia
Stephen Kobourov	University of Arizona, USA
Jan Kratochvíl	Charles University, Czech Republic
Giuseppe Liotta	University of Perugia, Italy
Debajyoti Mondal	University of Saskatchewan, Canada
Petra Mutzel (Chair)	University of Bonn, Germany
Shin-Ichi Nakano	Gunma University, Japan
Maurizio Patrignani	Roma Tre University, Italy
M. Sohel Rahman	BUET, Bangladesh
Md. Saidur Rahman (Chair)	BUET, Bangladesh
Slamin (Chair)	Universitas Jember, Indonesia
Kiki A. Sugeng	Universitas Indonesia, Indonesia
Wing-Kin Sung	National University of Singapore, Singapore
Etsuji Tomita	The University of Electro-Communications, Japan
Ryuhei Uehara	Japan Advanced Institute of Science and Technology, Japan
Sue Whitesides	University of Victoria, Canada
Mingyu Xiao	University of Electronic Science and Technology of China, China
Hsu-Chun Yen	National Taiwan University, Taiwan
Xiao Zhou	Tohoku University, Japan

We deeply mourn the sudden death of Prof. Etsuji Tomita while he was active in the Program Committee of WALCOM 2022 and gratefully remember his contribution to WALCOM.

Organizing Committee

Robiatul Adawiyah	Universitas Jember, Indonesia
Dafik (Chair)	Universitas Jember, Indonesia
Tio Dharmawan	Universitas Jember, Indonesia
Ikhsanul Halikin	Universitas Jember, Indonesia
Elia Musrofa	Universitas Jember, Indonesia
Antonius Cahya Prihandoko (Chair)	Universitas Jember, Indonesia
Rafiantika Megahnia Prihandini	Universitas Jember, Indonesia
Januar Adi Putra	Universitas Jember, Indonesia
Qurrota A'yuni Ar Ruhimat	Universitas Jember, Indonesia
Slamin (Chair)	Universitas Jember, Indonesia
Kristiana Wijaya	Universitas Jember, Indonesia
Kholifatul Zahro	Universitas Jember, Indonesia

Technical Co-sponsors

Information Processing Society of Japan (IPSJ), Japan
The Institute of Electronics, Information and Communication Engineers (IEICE), Japan
Japan Chapter of the European Association of Theoretical Computer Science (EATCS Japan), Japan

External Reviewers

Acharjee, Sajib	Bulteau, Laurent
Ahmed, Abu Reyan	Chaplick, Steven
Ahmed, Shareef	Chiu, Man Kwun
Ahn, Hee-Kap	Chopin, Morgan
Alegría, Carlos	Choudhary, Pratibha
Angelini, Patrizio	Cloteaux, Brian
Arrighi, Emmanuel	Da Lozzo, Giordano
Asahiro, Yuichi	Das, Bireswar
Aziz, Haris	Das, Syamantak
Bača, Martin	Dev, Subhadeep
Bai, Tian	Durocher, Stephane
Bandopadhyay, Susobhan	Escoffier, Bruno
Bandyapadhyay, Sayan	Evans, William
Bazgan, Cristina	Faliszewski, Piotr
Bhattacharya, Anup	Fekete, Sándor
Bhore, Sujoy	Ghosh, Arijit
Blais, Eric	Gomez, Renzo
Bodwin, Greg	Gonçalves, Daniel

Grilli, Luca
Grosso, Fabrizio
Gupta, Manoj
Gupta, Siddharth
Hakim, Sheikh Azizul
Haque, Md. Redwanul
Hasan, Md. Manzurul
Hashem, Tanzima
Heeger, Klaus
Higashikawa, Yuya
Horiyama, Takashi
Huang, Haoqiang
Jain, Pallavi
Johnson, Matthew
Kanesh, Lawqueen
Karthik C. S.
Kaykobad, Mohammad
Kaykobad, Tanvir
Khan, Samin Rahman
Knop, Dušan
Komusiewicz, Christian
Kristiana, Arika Indah
Krizanc, Danny
Lahiri, Abhiruk
Leucci, Stefano
Liu, Daphne
Lubiw, Anna
Luchsinger, Austin
Ma, Mengfan
Majumdar, Diptapriyo
Manea, Florin
Mihalák, Matúš
Mishra, Gopinath
Mouawad, Amer

Nandy, Subhas
Ochem, Pascal
Otachi, Yota
Papan, Bishal Basak
Paul, Subhabrata
Penna, Paolo
Pergel, Martin
Pizzonia, Maurizio
Pupyrev, Sergey
Rahman, Atif
Raju, Iqbal Hossain
Saitoh, Toshiki
Sakib, Sadman
Schnider, Patrick
Sen, Sayantan
Seto, Kazuhisa
Shur, Arseny
Susilowati, Liliek
Suzuki, Akira
Takaoka, Asahi
Takenaga, Yasuhiko
Tamura, Yuma
Tappini, Alessandra
Tushar, Abdur Rashid
Valtr, Pavel
Wang, Haitao
Wellnitz, Philip
Wijaya, Kristiana
Wismath, Steve
Yanhaona, Muhammad Nur
Zeman, Peter
Zhao, Jingyang
Zhou, Xiao

A Tribute to Professor Takao Nishizeki

With great sorrow we are sharing the sad news that one of the advisory board members of WALCOM and a founding steering committee member of WALCOM Prof. Dr. Takao Nishizeki, Professor Emeritus at Tohoku University, passed away on January 30, 2022. Prof. Nishizeki was born in 1947 in Fukushima, and was a student at Tohoku University, earning a bachelor's, a master's and a doctorate degree in 1969, 1971 and 1974 respectively. He continued at Tohoku University as a faculty member, and became a full professor there in 1988. He retired in 2010, becoming a Professor Emeritus at Tohoku University.

Professor Nishizeki established himself as a world leader in computer science, in particular, algorithms for planar graphs, edge coloring, network flows, very large-scale integration (VLSI) routing, graph drawing and cryptography. He is the co-author of two books "Planar Graphs: Theory and Algorithms" and "Planar Graph Drawing". Both books are considered as the most valuable pioneering work on planar graphs and planar graph drawings and have been widely distributed over the world. Professor Nishizeki served in the editorial boards of Algorithmica, Journal of Combinatorial Optimization, Discrete Mathematics and Theoretical Computer Science, Journal of Information Processing, Journal of Graph Algorithms and Applications, Transactions of IEICEJ, and Journal of IEICJ. In 1996, he became a life fellow of the IEEE "for contributions to graph algorithms with applications to physical design of electronic systems." In 1996 he was selected as a fellow of the Association for Computing Machinery "for contributions to the design and analysis of efficient algorithms for planar graphs, network flows and VLSI routing". Prof. Nishizeki was also a foreign fellow of the Bangladesh Academy of Sciences and contributed significantly to the computer science education and research in Bangladesh. For his great achievements in computer science, Professor Nishizeki was awarded the ICF Best Research Award by the International Communications Foundation in 2006. He received the prestigious Information Science Promotion Award by Funai Foundation for Information Technology in 2003 and the TELECOM Technology Award by the Telecommunication Advancement Foundation in 1998.

As the most active and renowned computer scientist in Asia, Professor Nishizeki also made a great contribution to build a community of computer science in the Asia and Pacific region. In 1990 he founded the International Symposium on Algorithms and Computation (ISAAC) for the purpose of expanding the research community among the Asian and Pacific countries. He was a founding steering committee member of Graph Drawing (GD) and the International Conference and Workshops on Algorithms and Computation (WALCOM). It is worth mentioning that WALCOM was started to celebrate the 60th birthday of Professor Takao Nishizeki on February 12, 2007.

Professor Nishizeki was very popular among his students for his innovative teaching methodology. He awarded 28 PhD degrees and many of his students have established themselves in academia and industry. Professor Nishizeki is no longer with us, but his values, teachings and research contributions will survive years after years for the benefit of mankind.

Contents

Computational Geometry

Computational Complexity

Online and Property Testing

Parameterized Complexity

Graph Algorithms

Approximation Algorithms

Invited Talks

Some Problems Related to the Space of Optimal Tree Reconciliations

(Invited Talk)

Tiziana Calamoneri[1]([✉])([iD]) and Blerina Sinaimeri[2,3]([iD])

[1] Sapienza University of Rome, Rome, Italy
calamo@di.uniroma1.it
[2] LUISS University, Rome, Italy
bsinaimeri@luiss.it
[3] Erable, INRIA Grenoble Rhône-Alpes, Montbonnot-Saint-Martin, France

Abstract. Tree reconciliation is a general framework for investigating the evolution of strongly dependent systems as hosts and parasites or genes and species, based on their phylogenetic information. Indeed, informally speaking, it reconciles any differences between two phylogenetic trees by means of biological events. Tree reconciliation is usually computed according to the parsimony principle, that is, to each evolutionary event a cost is assigned and the goal is to find tree reconciliations of minimum total cost. Unfortunately, the number of optimal reconciliations is usually huge and many biological applications require to enumerate and to examine all of them, so it is necessary to handle them.

In this paper we list some problems connected with the management of such a big space of tree reconciliations and, for each of them, discuss some known solutions.

Keywords: Tree Reconciliation · Enumeration Algorithms · Visualization Algorithms · Clustering

1 Introduction

Tree reconciliation is a general framework for investigating the co-evolution of related biological systems as for example hosts and their parasites [9], genes and their corresponding species [15], organisms and their living areas (biogeography) [4], or species and their geological history [27]. The similarity between all these classes of problems was pointed out since 1994 [25, 26] and in this paper we will use the terminology of host/parasite context.

Given two phylogenetic trees for two sets of organisms, hosts and parasites, denoted by H and P respectively, together with a mapping ϕ of the leaves of P to the leaves of H (ϕ represents the nowadays infections), a *reconciliation* is

Supported by *Sapienza* University of Rome, projects "Comparative Analysis of Phylogenies" (no. RM1181642702045E), "A deep study of phylogenetic tree reconciliations" (no. RM11916B462574AD) and "Measuring the similarity of biological and medical structures through graph isomorphism" (no. RM120172A3F313FE).

P. Mutzel et al. (Eds.): WALCOM 2022, LNCS 13174, pp. 3–14, 2022.
https://doi.org/10.1007/978-3-030-96731-4_1

a mapping ρ of the internal vertices of P to the vertices of H which extends ϕ under some constraints.

Informally, it attempts to reconcile any differences between the phylogenetic histories of the parasite with that of their hosts and explaining each diversification with a possible evolutionary event.

Tree reconciliations are generally computed according to the *parsimony* principle, that is, to each evolutionary event is assigned a cost and the goal is to find a reconciliation of minimum total cost. The resulting optimization problem is denoted as the *reconciliation problem.*

Given a cost for each evolutionary event, for many data sets there may be a huge number (possibly exponential in the dimension of the trees) of optimal reconciliations. Each one of them could represent a different biological scenario and thus without further information, any biological interpretation of the underlying co-evolution would require all optimal solutions to be enumerated and examined, which is often unfeasible in practice.

In this paper we survey some problems connected with the management of such a huge space of tree reconciliations; more in detail, we will describe the problems of enumerating all reconciliations, of visualizing them and of reducing the number of those to be examined.

2   Definitions and Notations

Given a tree T, we denote its *vertex, edge* and *leaf* sets by $V(T)$, $E(T)$ and $L(T)$, respectively. If T is rooted, its *root* is denoted by $r(T)$, and the *subtree in T rooted at v* by $T(v)$. For any $v, w \in V(T)$, $P(v, w)$ is the set of vertices on the unique *path connecting v to w*, and their *distance* is defined as $d_T(v, w) = |P(v, w)| - 1$.

The rooting of T induces a partial order on the vertices of $V(T)$: w is an *ancestor* (*descendant*, respectively) of v if w is a vertex on the path between $r(T)$ and v (v is on the path between $r(T)$ and w); in both cases, v and w are called *comparable*, otherwise they are *incomparable*.

The *lowest common ancestor* of two vertices $v, w \in V(T)$, denoted *lca* (v, w), is the (unique) least common vertex of the two paths leading from v and w up to $r(T)$.

A *phylogenetic tree T* is a leaf-labelled rooted full binary (*i.e.*, all of its internal vertices have exactly two children) tree that models the evolution of a set of taxa (placed at the leaves) from their most recent common ancestor (placed at the root). Let H and P be the phylogenetic trees for the host and parasite species, respectively. Function ϕ (defined from the leaves of P to the leaves of H) indicates the association between currently living host and parasite species.

A *reconciliation ρ* is a function from the set of vertices of P to the set of vertices of H that extends the mapping ϕ of the leaves under some constraints. Note that each internal vertex of P can be associated to an event among: *cospeciation* (when both the parasite and the host speciate), *duplication* (when the parasite speciates but not the host) and *host switch* (when the parasite speciates and one of its children is associated to an incomparable host), while each arc

(u, v) of P is associated to a certain number of *loss* events $l_{(u,v)} \geq 0$ that is equal to the length of $path_H(\rho(u), \rho(v))$ if u is an ancestor of v. It is therefore possible to associate to each reconciliation ρ a vector $E_\rho = \langle e_c, e_d, e_s, e_l \rangle$, called *event vector*, where e_c, e_d, e_s and e_l denote the number of cospeciations, duplications, host switches and losses, respectively, that are identified by ρ.

Given a vector $C = \langle c_c, c_d, c_s, c_l \rangle$ of real values that correspond to the cost of each type of event, the most parsimonious (or optimal) reconciliations are the ones that minimize the total cost, *i.e.* that minimize $cost(\rho) = \sum_{i \in \{c,d,s,l\}} e_i c_i$. Note that it is usual to assume $c_c < c_d$ and $c_l > 0$; in the following we adopt these assumptions.

Finally, in our framework we assume the host tree is undated (*i.e.* a total order of the internal vertices is not known). This because dating a phylogenetic tree is usually a hard task and often unreliable. Working with undated trees increases the difficulty of the reconciliation problem as sometimes the inferred host switches may induce contradictory time constraints on the internal vertices of the host. Thus, finding an optimal time-consistent reconciliation is NP-hard [24, 34].

Nevertheless, it is possible to detect in polynomial time, whether a given tree reconciliation is time-consistent or not. Concerning the definition and the checking of time-consistency we refer the reader to [33].

3   Enumerating Reconciliations

It is well-known that finding an optimal time-consistent reconciliation is NP-hard (see for example [24, 34]). However, if the time-consistent constraint is dropped, the problem can be solved efficiently in polynomial time using a dynamic programming algorithm [21, 34]. It has been observed in practice that often there exists a time-consistent reconciliation of optimal cost. This lead to the design of efficient enumeration algorithms for the reconciliation problem. However, the main reason why one wants to enumerate all the optimal solutions is because –even if we guarantee the time-consistency– different optimal reconciliations can exist. Though these reconciliations are all optimal and hence have all the same (minimum) total cost, they can be quite different in terms of the mapping function or even of the number of events. For example in Fig. 5 a-b we present two reconciliations with the same events (and the same total cost) but which correspond to very different mappings and thus to two different biological scenarios. To this purpose, most of the software proposed in the literature do not rely on only one optimal solution but enumerate (*i.e.* list) *all* of them (see *e.g.* [11, 18, 20, 29, 33, 36]). The general enumeration algorithm underlying these tools employs a fairy simple approach. It is based on the dynamic programming (DP) technique for computing a single optimal solution (see *e.g.* [11, 21, 33, 34]) with some additional information useful for an exhaustive traceback stored in a matrix. The dynamic programming matrix has size $O(|V(P)| \times |V(H)|)$; each cell labeled by a parasite/host association $(p : h)$ contains the information needed for the reconstruction of all reconciliations of minimum cost between the subtree of P rooted at vertex p and the host tree, such that p is mapped to h. Once the matrix has been filled, the optimal solution can be found in the last row of the matrix in correspondence of the mapping of the root of P. It is possible then, starting from the root of P and using the backtrack arcs to traverse in a depth-first search manner the matrix and getting thus all the optimal solutions. It is not difficult to see that this algorithm takes $O(|V(H)|^2|V(P)|)$ time to fill the matrix and then only $O(|V(P)|)$ to output each subsequent optimal reconciliation. Therefore, there is an algorithm with a $O(|V(P)||V(H)|^2)$ time pre-processing step and $O(|V(P)|)$ time delay for enumerating the optimal reconciliations (see [11, 36]).

In various applications one is interested not only in optimal solutions but also in sub-optimal ones, that is, those having a cost strictly larger than the minimum. In the context of reconciliations, for a given input, one can *hypothetically* consider all the possible reconciliations in an increasing order based on their costs (the ordering between solutions having the same cost is arbitrary). Given an integer K, the goal is to output the first K reconciliations in this order. If K is larger than the number of optimal reconciliations, then the output should contain all optimal reconciliations and also a number of sub-optimal ones. To the best of our knowledge, [36] is the only paper that handles sub-optimal reconciliations, and is based on an algorithm that is a non trivial extension of the dynamic programming algorithm for listing all optimal reconciliations.

A variation of the problem involves restricting the reconciliations to have all host-switches bounded by some fixed distance k. The distance of a host-switch is defined as the distance in the host tree of the species involved in the switch of a parasite from one host to another. The biological motivation behind this constraint is that it may be related to the ability of the parasite to invade "very different" species. Such constraint has already been included in the parsimonious framework developed in [11] by requiring that host switches be allowed to happen only between "closely" related species, *i.e.* species that are within some fixed distance in the host tree. To the best of our knowledge, [7] is the only paper that considers this problem for low values of k ($k = 2$).

Open Problems. First, it is still open whether an optimal reconciliation where the distance of the host-switches is bounded by k (k is not part of the input) can be computed in polynomial time. In another direction, it is possible to analyse whether an optimal time-consistent reconciliation can be found in polynomial time for some particular topologies of trees. Some preliminary results on this problem can be found in [6].

4   Visualizing Reconciliations

Producing readable and compact representations of trees has a long tradition in the graph drawing research field. Besides the standard vertex-link diagrams (*e.g.* layered trees, radial trees, hv-drawings, etc.), trees can be visualized via the so-called space-filling metaphors (*e.g.* circular and rectangular treemaps, sunbursts, icicles, sunrays, icerays, etc.) [30].

It is crucial for biologists to unambiguously and effectively visualize tree reconciliations, in order to study and compare them quickly. So, in the literature, there are a number of papers and tools going in this direction. We classify into three families, schematically represented in Fig. 1, the representation conventions commonly adopted.

The simplest strategy represents the two trees by adopting the traditional vertex-link metaphor, where the vertices of P are drawn close to the vertices of H they are associated to through the reconciliation (see Fig. 1(a)). The advantage of this strategy lays in its simplicity. However, the drawing tends to become cluttered and with a high number of crossings.

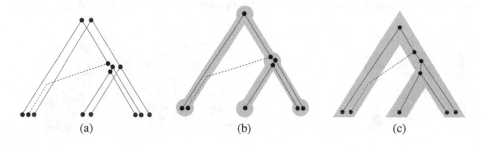

(a)   (b)   (c)

Fig. 1. Three visualization strategies for representing reconciliations.

Figure 2(a) shows an example of a straight-line representation obtained with CoRe-PA [38], adopting this strategy, while Fig. 2(b) shows a rightward orthogonal representation of a reconciliation obtained with Jane 4 [19].

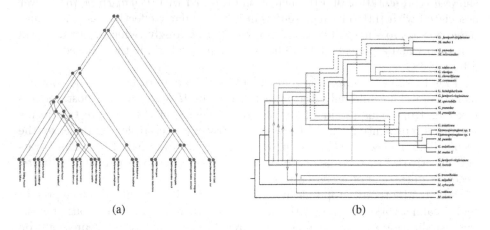

(a)   (b)

Fig. 2. (a) Reconciliation of Seabirds-Chewing Lice co-evolution trees visualized with CoRe-PA [38]. (b) Reconciliation of Gymnosporangium-Malus pathosystem co-evolution trees visualized with Jane 4 [19] (picture from [39]).

An alternative strategy consists in representing H as a background shape, such that its vertices are shaded disks and its arcs are pipes, while P is contained

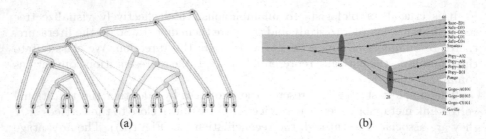

Fig. 3. (a) An example of co-phylogenetic tree drawn by *CophyTrees*, the viewer associated with [11]. (b) Reconciliation of Major Histocompatibility Complex class I in Gorilla, Orangutan-Tamarin visualized with *Primetv* [32] (picture from [32]).

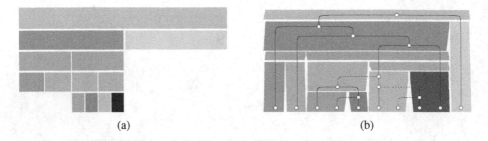

Fig. 4. (a) An icicle. (b) The representation adopted for trees H and P.

in H and drawn in the traditional vertex-link style (see Fig. 1(b)). Figure 3(a) shows a representation of a reconciliation obtained with *CophyTrees*, the viewer associated with [11]. The representation is particularly effective, as it is unambiguous and crossings between the two trees are strongly reduced, but it is still cluttered when a parasite subtree has to be squeezed inside the reduced area of a host vertex.

Finally, some visualization tools adopt the strategy of keeping the containment metaphor while only drawing thick arcs of H and omitting host vertices (see Fig. 1(c)). This produces a vertex-link drawing of the parasite tree inside the pipes representing the host tree. This strategy is used, for example, by the reconciliation viewer *SylvX* [10] or by *Primetv* [32] (see Fig. 3(b)). This strategy, effective when there is uncertainty in the association of parasites to hosts, becomes unclear in other situations.

Inspired by recent proposals of adopting space-filling techniques to represent biological networks [35], and with the aim of overcoming the limitations of existing visualization strategies, in [8] a new hybrid metaphor for the representation of reconciliations is introduced. A space-filling approach is used to represent H, while tree P maintains the traditional vertex-link representation. The reconciliation is unambiguously conveyed by placing parasite vertices inside the regions associated with the hosts they are mapped to.

More specifically, the representation of tree H is a variant of a representation known in the literature with the name of *icicle*. An *icicle* is a space-filling representation of hierarchical information in which vertices are represented by rectangles and arcs are represented by the contact of rectangles, such that the bottom side of the rectangle representing a vertex touches the top sides of the rectangles representing its children (see Fig. 4(a)). In this model, in order to contain parasite subtrees of different depths, rectangles of different height are allowed. Also, all leaves of H (*i.e.* present-day hosts) are forced to share the same bottom line that intuitively represents current time. This is the first representation guaranteeing the downwardness of P when time-consistent reconciliations are considered.

In the same paper [8], the problem of minimizing the number of crossings in the representation is addressed, proving that it is NP-complete; moreover, a characterization of reconciliations that can be planarly drawn is given, so putting, for the first time, in relation properties of the reconciliation and of its visualization.

Open Problems. A tanglegram drawing consists of a pair of plane trees whose leaves are connected by straight-line edges and they are very studied (*e.g.* in [1,5,23,31]). It is known that finding a tanglegram drawing with the minimum number of crossings is NP-hard; we would like to adapt heuristics for the reduction of the crossings of tanglegram drawings, such as those in [3,23,31], to our problem of clearly visualizing reconciliations.

Moreover, there are some situations in which the taxa are associated with some further information (such as, for example, the geographical zone where they live [2]). It would be interesting to be able to visualizing also this additional information inside the drawing; an idea could be to use colors, and in this case a drawing requirement could be to put close vertices associated to the same geographical zone.

5   Reducing the Number of Reconciliations

As already mentioned, the number of optimal solutions can be exponential in the size of the trees and hence, in many real cases, the current cophylogeny model may lead to a number of optimal solutions that is unrealistically large [11,12,36,37], making it practically impossible to analyze each one of them separately. To address this issue, different directions have been proposed in the literature. Some of them involve clustering and random sampling of the space of optimal solutions. However, as the optimal reconciliation space can be both large and heterogeneous [14], this does not guarantee that important information is not lost.

Here we focus on two main approaches: the first one involves the definition of a similarity/distance measure defined on the set of optimal reconciliations. Then the problem would be to find a subset of reconciliations that is representative of the whole set. More formally, we want to find a subset of optimal reconciliations S, such that each of the optimal reconciliations is at distance at most d from at least one of the reconciliations in S. Many methods have been designed in this direction, as computing in polynomial time the pairwise distance among the optimal reconciliations [28] or finding a single reconciliation (e.g. a "median" reconciliation) to represent the whole space of optimal ones [13,17,22]. However, the results presented in [13,14,16,28] show that the space can be very diverse and making inferences from a single reconciliation might lead to conclusions that can be contradicted by other optimal reconciliations.

Another direction is to define equivalence classes for grouping the reconciliations that may be considered *biologically equivalent*. Once this notion of equivalence is defined, we could group the optimal solutions in equivalence classes and output a single solution for each class. Intuitively, this would allow to generate solutions that are sufficiently different and would provide a first information on the space of optimal solutions.

For example one notion of equivalence is based on the comparison of the number of each one of the four events (cospeciation, duplication, loss and host switch): two reconciliations are considered equivalent, and hence put in a same class, if they have the same number of each event, *i.e.* if they have the same event vector. Notice that, while the number of optimal solutions can be exponential, the number of event vectors is bounded by $O(|H||P|^3)$ and hence can be enumerated efficiently in polynomial time [36,37]. The set of the event-vector classes provides already a first information about the co-evolutionary history of the hosts and their parasites. Indeed, a high number of cospeciations may indicate that hosts and parasites evolved together, while a high number of host-switches may indicate that the parasites are able to infect different host species. However, as shown in Fig. 5, two reconciliations may have the same event vector and still be very different from a biological point of view.

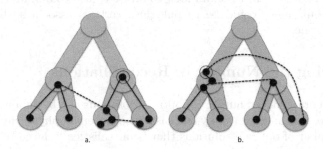

Fig. 5. a. and b. Two reconciliations with the same event vector that nevertheless are rather different. The grey tubes represent the host tree, while the black (plain or dotted) lines inside the tubes represent the parasite tree. The roots of the parasite trees are double lined to facilitate their recognition (picture from [12]).

Some more interesting notions of equivalence were first considered in [12]. One of the equivalences defined by the authors was inspired by the following theoretical result: once the set of vertices of P that are associated to switches is fixed, an optimal reconciliation can be easily identified using the least common ancestor mapping. Hence, the authors consider two reconciliations ρ, ρ' as equivalent if for each vertex $v \in V(P)$ which is not associated to a host-switch, we have $\rho(v) = \rho(v')$ and for each host switch edge (u, v) in the parasite tree, map u on a different vertex of H lying on the path between the vertices reconciled with the parent of u and with the sibling of v. The latter is called a *sliding path* to highlight the idea that u can be moved anywhere inside this path without modifying the cost of the reconciliation. The partition of reconciliations induced by this equivalence relation is finer than the partition induced by the event vector, since two reconciliations that are equivalent w.r.t. it are surely equivalent w.r.t. to the event vector partition, but the opposite is not true, and this is in agreement with the fact that two reconciliations with the same event vector can be very different: in such a case, this equivalence distinguishes them.

However, the method presented in [12] requires first the listing (*i.e.* the enumeration) of all the optimal solutions and then clustering them according to the equivalence notion. This represents a difficulty as when the number of reconciliations is too large, for example, $> 10^{42}$ [12,36,37], listing all the solutions is not feasible. Hence, the ultimate goal would be to extract as much information as possible about the space of optimal reconciliations, *without* considering all of the elements of this space. This has been solved recently in [37] where for three different notions of equivalence, a polynomial delay algorithm has been proposed in order to enumerate only one representative for equivalence class. The authors consider three different equivalence relations; one of them is the *event-partition equivalence*: two reconciliations are event-partition equivalent if, for each internal vertex in the parasite tree, the event (cospeciation, duplication, host-switch) assigned by each of the two reconciliations is the same. Notice that the difference with the event-vector equivalence is that in the event-partition equivalence the interest is not only in the number of the events but also in where these events have taken place in the parasite tree.

Open Problems. First it would be interesting to explore the connections between the equivalences defined in [12] and the ones defined in [37] and determine whether there exist polynomial delay algorithms enumerating the representative reconciliations of the equivalence classes defined in [12]. More generally, it would be interesting to define other (biologically meaningful) notions of equivalence for which such algorithms exist.

References

1. Bansal, M.S., Chang, W.-C., Eulenstein, O., Fernández-Baca, D.: Generalized binary tanglegrams: algorithms and applications. In: Rajasekaran, S. (ed.) BICoB 2009. LNCS, vol. 5462, pp. 114–125. Springer, Heidelberg (2009). https://doi.org/10.1007/978-3-642-00727-9_13
2. Berry, V., Chevenet, F., Doyon, J.P., Jousselin, E.: A geography-aware reconciliation method to investigate diversification patterns in host/parasite interactions. Mol. Ecol. Resour. **18**(5), 1173–1184 (2018). https://doi.org/10.1111/1755-0998.12897

3. Böcker, S., Hüffner, F., Truss, A., Wahlström, M.: A faster fixed-parameter approach to drawing binary tanglegrams. In: Chen, J., Fomin, F.V. (eds.) IWPEC 2009. LNCS, vol. 5917, pp. 38–49. Springer, Heidelberg (2009). https://doi.org/10.1007/978-3-642-11269-0_3

4. Brooks, D.R., Ferrao, A.L.: The historical biogeography of co-evolution: emerging infectious diseases are evolutionary accidents waiting to happen. J. Biogeogr. **32**, 1291–1299 (2005)

5. Buchin, K., et al.: Drawing (complete) binary tanglegrams. Algorithmica **62**(1–2), 309–332 (2012). https://doi.org/10.1007/s00453-010-9456-3

6. Calamoneri, T., Monti, A., Sinaimeri, B.: Co-divergence and tree topology. J. Math. Biol. **79**(3), 1149–1167 (2019). https://doi.org/10.1007/s00285-019-01385-w

7. Calamoneri, T., Tavernelli, D., Vocca, P.: Linear time reconciliation with bounded transfers of genes. IEEE/ACM Trans. Comput. Biol. Bioinform. (2020). https://doi.org/10.1109/TCBB.2020.3027207. https://ieeexplore.ieee.org/document/9207859

8. Calamoneri, T., Donato, V.D., Mariottini, D., Patrignani, M.: Visualizing cophylogenetic reconciliations. Theoret. Comput. Sci. **815**, 228–245 (2020). https://doi.org/10.1016/j.tcs.2019.12.024

9. Charleston, M.: Jungles: a new solution to the host/parasite phylogeny reconciliation problem. Math. Biosci. **149**, 191–223 (1998)

10. Chevenet, F., Doyon, J.P., Scornavacca, C., Jacox, E., Jousselin, E., Berry, V.: SylvX: a viewer for phylogenetic tree reconciliations. Bioinformatics **32**(4), 608–610 (2016). https://doi.org/10.1093/bioinformatics/btv625

11. Donati, B., Baudet, C., Sinaimeri, B., Crescenzi, P., Sagot, M.F.: EUCALYPT: efficient tree reconciliation enumerator. Algorithms Mol. Biol. **10**(3), 1–11 (2015). https://doi.org/10.1186/s13015-014-0031-3

12. Gastaldello, M., Calamoneri, T., Sagot, M.-F.: Extracting few representative reconciliations with host switches. In: Bartoletti, M., et al. (eds.) CIBB 2017. LNCS, vol. 10834, pp. 9–18. Springer, Cham (2019). https://doi.org/10.1007/978-3-030-14160-8_2

13. Grueter, M., Duran, K., Ramalingam, R., Libeskind-Hadas, R.: Reconciliation reconsidered: in search of a most representative reconciliation in the duplication-transfer-loss model. IEEE/ACM Trans. Comput. Biol. Bioinform. **18**(6), 2136–2143 (2019). https://doi.org/10.1109/TCBB.2019.2942015

14. Haack, J., Zupke, E., Ramirez, A., Wu, Y.C., Libeskind-Hadas, R.: Computing the diameter of the space of maximum parsimony reconciliations in the duplication-transfer-loss model. IEEE/ACM Trans. Comput. Biol. Bioinf. **16**(1), 14–22 (2019). https://doi.org/10.1109/TCBB.2018.2849732

15. Hallett, M.T., Lagergren, J.: Efficient algorithms for lateral gene transfer problems. In: Proceedings of the Fifth Annual International Conference on Computational Biology (RECOMB 2001), pp. 149–156 (2001)

16. Huber, K.T., Moulton, V., Sagot, M.F., Sinaimeri, B.: Exploring and visualizing spaces of tree reconciliations. Syst. Biol. **68**(4), 607–618 (2018). https://doi.org/10.1093/sysbio/syy075

17. Huber, K.T., Moulton, V., Sagot, M., Sinaimeri, B.: Geometric medians in reconciliation spaces of phylogenetic trees. Inf. Process. Lett. **136**, 96–101 (2018). https://doi.org/10.1016/j.ipl.2018.04.001

18. Jacox, E., Chauve, C., Szöllősi, G.J., Ponty, Y., Scornavacca, C.: ecceTERA: comprehensive gene tree-species tree reconciliation using parsimony. Bioinformatics (2016). https://doi.org/10.1093/bioinformatics/btw105

19. Libeskind-Hadas, R.: Jane 4 - a software tool for the cophylogeny reconstruction problem. https://www.cs.hmc.edu/~hadas/jane/

20. Merkle, D., Middendorf, M., Wieseke, N.: A parameter-adaptive dynamic programming approach for inferring cophylogenies. BMC Bioinformatics **11**(Suppl. 1), 1–10 (2010). https://doi.org/10.1186/1471-2105-11-S1-S60

21. Bansal, M.S., Alm, E.J., Kellis, M.: Efficient algorithms for the reconciliation problem with gene duplication, horizontal transfer and loss. Bioinformatics **28**, 283–291 (2012). https://doi.org/10.1093/bioinformatics/bts225

22. Nguyen, T.H., Ranwez, V., Berry, V., Scornavacca, C.: Support measures to estimate the reliability of evolutionary events predicted by reconciliation methods. PLoS ONE **8**(10), 1–14 (2013). https://doi.org/10.1371/journal.pone.0073667

23. Nöllenburg, M., Völker, M., Wolff, A., Holten, D.: Drawing binary tanglegrams: an experimental evaluation. In: Finocchi, I., Hershberger, J. (eds.) ALENEX 2009, pp. 106–119. SIAM (2009). https://doi.org/10.1137/1.9781611972894.11

24. Ovadia, Y.J., Fielder, D., Conow, C., Libeskind-Hadas, R.: The cophylogeny reconstruction problem is np-complete. J. Comput. Biol. **18**(1), 59–65 (2011). https://doi.org/10.1089/cmb.2009.0240

25. Page, R.D.M.: Maps between trees and cladistic analysis of historical associations among genes, organisms, and areas. Syst. Biol. **43**, 58–77 (1994)

26. Ronquist, F.: Parsimony analysis of coevolving species associations. In: Tangled Trees: Phylogeny, Cospeciation, and Coevolution (2002)

27. Rosen, D.E.: Vicariant patterns and historical explanation in biogeography. Syst. Biol. **27**, 159–188 (1978)

28. Santichaivekin, S., Mawhorter, R., Libeskind-Hadas, R.: An efficient exact algorithm for computing all pairwise distances between reconciliations in the duplication-transfer-loss model. BMC Bioinformatics **20**(20), 636 (2019). https://doi.org/10.1186/s12859-019-3203-9

29. Santichaivekin, S., et al.: eMPRess: a systematic cophylogeny reconciliation tool. Bioinformatics **37**(16), 2481–2482 (2020). https://doi.org/10.1093/bioinformatics/btaa978

30. Schulz, H.J.: Treevis.net: a tree visualization reference. IEEE Comput. Graphics Appl. **31**(6), 11–15 (2011). https://doi.org/10.1109/MCG.2011.103

31. Scornavacca, C., Zickmann, F., Huson, D.H.: Tanglegrams for rooted phylogenetic trees and networks. Bioinformatics **13**(27), i248–i256 (2011). https://doi.org/10.1093/bioinformatics/btr210

32. Sennblad, B., Schreil, E., Sonnhammer, A.C.B., Lagergren, J., Arvestad, L.: primetv: a viewer for reconciled trees. BMC Bioinformatics **8**(1), 148 (2007)

33. Stolzer, M., Lai, H., Xu, M., Sathaye, D., Vernot, B., Durand, D.: Inferring duplications, losses, transfers and incomplete lineage sorting with nonbinary species trees. Bioinformatics **28**(18), i409–i415 (2012). https://doi.org/10.1093/bioinformatics/bts386

34. Tofigh, A., Hallett, M., Lagergren, J.: Simultaneous identification of duplications and lateral gene transfers. IEEE/ACM Trans. Comput. Biol. Bioinf. **8**(2), 517–535 (2011). https://doi.org/10.1109/TCBB.2010.14

35. Tollis, I.G., Kakoulis, K.G.: Algorithms for visualizing phylogenetic networks. In: Hu, Y., Nöllenburg, M. (eds.) GD 2016. LNCS, vol. 9801, pp. 183–195. Springer, Cham (2016). https://doi.org/10.1007/978-3-319-50106-2_15

36. Wang, Y., Mary, A., Sagot, M., Sinaimeri, B.: Capybara: equivalence class enumeration of cophylogeny event-based reconciliations. Bioinformatics **36**(14), 4197–4199 (2020). https://doi.org/10.1093/bioinformatics/btaa498

37. Wang, Y., Mary, A., Sagot, M., Sinaimeri, B.: Making sense of a cophylogeny output: efficient listing of representative reconciliations. In: 21st International Workshop on Algorithms in Bioinformatics, WABI 2021, 2–4 August 2021, Virtual Conference, pp. 3:1–3:18 (2021). https://doi.org/10.4230/LIPIcs.WABI.2021.3
38. Wieseke, N., Hartmann, T., Bernt, M., Middendorf, M.: Cophylogenetic reconciliation with ILP. IEEE/ACM Trans. Comput. Biol. Bioinform. **12**(6), 1227–1235 (2015). https://doi.org/10.1109/TCBB.2015.2430336
39. Zhao, P., Liu, F., Li, Y.M., Cai, L.: Inferring phylogeny and speciation of Gymnosporangium species, and their coevolution with host plants. Sci. Rep. **6**, Article No. 29339 (2016). https://doi.org/10.1038/srep29339

From the W-hierarchy to XNLP
Classes of Fixed Parameter Intractability

Hans L. Bodlaender$^{(\boxtimes)}$ (iD)

Department of Information and Computing Sciences,
Utrecht University, Utrecht, The Netherlands
h.l.bodlaender@uu.nl

Abstract. In this short survey, a number of old and new notions from parameterized complexity are discussed. We start with looking at the W-hierarchy, including the classes $W[1]$, $W[2]$, $W[P]$. Then, a recent development where problems are shown to be complete for simultaneously non-deterministic time of the form $f(k)n^c$ and space of the form $f(k)\log n$, is discussed. Some consequences and other notions are briefly explored.

Keywords: Parameterized complexity · W-hierarchy · XP · XNLP

1 Introduction

The study of parameterized algorithms and complexity starts at the insight that many computationally hard problems become easier when a parameter of the input can be assumed small. Suppose we are to solve a facility location problem, e.g., we have to place as few as possible fire stations in a city, such that each house in the city is at most a 15 min drive away from a fire station. It is not hard to observe that this is an NP-hard problem. However, if we know that we have only funds available for three fire stations, then an exhaustive search for all possible combinations of at most three locations gives a tractable (polynomial time) algorithm to solve the problem.

In the theory of parameterized algorithms and complexity, we look at parameterized problems; i.e., we identify some aspect of the input as the parameter. Then we ask: when this parameter is a constant, is there a polynomial time algorithm. And if so, does the degree of the polynomial depend on the parameter. The theory started with work by Fellows and Langston at the late 1980s (e.g., [21,22], with some central notions first identified by Abrahamson et al. in 1989 [2], and much foundational work done in the 1990s by Downey and Fellows (e.g., [14–16] and [17].)

Throughout this paper, we view a parameterized problem as a subset of $\Sigma^* \times \mathbb{N}$, with Σ some finite alphabet. We are interested in the algorithmic complexity of parameterized problems for which its 'classic' variant (i.e., where the parameter is just part of the input) is intractable, e.g., NP-hard. Many parameterized problems fall in one of the following three categories (in order of increasing desirability):

© Springer Nature Switzerland AG 2022
P. Mutzel et al. (Eds.): WALCOM 2022, LNCS 13174, pp. 15–25, 2022.
https://doi.org/10.1007/978-3-030-96731-4_2

- There is a value of the parameter for which the problem is NP-hard. E.g., if we consider GRAPH COLOURING where the number of colours is the parameter, then this problem is NP-hard with the parameter (number of colours) equal to 3 [27]. Parameterized problems which are NP-hard for some fixed value of the parameter are called *para-NP-hard*.
- There is an algorithm that solves the problem for inputs of the form (x, k) in $O(n^{f(k)})$ time, where $n = |x|$ is the size of the input, k the parameter, and f a (computable) function. The class of problems with such an algorithm is called XP.
- There is an algorithm that solves the problem for inputs of the form (x, k) in $O(f(k)n^c)$ time, with again $n = |x|$ the size of the input, k the parameter, f a computable function, and c a constant. Problems of this type are called *fixed parameter tractable*, and the class of such problems is called FPT.

One can distinguish different flavours of FPT (and XP), namely *non-uniform* (for each value of k, there is an algorithm of the stated running time), *uniform* (there is one algorithm working for all values of k, but we do not require that f is computable), and *strongly uniform* (as above: we have one algorithm for all values of k, and f is computable). Examples of non-uniform fixed parameter tractability can be obtained with help of well quasi orderings: if we have a graph parameter h which cannot increase by taking a minor of a graph, then from Robertson-Seymour graph theory, we obtain a non-constructive proof tells us that for each k, there is an $O(n^2)$ algorithm that decides for a given graph if $h(G) \leq k$. (See e.g., [13, Section 6.3] with [29].) But, we may not be able to construct the algorithms and thus only know that for each k there exists a (separate) algorithm. See the discussion in [17, Chapter 19]. In the remainder, we only look at strongly uniform cases.

NP-completeness theory tells us when a problem is para-NP-hard. Assuming P$\neq$NP, para-NP-hard problems do not belong to XP (or FPT). Thus an NP-hardness result for a specific value of a parameter gives evidence that the problem at hand is not likely to belong to XP. To give similar evidence to tell for studied problems that they are not fixed parameter tractable, a number of complexity classes have been introduced, all which are assumed to be not a subset of or equal to FPT. Thus, hardness of a problem for such a class tells that it is unlikely that the problem belongs to FPT.

For a few problems, an unconditional proof that they do not belong to XP is known. Diagonalisation gives that FPT is a proper subset of XP [17, Proposition 27.1.1]. A few problems (formulated in terms of games) are known to be XP-complete [3,4], and thus, these cannot belong to FPT. (See also [18, Chapter 27].)

This short (and incomplete) survey reviews a number of classes of problems assumed not to be fixed parameter tractable, with some classic results from the field, and some recent developments. The theory in this field is rich (much richer than this survey can show); the focus in this short survey is on classes that contain complete problems that are studied in combinatorial optimisation algorithms. Much information can also be found in a number of excellent text books that on parameterized algorithms and complexity [13, 17, 18, 24, 31] and on the topic of kernelization (a subtopic in the field, not discussed in this survey) [26].

2 Reductions

Hardness and completeness for classes is as usual defined with help of reductions between problems.

A *parameterized reduction* from parameterized problem Q to parameterized problem R is an algorithm A that maps inputs for Q to inputs for R, such that $(x, k) \in Q \Leftrightarrow A((x, k)) \in R$, A uses time $f(k)n^c$ for a computable function f and constant c, and if $A(x, k) = (x', k')$, then $k' \leq g(k)$ for a computable function g.

We also look at *parameterized logspace reductions* (or pl-reductions), where we additionally require that the space used by the reduction algorithm is $O(h(k) + \log n)$. Different types of reductions also are used in the field of parameterized complexity, but these will not be discussed here.

3 The W-hierarchy

Downey and Fellows (see e.g. [17]) have introduced the W-hierarchy: a hierarchy of complexity classes of parameterized problems. The hierarchy contains a class $W[i]$ for each positive integer i, the class $W[SAT]$, and the class $W[P]$. Together with FPT and XP, we have the following inclusions.

$$FPT \subseteq W[1] \subseteq W[2] \subseteq W[3] \cdots \subseteq W[SAT] \subseteq W[P] \subseteq XP$$

It is conjectured [17, Chapter 12] that each inclusion is proper. In particular, when $FPT = W[1]$, then the Exponential Time Hypothesis would not hold [10].

$W[i]$ is defined with help of combinatorial circuits. Consider a circuit, with n Boolean input gates, and one output node. Take some fixed constant c. (The choice of c does not matter for the results, e.g., we can set $c = 2$.) The *weft* of the circuit is the maximum number of internal nodes with indegree more than c on a path from an input gate to the output node. Now, $W[i]$ is defined as the parameterized problems with a parameterized reduction to the problem to decide for a given circuit, if we can set k input gates to true and all other input gates to false, such that the circuit outputs true. (k is the parameter of the problem.)

An alternative definition, that can be of help to prove $W[i]$-hardness, is in terms of Boolean formulas. Consider a formula on n Boolean variables. We say the formula is i-normalised, if its is the conjunction of the disjunction of the conjunction of ... of literals, with i alternations between conjunction and disjunction. $W[i]$ can also be defined as the problems with a parameterized reduction to the problem to decide if a given i-normalised formula can be satisfied by setting exactly k variables to true, and all others to false. Again, k is the parameter. (Equivalently, we can ask to set at most k variables to true.)

For $i = 1$, we obtain the $W[1]$-complete problem, for each fixed integer q, WEIGHTED q-CNF SATISFIABILITY. Given is a Boolean formula in Conjunctive Normal Form, with each clause having at most q literals, and we ask if we can satisfy it by setting exactly k (the parameter) variables to true.

Important examples of complete problems are CLIQUE and INDEPENDENT SET, who are $W[1]$-complete, and DOMINATING SET, which is $W[2]$-complete.

As we assume that the hierarchy is proper, this implies that it is unlikely that CLIQUE, INDEPENDENT SET, and DOMINATING SET are in FPT.

Intuition why CLIQUE and INDEPENDENT SET are in $W[1]$, while DOMINATING SET is not, is the following. We can take an input gate (or a Boolean variable) for each vertex of the graph, which is true iff the vertex is in the solution set. To verify that this set forms a independent set, we need to perform a polynomial number of tests (one for each pair of nonadjacent vertices), where each such test looks at two variables (checking that at least one of these is false)—this corresponds to a circuit of weft one. (The same type of argument works for CLIQUE.) To verify that we have a dominating set, we need to perform a polynomial number of tests (one for each vertex), but each of these tests can involve a large number of variables (we check that the vertex is dominated, thus need to look at the variables of the vertex and its neighbours)—this corresponds to a circuit of weft two.

$W[SAT]$ is defined in the same manner as the classes $W[i]$, but now we can use any Boolean formula of polynomial size, and $W[P]$ is defined with combinatorial circuits of polynomial size (without weft restrictions).

In the parameterized algorithms and complexity literature, a large number of problems from various applications have been shown to be hard or complete for classes in the W-hierarchy. In particular, the classes $W[1]$ and $W[2]$ play an important role.

4   Logarithmic Space

4.1   The Story of BANDWIDTH

There are several problems that are shown to be hard for $W[1]$, for $W[2]$, or for all classes $W[i]$ for all integers $i \in \mathbb{N}$, but which are not known to be member of $W[P]$, i.e., we do not know whether they belong to a class in the W-hierarchy.

As a central example, we look at the BANDWIDTH problem. Given here is an undirected graph $G = (V, E)$, and the integer parameter k, and we want to decide if there is a bijective function $f : V \to \{1, \ldots, |V|\}$, such that for all edges $\{v, w\} \in E$, $|f(v) - f(w)| \leq k$. BANDWIDTH is a long studied problem—amongst others, because it is equivalent to asking for a symmetric matrix whether we can permute rows and columns simultaneously, such that all non-zero elements are at a band of width k around the main diagonal.

Already in 1980, Saxe [34] showed that BANDWIDTH can be solved in $O(n^{k+1})$ time, thus belongs to XP; this was later improved to $O(n^k)$ by Gurari and Sudborough [28]. In 1994, Bodlaender et al. [7] claimed that BANDWIDTH was hard for all classes $W[i]$, $i \in \mathbb{N}$, but it took till 2020 till a proof of this fact was written down [5]. In 2014, Dregi and Lokshtanov [19] showed that BANDWIDTH is W[1]-hard for trees of pathwidth at most two.

Each of these results showed *hardness* for classes in the W-hierarchy, but membership. This gives the question: is BANDWIDTH member of a class in the W-hierarchy, and can we find a class for which this problem is complete? The same question can be asked for many other problems, that are known to be hard for $W[1]$, but not known to reside in the W-hierarchy.

In the midst of the 1990s, Hallett gave an argument why it is unlikely that BANDWIDTH belongs to $W[P]$; the argument is discussed in [23]. The argument is as follows: certificates for problems in $W[P]$ have size $O(k \log n)$: we use $\log n$ bits for each of the k input gates that is true to give its index. However, one expects that BANDWIDTH cannot have such small certificates; for instance, we can have a graph with many connected components; one would expect to need certificates of size at least (but probably much larger than) the number of connected components. The argument resembles the later development of compositionality arguments for showing lower bounds for kernels [6].

So, if BANDWIDTH is not (likely) in $W[P]$, where is it?

We can go back to the first dynamic programming algorithm by Saxe [34] for BANDWIDTH. In this algorithm, we build n tables: each table entry of the i table gives 'essential information' of an ordering f of a set S with i vertices. The essential information gives all that is needed to remember of f and S to later determine if there is an ordering of V that starts with f, and then gives the vertices in $V \setminus S$ in some order. A simpler (slower) algorithm is obtained by taking as essential information the last $2k$ vertices of S with their order. One can turn this dynamic programming algorithm into a non-deterministic algorithm, where we do not store all elements of a table, but just guess one entry. We then have the following, non-deterministic algorithm for BANDWIDTH: repeatedly guess the next vertex in the order, and keep in memory the last $2k$ vertices. (We need to check that we never guess a vertex that is already ordered, but this verification can be done with a dfs search with help of the $2k$ stored vertices; we leave the details as a simple puzzle for the reader.)

What we now have is a non-deterministic algorithm for BANDWIDTH; the algorithm uses polynomial time, n non-deterministic guesses of a vertex, $O(k \log n)$ memory (as we remember $O(k)$ vertices with order).

In 2015, Elberfeld et al. [20] introduced a number of different classes of parameterized problems, including several subclasses of FPT and of XP, characterising the use of time, space, size of kernels, and more. One of these subclasses is the class, which we call here XNLP (and was called $N[fpoly, flog]$ in [20]). XNLP is the class of parameterized problems that can be recognized by a non-deterministic algorithm that simultaneously use $O(f(k)n^c)$ time and $O(f(k) \log n)$ memory, with f a computable function, and c a constant.

The non-deterministic algorithm for BANDWIDTH sketched above shows that it belongs to this class XNLP. Interestingly, it is possible to show that

BANDWIDTH is XNLP-complete [8]. For XNLP-completeness, we need to use parameterized logspace reductions.

4.2 XNLP-complete Problems

To show that problems are XNLP-hard, we use parameterized logspace reductions from known XNLP-hard problems. Recently, several problems have been shown to be XNLP-complete [8,20]. Several need a chain of reductions. Useful intermediate XNLP-complete problems are, amongst others:

- TIMED ACCEPTING NON-DETERMINISTIC LINEAR CELLULAR AUTOMATON [20]. We have a linear cellular automaton: a row of k cells, each having at each time step a value (state) from an alphabet (which can be of linear size, so we use $O(\log n)$ bits to denote an element from the alphabet). At each time step, each cell receives a value, non-deterministically depending on its value at that of the neighbouring cell(s). One state is said to be accepting, and the question is whether there exists a run where after t (given in unary) time steps, a cell has the accepting state. k is the parameter.
- CHAINED WEIGHTED CNF-SATISFIABILITY [8]. We have n sets of Boolean variables $X^1, \ldots, X^n$, each of size r, and a Boolean formula in Conjunctive Normal Form F, and a parameter k. The question is to set of each set X^i exactly k variables to true, and all others to false, such that the following formula is satisfied:

$$\wedge_{1 \leq i < n} F(X_i, X_{i+1})$$

Several special cases are also shown to be XNLP-complete in [8]. The hardness proof is of a similar vein as the Cooks proof of the NP-hardness of SATISFIABILITY [12]: the logic formula describes the working of the automaton.
- CHAINED CLIQUE [8]. Given is a graph $G = (V, E)$, where V is partitioned into n subsets $V_1, \ldots, V_n$, and the parameter $k \in \mathbb{N}$. Question is whether we can choose from each set V_i a subset $S_i \subseteq V_i$ of k vertices, such that for each pair of successive sets, $S_i \cup S_{i+1}$ $(1 \leq i < n)$ forms a clique of size $2k$.
- ACCEPTING NNCC MACHINE [8]. The XNLP-completeness of the problem whether the following non-deterministic machine has an accepting run has been proven to be a very helpful tool to show several problems XNLP-hard. The machine has k integer counters, which start at 0. At each time step, all counters can be increased non-deterministically to an integer that is at most n. There is a series of tests: each test looks at two counters, and has two integers from $[0, n]$. If the first counter equals the first of these integers, and the second counter equals the second integer, then the machine halts and rejects. If all tests succeed, the machine accepts.

From the XNLP-hardness of ACCEPTING NNCC MACHINE, we can (with one intermediate step) obtain the XNLP-hardness of BANDWIDTH, but also XNLP-hardness of SCHEDULING WITH PRECEDENCE CONSTRAINTS, parameterized by the number of machines and thickness. Other XNLP-complete problems include LONGEST COMMON SUBSEQUENCE [20], LIST COLOURING with the pathwidth of the graph as parameter [8], and INDEPENDENT SET and DOMINATING SET on graphs of pathwidth $k \log n$, where again k is the parameter.

XNLP-completeness has two interesting consequences. First, it implies hardness for all classes $W[i]$ for all $i \in \mathbb{N}$. Interestingly, often the XNLP-hardness proofs are easier than the earlier proofs of $W[i]$-hardness for all i. Second, a conjecture of Pilipczuk and Wrochna [32] for LONGEST COMMON SUBSEQUENCE implies the same conjecture for all XNLP-hard problems.

Conjecture 1 (Pilipczuk and Wrochna [32]). Suppose parameterized problem Q is XNLP-hard. Then Q has no algorithm that runs in $n^{f(k)}$ time and $f(k)n^c$ space, for a computable function f and constant c, with k the parameter, and n the total input size.

XNLP is a subset of XP (instead of making non-deterministic guesses, we tabulate all reachable states of the memory), but from Conjecture 1, we obtain that it is unlikely that an XNLP-complete problem has an XP algorithm that uses little space ('fpt space').

4.3   Other Classes with Logarithmic Space and Reconfiguration

Well known in classic complexity theory are the classes L and NL: problems solvable in logarithmic spacew with a deterministic, respectively non-deterministic algorithm. An interesting class is SL (with the S an abbreviation of *'symmetric'*), which allows to 'reverse' computations. Reingold [33] showed that L=SL, which is used in a result discussed below.

The parameterized counterparts of L and NL are respectively XL (parameterized problems solvable in $f(k) \log n$ space), and XNL (parameterized problems solvable with a non-deterministic algorithm in $f(k) \log n$ space). See e.g., [11].

Recently, Bodlaender et al. [9] explored the complexity of INDEPENDENT SET and DOMINATING SET reconfiguration. Given are two sets S_1 and S_2, which are both independent sets of G (or, respectively, both dominating sets). We want to change S_1 into S_2 in a number of moves, where each move changes one vertex of the set to another one, while each intermediate set still must be an independent (or dominating) set. We look at the problem if such a move sequence exists, or such a move sequence with t moves exists. The sizes $|S_1| = |S_2|$ are the parameter of the problem. The complexities of these questions depend on whether t is not given, a second parameter[1], given in unary, or given in binary. Table 1 summarises the different results. The XL-completeness for the case where there is no bound on the number of moves uses an interesting argument, with the following intuition: when we can use arbitrary many moves, we can always reverse any move. That corresponds (via the reductions) to a computation on a symmetric Turing Machine, which yields XSL-completeness, where XSL is the parameterized counterpart of SL. But, by Reingolds result [33], SL = L, which implies XSL = XL, thus the problems without a bound on the number of moves are XL-complete.

---

[1] Formally, instead of giving a problem two parameters, we can take the sum of these two values as parameter.

Table 1. Complexity of reconfiguration problems, with set sizes as (one of the) parameter(s)

Nb of steps	INDEPENDENT SET	DOMINATING SET	References
parameter	$W[1]$-complete	$W[2]$-complete	[9,30]
unary	XNLP-complete	XNLP-complete	[8]
binary	XNL-complete	XNL-complete	[9]
not bounded	XL-complete	XL-complete	[9]

5   Other Classes of Hard Parameterized Problems

There are a several other important classes of parameterized problems, which are assumed not to be fixed parameter tractable. The following brief overview mentions just a few of these, and is far from complete.

The A-hierarchy. Flum and Grohe [25] introduced the **A-hierarchy**: parameterized equivalences of the classes in the polynomial time hierarchy. The hierarchy contains classes $A[1], A[2], \ldots$ While $A[1] = W[1]$, classes higher in the hierarchy contain their W-counterparts as (likely proper) subsets. We will not give the formal definitions here; intuitively, each level in the A-hierarchy adds one alternation between universal and existential quantification.

One such alternation can be seen in the CLIQUE DOMINATING SET problem. Given is an undirected graph $G = (V, E)$, and integers k and ℓ, which both are parameters of the problem. (Or, more precisely, we take $k + \ell$ as parameter.) The question is if there is a set S of k vertices that dominates all cliques with ℓ vertices. (I.e., for every clique C of size ℓ, C contains a vertex that is in S or has a neighbour in S.) CLIQUE DOMINATING SET is an example of an $A[2]$-complete problem [25, Theorem 8.20].

The AW-hierarchy. Alternation is also a key element in the classes defined in the AW-hierarchy [1], see also [17, Chapter 14]. The classes can be defined with help of weighted variants of QUANTIFIED BOOLEAN FORMULAS. Several complete problems for these classes are defined in terms of combinatorial games, where the problem is whether there is a winning strategy for the first player in a given position in at most k moves, where this number of moves k is taken as parameter.

Counting Problems. Let us now consider counting problems, i.e., we want to determine the number of solutions to a problem. In classic complexity theory, many complexity classes have counting variants, with $\#P$ (the class which map the input to the number of accepting paths of a non-deterministic Turing Machine) of central importance. Typical $\#P$-complete problems are: given a Boolean formula, how many satisfying truth assignments does it have; given a graph, how many Hamiltonian circuits does it have? Flum and Grohe [24] introduced parameterized classes for counting problems, including counting variants of the classes in the W-hierarchy.

An interesting example of a $\#W[1]$-complete problem is that of counting the number of paths of length k in a given graph G; k is again the parameter [24]. In contrast, deciding if there is at least one path of length k is fixed parameter tractable. The difference between the complexity of deciding and counting can here be explained by the fact that negative inputs (graphs that do not have a path of length k) have a special structure (e.g., they have treewidth at most k), and such structure can be exploited algorithmically. In contrast, when counting we cannot make assumptions on the graph's structure.

6   Conclusions

In the study of parameterized algorithms, many parameterized problems are known to be hard for a complexity class that is assumed not to be a subset of FPT, and thus, are believed not to be in FPT. For a subset of these problems, completeness for a parameterized class is known. Still, there are many problems where only hardness for some class has been proved, but membership in that class is not known, and sometimes not expected.

Thus, the situation is much less clear than in classic NP-completeness theory. There, for many problems, both NP-hardness and membership in NP is known. The parameterized counterparts of those problems often reside in different classes, and their precise complexity in the hierarchies has often not yet been established. To gain more understanding and give precise characterisations of the parameterized complexity of well known combinatorial problems gives a large number of intriguing open problems. The discussion on XNLP shows that such results can have wider consequences, e.g., give more information on the use of additional resources like memory by the algorithms.

References

1. Abrahamson, K.A., Downey, R.G., Fellows, M.R.: Fixed-parameter tractability and completeness IV: On completeness for $W[P]$ and PSPACE analogues. Ann. Pure Appl. Logic **73**, 235–276 (1995)
2. Abrahamson, K.R., Ellis, J.A., Fellows, M.R., Mata, M.E.: On the complexity of fixed-parameter problems. In: Proceedings of the 30th Annual Symposium on Foundations of Computer Science, FOCS 1989, pp. 210–215 (1989)
3. Adachi, A., Iwata, S., Kasai, T.: Classes of pebble games and complete problems. SIAM J. Comput. **8**(4), 576–586 (1979)
4. Adachi, A., Iwata, S., Kasai, T.: Some combinatorial game problems require $\Omega(n^k)$ time. J. ACM **31**(2), 361–376 (1984)
5. Bodlaender, H.L.: Parameterized complexity of BANDWIDTH of caterpillars and WEIGHTED PATH EMULATION. In: Kowalik, Ł, Pilipczuk, M., Rzążewski, P. (eds.) WG 2021. LNCS, vol. 12911, pp. 15–27. Springer, Cham (2021). https://doi.org/10.1007/978-3-030-86838-3_2
6. Bodlaender, H.L., Downey, R.G., Fellows, M.R., Hermelin, D.: On problems without polynomial kernels. J. Comput. Syst. Sci. **75**, 423–434 (2009)

7. Bodlaender, H.L., Fellows, M.R., Hallett, M.: Beyond NP-completeness for problems of bounded width: Hardness for the W hierarchy. In: Proceedings of the 26th Annual Symposium on Theory of Computing, STOC 1994, pp. 449–458. ACM Press, New York (1994)
8. Bodlaender, H.L., Groenland, C., Nederlof, J., Swennenhuis, C.M.F.: Parameterized problems complete for nondeterministic FPT time and logarithmic space. arXiv abs/2105.14882 (2021). https://arxiv.org/abs/2105.14882. To appear in proceedings FOCS 2021
9. Bodlaender, H.L., Groenland, C., Swennenhuis, C.M.F.: Parameterized complexities of dominating and independent set reconfiguration. In: Golovach, P.A., Zehavi, M. (eds.) 16th International Symposium on Parameterized and Exact Computation, IPEC 2021. LIPIcs, vol. 214, pp. 9:1–9:16. Schloss Dagstuhl - Leibniz-Zentrum für Informatik (2021). https://doi.org/10.4230/LIPIcs.IPEC.2021.9
10. Chen, J., Huang, X., Kanj, I.A., Xia, G.: Strong computational lower bounds via parameterized complexity. J. Comput. Syst. Sci. **72**, 1346–1367 (2006)
11. Chen, Y., Flum, J., Grohe, M.: Bounded nondeterminism and alternation in parameterized complexity theory. In: 18th Annual IEEE Conference on Computational Complexity (Complexity 2003), Aarhus, Denmark, 7–10 July 2003, pp. 13–29. IEEE Computer Society (2003). https://doi.org/10.1109/CCC.2003.1214407
12. Cook, S.A.: The complexity of theorem-proving procedures. In: Proceedings of the 3rd Annual Symposium on Theory of Computing, STOC 1971, pp. 151–158. ACM, New York (1971)
13. Cygan, M., et al.: Parameterized Algorithms. Springer, Cham (2015). https://doi.org/10.1007/978-3-319-21275-3
14. Downey, R.G., Fellows, M.R.: Fixed-parameter tractability and completeness III: Some structural aspects of the W hierarchy. In: Ambos-Spies, K., Homer, S., Schöning, U. (eds.) Complexity Theory, pp. 191–226. Cambridge University Press, Cambridge (1993)
15. Downey, R.G., Fellows, M.R.: Fixed-parameter tractability and completeness I: Basic results. SIAM J. Comput. **24**, 873–921 (1995)
16. Downey, R.G., Fellows, M.R.: Fixed-parameter tractability and completeness II: On completeness for $W[1]$. Theoret. Comput. Sci. **141**, 109–131 (1995)
17. Downey, R.G., Fellows, M.R.: Parameterized Complexity. Springer, New York (1999). https://doi.org/10.1007/978-1-4612-0515-9
18. Downey, R.G., Fellows, M.R.: Fundamentals of Parameterized Complexity. TCS, Springer, London (2013). https://doi.org/10.1007/978-1-4471-5559-1
19. Dregi, M.S., Lokshtanov, D.: Parameterized complexity of bandwidth on trees. In: Esparza, J., Fraigniaud, P., Husfeldt, T., Koutsoupias, E. (eds.) ICALP 2014. LNCS, vol. 8572, pp. 405–416. Springer, Heidelberg (2014). https://doi.org/10.1007/978-3-662-43948-7_34
20. Elberfeld, M., Stockhusen, C., Tantau, T.: On the space and circuit complexity of parameterized problems: classes and completeness. Algorithmica **71**(3), 661–701 (2014). https://doi.org/10.1007/s00453-014-9944-y
21. Fellows, M.R., Langston, M.A.: Nonconstructive advances in polynomial-time complexity. Inf. Process. Lett. **26**, 157–162 (1987)
22. Fellows, M.R., Langston, M.A.: Nonconstructive tools for proving polynomial-time decidability. J. ACM **35**, 727–739 (1988)
23. Fellows, M.R., Rosamond, F.A.: Collaborating with Hans: Some remaining wonderments. In: Fomin, F.V., Kratsch, S., van Leeuwen, E.J. (eds.) Treewidth, Kernels, and Algorithms. LNCS, vol. 12160, pp. 7–17. Springer, Cham (2020). https://doi.org/10.1007/978-3-030-42071-0_2

24. Flum, J., Grohe, M.: The parameterized complexity of counting problems. SIAM J. Comput. **33**(4), 892–922 (2004). https://doi.org/10.1137/S0097539703427203

25. Flum, J., Grohe, M.: Parameterized Complexity Theory. TTCSAES, Springer, Heidelberg (2006). https://doi.org/10.1007/3-540-29953-X

26. Fomin, F., Loksthanov, D., Saurabh, S., Zehavi, M.: Kernelization - Theory of Parameterized Preprocessing. Cambridge University Press, Cambridge (2019)

27. Garey, M.R., Johnson, D.S.: Computers and Intractability, A Guide to the Theory of NP-Completeness. W.H. Freeman and Company, New York (1979)

28. Gurari, E.M., Sudborough, I.H.: Improved dynamic programming algorithms for bandwidth minimization and the mincut linear arrangement problem. J. Algorithms **5**, 531–546 (1984)

29. Kawarabayashi, K., Kobayashi, Y., Reed, B.A.: The disjoint paths problem in quadratic time. J. Comb. Theory Ser. B **102**(2), 424–435 (2012). https://doi.org/10.1016/j.jctb.2011.07.004

30. Mouawad, A.E., Nishimura, N., Raman, V., Simjour, N., Suzuki, A.: On the parameterized complexity of reconfiguration problems. Algorithmica **78**(1), 274–297 (2016). https://doi.org/10.1007/s00453-016-0159-2

31. Niedermeier, R.: Invitation to Fixed-Parameter Algorithms. Oxford Lecture Series in Mathematics and Its Applications, Oxford University Press, Oxford (2006)

32. Pilipczuk, M., Wrochna, M.: On space efficiency of algorithms working on structural decompositions of graphs. ACM Trans. Comput. Theory **9**(4), 18:1-18:36 (2018). https://doi.org/10.1145/3154856

33. Reingold, O.: Undirected connectivity in log-space. J. ACM **55**(4), 1–24 (2008)

34. Saxe, J.B.: Dynamic programming algorithms for recognizing small-bandwidth graphs in polynomial time. SIAM J. Algebraic Discrete Methods **1**, 363–369 (1980)

Invitation to Combinatorial Reconfiguration

Takehiro Ito[✉]

Graduate School of Information Sciences, Tohoku University, Sendai, Japan
takehiro@tohoku.ac.jp

Abstract. Combinatorial reconfiguration arises when we wish to find a step-by-step transformation on the solution space formed by feasible solutions of an instance of a search problem. Many reconfiguration problems have been shown PSPACE-complete, while several algorithmic techniques have been developed. In this talk, I will give a broad introduction of combinatorial reconfiguration.

1 Introduction

Combinatorial reconfiguration [11,14,18] studies the reachability/connectivity of the solution space formed by feasible solutions of an instance of a search problem. A familiar example of combinatorial reconfiguration appears in puzzles. In the 15-puzzle, the tiles with numbers from 1 to 15 are given on a board, and a tile can be slid to an empty spot. The goal is to rearrange the tiles to their target positions. (See Fig. 1.) In the 15-puzzle, the number of possible board configurations is 16! (approximately ten trillion). This means that the solution space of the 15-puzzle consists of approximately ten trillion board configurations, and we are asked to find a step-by-step transformation (namely, a sliding procedure) on the solution space from the initial board configuration to the target one.

The complexity status of the $(n^2 - 1)$-puzzle shows interesting behavior, where the $(n^2 - 1)$-puzzle is a generalization of the 15-puzzle such that the tiles are numbered from 1 to $n^2 - 1$. Given two board configurations of the $(n^2 - 1)$-puzzle, it can be checked in polynomial time whether or not they are reachable each other (see [1] for example). On the other hand, it is NP-hard to minimize the number of tile slides during a transformation between two given board configurations [8, 19].

Fig. 1. Example of the 15-puzzle.

P. Mutzel et al. (Eds.): WALCOM 2022, LNCS 13174, pp. 26–31, 2022.
https://doi.org/10.1007/978-3-030-96731-4_3

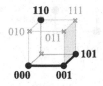

Fig. 2. Solution space for a SAT formula $\phi(x, y, z) = (x \vee \bar{y}) \wedge (\bar{x} \vee y \vee z) \wedge (\bar{y} \vee \bar{z})$, where two truth assignments are adjacent if their Hamming distance is one.

1.1 Framework and Algorithmic Challenge

In general, reconfiguration problems can be defined as follows: For a search problem $\mathcal{P}$, we introduce a *reconfiguration rule* A on the set of feasible solutions which defines whether or not two feasible solutions of an instance I of $\mathcal{P}$ are "adjacent." Then, the *solution space* for I under A is a graph whose node set is the set of all feasible solutions of I and there is an edge joining two nodes if and only if their corresponding solutions are adjacent under A. (See Fig. 2 for example.) We note that the solution space is not given explicitly in reconfiguration problems, but is given implicitly as an instance I of the underlying problem $\mathcal{P}$.

On such a solution space, several natural questions can arise. In the *reachability variant*, we are given two nodes of the solution space, and the goal is to determine whether or not there is a path between the two nodes in the solution space. In the *shortest variant*, the goal is to compute the shortest length of a path between two given nodes in the solution space. Note that the reachability and shortest variants do not ask for an actual path as an output; the reachability variant is a decision problem asking for the existence of a path in the solution space, and the shortest variant outputs simply a shortest length (integer). In the *connectivity variant*, the goal is to determine whether or not the solution space is a connected graph.

In this decade, many reconfiguration problems have been studied from the algorithmic viewpoints, and clarified their complexity status. In particular, reachability variants have been studied intensively for many central combinatorial problems, such as SATISFIABILITY, INDEPENDENT SET and COLORING, and have been shown PSPACE-complete in general. (See, e.g., [11,14,18].) The PSPACE-hardness implies that, unless NP = PSPACE, there exists an instance in the reachability variant which requires a super-polynomial length even in a shortest path in the solution space; in other words, the diameter of the solution space has a super-polynomial length. Thus, the main challenge for solving reconfiguration problems efficiently is to develop smart search methods which need not construct the solution space directly.

1.2 Motivation

The study of reconfiguration problems has motivation from a variety of fields such as puzzles, discrete geometry, and statistical physics. Conversely, there are

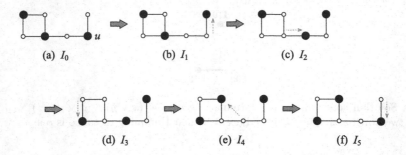

(a) I_0 (b) I_1 (c) I_2

(d) I_3 (e) I_4 (f) I_5

Fig. 3. A sequence $\langle I_0, I_1, \ldots, I_5 \rangle$ of independent sets in the same graph under TJ rule, where the vertices in independent sets are depicted by large black circles (tokens).

some examples such that the algorithmic research of reconfiguration problems gives a new proof to a known result. (For example, see the proof of Theorem 19 in [6].)

Combinatorial reconfiguration also appears in industry. In particular, applications to "24/7 systems" such as power distribution systems are prominent. A power distribution network is designed as supplying electricity via multiple numbers of routes, so as to reduce the blackout duration when failure happens. For example, the Japan standard benchmark model of power distribution networks has approximately 10^{58} alternatives for the choice of network configurations [13]. Even the computation of a single optimal network configuration among them is quite difficult. Furthermore, even if we may compute a single optimal network configuration, we encounter another issue that is characteristic of 24/7 systems. Namely, upon a switching procedure to reconfigure the current configuration to the optimal one, we may not allow any power failure during the process.

2   Independent Set Reconfiguration

In this decade, various reconfiguration problems have been studied for many central combinatorial problems especially on graphs [11,14,18]. As an example, we here explain the reachability variant for independent sets, which is one of the most well-studied reconfiguration problems.

2.1   Three Reconfiguration Rules

Recall that an *independent set* of a graph G is a vertex subset of G in which no two vertices are adjacent, and imagine that a token (coin) is placed on each vertex in the independent set. (Figure 3 depicts six different independent sets in the same graph.) For independent sets, the following three reconfiguration rules have been studied, which are now considered as the most basic reconfiguration rules on graphs.

- *Token Jumping* (TJ rule): We can move a single token to any vertex if it results in an independent set. Formally, two independent sets I and I' of G are adjacent if $|I \setminus I'| = |I' \setminus I| = 1$.

- *Token Sliding* (TS rule): We can slide a single token along an edge of a graph if it results in an independent set. Formally, two independent sets I and I' of G are adjacent if $I \setminus I' = \{v\}$ and $I' \setminus I = \{w\}$ for an edge vw of G.

- *Token Addition and Removal* (TAR rule): We can either add or remove a single token at a time if it results in an independent set of cardinality at least a given threshold.

For example, Fig. 3 shows a sequence of independent sets (i.e., a path in the solution space) under TJ rule. I_3 and I_4 are adjacent only under TJ rule, but the other two consecutive independent sets in the sequence are adjacent also under TS rule. However, there is no path between I_0 and I_5 in the solution space under TS rule. Thus, Fig. 3 is a yes-instance under TJ rule, but is a no-instance under TS rule. In this way, the existence of a desired path depends deeply on the reconfiguration rules. Kamiński et al. [15] showed that TJ and TAR rules are essentially the same, and hence we consider only TJ and TS rules below.

2.2 Complexity Status

The reachability variant for independent sets under TS rule was originally introduced by Hearn and Demaine [9,10] as a one-player game. They proved that INDEPENDENT SET REACHABILITY under TS rule is PSPACE-complete for planar graphs, by a reduction from the problem NONDETERMINISTIC CONSTRAINT LOGIC. We note in passing that NONDETERMINISTIC CONSTRAINT LOGIC plays a very important role in combinatorial reconfiguration, because several PSPACE-hardness of reconfiguration problems have been proved using reductions from this problem.

The complexity status of INDEPENDENT SET REACHABILITY has been analyzed very precisely, in terms of graph classes and graph width parameters.

Surprisingly, Wrochna [20] proved that the problem is PSPACE-complete under both TJ and TS rules even for graphs with bounded bandwidth. Note that the bandwidth of a graph gives an upper bound on the pathwidth and treewidth of the graph. van der Zanden [21] proved that the problem remains PSPACE-complete under both TJ and TS rules even for planar graphs of maximum degree three and bounded bandwidth. On the other hand, Belmonte et al. [2] proved that the problem is fixed-parameter tractable under both TJ and TS rules when parameterized by the modular-width of graphs.

We then summarize the complexity status of the problem from the viewpoint of graph classes. (See Fig. 4.) The problem is PSPACE-complete under both TJ and TS rules for planar graphs [9,10], and for perfect graphs [15]. The problem under TJ rule is NP-complete (surprisingly, it belongs to NP) for bipartite graphs [16], while it can be solved in polynomial time for even-hole-free graphs [15], for cographs [5], and for cactus graphs [17]. Under TS rule, the

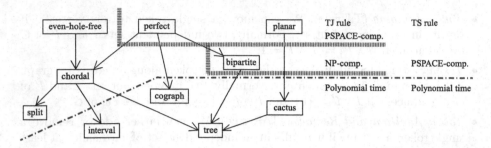

Fig. 4. Complexity status of INDEPENDENT SET REACHABILITY under TJ and TS rules, where each arrow represents the inclusion relationship between graph classes: $A \to B$ indicates that the graph class B is properly included in the graph class A. Thick dotted lines indicate complexity boundaries under TJ rule, and the dash-dotted line indicates the boundary under TS rule.

problem remains PSPACE-complete for bipartite graphs [16] and even for split graphs [3], while it can be solved in polynomial time for cographs [15], for interval graphs [4,7], and for cactus graphs [12].

Acknowledgments. The author thanks Ryuhei Uehara and Yota Otachi for their helpful suggestions. This work is partially supported by JSPS KAKENHI Grant Numbers JP18H04091, JP19K11814 and JP20H05793, Japan.

References

1. Archer, A.F.: A modern treatment of the 15 puzzle. Am. Math. Mon. **106**(9), 793–799 (1999)
2. Belmonte, R., Hanaka, T., Lampis, M., Ono, H., Otachi, Y.: Independent set reconfiguration parameterized by modular-width. Algorithmica **82**(9), 2586–2605 (2020)
3. Belmonte, R., Kim, E.J., Lampis, M., Mitsou, V., Otachi, Y., Sikora, F.: Token sliding on split graphs. Theory Comput. Syst. **65**(4), 662–686 (2021)
4. Bonamy, M., Bousquet, N.: Token sliding on chordal graphs. In: Bodlaender, H.L., Woeginger, G.J. (eds.) WG 2017. LNCS, vol. 10520, pp. 127–139. Springer, Cham (2017). https://doi.org/10.1007/978-3-319-68705-6_10
5. Bonsma, P.: Independent set reconfiguration in cographs and their generalizations. J. Graph Theory **83**(2), 164–195 (2016)
6. Brewster, R.C., McGuinness, S., Moore, B., Noel, J.A.: A dichotomy theorem for circular colouring reconfiguration. Theoret. Comput. Sci. **639**, 1–13 (2016)
7. Briański, M., Felsner, S., Hodor, J., Micek, P.: Reconfiguring independent sets on interval graphs. In: Bonchi, F., Puglisi, S.J. (eds.) 46th International Symposium on Mathematical Foundations of Computer Science (MFCS 2021). Leibniz International Proceedings in Informatics, Dagstuhl, Germany, vol. 202, pp. 23:1–23:14. Schloss Dagstuhl - Leibniz-Zentrum für Informatik (2021)
8. Demaine, E.D., Rudoy, M.: A simple proof that the (n^2-1)-puzzle is hard. Theoret. Comput. Sci. **732**, 80–84 (2018)

9. Hearn, R.A., Demaine, E.D.: PSPACE-completeness of sliding-block puzzles and other problems through the nondeterministic constraint logic model of computation. Theoret. Comput. Sci. **343**(1–2), 72–96 (2005)

10. Hearn, R.A., Demaine, E.D.: Games, Puzzles, and Computation. A. K. Peters Ltd, Natick (2009)

11. van den Heuvel, J.: The complexity of change. In: Blackburn, S.R., Gerke, S., Wildon, M. (eds.) Surveys in Combinatorics 2013. London Mathematical Society Lecture Note Series, vol. 409, pp. 127–160. Cambridge University Press, Cambridge (2013)

12. Hoang, D.A., Uehara, R.: Sliding tokens on a cactus. In: Hong, S.-H. (ed.) 27th International Symposium on Algorithms and Computation (ISAAC 2016). Leibniz International Proceedings in Informatics, Dagstuhl, Germany, vol. 64, pp. 37:1–37:26. Schloss Dagstuhl-Leibniz-Zentrum fuer Informatik (2016)

13. Inoue, T., et al.: Distribution loss minimization with guaranteed error bound. IEEE Trans. Smart Grid **5**(1), 102–111 (2014)

14. Ito, T., et al.: On the complexity of reconfiguration problems. Theoret. Comput. Sci. **412**(12–14), 1054–1065 (2011)

15. Kamiński, M., Medvedev, P., Milanič, M.: Complexity of independent set reconfigurability problems. Theoret. Comput. Sci. **439**, 9–15 (2012)

16. Lokshtanov, D., Mouawad, A.E.: The complexity of independent set reconfiguration on bipartite graphs. ACM Trans. Algorithms **15**(1), 7:1-7:19 (2019)

17. Mouawad, A.E., Nishimura, N., Raman, V., Siebertz, S.: Vertex cover reconfiguration and beyond. Algorithms **11**(2), Paper ID 20 (2018)

18. Nishimura, N.: Introduction to reconfiguration. Algorithms **11**(4), Paper ID 52 (2018)

19. Ratner, D., Warmuth, M.: The $(n^2 - 1)$-puzzle and related relocation problems. J. Symb. Comput. **10**(2), 111–137 (1990)

20. Wrochna, M.: Reconfiguration in bounded bandwidth and tree-depth. J. Comput. Syst. Sci. **93**, 1–10 (2018)

21. van der Zanden, T.C.: Parameterized complexity of graph constraint logic. In: Husfeldt, T., Kanj, I. (eds.) 10th International Symposium on Parameterized and Exact Computation (IPEC 2015). Leibniz International Proceedings in Informatics, Dagstuhl, Germany, vol. 43, pp. 282–293. Schloss Dagstuhl-Leibniz-Zentrum fuer Informatik (2015)

Combinatorial Reconfiguration

Reconfiguration of Regular Induced Subgraphs

Hiroshi Eto[1]([✉]), Takehiro Ito[1] [iD], Yasuaki Kobayashi[2] [iD], Yota Otachi[3] [iD], and Kunihiro Wasa[4] [iD]

[1] Graduate School of Information Sciences, Tohoku University, Sendai, Japan
{hiroshi.eto.b4,takehiro}@tohoku.ac.jp
[2] Graduate School of Informatics, Kyoto University, Kyoto, Japan
kobayashi@iip.ist.i.kyoto-u.ac.jp
[3] Nagoya University, Nagoya, Japan
otachi@nagoya-u.jp
[4] Department of Computer Science and Engineering, Toyohashi University of Technology, Toyohashi, Japan
wasa@cs.tut.ac.jp

Abstract. We study the problem of checking the existence of a step-by-step transformation of d-regular induced subgraphs in a graph, where $d \geq 0$ and each step in the transformation must follow a fixed reconfiguration rule. Our problem for $d = 0$ is equivalent to INDEPENDENT SET RECONFIGURATION, which is one of the most well-studied reconfiguration problems. In this paper, we systematically investigate the complexity of the problem, in particular, on chordal graphs and bipartite graphs. Our results give interesting contrasts to known ones for INDEPENDENT SET RECONFIGURATION.

Keywords: Combinatorial reconfiguration · Regular induced subgraph · Computational complexity

1 Introduction

Combinatorial reconfiguration [9,10,16] studies the reachability in the solution space formed by feasible solutions of an instance of a search problem. In a reconfiguration problem, we are given two feasible solutions of a search problem and are asked to determine whether we can modify one to the other by repeatedly applying a prescribed reconfiguration rule while keeping the feasibility. Such problems arise in many applications and studying them is important also for understanding the underlying problems deeper (see the surveys [9,16] and the references therein) (Fig. 1).

Partially supported by JSPS KAKENHI Grant Numbers JP18H04091, JP18K11168, JP18K11169, JP19K11814, JP19K20350, JP20H05793, JP20K19742, JP21H03499 and JP21K11752, and JST CREST Grant Number JPMJCR18K3, Japan.

© Springer Nature Switzerland AG 2022
P. Mutzel et al. (Eds.): WALCOM 2022, LNCS 13174, pp. 35–46, 2022.
https://doi.org/10.1007/978-3-030-96731-4_4

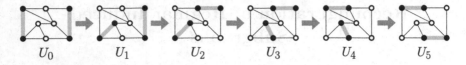

$$U_0 \qquad U_1 \qquad U_2 \qquad U_3 \qquad U_4 \qquad U_5$$

Fig. 1. Example of a TJ-sequence for $d = 1$, where each vertex in a d-regular set U_i is colored with black. Under TS, only $U_2 \leftrightarrow U_3$ holds.

1.1 Our Problems

In this paper, we take d-regular induced subgraphs in a graph as feasible solutions of the solution space. Recall that a graph is *d-regular* if every vertex in the graph is of degree d. By the choice of d, we can represent some well-known graph properties: $d = 0$ corresponds to an independent set of a graph G, and $d = 1$ corresponds to an induced matching of G. If we require d-regular induced subgraphs to be connected, then $d = 2$ corresponds to an induced cycle of G.

We then define two reconfiguration rules on the d-regular induced subgraphs. Since we consider only induced subgraphs of a graph G, each feasible solution can be represented by a vertex subset U of G. We denote by $G[U]$ the subgraph of G induced by U. We say that a vertex subset U of a graph G is a *d-regular set of G* if $G[U]$ is d-regular. Then, there are two well-studied reconfiguration rules [13], called Token Jumping rule (TJ rule for short) and Token Sliding rule (TS rule for short).[1] Let U and U' be two d-regular sets of G. Then, we write

- $U \leftrightarrow U'$ *under* TJ if $|U \setminus U'| = |U' \setminus U| = 1$; and
- $U \leftrightarrow U'$ *under* TS if $U \setminus U' = \{v\}$, $U' \setminus U = \{w\}$, and $vw \in E(G)$.

A sequence $\langle U_0, U_1, \ldots, U_\ell \rangle$ of d-regular sets of G is called a *reconfiguration sequence under* TJ (or TS) between two d-regular sets U_0 and U_ℓ if $U_{i-1} \leftrightarrow U_i$ holds under TJ (resp., TS) for all $i \in \{1, 2, \ldots, \ell\}$. A reconfiguration sequence under TJ (or TS) is simply called a *TJ-sequence* (resp., *TS-sequence*). Note that all d-regular sets in the sequence have the same cardinality.

We now define the problem for a rule $\mathsf{R} \in \{\mathsf{TJ}, \mathsf{TS}\}$, as follows:

---

d-REGULAR INDUCED SUBGRAPH RECONFIGURATION under R (abbreviated as RISR$_d$)

Input: A graph G and d-regular sets U^s and U^t of G.
Question: Is there an R-sequence between U^s and U^t?

A d-regular set U of G is *connected* if $G[U]$ is connected. We also consider the following special case of RISR$_d$, which only allows connected sets as the initial and target d-regular sets:

---

[1] There is another well-studied rule, Token Addition and Removal (TAR) [13]. We are not going to consider this rule as it cannot keep d-regularity unless $d = 0$.

> CONNECTED d-REGULAR INDUCED SUBGRAPH RECONFIGURATION
> under R (abbreviated as CRISR$_d$)
> **Input:** A graph G and connected d-regular sets U^s and U^t of G.
> **Question:** Is there an R-sequence between U^s and U^t?

Although CRISR$_d$ does not explicitly ask intermediate d-regular sets to be connected, it is actually forced by the d-regularity and connectivity of the initial set (see Sect. 2).

1.2 Known and Related Results

Hanaka et al. [8] introduced SUBGRAPH RECONFIGURATION, which unifies several reconfiguration problems where feasible solutions are defined as (induced or non-induced) subgraphs in a graph satisfying a specific property. They considered several graph properties for defining feasible solutions, and one of their results shows that CRISR$_2$ is PSPACE-complete under TS and TJ.

As related work, Ito et al. [12] introduced CLIQUE RECONFIGURATION, which can be seen as a special case of d-regular induced subgraphs. The problem is solvable in polynomial time on even-hole-free graphs (and hence on chordal graphs) under TJ and TS [12]. Mühlenthaler [15] proved that SUBGRAPH RECONFIGURATION is solvable in polynomial time when (not necessarily induced) regular graphs are taken as feasible solutions. His result can be generalized to degree-constrained subgraphs where each vertex has lower and upper bounds for its degree.

INDEPENDENT SET RECONFIGURATION, equivalent to RISR$_0$, is one of the most well-studied reconfiguration problems. RISR$_0$ under TJ is PSPACE-complete on perfect graphs [13], and is NP-complete on bipartite graphs [14], whereas it is solvable in polynomial time on even-hole-free graphs [13], claw-free graphs [6], and cographs [5]. On the other hand, RISR$_0$ under TS is PSPACE-complete on bipartite graphs [14], and on split graphs [3].

These precise complexity analyses of RISR$_0$ show interesting contrast with respect to the reconfiguration rules TS and TJ. (See Table 1.) On chordal graphs, tractability of RISR$_0$ depends on the choice of reconfiguration rules. On bipartite graphs, the complexity of RISR$_0$ shows arguably the most surprising behavior depending on the reconfiguration rules. Lokshtanov and Mouawad [14] showed that, on bipartite graphs, RISR$_0$ is PSPACE-complete under TS but NP-complete under TJ [14]. That is, RISR$_0$ is intractable under both rules but in different senses.

1.3 Our Contribution

In this paper, we investigate the complexity of RISR$_d$ and CRISR$_d$ systematically. In particular, we focus on chordal graphs and bipartite graphs, where the complexity of RISR$_0$ (i.e., INDEPENDENT SET RECONFIGURATION) is known to show interesting behavior depending on the reconfiguration rules. Our results are summarized in Table 1 together with known results for RISR$_0$.

Table 1. Summary of the results.

| | RISR$_d$ | | CRISR$_d$ $(d \geq 2)$ | |
	TS	TJ	TS	TJ
constant bandwidth	PSPACE-c [Corollary 5]		PSPACE-c [Theorem 9]	
chordal	$d = 0$: PSPACE-c [3] $d \geq 1$: PSPACE-c [Theorem 4]	$d = 0$: P [13] $d \geq 1$: PSPACE-c [Theorem 4]	P [12]	
bipartite	$d = 0$: PSPACE-c [14] $d \geq 1$: P [Observation 7]	$d = 0$: NP-c [14] $d \geq 1$: PSPACE-c [Theorem 10]	P [Observation 7]	PSPACE-c [Theorem 8]

We note that our results for $d \geq 1$ give interesting contrasts to known ones for $d = 0$. On chordal graphs, both TS and TJ rules have the same complexity for $d \geq 1$, whereas they are different for $d = 0$. On bipartite graphs, the complexity of RISR$_d$ for $d \geq 1$ shows a kind of reverse phenomenon from $d = 0$: the problem becomes easier (indeed, becomes a trivial case as in Observation 7) under TS, but becomes harder under TJ.

Due to the space limitation, the proofs of the statements marked with $\star$ are deferred to the full version [7].

2   Preliminaries

In this paper, we only consider simple and undirected graphs. Let $G = (V, E)$ be a graph. We sometimes denote by $V(G)$ and $E(G)$ the vertex and edge sets of G, respectively. For a vertex subset U of G, we denote by $G[U]$ the subgraph of G induced by U. We say that U is *connected* if $G[U]$ is connected.

In the introduction, we have defined the problems RISR$_d$ and CRISR$_d$ under R for a rule R $\in \{$TJ, TS$\}$. We denote by $\langle G, U^{\mathsf{s}}, U^{\mathsf{t}} \rangle$ an instance of these problems. Since the problems clearly belong to PSPACE, we will only show PSPACE-hardness for proving PSPACE-completeness.

Note that although CRISR$_d$ only asks the input sets U^{s} and U^{t} to be connected, it is actually required that all sets in a reconfiguration sequence are connected by Lemma 1 below.

Lemma 1 ($\star$). *Let U and U' be d-regular sets of a graph G. If $U \leftrightarrow U'$ under* TJ *or* TS, *then $G[U]$ and $G[U']$ are isomorphic.*

For a positive integer k, we define $[k] = \{d \in \mathbb{Z} \mid 1 \leq d \leq k\}$. For an n-vertex graph $G = (V, E)$, the *width* of a bijection $\pi \colon V \to [n]$ is $\max_{\{u,v\} \in E} |\pi(u) - \pi(v)|$. The *bandwidth* of G is the minimum width over all bijections $\pi \colon V \to [n]$. The graphs of constant bandwidth are quite restricted in the sense that the bandwidth of a graph is an upper bound of pathwidth (and of treewidth) [17]. For example, a path has bandwidth 1 and a cycle has bandwidth 2.

For a positive integer t, a graph H is a *t-sketch* of a graph G if there exists a mapping $f \colon V(G) \to V(H)$ such that $|f^{-1}(v)| \leq t$ for every $v \in V(H)$ and

$\{u, v\} \in E(G)$ implies either $f(u) = f(v)$ or $\{f(u), f(v)\} \in E(H)$. The following lemma is useful for bounding the bandwidth of a graph.

Lemma 2 ($\star$). *Let H be a graph of bandwidth at most b. If H is a t-sketch of G, then the bandwidth of G is at most $t(b + 1)$.*

3 Complexity of RISR and CRISR

In this section, we show that RISR and CRISR are intractable on the classes of chordal graphs, bipartite graphs, and graphs of bounded bandwidth. We also observe that some cases are tractable.

Observe that a connected 0-regular graph is a single vertex and a connected 1-regular graph is a single edge. This implies that, for $d \leq 1$, CRISR_d is polynomial-time solvable under TJ and TS. Hence, we only consider the case of $d \geq 2$ for CRISR_d. On the other hand, we cannot put such an assumption for the more general problem RISR_d as RISR_0 is PSPACE-complete under TJ and TS [13].

3.1 Chordal Graphs

For chordal graphs, let us first consider CRISR_d, which is actually easy. Since every connected regular induced subgraph in a chordal graph is a complete graph [1], CRISR_d on chordal graphs can be solved by using a linear-time algorithm for CLIQUE RECONFIGURATION on chordal graphs [12]. Also, we can use the same algorithm for the general RISR_d with $d \geq 1$ on split graphs since a split graph can have at most one nontrivial connected component.

Observation 3. CRISR_d *with* $d \geq 0$ *on chordal graphs and* RISR_d *with* $d \geq 1$ *on split graphs can be solved in linear time.*

Now let us turn our attention to the general problem RISR_d on chordal graphs. When $d = 0$, the complexity depends on the reconfiguration rules. On chordal graphs, RISR_0 is PSPACE-complete under TS [3], while it is trivially polynomial-time solvable under TJ as it does not have a no-instance [13]. For $d \geq 1$, we show that the problem is PSPACE-complete under both rules.

Theorem 4. *For every constant $d \geq 1$, RISR_d is PSPACE-complete on chordal graphs under TJ and TS.*

Proof. We give a polynomial-time reduction from INDEPENDENT SET RECONFIGURATION on chordal graphs under TS. Let $\langle H, I, I' \rangle$ be an instance of INDEPENDENT SET RECONFIGURATION under TS on chordal graphs. From this instance, we construct an instance $\langle G, U^s, U^t \rangle$ of RISR_d under TJ and TS on chordal graphs as follows. For each $v \in V(H)$, we take a set X_v of $d+1$ vertices. We set $V(G) = \bigcup_{v \in V(H)} X_v$. Each X_v is a clique in G. We add all possible edges between X_u and X_v if $\{u, v\} \in E(H)$. We set $U^s = \bigcup_{v \in I} X_v$ and $U^t = \bigcup_{v \in I'} X_v$. This reduction can be seen as repeated additions of true twin vertices,[2] which do not break the chordality, and thus G is chordal.

---

[2] Two vertices u, v are *true twins* if $N[u] = N[v]$.

We now show that $\langle H, I, I' \rangle$ is a yes-instance of INDEPENDENT SET RECONFIGURATION under TS if and only if $\langle G, U^s, U^t \rangle$ is a yes-instance of RISR$_d$ under TJ and TS.

To prove the only-if direction, we assume that there exists a TS-sequence $\langle I_0, \ldots, I_\ell \rangle$ between I and I'. For each i with $0 \leq i \leq \ell$, we set $R_i = \bigcup_{v \in I_i} X_v$. Observe that each R_i is a d-regular set, $R_0 = U^s$, and $R_\ell = U^t$. Hence, it suffices to show that there is a TS-sequence (and thus a TJ-sequence as well) from R_i to R_{i+1} for all $1 \leq i \leq \ell - 1$. Let $I_i \setminus I_{i+1} = \{p\}$ and $I_{i+1} \setminus I_i = \{q\}$, and thus $R_i \setminus R_{i+1} = X_p$ and $R_{i+1} \setminus R_i = X_q$. Let $X_p = \{p_h \mid 1 \leq h \leq d+1\}$ and $X_q = \{q_h \mid 1 \leq h \leq d+1\}$. We set $T^0 = R_i$, and for each j with $1 \leq j \leq d+1$, we define $T^j = T^{j-1} \setminus \{p_j\} \cup \{q_j\}$. Observe that each T^j is d-regular and $T^{d+1} = R_{i+1}$. We can see that the exchanged vertices p_j and q_j are adjacent in G since p and q are adjacent in H. This implies that $\langle T^0, \ldots, T^{d+1} \rangle$ is a TS-sequence between R_i and R_{i+1}.

To show the if direction, we first show that the existence of a TJ-sequence between U^s and U^t implies the existences of a TS-sequence between them as well. This allows us to start the proof of this direction with the stronger assumption of having a TS-sequence.

Let $\langle R_0, \ldots, R_\ell \rangle$ be a TJ-sequence between U^s and U^t. For each $i \in [\ell]$, we show that there is a TS-sequence between R_{i-1} and R_i. Let $\{u\} = R_{i-1} \setminus R_i$ and $\{v\} = R_i \setminus R_{i-1}$. If $\{u, v\} \in E(G)$, then $R_{i-1} \leftrightarrow R_i$ under TS. Assume that $\{u, v\} \notin E(G)$. Let $w \in R_{i-1} \cap R_i$ be a common neighbor of u and v. Such a vertex exists since $d \geq 1$. Let z be the vertex of H such that $w \in X_z$. Observe that $u, v \notin X_z$ as $\{u, v\} \notin E(G)$. Now $X_z \not\subseteq R_{i-1} \cap R_i \ (= R_{i-1} \setminus \{u\})$ holds since otherwise u has at least $|X_z| = d+1$ neighbors in $G[R_{i-1}]$ as X_z is a set of true twins. Since $u, v \notin X_z$, it holds that $X_z \not\subseteq R_{i-1} \cup R_i$. Thus there exists $w' \in X_z$ that does not belong to $R_{i-1} \cup R_i$. We claim that $R_{i-1} \leftrightarrow (R_{i-1} \setminus \{u\} \cup \{w'\}) = (R_{i-1} \setminus \{v\} \cup \{w'\}) \leftrightarrow R_i$ under TS. We only need to show that the intermediate set $R_{i-1} \setminus \{u\} \cup \{w'\}$ is d-regular. By Lemma 1 and the definition of R_{i-1}, each connected component in $G[R_{i-1}]$ is a $(d+1)$-clique. Let C be the $(d+1)$-clique $G[R_{i-1}]$ that includes u and w. Since w and w' are true twins, the $(d+1)$-clique $C \setminus \{u\} \cup \{w'\}$ is a connected component of $G[R_{i-1} \setminus \{u\} \cup \{w'\}]$. Thus, $R_{i-1} \setminus \{u\} \cup \{w'\}$ is d-regular.

Now assume that there is a TS-sequence $\langle R_0, R_1, \ldots, R_\ell \rangle$ between U^s and U^t. From this sequence, we construct a sequence $\langle S_0, S_1, \ldots, S_\ell \rangle$ of independent sets in G as follows. Recall that $R_0 = \bigcup_{v \in I_0} X_v$. To construct S_0, we pick one arbitrary vertex from X_v for each $v \in I_0$. That is, S_0 is a set such that $|S_0| = |I|$ and $|S_0 \cap X_v| = 1$ for each $v \in I_0$. For $0 \leq i \leq \ell - 1$, we define

$$S_{i+1} = \begin{cases} S_i & S_i \subseteq R_{i+1}, \\ S_i \setminus \{u\} \cup \{v\} & R_i \setminus R_{i+1} = \{u\} \subseteq S_i, \ R_{i+1} \setminus R_i = \{v\}. \end{cases}$$

Intuitively, $S_0 \subseteq R_0$ can be seen as the set of representatives of all X_v contained in R_0. For $i \geq 1$, the set $S_i \subseteq R_i$ traces the tokens started on the representatives chosen to S_0. Each S_i is an independent set of G with size $|I|$. For $0 \leq i \leq \ell$, we now construct I_i by projecting the vertices in S_i onto $V(H)$; that is, we define

$$I_i = \{v \in V(H) \mid X_v \cap S_i \neq \emptyset\}.$$

Since each S_i is an independent set, each I_i is an independent set too. Assume that $I_i \neq I_{i+1}$ for some i, $I_i \setminus I_{i+1} = \{u\}$, and $I_{i+1} \setminus I_i = \{v\}$. Then, $S_i \neq S_{i+1}$ holds. In particular, $S_i \setminus S_{i+1} \subseteq X_u$ and $S_{i+1} \setminus S_i \subseteq X_v$. By the definition of the sequence $\langle S_0, \ldots, S_\ell \rangle$, this further implies that $R_i \neq R_{i+1}$, $R_i \setminus R_{i+1} \subseteq X_u$, and $R_{i+1} \setminus R_i \subseteq X_v$. Since $\langle R_0, \ldots, R_\ell \rangle$ is a TS-sequence in G, the vertices u and v are adjacent in H by the definition of G. Hence, $I_i \leftrightarrow I_{i+1}$ under TS. Therefore, by skipping the sets I_i with $I_i = I_{i+1}$ in $\langle I_0, \ldots, I_\ell \rangle$, we have a TS-sequence from I to I'. □

We can use the same reduction in the proof of Theorem 4 from the same problem but on a different graph class to show the PSPACE-completeness on graphs of bounded bandwidth.

Corollary 5. *For every constant $d \geq 0$, there is a constant b_d depending only on d such that RISR_d is PSPACE-complete under TJ and TS on graphs of bandwidth at most b_d.*

Proof. Wrochna [18] showed that there exists a constant b_0 such that RISR_0 is PSPACE-complete under TJ and TS on graphs of bandwidth at most b_0. Let H be a graph of bandwidth at most b_0. By applying the reduction in the proof of Theorem 4 to H, we obtain a graph G of bandwidth at most $b_d := (d+1)(b_0+1)$. This upper bound of the bandwidth follows from the observation that H is a $(d+1)$-sketch of G and by Lemma 2. □

Similarly, we can show the W[1]-hardness of RISR_d under TJ and TS parameterized by the natural parameter $|U^a|$, which is often called as the *solution size*. For $d = 1$, the W[1]-hardness is known under both TJ [11] and TS [2]. For an instance of RISR_0 under TS parameterized by the solution size k, we apply the reduction in the proof of Theorem 4. The obtained equivalent instance of RISR_d under TJ and TS has solution size $k(d+1)$, and thus the following holds.

Corollary 6. *For every constant $d \geq 0$, RISR_d is W[1]-hard parameterized by the solution size under TJ and TS.*

3.2 Bipartite Graphs

As mentioned in Introduction, on bipartite graphs, RISR_0 is PSPACE-complete under TS but NP-complete under TJ [14]. That is, RISR_0 is intractable under both rules but in different senses. In this section, we study the complexity of RISR_d on bipartite graphs for $d \geq 1$ and show a kind of reverse phenomenon: the problem becomes trivial with TS but harder with TJ.

First we observe the triviality under TS. Observe that if $U \leftrightarrow U'$ under TS, then there exist adjacent vertices $u \in U \setminus U'$ and $v \in U' \setminus U$ such that u and v have a common neighbor $w \in U \cap U'$ as $d \geq 1$. Thus the graph contains a triangle formed by u, v, w. Therefore, in triangle-free graphs (and thus in bipartite graphs), no nontrivial TS-sequence exists.

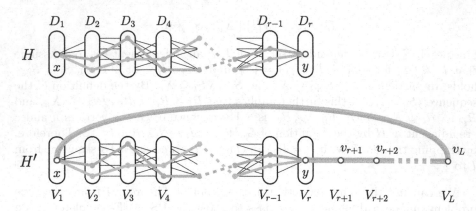

Fig. 2. Reduction from SPR to $CRISR_2$ on bipartite graphs.

Observation 7. *For every $d \geq 1$, $RISR_d$ under* TS *on triangle-free graphs is polynomial-time solvable.*

Under the TJ rule, we first show that the restricted version $CRISR_d$ for $d \geq 2$ is already PSPACE-complete on bipartite graphs. It is known that $CRISR_2$ is PSPACE-complete on general graphs [8]. Our proof for $d = 2$ basically follows their reduction from SHORTEST PATH RECONFIGURATION (SPR, for short) to $CRISR_2$ on general graphs. We start with a reduction for the case of $d = 2$ that has some additional properties and then increase d by using the properties.

An instance of SPR consists of a graph H and the vertex sets P and P' of two shortest x–y paths in H. SPR asks whether there is a sequence $\langle P_0, P_1, \ldots, P_q \rangle$ such that $P_0 = P$, $P_q = P'$, each P_i is the vertex set of a shortest x–y path, and $|P_i \setminus P_{i-1}| = |P_{i-1} \setminus P_i| = 1$ for $1 \leq i \leq q$. Observe that in polynomial time, we can remove all vertices and edges that are not on any shortest x–y path. Hence, we can assume that the vertex set of H is partitioned into independent sets $D_1, D_2, \ldots, D_r$ such that $D_1 = \{x\}$, $D_r = \{y\}$, and each D_i is the set of vertices of distance $i - 1$ from x (see Fig. 2). Bonsma [4] showed that SPR is PSPACE-complete.

Theorem 8. *For every constant $d \geq 2$, $CRISR_d$ is PSPACE-complete on bipartite graphs under* TJ.

Proof. Let $\langle H, P, P' \rangle$ be an instance of SPR with the partition $D_1, D_2, \ldots, D_r$ of the vertex set defined as above. We assume that all sets D_i are independent sets. Let L be the smallest multiple of $2d$ larger than or equal to $\max\{r, 6\}$. To the graph H, we add $L - r$ vertices $v_{r+1}, \ldots, v_L$ and $L - r + 1$ edges $\{y, v_{r+1}\}$, $\{v_L, x\}$, and $\{v_i, v_{i+1}\}$ for $r + 1 \leq i \leq L - 1$. We set $S = P \cup \{v_{r+1}, \ldots, v_L\}$ and $S' = P' \cup \{v_{r+1}, \ldots, v_L\}$. For $1 \leq i \leq r$, we set $V_i = D_i$, and for $r + 1 \leq i \leq L$, we set $V_i = \{v_i\}$. We call the graph constructed H'. See Fig. 2.

The construction is done if $d = 2$. For the cases of $d \geq 3$, we need some additional parts to increase the degree. For each i with $1 \leq i \leq L/(2d)$, we attach the

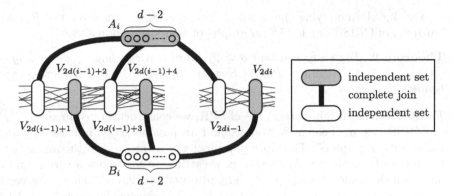

Fig. 3. Increasing the degree from 2 to d.

gadget in Fig. 3 to the consecutive $2d$ sets $V_{2d(i-1)+1}, V_{2d(i-1)+2}, \ldots, V_{2di}$. That is, we add two size-$(d-2)$ independent sets A_i and B_i, and add all possible edges between A_i and $V_{2d(i-1)+(2j-1)}$ for $1 \le j \le d$ and between B_i and $V_{2d(i-1)+(2j)}$ for $1 \le j \le d$. We call the resultant bipartite graph G. Let I denote the set of vertices added, i.e., $I = \bigcup_{1 \le i \le L/(2d)} (A_i \cup B_i)$. We set $U^s = S \cup I$ and $U^t = S' \cup I$. Clearly, U^s and U^t are connected d-regular sets of G.

We now show that $\langle G, U^s, U^t \rangle$ is a yes-instance of CRISR$_d$ if and only if $\langle H, P, P' \rangle$ is a yes-instance of SPR.

To show the if direction, assume that there is a reconfiguration sequence $\langle P_0, \ldots, P_q \rangle$ between P and P'. For $0 \le i \le q$, we set $U_i = P_i \cup \{v_{r+1}, \ldots, v_L\} \cup I$. Observe that U_i is a connected d-regular set: $P_i \cup \{v_{r+1}, \ldots, v_L\}$ induces a cycle, each vertex in the cycle has $d - 2$ neighbors in I, and each vertex in I has d neighbors in the cycle. Since $P_i \setminus P_{i-1} = U_i \setminus U_{i-1}$ and $P_{i-1} \setminus P_i = U_{i-1} \setminus U_i$ for $1 \le i \le q$, $\langle U_0, \ldots, U_q \rangle$ is a TJ-sequence between U^s and U^t.

To show the only-if direction, assume that there is a TJ-sequence $\langle U_0, \ldots, U_q \rangle$ between U^s and U^t. We first show the following fact.

Claim. $I \subseteq U_i$ for each i and $|U_i \cap V_j| = 1$ for each i and j.

Proof of Claim. The claim is true for $U_0 = U^s$. Assume that the claim holds for some i with $0 \le i < q$. Let $U_i \setminus U_{i+1} = \{u\}$ and $U_{i+1} \setminus U_i = \{v\}$. Suppose that $u \in I$. The d neighbors of u in $G[U_i]$ belong to d different sets $V_p, V_{p+2}, \ldots, V_{p+2d-2}$ for some p. Since v has the same neighborhood in $G[U_{i+1}]$, it has to belong to I as well. This contradicts the assumption that $v \notin U_i$ and $I \subseteq U_i$. Thus u belongs to some V_j. In $G[U_i]$, u has neighbors in I, V_{j-1}, and V_{j+1} (where $V_0 = V_L$ and $V_{L+1} = V_1$). Since $L \ge 6$, v has to belong to V_j. Thus the claim holds. ◁

By the claim above, each U_i includes the vertices $v_{r+1}, \ldots, v_L$. Let $P_i = U_i \setminus (I \cup \{v_{r+1}, \ldots, v_L\})$. For $2 \le j \le r - 1$, the unique vertex in $P_i \cap V_j$ has exactly two neighbors in $H[P_i]$; one in V_{j-1} and the other in V_{j+1}. That is, P_i is a shortest x–y path in H. Therefore, $\langle P_0, P_1, \ldots, P_q \rangle$ is a reconfiguration sequence from P to P'. □

By slightly modifying the proof of Theorem 8, we can show the PSPACE-hardness of CRISR_d under TS on graphs of bounded bandwidth.

Theorem 9. *For every constant $d \geq 2$, there is a constant b_d depending only on d such that CRISR_d is PSPACE-complete under TJ and TS on graphs of bandwidth at most b_d.*

Proof. From an instance $\langle H, P, P' \rangle$ of SPR, we construct an instance $\langle G, U^{\text{s}}, U^{\text{t}} \rangle$ of CRISR_d as in Theorem 8. Now we add all possible edges in each V_i and call the resultant graph G'. This does not affect the correctness of the arguments in the proof of Theorem 8. At each step, we still have to remove a vertex and add one from the same set V_i. Since we add all possible edges in each V_i, the vertices involved in this step are adjacent. That is, a TJ-sequence between U^{s} and U^{t} in G is a TS-sequence between U^{s} and U^{t} in G', and vice versa.

Now we prove the claim on the bandwidth. Wrochna [18] showed that there is a constant b such that SPR is PSPACE-complete even if each D_i has size at most b. Let us start the reduction in the proof of Theorem 8 with this restricted version. Let $W_i = A_i \cup B_i \cup \bigcup_{1 \leq j \leq 2d} V_{2d(i-1)+j}$ for each i. We can see that $|W_i| \leq 2db + 2(d-2)$ and that each edge is either in W_i for some i or connecting W_i and W_{i+1}, where $W_{L+1} = W_1$. This implies that a cycle is a $(2db + 2(d-2))$-sketch of G, and thus G has bandwidth at most $b_d := 2(2db + 2(d-2))$ by Lemma 2.   □

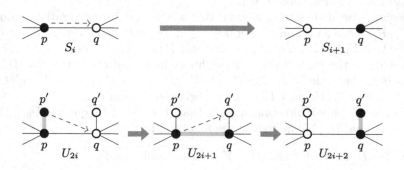

Fig. 4. Simulating a token sliding by two token jumps.

Now we turn our attention back to bipartite graphs and complete the map of complexity of RISR_d.

Theorem 10. *For every constant $d \geq 1$, RISR_d is PSPACE-complete on bipartite graphs under TJ.*

Proof. By Theorem 8, it suffices to show that RISR_1 on bipartite graphs is PSPACE-complete under TJ.

We prove by a reduction from RISR_0 on bipartite graphs under TS, which is PSPACE-complete [14]. Let $\langle H, S, T \rangle$ be an instance of RISR_0 under TS, where

H is bipartite. We obtain a graph G from H by attaching a pendant vertex v' to each vertex v of H. Formally, we set $V(G) = V(H) \cup \{v' \mid v \in V(H)\}$ and $E(G) = E(H) \cup \{\{v, v'\} \mid v \in V(H)\}$. We set $U^s = S \cup \{v' \mid v \in S\}$ and $U^t = T \cup \{v' \mid v \in T\}$. This completes the construction. Note that U^s and U^t are 1-regular sets of G. We show that $\langle G, U^s, U^t \rangle$ is a yes-instance of RISR_1 under TJ if and only if $\langle H, S, T \rangle$ is a yes-instance of RISR_0 under TS.

To show the if direction, assume that $\langle S_0, \ldots, S_\ell \rangle$ is a TS-sequence from S to T. For each i with $0 \le i < \ell$, we set $U_{2i} = S_i \cup \{v' \mid v \in S_i\}$ and $U_{2i+1} = U_{2i} \setminus \{p'\} \cup \{q\}$, where $S_i \setminus S_{i+1} = \{p\}$ and $S_{i+1} \setminus S_i = \{q\}$. That is, when S_{i+1} is obtained from S_i by sliding a token from p to a neighbor q, we obtain U_{2i+2} from U_{2i} by first jumping the token from p' to q (we obtain U_{2i+1}), and then the other token from p to q'. See Fig. 4. Since p, q are adjacent in H, the sets U_{2i}, U_{2i+1}, and U_{2i+2} are 1-regular in G. Therefore, $\langle U_0, \ldots, U_{2\ell} \rangle$ is a TJ-sequence from $U_0 = U^s$ and $U_{2\ell} = U^t$.

To show the only-if direction, assume that $\langle U_0, \ldots, U_\ell \rangle$ is a TJ-sequence from U^s to U^t. Let (A, B) be a partition of $V(H)$ into independent sets of H. From each U_i, we define

$$R_i = \{v \in V(H) \mid \{v, v'\} \subseteq U_i\} \cup \{u \in A \mid \{u, v\} \subseteq U_i,\ \{u, v\} \in E(H)\}.$$

That is, R_i is obtained by projecting U_i onto $V(H)$ and then further replacing two adjacent vertices in $U_i \cap V(H)$ with the one in A. Clearly, each R_i is an independent set of H with size $|U_i|/2 = |S|$. In particular, $R_0 = S$ and $R_\ell = T$. Hence, it suffices to show that there is a TS-sequence from R_i to R_{i+1} for $0 \le i < \ell$. By the definition of R_i, the size of the symmetric difference of R_i and R_{i+1} is at most two. Assume that $R_i \neq R_{i+1}$ and that $u \in R_i \setminus R_{i+1}$ and $v \in R_{i+1} \setminus R_i$ are not adjacent, since otherwise we are done. Now there must be a common neighbor $w \notin R_i \cup R_{i+1}$ of u and v such that $R_i \setminus \{u\} \cup \{w\} = R_{i+1} \setminus \{v\} \cup \{w\}$ is an independent set. That is, $\langle R_i, R_i \setminus \{u\} \cup \{w\}, R_{i+1} \rangle$ is a TS-sequence. $\square$

4 Conclusion

In this paper, we have investigated the computational complexity of d-REGULAR INDUCED SUBGRAPH RECONFIGURATION (RISR_d) and CONNECTED d-REGULAR INDUCED SUBGRAPH RECONFIGURATION (CRISR_d). We have shown that RISR_d is PSPACE-complete for any fixed $d \ge 1$ even on chordal graphs and on bipartite graphs under two well-studied reconfiguration rules, Token Jumping and Token Sliding, except for some trivial cases. The results give interesting contrasts to known results for $d = 0$, namely INDEPENDENT SET RECONFIGURATION. On chordal graphs, the two reconfiguration rules do not make a difference in the complexity of RISR_d for $d \ge 1$, whereas they do make a significant difference for $d = 0$, and on bipartite graphs, the complexity for $d \ge 1$ shows a reverse phenomenon from $d = 0$. For any fixed $d \ge 2$, CRISR_d on bipartite graphs is PSPACE-complete under Token Sliding rule and polynomial-time solvable under Token Jumping rule.

References

1. Asahiro, Y., Eto, H., Ito, T., Miyano, E.: Complexity of finding maximum regular induced subgraphs with prescribed degree. Theor. Comput. Sci. **550**, 21–35 (2014). https://doi.org/10.1016/j.tcs.2014.07.008
2. Bartier, V., Bousquet, N., Dallard, C., Lomer, K., Mouawad, A.E.: On girth and the parameterized complexity of token sliding and token jumping. Algorithmica **83**(9), 2914–2951 (2021). https://doi.org/10.1007/s00453-021-00848-1
3. Belmonte, R., Kim, E.J., Lampis, M., Mitsou, V., Otachi, Y., Sikora, F.: Token sliding on split graphs. Theory Comput. Syst. **65**(4), 662–686 (2020). https://doi.org/10.1007/s00224-020-09967-8
4. Bonsma, P.S.: The complexity of rerouting shortest paths. Theor. Comput. Sci. **510**, 1–12 (2013). https://doi.org/10.1016/j.tcs.2013.09.012
5. Bonsma, P.S.: Independent set reconfiguration in cographs and their generalizations. J. Graph Theory **83**(2), 164–195 (2016). https://doi.org/10.1002/jgt.21992
6. Bonsma, P., Kamiński, M., Wrochna, M.: Reconfiguring independent sets in claw-free graphs. In: Ravi, R., Gørtz, I.L. (eds.) SWAT 2014. LNCS, vol. 8503, pp. 86–97. Springer, Cham (2014). https://doi.org/10.1007/978-3-319-08404-6_8
7. Eto, H., Ito, T., Kobayashi, Y., Otachi, Y., Wasa, K.: Reconfiguration of regular induced subgraphs. CoRR abs/2111.13476 (2021)
8. Hanaka, T., et al.: Reconfiguring spanning and induced subgraphs. Theor. Comput. Sci. **806**, 553–566 (2020). https://doi.org/10.1016/j.tcs.2019.09.018
9. van den Heuvel, J.: The complexity of change. In: Surveys in Combinatorics 2013, pp. 127–160 (2013). https://doi.org/10.1017/CBO9781139506748.005
10. Ito, T., et al.: On the complexity of reconfiguration problems. Theor. Comput. Sci. **412**(12–14), 1054–1065 (2011). https://doi.org/10.1016/j.tcs.2010.12.005
11. Ito, T., Kaminski, M.J., Ono, H., Suzuki, A., Uehara, R., Yamanaka, K.: Parameterized complexity of independent set reconfiguration problems. Discret. Appl. Math. **283**, 336–345 (2020). https://doi.org/10.1016/j.dam.2020.01.022
12. Ito, T., Ono, H., Otachi, Y.: Reconfiguration of cliques in a graph. In: Jain, R., Jain, S., Stephan, F. (eds.) TAMC 2015. LNCS, vol. 9076, pp. 212–223. Springer, Cham (2015). https://doi.org/10.1007/978-3-319-17142-5_19
13. Kamiński, M., Medvedev, P., Milanič, M.: Complexity of independent set reconfigurability problems. Theor. Comput. Sci. **439**, 9–15 (2012). https://doi.org/10.1016/j.tcs.2012.03.004
14. Lokshtanov, D., Mouawad, A.E.: The complexity of independent set reconfiguration on bipartite graphs. ACM Trans. Algorithms **15**(1), 7:1-7:19 (2019). https://doi.org/10.1145/3280825
15. Mühlenthaler, M.: Degree-constrained subgraph reconfiguration is in P. In: Italiano, G.F., Pighizzini, G., Sannella, D.T. (eds.) MFCS 2015. LNCS, vol. 9235, pp. 505–516. Springer, Heidelberg (2015). https://doi.org/10.1007/978-3-662-48054-0_42
16. Nishimura, N.: Introduction to reconfiguration. Algorithms **11**(4), 52 (2018). https://doi.org/10.3390/a11040052
17. Sorge, M., Weller, M.: The graph parameter hierarchy (2019). https://manyu.pro/assets/parameter-hierarchy.pdf
18. Wrochna, M.: Reconfiguration in bounded bandwidth and tree-depth. J. Comput. Syst. Sci. **93**, 1–10 (2018). https://doi.org/10.1016/j.jcss.2017.11.003

Traversability, Reconfiguration, and Reachability in the Gadget Framework

Joshua Ani[1], Erik D. Demaine[1], Yevhenii Diomidov[1], Dylan Hendrickson[1], and Jayson Lynch[2(✉)]

[1] Massachusetts Institute of Technology, Cambridge, MA, USA
{joshuaa,edemaine,diomidov,dylanhen}@mit.edu
[2] Cheriton School of Computer Science, University of Waterloo,
Waterloo, ON, Canada
jayson.lynch@uwaterloo.ca

Abstract. Consider an agent traversing a graph of "gadgets", each with local state that changes with each traversal by the agent. Prior work has studied the computational complexity of deciding whether the agent can reach a target location given a graph containing many copies of a given type of gadget. This paper introduces new goals and studies examples where the computational complexity of these problems are the same or differ from the original relocation goal. For several classes of gadgets—DAG gadgets, one-state gadgets, and reversible deterministic gadgets—we give a partial characterization of their complexity when the goal is to traverse every gadget at least once. We also study the complexity of reconfiguration, where the goal is to bring the entire system of gadgets to a specified state. We give examples where reconfiguration is a strictly harder problem than relocating the agent, and also examples where relocation is strictly harder. We also give a partial characterization of the complexity of reconfiguration with reversible deterministic gadgets.

1 Introduction

The *motion-planning-through-gadgets framework*, introduced in [3] and further developed in [4], captures a broad range of combinatorial motion-planning problems. It also serves as a powerful tool for proving hardness of games and puzzles that involve an agent moving in and interacting with an environment where the goal is to reach a specified location. Prior work [4] fully characterizes the complexity of 1-player motion planning with two natural classes of gadgets: *DAG k-tunnel* gadgets, which naturally lead to bounded games, and *reversible deterministic k-tunnel* gadgets, which naturally lead to unbounded games. Section 2 reviews these and other important definitions.

© Springer Nature Switzerland AG 2022
P. Mutzel et al. (Eds.): WALCOM 2022, LNCS 13174, pp. 47–58, 2022.
https://doi.org/10.1007/978-3-030-96731-4_5

All of the prior work considers **reachability**, where the decision problem is whether the agent can reach the target location.[1] In this paper, we begin extending the gadget model to victory conditions other than reaching a target location. In particular we examine the complexity of reconfiguring a system of gadgets and of visiting every single gadget. These extensions seem natural and interesting, but are also motivated by the fact that this model has been used to show hardness of reconfiguration problems and problems with Hamiltonian Path like constraints.

We consider the **universal traversal** problem of whether the agent can visit every gadget. In Sect. 3, we characterize the complexity of this problem for three classes of k-tunnel gadgets: DAG gadgets, one-state gadgets, and reversible deterministic gadgets. Of particular note is that universal traversal can be harder than reachability for the same gadget. In particular, there are DAG k-tunnel gadgets for which reachability is in P but universal traversal is NP-complete. Additionally, reachability for one-state gadgets is always in NL, but universal traversal can be NP-complete.

In Sect. 4 we consider the **reconfiguration** problem of whether the agent can cause the entire system of gadgets to reach a target configuration. We exhibit a gadget with non-interacting tunnels for which reconfiguration is PSPACE-complete, but reachability is in P. We also show that for reversible gadgets, reconfiguration is at least as hard as reachability. In contrast, we exhibit a nonreversible gadget for which the reconfiguration is contained in P while reachability is NP-complete. The gadgets framework has already been used to prove complexity results about reconfiguration problems related to swarm [2] and modular robotics [1], so understanding reconfiguration in the gadgets model may provide an easier and more powerful base for such applications.

2   Gadget Model

We now define the gadget model of motion planning, introduced in [3].

A **gadget** consists of a finite number of **locations** (entrances/exits) and a finite number of **states**. Each state S of the gadget defines a labeled directed graph on the locations, where a directed edge (a, b) with label S' means that an agent can enter the gadget at location a and exit at location b, changing the state of the gadget from S to S'. Each of these arcs is called a **transition**. Sometimes we will discuss a **traversal** from some location a to location b which refers to any possible transition from a to b in state s. Different states in a gadget can have different transitions while having the same traversability, because the transitions in those different states go from the same entrances to the same exits. Equivalently, a gadget is specified by its **transition graph**, a directed graph whose vertices are state/location pairs, where a directed edge from (S, a) to (S', b) represents that the agent can traverse the gadget from a to b if it is

---

[1] Assembly and motion planning literature often use the term reachability to refer to whether an agent can reach a target location. However, reconfiguration literature uses the term to refer to whether a target location in the configuration space is reachable from another. This would be equivalent to our reconfiguration problem which also specifies a target location for the agent.

in state S, and that such traversal will change the gadget's state to S'. Gadgets are *local* in the sense that traversing a gadget does not change the state of any other gadgets. An example can be seen in Fig. 1.

A *system of gadgets* consists of gadgets, the initial state of each gadget, and an undirected **connection graph** on the gadgets' locations. If two locations a and b of two gadgets (possibly the same gadget) are connected by a path in the connection graph, then an agent can traverse freely between a and b along the connection graph. The **configuration** of a system of gadgets is that system of gadgets along with a state for each of the gadgets in the system. (Equivalently, we can think of locations a and b as being identified, effectively contracting connected components of the connection graph.) These are all the ways that the agent can move: exterior to gadgets using the connection graph, and traversing gadgets according to their current states. An agent's **path** is a sequence of valid transitions through gadgets and moves in the connection graph.

Definition 1. *For a finite set of gadgets F, reachability for F is the following decision problem. Given a system of gadgets consisting of n copies of gadgets in F, and a starting location and a win location in that system of gadgets, is there a path the agent can take from the starting location to the win location?*

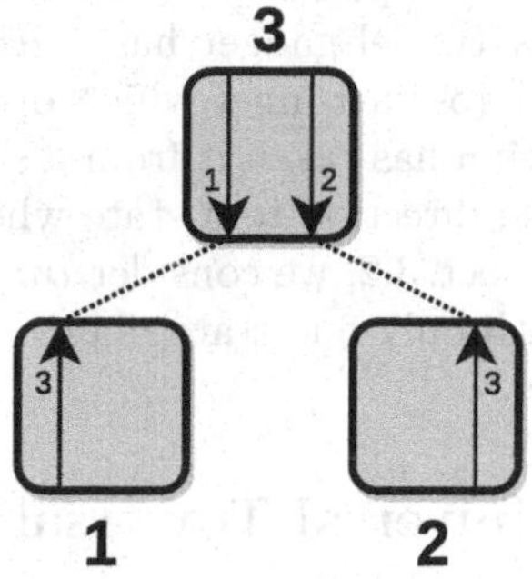

Fig. 1. A diagram describing the locking 2-toggle gadget. Each box represents the gadget in a different state, in this case labeled with the numbers $1, 2, 3$. Arrows represent transitions in the gadget and are labeled with the states to which those transition take the gadget. In the top state 3, the agent can traverse either tunnel going down, which blocks off the other tunnel until the agent reverses that traversal. Dotted lines help visualize the associated transitions between states.

We will consider several specific classes of gadgets.

A *k-tunnel* gadget has $2k$ locations, which are partitioned into k pairs called *tunnels*, such that every transition is between two locations in the same tunnel. The *state-transition graph* of a gadget is the directed graph which has a vertex for each state, and an edge $S \to S'$ for each transition from state S to S'. A **DAG** gadget is a gadget whose state-transition graph is acyclic. DAG gadgets naturally lead to problems with a polynomially bounded number of transitions, since each gadget can be traversed a bounded number of times. The complexity of the reachability problem for DAG k-tunnel gadgets, as well as the 2-player and team games, is characterized in [4].

A gadget is **deterministic** if every traversal goes to only one state and every location has at most 1 traversal from it. More precisely, its transition graph has maximum out-degree 1.

A gadget is **reversible** if every transition can be reversed. More precisely, its transition graph is undirected.

Reversible deterministic gadgets are gadgets whose transition graphs are partial matchings, and they naturally lead to unbounded problems. Prior work [4] characterizes the complexity of reachability for reversible deterministic k-tunnel gadgets and partially characterizes the complexity of 2-player and team games.

A k-tunnel gadget has a **distant opening** if there is a transition in some state across a tunnel which opens a different tunnel. A tunnel is **opened** if a transition has taken it from a state where the tunnel did not have traversability in some direction to a state where it is now traversable.

In Sect. 3.2, we consider **one-state**, k-tunnel gadgets. A transition in a gadget with only one state does not change the state, so the legal traversals never change.

3   Universal Traversal

In this section, we consider whether an agent in a system of gadgets can make a traversal across every gadget, called the **universal traversal** problem.

Definition 2. *For a finite set of gadgets F, **universal traversal for F** is the following decision problem. Given a system of gadgets consisting of n copies of gadgets in F, and a starting location and a win location in that system of gadgets is there a path the agent can take from the starting location which makes at least one traversal in every gadget?*

We provide a full characterization for the complexity of this problem for three classes of gadgets. In Sect. 3.1, we characterize DAG k-tunnel gadgets. Universal traversal is NP-hard for some DAG gadgets where reachability is in P. This is somewhat similar to the distinction between finding paths and finding Hamiltonian paths. In Sect. 3.2, we further emphasize this difference by characterizing one-state k-tunnel gadgets. Reachability is always in NL for one-state gadgets, but we find that universal traversal is often NP-complete. Finally, in the full version of the paper we consider the unbounded case by characterizing universal traversal for reversible deterministic k-tunnel gadgets. In this case, the dichotomy is the same as for reachability.

3.1   DAG Gadgets

In this subsection, we consider universal traversal for k-tunnel DAG gadgets and show this problem is NP-hard for any DAG gadget which has and actually uses at least 2 tunnels, in the sense defined below. For some simple 1-tunnel DAG gadgets, universal traversal is analogous to finding Eulerian paths and is thus in P; however, more complex 1-tunnel gadgets can not easily be converted to an Eulerian path problem. For example the 1-toggle which switches direction after each transition or a gadget which can be traversed at most twice. We leave the case of 1-tunnel DAG gadgets open.

Open Problem 1. *Is universal traversal with any 1-tunnel DAG gadget in P? Are there 1-tunnel DAG gadgets for which universal traversal is NP-complete?*

Some k-tunnel DAG gadgets with $k >$ 1 act like 1-tunnel gadgets in that it is never possible to make use of multiple tunnels. A simple example is shown in Fig. 2. We formalize this notion in the following definition.

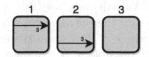

Fig. 2. A 2-tunnel DAG gadget which is not true 2-tunnel.

Definition 3. *A state of a k-tunnel gadget is **true 2-tunnel** if there are at least two tunnels, each of which is traversable in some state reachable (through any number of transitions) from that state. A gadget is **true 2-tunnel** if it is a k-tunnel gadget and has a true 2-tunnel state.*

Note that a k-tunnel gadget does not need multiple tunnels traversable in the same state to be true 2-tunnel: perhaps traversing the single traversable tunnel opens another tunnel. To justify this definition, we prove the following result.

Theorem 1. *Let G be a k-tunnel which is not true 2-tunnel. Then there is a 1-tunnel gadget G' and a bijection between states of G to states of G' such that replacing each copy of G in a system of gadgets with a copy of G' in the corresponding state gives an equivalent system of gadgets with respect to reachability and universal traversal.*

We will use the fact that every nontrivial DAG gadget simulates either a directed or an undirected single-use path, since we can take a final nontrivial state of the gadget [4]. The rest of this subsection is devoted to proving NP-completeness for universal traversal for true 2-tunnel DAG gadgets.

Theorem 2. *Universal traversal with any true 2-tunnel DAG gadget is NP-complete.*

To prove Theorem 2, we will focus on a **final** true 2-tunnel state of a DAG gadget, and only use the two tunnels which make this state true 2-tunnel. A final true 2-tunnel state is a true 2-tunnel state from which no other true 2-tunnel state can be reached. Such a state exists because the state-graph is a DAG. After making a traversal in this state, any resulting state is not true 2-tunnel, so only one of the two tunnels can be traversed in the future. If the gadget is nondeterministic, the agent may be able to choose which of the two tunnels this is. We consider several cases for the form of the last true 2-tunnel state, and show NP-hardness for each one. Most proofs are left to the full version of the paper.

The first case we consider is when the final true 2-tunnel state being considered has a distant opening.

Lemma 1. *Let G be a true 2-tunnel gadget and let S be a final true 2-tunnel state of G. If there exists a transition from S across one tunnel which opens a traversal across another tunnel, then universal traversal for G is NP-hard.*

proof. We will only use the two tunnels involved in the opening transition from S to S' where S' has some traversal which was not possible in S. Suppose traversing the top tunnel from left to right allows the agent to open the left-to-right traversal on the bottom tunnel. Then state S has one of the two forms shown in Fig. 3, depending on whether the bottom tunnel can be traversed right to left in S. In either case, the top tunnel may or may not be traversable from right to left in S. Since S is a final true 2-tunnel state, only the bottom tunnel is traversable in S'.

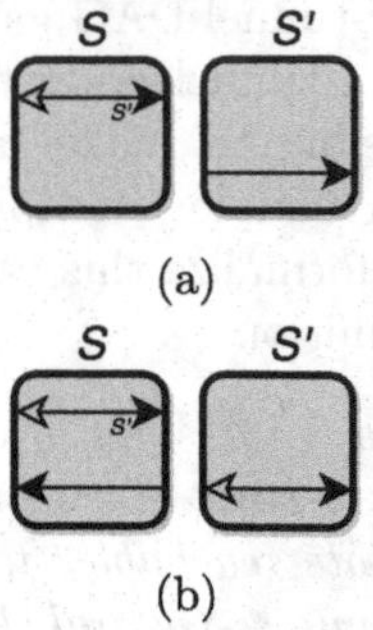

To show NP-hardness of universal traversal with true 2-tunnel gadget G, we reduce from reachability for G. Since the gadget has a distant opening, reachability is NP-complete [4]. We modify the system of gadgets in an instance of the reachability problem by adding a construction to each gadget which allows the agent to go back and make a traversal in it after reaching the win location. If the agent can reach the win location, it can then use any gadgets it did not already use, and if it cannot reach the win location, it cannot use the gadgets in this construction.

The construction is slightly different depending on whether the bottom tunnel can be traversed from right to left in state S. We use the construction in either Fig. 4 or Fig. 5. In either case, the agent cannot use the newly added gadgets until it first reaches the win location. Once it reaches the win location, it can

Fig. 3. Two cases for the form of the gadget in Lemma 1, assuming traversing the top tunnel to the right opens the bottom tunnel to the right. In (a) the bottom tunnel is not traversable to the left in state S and in (b) it is. Unfilled arrows are traversals that may or may not exist depending on the gadget. Unlabled transitions may be to arbitrary states not specified here.

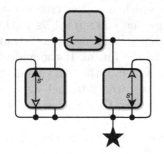

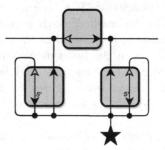

Fig. 4. The construction to allow the agent to use a gadget after reaching the win location (the star), when the bottom tunnel isn't traversable in state S (the case of Fig. 3a).

Fig. 5. The construction to allow the agent to use a gadget after reaching the win location (the star), when the bottom tunnel is traversable from right to left in state S (the case of Fig. 3b).

open tunnels in the added gadgets, traverse the (top) gadget the construction is attached to, and return. If the agent already used the gadget this is attached to, it can instead use a traversal in each added gadget without visiting that gadget. So it is possible to make a traversal in every gadget if and only if the original reachability problem is solvable. □

Now we will assume the final true 2-tunnel state has no distant opening. If only one tunnel is traversable in this state, then it cannot be true 2-tunnel because the other tunnel will never become traversable. So both tunnels are traversable, and after making any traversal, there is only one tunnel which will ever be traversable. With no distant opening, we first consider the case where at least one of the tunnels is directed in the final true 2-tunnel state.

Lemma 2. *Let G be a true 2-tunnel gadget and let S be a final true 2-tunnel state of G. Suppose no transition from S across one tunnel opens a traversal across the other tunnel. If, in S, some tunnel can be traversed in one direction but not in the other, then universal traversal for G is NP-hard.*

The remaining case is when, in the final true 2-tunnel state, there is no distant opening and all tunnels are undirected. We branch into two cases one last time, based on whether traversing one tunnel requires closing the other tunnel. These can be found in the full version of the paper.

Open Problem 2. *Is universal traversal restricted to planar systems of gadgets NP-hard for all true 2-tunnel DAG gadgets?*

3.2 One-State Gadgets

In this subsection, we consider universal traversal for k-tunnel gadgets with only one state. The reachability problem is clearly in NL for such gadgets, but we will see that universal traversal is often NP-complete.

A one-state k-tunnel gadget consists of directed and undirected tunnels, and is determined by the number of each type; we assume there is no untraversable tunnel since such a tunnel can be removed without affecting the problem. We fully characterize the complexity of universal traversal for such gadgets. We only prove a key lemma here, the rest can be found in the full version of the paper.

Theorem 3. *Let G be a one-state k-tunnel gadget. If G has no directed tunnels, then universal traversal for G is in L. Otherwise, if $k \leq 2$ universal traversal for G is NL-complete and if $k \geq 3$ universal traversal for G is NP-complete.*

Lemma 3. *Universal traversal with any one-state k-tunnel gadget is in NL if $k \leq 2$.*

Proof. We provide an algorithm which runs in nondeterministic logarithmic space with an oracle for reachability in directed graphs. This shows that the universal traversal problem is in NL^{NL}. The algorithm can be adapted to run in NL by first using the oracle to convert the problem to an instance of 2SAT. It then solves this instance, since 2SAT is in NL.

The 2SAT formula has a variable for each tunnel in the system of gadgets; a satisfying assignment will provide a set of tunnels we can traverse to solve the universal traversal problem. For each gadget with tunnels x_1 and x_2, we have a clause $x_1 \vee x_2$ (if the gadget has only one tunnel, $x_1 = x_2$). For each pair of distinct tunnels x and y, we query the reachability oracle to determine whether there is a path from the exit of x to the entrance of y or from the exit of y to the entrance of x (if x or y is undirected, we can use either location as the entrance or exit). If there is no path in either direction, we have a clause $\neg x \vee \neg y$.

We prove that this algorithm works, and then adapt it to an L^{NL} algorithm which is known to equal NL [6]. Suppose the universal traversal problem is solvable, and consider the assignment which contains the tunnels which are used in the solution. Since the solution must use a tunnel in every gadget, each clause $x_1 \vee x_2$ is satisfied. If the solution uses both tunnels x and y, there must be a path in some direction between x and y, namely the path the agent takes between the two tunnels. For each clause $\neg x \vee \neg y$ in the formula, there is no such path, so the solution does not use both tunnels x and y, so the clause is satisfied.

Now suppose the 2SAT formula is satisfiable, and consider the set T of tunnels corresponding to true variables in a satisfying assignment. Because of the clauses $x_1 \vee x_2$, T must contain a tunnel in each gadget. We define a relation $\rightarrow$ on T where $x \rightarrow y$ if there is a path from the exit of x to the entrance of y. As suggested by the notation, this relation is transitive: if $x \rightarrow y \rightarrow z$, there is a path from the exit of x to the entrance of y, across y, and then to the entrance of z, so $x \rightarrow z$. Since each clause $\neg x \vee \neg y$ is satisfied, for any distinct $x, y \in T$ we have $x \rightarrow y$ or $y \rightarrow x$. That is, $\rightarrow$ is a strict total pre-order.

Then there must be a (strict) total order $\prec$ on T such that $x \prec y \implies x \rightarrow y$: define another relation $\sim$ where $x \sim y$ if $x = y$ or both $x \rightarrow y$ and $y \rightarrow x$. Then $\sim$ is clearly an equivalence relation, and $\rightarrow$ is a total order on $T/\sim$. We can construct $\prec$ by putting the equivalence classes under $\sim$ in order according to $\rightarrow$, and arbitrarily ordering the elements of each equivalence class.

This now shows that there exists a set of locations from which a universal traversal is possible. The last step is to nondeterministically check that the start location of the agent has no strict predecessor in the preorder. This can be done by checking that the starting location is in the same equivalence class as the minimal element in our chosen total ordering. The agent can traverse the tunnels in T in the order described by $\prec$. This is a solution to the universal traversal problem.

We run the algorithm in nondeterministic logarithmic space as follows. Begin with an NL algorithm that solves 2SAT, and assume the input is given in a format where we can check whether a clause $a \vee b$ is in the formula by checking a single bit for literals a and b. For example, the input can be given as a matrix with a row and column for each literal. We run this nondeterministic 2SAT algorithm, except that whenever we would read a bit of the input, we perform a procedure to determine whether that clause is in the formula.

Suppose the algorithm to solve universal traversal wants to know whether $a \vee b$ is in the formula. If a and b are both positive literals, we simply check whether they correspond to tunnels in the same gadget. If a and b have different signs, the clause is not in the formula. The interesting case is when $a = \neg x$ and $b = \neg y$ for tunnels x and y, where we need to determine whether there is a path from the exit of x to the entrance of y or vice-versa.

In this case, we nondeterministically guess whether the clause exists, and then check whether the guess was correct. If we guess it does exist, we run a coNL algorithm to verify that there is no path from the exit of x to the entrance of y or vice versa; this can be converted to an NL algorithm. If the verification succeeds, we proceed; if it fails, we halt and reject. Similarly, if we guess the clause does not exist, we run an NL algorithm to verify that there is such a path, proceeding on success and rejecting on failure.

Consider the computation branches which have not rejected after this process. If the clause exists, the branch which attempted to verify it does not exist has entirely rejected, and the branch which attempted to verify it does exist has succeeded in at least one branch. So there is at least one continuing branch, and every such branch believes that the clause exists. Similarly if the clause does not exist, we end up with only branches which guessed that it does not exist.   □

4   Gadget Reconfiguration

In this section we study the question of whether an agent has a series of moves after which the system of gadgets will be in some target configuration. In Sect. 4.1 we show that for reversible deterministic gadgets the reconfiguration problem is always PSPACE-complete if the reachability problem is PSPACE-complete. Section 4.2 shows some methods for constructing new PSPACE-complete gadgets from known ones and shows the reconfiguration problem can be PSPACE-complete even when a gadget does not change traversability. Finally, in Sect. 4.3, we show an interesting connection between reconfiguration problems and bounded reachability problems, expanding the classes of gadgets known to be in NP. We also exhibit, a gadget for which the reachability question is NP-complete but the reconfiguration question is in P.

Definition 4. *For a finite set of gadgets F, **reconfiguration for F** is the following decision problem. Given a system of gadgets consisting of n copies of gadgets in F, a target configuration for that system of gadgets, and a starting location is there a path the agent can take from the starting location which makes the configuration of the system of gadgets equal to the target configuration?*

4.1   Reconfiguring Reversible Gadgets

Theorem 4. *For any set of reversible gadgets containing at least one gadget which is able to change state, there is a polynomial-time reduction from reachability with those gadgets to reconfiguration with those gadgets.*

Proof. We use the same technique as used to show that reconfiguration Nondeterministic Constraint Logic is PSPACE-complete [5]. We are given an instance of the reachability problem, which is a network of gadgets with a target location. At the target location, add a loop with a single gadget which permits a traversal which changes its state. For the reconfiguration problem, we set the target states of all but the newly added gadget to be the same as the initial states, and we set the target state of the added gadget to be one reachable by making a traversal in the loop. If the reachability problem is solvable, the reconfiguration problem can be solved by navigating to the target location, traversing the loop through the added gadget, and taking the inverse transitions of the path taken to the target location to restore all other gadgets to the initial state. If the reconfiguration problem is solvable, its solution must involve visiting the added gadget, so the reachability problem is solvable. □

4.2 Verified Gadgets and Shadow Gadgets

In this section we will discuss a technique for generating gadgets for which the reconfiguration and reachability problems are computationally hard. The main idea is constructing a gadget which behaves well when used like a gadget with a known hardness reduction, but might also have other transitions which are allowed but put the gadget into some undesirable state.

First, we will pick some base gadget which we want to modify. Next we will add additional **shadow states** to the gadget and additional transitions with the restriction that all newly added transitions must take the gadget to a shadow state. We call such a construction a **shadow gadget** of the base gadget. This has the nice property that if the agent takes any transition not be allowed in the base gadget, then the gadget will always stay in a shadow state after that transition.

Theorem 5. *Reconfiguration with a shadow gadget is at least as hard as reconfiguration with the base gadget.*

Corollary 1. *There is a gadget which never changes its traversability but with which reconfiguration is PSPACE-complete.*

Figure 6 contains a diagram of the 2-toggle which is PSPACE-complete for reachability [3] and an example of a shadow 2-toggle which is PSPACE-complete for reconfiguration and is a gadget which never changes traversability (if there is a transition from some location a to another location b in any state, there must be a transition from a to b in every state). In fact all tunnels are always traversable in both directions. Finally, the figure shows a verified 2-toggle which is PSPACE-complete for reachability and a construction we will discuss next.

A **verified gadget** is a shadow gadget with some additional structure. From a shadow gadget we add two more locations, the **verifying locations** to the gadget. We may also add **verified states** which can only be reached by transitions from the added locations while the gadget is in normal states. We now add transitions among the verifying locations such that these locations can be connected in a series, there is always a traversal from the first to the last location if the gadget is in a normal state, and there is no such traversal if the gadget is in a shadow state. We call this added traversal the **verification traversal**.

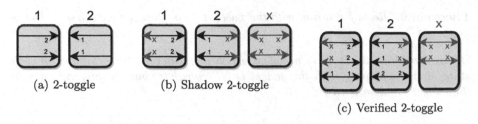

(a) 2-toggle   (b) Shadow 2-toggle

(c) Verified 2-toggle

Fig. 6. Examples of a PSPACE-complete gadget and constructions of a shadow gadget and verified gadget based on it.

Theorem 6. *Reachability with a verified gadget is at least as hard as reachability with the base gadget.*

4.3 Reconfiguration and DAG-like Gadgets

In [4] we study DAG gadgets as a naturally bounded class of gadgets. We now consider a generalized class of gadgets and describe cases in which the reachability question remains in NP.

For a finite family of gadgets F, we call a gadget ***F-DAG-like*** if its state graph can be decomposed into disjoint subgraphs for which those subgraphs are gadgets in F and the transitions between these subgraphs are acyclic. We call the transitions between the subgraphs ***F-DAG-like transitions***.

With this notion, one may wonder what gadgets can be used in an F-DAG-like gadget and have the resulting reconfiguration or rechability problem with that gadget still be in NP. We then show that if F is a family of gadgets for which the reconfiguration problem is in NP, then the reconfiguration and reachability problems for F and for F-DAG-like gadgets are also in NP. We call a finite set of gadgets ***NPReDAG*** if they are all F-DAG-like for some fixed family F for which the reconfiguration problem is in NP. Proofs can be found in the full version of the paper.

Theorem 7. *Reconfiguration with any NPReDAG set of gadgets is in NP.*

Theorem 8. *If reconfiguration with some set of gadgets is in NP, than reachability is also in NP.*

4.4 Reconfiguration Can Be Easier

In this section we introduce the Labeled Two-Tunnel Single-Use gadget for which the reachability question is harder than the reconfiguration problem. The Labeled Two-Tunnel Single-Use gadget is a DAG gadget where going through either tunnel closes both of them; however, the states are distinguished based on which tunnel was traversed. This is a DAG gadget with a forced distant door closing, so it is NP-complete by Theorem 22 in [4]. We give a polynomial time algorithm for the reconfiguration problem in the full version of the paper.

Theorem 9. *Reconfiguration with the Labeled Two-Tunnel Single-Use gadget is in P.*

Open Problem 3. *Is there a gadget which has different traversability in each state, and reachability with the gadget is NP-complete, but reconfiguration with the gadget problem is in P?*

4.5   Reconfiguration with 1-tunnel

Just as with universal traversal, moving to reconfiguration can also be NP-hard with 1-tunnel gadgets. As an example we show that reconfiguration with a diode and a single-use gadget is NP-complete. A diode is a 1-state gadget which only allows traversals from location A to location B. A single-use gadget is a two state gadget in which state 1 allows a traversal between locations A and B which changes the gadget to state 2. State 2 has no traversals. The reduction is from Hamiltonian path.

Theorem 10. *Reconfiguration with the diode and the single-use gadget is NP-complete.*

Open Problem 4. *For what classes of 1-tunnel gadgets is reconfiguration NP-complete? What about PSPACE-complete?*

Acknowledgments. This work grew out of an open problem session and a final project from MIT class on Algorithmic Lower Bounds: Fun with Hardness Proofs (6.892) from Spring 2019.

References

1. Akitaya, H.A., et al.: Characterizing universal reconfigurability of modular pivoting robots. In: 37th International Symposium on Computational Geometry (2021)
2. Balanza-Martinez, J., et al.: Full tilt: universal constructors for general shapes with uniform external forces. In: Proceedings of the Thirtieth Annual ACM-SIAM Symposium on Discrete Algorithms, pp. 2689–2708. SIAM (2019)
3. Demaine, E.D., Grosof, I., Lynch, J., Rudoy, M.: Computational complexity of motion planning of a robot through simple gadgets. In: Proceedings of the 9th International Conference on Fun with Algorithms (FUN 2018), La Maddalena, Italy, pp. 18:1–18:21, June 2018
4. Demaine, E.D., Hendrickson, D., Lynch, J.: Toward a general theory of motion planning complexity: characterizing which gadgets make games hard. In: Proceedings of the 11th Conference on Innovations in Theoretical Computer Science (ITCS 2020), Seattle, Washington, pp. 62:1–62:42, January 2020
5. Hearn, R.A., Demaine, E.D.: PSPACE-completeness of sliding-block puzzles and other problems through the nondeterministic constraint logic model of computation. Theoret. Comput. Sci. **343**(1–2), 72–96 (2005)
6. Immerman, N.: Nondeterministic space is closed under complementation. SIAM J. Comput. **17**(5), 935–938 (1988)

1-Complex s, t Hamiltonian Paths: Structure and Reconfiguration in Rectangular Grids

Rahnuma Islam Nishat[1]([✉]), Venkatesh Srinivasan[2], and Sue Whitesides[2]

[1] Department of Computer Science, Ryerson University, Toronto, ON, Canada
rnishat@ryerson.ca
[2] Department of Computer Science, University of Victoria, Victoria, BC, Canada
{srinivas,sue}@uvic.ca

Abstract. We give a complete structure theorem for 1-complex s, t Hamiltonian paths in rectangular grid graphs. We use the structure theorem to design an algorithm to reconfigure one such path into any other in linear time, making a linear number of *switch* operations in grid cells.

1 Introduction

Let $\mathbb{G}$ be an $m \times n$ *rectangular grid graph*, which is an induced, embedded subgraph of the infinite integer grid and has m rows and n columns in an $(m - 1) \times (n - 1)$ rectangle $R_{\mathbb{G}}$. Let s and t be the top left and bottom right corners of $R_{\mathbb{G}}$. We define a class of s, t Hamiltonian paths (s and t are the endpoints of the Hamiltonian path) that we call *1-complex paths*. A 1-complex path is an s, t Hamiltonian path of $\mathbb{G}$, where each vertex of $\mathbb{G}$ can be connected to a node on one of the sides $\mathcal{E}, \mathcal{W}, \mathcal{N}, \mathcal{S}$ of $R_{\mathbb{G}}$ by a straight line segment on the path, as in Fig. 1(a). We reconfigure such paths using *switches* in 1×1 grid *cells*. Let $R_{\mathbb{G}}$ be covered by an s, t path and 0 or more disjoint cycles (e.g., initially $R_{\mathbb{G}}$ is covered by an s, t Hamiltonian path). If a cell has two parallel grid edges that belong to the cover and two parallel grid edges that do not, then a *switch* exchanges the

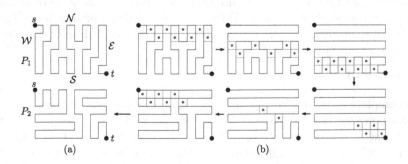

(a)   (b)

Fig. 1. (a) Two 1-complex s, t Hamiltonian paths P_1 and P_2. Sides of $R_{\mathbb{G}}$ are $\mathcal{N}, \mathcal{S}, \mathcal{W}, \mathcal{E}$. (b) A sequence of *switches* (shown by red dots) taking P_1 to P_2. (Color figure online)

© Springer Nature Switzerland AG 2022
P. Mutzel et al. (Eds.): WALCOM 2022, LNCS 13174, pp. 59–70, 2022.
https://doi.org/10.1007/978-3-030-96731-4_6

edges in that cell that belong to the cover for the two edges not in the cover (see Fig. 1(b)). A switch produces a new cover comprising an s, t path and 0 or more disjoint cycles. Whether a cell may be switched depends on the cover.

The question we ask is this: given any two 1-complex paths of $\mathbb{G}$, can one of them be reconfigured to the other with only $O(|\mathbb{G}|)$ switches in grid cells, and if so, can the sequence of switches to be performed be computed efficiently?

As shown in Fig. 2, a "cross-separator" (i.e., a subpath η_i joining nodes on opposite sides of $R_{\mathbb{G}}$) may have many forms, depending on whether and where this subpath has bends. In previous work [9] we introduced a special case of 1-complex s, t Hamiltonian paths we called "simple". By definition, a simple s, t path has only straight cross-separators, so none of the subpaths in Fig. 2 that contain bent cross-separators can occur in simple paths.

A key contribution of this work is a complete structure theorem for 1-complex paths, building on the work in [9] for the special case of simple paths. Using our structure theorem for 1-complex paths, we achieve a linear time algorithm for reconfiguration of 1-complex s, t Hamiltonian paths $P_{s,t}$. Roughly, our algorithm uses the structure of a $P_{s,t}$ to define smaller sub-rectangles within $R_{\mathbb{G}}$. Path $P_{s,t}$ determines s', t' Hamiltonian paths $p'_{s',t'}$ within each sub-rectangle. We define the sub-rectangles so that our structure theorem as well as our new reconfiguration tools (Sect. 4) apply to each path $p'_{s',t'}$. See Fig. 2 and Sect. 5.

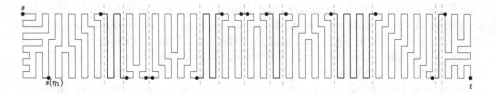

Fig. 2. The structure of the s, t path is used to break $\mathbb{G}$ into sub-rectangles.

The existence problem for Hamiltonian paths in rectangular grids [5] and non-rectangular grids [2,7] has long been studied, as well as enumeration [6] and generating functions [1] for such paths. In recent years, reconfiguration problems for Hamiltonian cycles [10,11,13] and for Hamiltonian paths [9] have attracted attention. Here we advance knowledge on structure and reconfiguration of Hamiltonian paths, motivated by the many applications of both Hamiltonian cycles and Hamiltonian paths in grid graphs (see e.g. [3,4,8,12]).

Our contributions:

(1) a complete structure for 1-complex s, t Hamiltonian paths (Sect. 3);
(2) two powerful new reconfiguration tools that find cells to switch and then sequence the switches to create straight path segments (Sect. 4); and
(3) an algorithm to reconfigure any 1-complex Hamiltonian path to any other such path in $O(|\mathbb{G}|)$ time, making $O(|\mathbb{G}|)$ switches in 1×1 grid *cells*, where $|\mathbb{G}|$ is the size of the grid graph $\mathbb{G}$ (Sects. 5 and 6).

2   Preliminaries

We define terminology and an assumption used throughout the paper. These have previously been defined in [9], but we repeat them here for completeness.

Assumption. $\mathbb{G}$ is an $m \times n$ grid graph, where $m, n \geq 4$, and α and β are the bottom left and top right corner vertices of $\mathbb{G}$. Without loss, *we assume the input 1-complex path $P_{s,t}$ visits α before β.* The target 1-complex path for the reconfiguration as well as intermediate configurations may visit β before α.

A vertex of $\mathbb{G}$ with coordinates (x, y) is denoted by $v_{x,y}$, where $0 \leq x \leq n-1$ and $0 \leq y \leq m - 1$. The top left corner vertex s of $\mathbb{G}$ is $(0, 0)$; the positive x-direction is rightward and the positive y-direction is downward. We use the terms *node* and *vertex* interchangeably.

Column x of $\mathbb{G}$ is the shortest path of $\mathbb{G}$ between $v_{x,0}$ and $v_{x,m-1}$, and *Row y* is the shortest path between $v_{0,y}$ and $v_{n-1,y}$. We call Columns 0 and $n - 1$ the *west* $(\mathcal{W})$ and *east* $(\mathcal{E})$ boundaries of $\mathbb{G}$ (i.e., sides of $R_{\mathbb{G}}$), respectively, and Rows 0 and $m - 1$ the *north* $(\mathcal{N})$ and *south* $(\mathcal{S})$ boundaries (sides).

Throughout this paper, a *1-complex path* means a 1-complex s, t Hamiltonian path P of $\mathbb{G}$; P visits each node of $\mathbb{G}$ exactly once and uses only edges in $\mathbb{G}$. We denote by $P_{u,v}$ the directed subpath of P from u to v. Straight subpaths of P are called *segments*, denoted $seg[u, v]$, where u and v are the segment endpoints.

Cookies and Separators. Every internal node v of $\mathbb{G}$ lies on an *internal subpath* of P, namely the subpath joining the first boundary vertices v_s and v_t met when travelling along P from v toward s and toward t, respectively. Such an internal subpath is called a *cookie* if v_s and v_t lie on the same boundary, a *corner separator* if they lie on adjacent boundaries, and a *cross-separator* if they lie on opposite boundaries. Because P is 1-complex, a cross-separator can either be *bent* and have two adjacent bends, or it can be *straight*.

A cookie c has one of four types, $\mathcal{N}, \mathcal{S}, \mathcal{E}$, and $\mathcal{W}$, depending on the boundary where c has its *base*. A cookie consists of three segments of P. The common length of the two parallel segments of c measures the *size* of c.

We say a corner separator *cuts off* a corner of $\mathbb{G}$. Traveling along $P_{s,t}$, we denote the i-th corner separator cutting off s by μ_i, where $0 \leq i \leq j$. We denote its internal bend by $b(\mu_i)$, and its endpoints by $s(\mu_i)$ and $t(\mu_i)$, where $s(\mu_i)$ is the first endpoint met. We use similar notation for the i-th corner separator ν_i cutting off t. A corner separator that has one of its endpoints connected to s or t by a segment of P is called a *corner cookie*. P can have at most two corner cookies, one at either end.

Similar to the corner separators, we denote the i-th cross-separator met along $P_{s,t}$ by η_i, and its endpoints by $s(\eta_i)$ and $t(\eta_i)$, where $s(\eta_i)$ is the first endpoint met. If η_i is bent, then $b(\eta_i)$ denotes the first bend met. In total, we have j corner separators μ_i cutting off s, and k cross-separators η_i, and ℓ corner separators ν_i cutting off t.

Runs of Cookies. A *run of cookies* is a subpath of P consisting of cookies of the same type, spaced one unit apart and joined by the single boundary edges between them, possibly extended at either end by an edge joining a cookie endpoint to an adjacent boundary vertex. A run of cookies is denoted $Run[u, v]$, where u and v belong to the same boundary and delimit the range of boundary vertices covered; $Run[u, v]$ may consist of a single boundary edge (u, v).

To describe the path structure, we define three types of runs, depending on the cookie sizes along the run: the sizes may remain the same, or be non-increasing (denoted $Run^{\geq}[u, v]$) or non-decreasing ($Run^{\leq}[u, v]$). Runs are assumed to have cookies of the same size unless specified otherwise.

Canonical Paths. A *canonical path* $\mathbb{P}$ is a 1-complex path with no bends at internal vertices. If m is odd, $\mathbb{P}$ can be type $\mathcal{E}$-$\mathcal{W}$, filling rows of $\mathbb{G}$ one by one; if n is odd, $\mathbb{P}$ can be of type $\mathcal{N}$-$\mathcal{S}$, filling columns. There are no other types.

Observation 1. [9] For P to be Hamiltonian, the subpath $P_{s,\alpha}$ of $P_{s,t}$ must cover all the $\mathcal{W}$ vertices, no corner separators can cut off α or β, and $P_{s,t}$ must visit β before visiting any other $\mathcal{E}$ vertices. It follows that the corner separators μ_i cutting off s must occur in $P_{s,t}$ before α, and that the corner separators ν_i cutting off t occur in $P_{s,t}$ after β. Furthermore, all cross-separators η_i occur on $P_{s,t}$ between α and β, and the number $k \geq 1$ of them must be odd.   □

3   Structure of 1-Complex Path P of $\mathbb{G}$

We regard P in its directed form $P_{s,t}$ as composed of an *initial* subpath $P_{s,s(\eta_1)}$, followed by a *middle* subpath $P_{s(\eta_1),t(\eta_k)}$, and a *final* subpath $P_{t(\eta_k),t}$. By reversing the edge directions, the final subpath of $P_{s,t}$ can be viewed as the initial subpath of the path $P_{t,s}$ from t to s. The grid $\mathbb{G}$ can be rotated by π to place t in the upper left corner and s in the lower right corner. Thus, apart from changes in notation, the structural possibilities for the final and initial subpaths of $P_{s,t}$ are the same, and we do not discuss the final subpath further.

We present a series of observations whose straightforward proofs may require some reflection. They lead to a *structure theorem* at the end of this section.

Initial Subpath. The structure of the initial subpath depends on the form of the first cross-separator η_1 of $P_{s,t}$. As η_1 may be bent or straight, there are five possible forms for it (in Fig. 3(a), see forms A, B, D, F, and G). We discuss and give the structure of the initial subpath for each of these five forms.

In the observations below, w is the vertex on $\mathcal{N}$ just one column west of $t(\eta_1)$, and a' is the vertex on $\mathcal{S}$ in the same column as w when η_1 has form G, and one column west of $s(\eta_1)$ otherwise. The vertex $v_{0,2}$ is denoted by a.

Separator η_1 has form B or G. The initial subpath has the same structure as the initial subpath of a *simple s, t path* described in [9].

Observation 2 (η_1 form B or G). (a) If $t(\eta_1)$ is in Column 1 next to s on the $\mathcal{N}$ boundary, then the initial subpath consists of two boundary segments $seg[s, \alpha]$ and $seg[\alpha, s(\eta_1)]$ on the $\mathcal{W}$ and $\mathcal{S}$ boundaries, respectively; see Fig. 3(b).

(b) If $j = 0$ and $x(t(\eta_1)) > 1$, $P_{s,s(\eta_1)}$ must have a corner cookie containing s and w, then next either $seg[a, \alpha]$ $Run[\alpha, a']$, or $seg[a, \alpha]$ $seg[\alpha, a']$, or $Run^{\geq}[a, \alpha]$

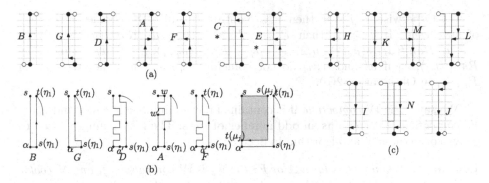

Fig. 3. (a) Seven forms for $\mathcal{S}$ to $\mathcal{N}$ cross-separators; η_1 cannot have forms E or C as any $\mathcal{S}$ cookie preceeding η_1 must be included in the initial subpath, not the middle subpath. Path P must contain the boundary segments shown. (b) Degenerate cases for different forms of η_1. (c) $\mathcal{N}$ to $\mathcal{S}$ forms for cross-separators.

$seg[\alpha, u]\ Run^{\le}[u, a']$, where u is at least two units from α on $\mathcal{S}$ or $Run^{\ge}[a, u']$
$seg[u', \alpha]\ Run^{\le}[\alpha, a']$, where u' is at least two units from α on $\mathcal{W}$; then ending with $seg[a', s(\eta_1)]$.

(c) If $j \ge 1$, then $P_{s,t(\mu_j)} = Run[s, s(\mu_1)]\ \mu_1\ Run[t(\mu_1), s(\mu_2)]\ \mu_2\ \ldots\ \mu_i$ $Run[t(\mu_i), s(\mu_{i+1})]\ \mu_{i+1}\ \ldots\ \mu_{j-1}\ Run[t(\mu_{j-1}), s(\mu_j)]\ \mu_j$. Since μ_j ends at $t(\mu_j)$ on $\mathcal{W}$, $s(\mu_1)$ must lie on $\mathcal{N}$ for j odd and on $\mathcal{W}$ for $j \ge 1$ even. Path $P_{t(\mu_j),s(\eta_1)}$ either has the structure of $P_{a,s(\eta_1)}$ in (b), or as shown in Fig. 3.(b), consists of edge $(t(\mu_j), \alpha)$ followed by $seg[\alpha, s(\eta_1)]$. □

Separator η_1 has form D. The node below w must be a bend (else P cannot visit all the vertices in the column of $t(\eta_1)$); this bend must connect to $\mathcal{W}$ forming a corner cookie. Thus $j = 0$.

Observation 3 (η_1 form D). (a) If $x(t(\eta_1)) = 1$, $P_{s,s(\eta_1)} = seg[s, a]\ Run[a, \alpha]$ $seg[\alpha, s(\eta_1)]$, where the $\mathcal{W}$ cookies have unit size. See Fig. 3.(b). (b) Otherwise, $P_{s,s(\eta_1)}$ consists of a corner cookie containing s and w, followed by $P_{a,s(\eta_1)}$ with the structure given in Observation 2(b). □

Separator η_1 has form A or F. We define two rectangular regions of $\mathbb{G}$ covered by the initial subpath, and use them in designing our algorithm in Sects. 5 and 6. Let w' denote the vertex on the $\mathcal{W}$ boundary in Row $y(b(\eta_1))$ for form F, and in Row $y(b(\eta_1)) - 1$ for form A. Let w'' be the vertex on the $\mathcal{W}$ boundary one row below w'. We denote by $\mathbb{R}_s$ the rectangular region of $\mathbb{G}$ that is delimited by Columns 0 and $x(w)$ and Rows 0 and $y(w')$; the rectangular region delimited by Columns 0 and $x(a')$ and Rows $y(w'')$ and $m - 1$ is denoted $\mathbb{R}_\alpha$.

Observation 4 (η_1 form A or F). (a) If $t(\eta_1)$ is in Column 1, η_1 has form F, and the initial subpath is $(\mathbb{R}_s = seg[s, w'])seg[w', w''](\mathbb{R}_\alpha = Run[w'', \alpha],$ $seg[\alpha, a'])\ seg[a', s(\eta_1)]$. If $s(\eta_1)$ is in Column 1, η_1 has form A, and the initial subpath is $(\mathbb{R}_s = Run[s, w'])seg[w', w''](\mathbb{R}_\alpha = seg[w'', \alpha], seg[\alpha, a'])$ $seg[a', s(\eta_1)]$. In both cases, the run contains unit size $\mathcal{W}$ cookies. See Fig. 3.(b).

(b) Otherwise, if $j = 0$, then $P_{s,w'}$ contains a run of $\mathcal{W}$ cookies $Run[s, w']$ of size $x(w)$. If $j \geq 1$, then $P_{s,w'} = Run[s, s(\mu_1)]\ \mu_1\ Run[t(\mu_1), s(\mu_2)]\ \mu_2$ $\ldots\ \mu_i\ Run[t(\mu_i), s(\mu_{i+1})]\ \mu_{i+1}\ \cdots\ \mu_{j-1}\ Run[t(\mu_{j-1}), s(\mu_j)]\ \mu_j$, followed by $Run[t(\mu_j), w']$ if w' and $t(\mu_j)$ do not coincide. $P_{w'',s(\eta_1)}$ has the structure of $P_{a,s(\eta_1)}$ in Observation 2(b).   □

We say $\mathbb{R}_s$ is $\mathcal{W}$ *compatible* if it contains an even number of rows, and $\mathbb{R}_\alpha$ is $\mathcal{W}$ *compatible* if it contains an odd number of rows. The next lemma is used to prove correctness of an algorithm in Sect. 5.

Lemma 1. *When η_1 has form A or F: (a) $\mathbb{R}_s$ is $\mathcal{W}$ compatible iff any $\mathcal{N}$ cookie in $\mathbb{R}_s$ has even size; $\mathbb{R}_\alpha$ is $\mathcal{W}$ compatible iff any $\mathcal{S}$ cookie in $\mathbb{R}_\alpha$ has even size. (b) At least one of $\mathbb{R}_s$ and $\mathbb{R}_\alpha$ must be $\mathcal{W}$ compatible.*

Middle Subpath. The middle subpath $P_{s(\eta_1),t(\eta_k)}$ by definition contains all the cross-separators. Since P is 1-complex, each internal grid point of $\mathbb{G}$ is connected to a boundary by a segment of P. Hence cross-separators are either straight, or bent toward $\mathcal{E}$ or $\mathcal{W}$ by one unit from their start nodes $s(\eta_i)$. Figure 3(a) and (c) contain the exhaustive list of $\mathcal{S}$ to $\mathcal{N}$ and $\mathcal{N}$ to $\mathcal{S}$ cross separators, respectively.

Observation 5 (constraints on η_i, η_{i+1}). (a) $x(t(\eta_i)) = x(s(\eta_i)) \pm 1$ for bent η_i, and $x(t(\eta_i)) = x(s(\eta_i))$ for straight η_i.
(b) $x(t(\eta_{i+1})) = 1 + x(s(\eta_i))$ because the roundtrip made by $P_{s(\eta_i),t(\eta_{i+1})}$ must cover all grid points between η_i and η_{i+1}.
(c) By (a) and (b), nodes $s(\eta_{i+1})$ and $t(\eta_i)$ are 1, 2, or 3 units apart $(i < k)$. Node $s(\eta_{i+1})$ lies to the right of $t(\eta_i)$, and when the nodes are 3 apart, P must have one cookie between them to cover the intervening points on the boundary.   □

To describe the middle subpath structure, we introduce four special cases E, D, C, and G for form F and four special cases L, I, N, and J for form H; see Fig. 3. Note that forms H and F both have two vertical segments of length at least two. Forms E and C describe the scenario that an $\mathcal{S}$ cookie occurs between η_{i-1} and η_i. Like form F, forms D and G turn to the left (toward $\mathcal{W}$) at the first bend, but unlike F, their internal bends occur one unit from the $\mathcal{N}$ or $\mathcal{S}$ boundary. The four special cases L, I, N, and J for form H are described similarly. Which forms can appear consecutively is determined by Observation 5.

Theorem 1 (1-complex Path Structure). *Let $P_{s,t}$ be a 1-complex path with k cross-separators and j and ℓ corner separators cutting off s and t, respectively.*
initial subpath $[P_{s,s(\eta_1)}]$ *Its structure is given by Observations 2–4.*
middle subpath $[P_{s(\eta_1),t(\eta_k)}]$ *Its structure is given by Observation 5 and the text following the observation.*
final subpath $[P_{t(\eta_k),t}]$ *As $P_{t,t(\eta_k)}$ is the initial subpath of $P_{t,s}$ (the reverse of $P_{s,t}$), the forms for the final and the initial subpaths are the same.*

4   Zip Operation

In this section, we define a *zip* operation Z (zip for short) that applies a sequence of switches to cells that appear on two sides of a line (row or column) of $\mathbb{G}$. The cells must be *switchable*, as described in Sect. 1. The line, called a *zipline* and denoted $l_z^{q_1,q_2}$ (the superscript may be omitted for short), is directed from node q_1 to node q_2 where q_1 and q_2 are not corners of $R_{\mathbb{G}}$.

As mentioned in Sect. 1, a *cycle-path cover* $\mathbb{P}$ of $\mathbb{G}$ is a set of cycles and paths that collectively cover all the vertices of $\mathbb{G}$.

The *zone* R_Z of a zip is a rectangle determined by the zipline $l_z^{q_1 q_2}$ and the two adjacent and parallel grid lines $l_a = [a_1, a_2]$ and $l_b = [b_1, b_2]$, where a_1 and b_1 are adjacent to q_1 on a boundary of $\mathbb{G}$ and a_2 and b_2 are adjacent to q_2 on the opposite boundary. Thus the corners of R_Z are a_1, a_2, b_1, b_2. We call the subgraph of $\mathbb{G}$ induced by l_a and l_z the *main track* tr of the zip, and the subgraph with sides l_z and l_b the *side track* tr'.

To describe the structure of a Hamiltonian path P inside the zone R_Z of a zip, we define the notion of a *local cookie*. See Fig. 4.

Definition 1 (local cookie). Let f' be a switchable cell for P in the side track tr' of the zone R_Z of a zipline l_z, such that the two sides (a, d) and (b, c) of f' that belong to P are perpendicular to l_z and l_b, and the other two sides (a, b) and (d, c) lie in l_z and l_b, respectively. Let C be a cycle of grid edges in tr such that (a, b) belongs to C and is the only edge of C not in P. Then the edges (a, d) and (b, c) of f' together with the edges of C except for (a, b) determine a subpath of P called a *local cookie* with base (d, c) on l_b. Depending on its shape, a local cookie has one of four types: I, T, q_1-facing and q_2-facing. For example, C and f_2' in Fig. 4 make a type T local cookie.

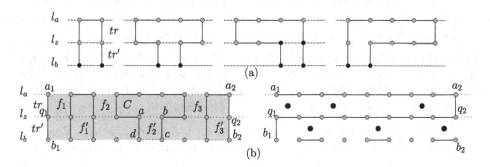

Fig. 4. (a) The types of local cookies. (b) Zip Z turns l_a and l_z into segments. *Left:* R_Z (in grey) before zip Z. Dotted edges do not belong to P. Track tr is covered locally. Cells in S_{tr} and $S_{tr'}$ are labeled f_i and f_i'. *Right:* The big dots mark the switched cells after the zip. On l_b, only edges of P in the f_i's are shown.

The goal of a zip operation is to create a new s, t Hamiltonian path P' that contains all of l_a and l_z as two segments in P' and joins them with a boundary edge of tr in P'. These two segments and the boundary edge that connects them can be viewed as a round trip from one boundary to the opposite boundary and back. This goal motivates the following definition.

Definition 2. *The main track tr is* locally covered *by P provided (i) P contains edge (a_1, q_1) or else contains a regular cookie whose base is the edge (a_1, q_1)– either way, P contains the edge (q_1, b_1), and (ii) P covers the remaining nodes of tr with local cookies with base in l_b (e.g., Fig. 4 (Left)).*

Note that here, either a_1 has degree 1 on path P and lies at a corner of $R_\mathbb{G}$, or else a_1 is incident to a boundary edge that belongs to P but lies outside tr.

Definition 3. *The zone R_Z of a zip Z is* zippable *provided the main track tr is locally covered by P (See Fig. 4 (Left)).*

Observation 6 (Special switchable cells). (a) By Definition 1, each grid cell f' of tr' that contains the base of a local cookie of R_Z is switchable. Switching any such f' creates a cycle-path cover $\mathbb{P} = \{P', C\}$ of $\mathbb{G}$ where cycle C lies in tr. (b) Each grid cell f in the main track tr of a zippable zone R_Z that does not lie inside a local cookie is switchable for P.

We now define two special sets of switchable cells S_{tr} in tr and $S_{tr'}$ in tr' for a zippable zone R_Z and tell how we index the cells.

The set S_{tr} consists of the following cells of tr: any cell that has one side in each of two distinct local cookies; any cell that has a_1q_1 as a side where (a_1, q_1) belongs to P, and has its parallel side in a local cookie; and any cell that has as one side the end of a cookie lying in tr with base a_1, q_1, and has for a parallel side an edge in a local cookie. We index the cells of S_{tr} $f_1 \ldots$ in their order of occurrence from the q_1 end to the q_2 end of tr. We define the set $S_{tr'}$ to be all the cells in tr' that have a side on l_b that is the base of a local cookie. We index the cells of $S_{tr'}$ $f'_1 \ldots$ in order of their occurrence in tr'. We note that $|S_{tr}| = |S_{tr'}|$. We index each local cookie according to the cell f'_i it encloses.

Definition 4 (Zip for zippable R_Z). *Let $l_z^{q_1, q_2}$ be a zipline of $\mathbb{G}$ whose zone R_Z is zippable, and let S_{tr} and $S_{tr'}$ be the special sets of switchable cells in the two tracks tr and tr' of R_Z. The zip operation $Z = zip(\mathbb{G}, P, l_z^{q_1, q_2}, tr)$ applies switch to all the cells of S_{tr} and $S_{tr'}$ in the following order: $f_1, f'_1, f_2, f'_2, \ldots$.*

Note that the zip operation is only defined for a zippable zone R_Z. Proofs of correctness of our algorithms will show that the zips are done in zippable zones.

Observation 7. Let P' be the s, t path resulting from a zip Z on an s, t path P of $\mathbb{G}$, where the zone R_Z is zippable. Then, the following hold: (1) Path P' is Hamiltonian and differs from P only in the cells of S_{tr} and $S_{tr'}$; (2) P' contains segments $seg[a_1, a_2]$ on l_a and $seg[q_2, q_1]$ on l_z and the boundary edge (a_2, q_2) joining their end points a_2 and q_2; (3) The boundary edge (q_1, b_1) is the only edge of P' that joins $seg[q_2, q_1]$ to l_b; and (4) P' can be obtained from P in $O(\max\{m, n\})$ switches.

In the next section, we will use path structure (Sect. 3) to show that doing a zip leaves the next pair of tracks zippable. When this condition holds, we will be able to apply a sequence of zips advancing the zipline by 2 units each time. We refer to such a sequence of zips as a *sweep*. Sweeps can be done in any of the four directions: up, down, left and right.

5   Reconfiguring 1-Complex Paths to Canonical Forms

We give an algorithm called 1COMPLEXTOCANONICAL to reconfigure any 1-complex s, t path P to a canonical path. In Sect. 6, we will use this algorithm to design another algorithm to reconfigure between any two 1-complex paths.

We first define some terminology in order to describe the algorithm. A *sub-rectangle* $\mathbb{G}'$ is an $m \times n'$, $n' \leq n$, subgrid of $\mathbb{G}$ such that the subpath P' of P covering the vertices of $\mathbb{G}'$ is a Hamiltonian path between two corner vertices of $\mathbb{G}'$; we call P' a *sub-rectangular path*. We use *sub-rectangle* and *sub-rectangular path* interchangeably for the rest of the section.

We now give a brief overview of the algorithm. It first breaks $\mathbb{G}$ into sub-rectangles using P and the *Structure Theorem* (Theorem 1), so that each sub-rectangle $\mathbb{G}'$ contains an s', t' Hamiltonian path $P'_{s',t'}$ that can be reconfigured to a canonical path of $\mathbb{G}'$ using at most two sweeps in $\mathbb{G}'$. We then merge all the canonical sub-rectangular paths into an s, t Hamiltonian path of $\mathbb{G}$; and using at most one sweep in $\mathbb{G}$, we reconfigure it to a canonical path of $\mathbb{G}$.

Breaking P into (extended) sub-rectangles.   We break P into sub-rectangles $\mathbb{G}_h$, $1 \leq h \leq Q$, by removing the following edges: all straight separators, and the edges on $\mathcal{N}$ and $\mathcal{S}$ preceding and following them; the edge between Columns x and $x + 1$ on $\mathcal{N}$ or $\mathcal{S}$, when the internal vertices of Column x are completely covered by a separator of form D or I, respectively, or the internal vertices of Column $x + 1$ are completely covered by a separator of form J or G. In the path in Fig. 2, removing the bold black edges will break the path into the $Q = 8$ sub-rectangles in Fig. 5(a). We call $\mathbb{G}_1$ and $\mathbb{G}_Q$ the *terminal* sub-rectangles, and the others the *middle* sub-rectangles.

Let s_h and t_h be the starting and ending points of the sub-rectangular path P_h of $\mathbb{G}_h$. If s_h is on $\mathcal{S}$, we flip $\mathbb{G}_h$ along $\mathcal{S}$ when $h < Q$. In case of $\mathbb{G}_Q$, we rotate it by π about its center. If t_h is on $\mathcal{N}$, we add a column to the east of $\mathbb{G}_h$, $1 \leq h \leq Q$, to create the *extended sub-rectangle* $\mathbb{G}'_h$, then connect t_h to the bottom-right corner t'_h of $\mathbb{G}'_h$ through the new edges to get an s_h, t'_h Hamiltonian path of $\mathbb{G}'_h$. From now on, we use $\mathbb{G}_h$ to denote the final sub-rectangle obtained after the optional flipping and/or extending steps. We now observe some properties of the middle and terminal sub-rectangles, then describe Algorithm RECONFIGSUBRECT, to reconfigure a sub-rectangle to a canonical form.

Middle Sub-rectangles.   As shown in Fig. 5, each middle sub-rectangle $\mathbb{G}_h$, $1 < h < Q$, must have a corner $\mathcal{W}$ cookie c of unit size. If an $\mathcal{S}$ cookie follows c, then it must be followed by a separator of form D and then the dummy $\mathcal{E}$ boundary. Otherwise, c is followed by a separator of form F, which is the start of the middle subpath of P_h.

Lemma 2. *Let P' be a middle sub-rectangular path. Then any S cookie of P' has the same parity as m and any N cookie has even size.*

Terminal Sub-rectangles. We limit our discussion of terminal sub-rectangles to $\mathbb{G}_1$, since $\mathbb{G}_Q$ has similar structure. If η_1 of the 1-complex path P of $\mathbb{G}$ has form B or G, then $\mathbb{G}_1$ contains only the initial subpath of P. Otherwise, $\mathbb{G}_1$ contains η_1 of P as the first cross-separator of P_1, where η_1 must have either form D, A or F. If η_1 has form D, then it must also be the last separator, and there are no corner separators or N cookies in $\mathbb{G}'_s$ by Observation 3. We now assume that η_1 has either form A or F and thus can have cookies in the middle subpath. We show that the size of those cookies depends on the W compatibility of $\mathbb{R}_s$ and $\mathbb{R}_\alpha$.

Lemma 3. *Let η_1 of $\mathbb{G}_1$ have form A or F. (a) If $\mathbb{R}_s$ is W compatible, then any N cookie in the middle subpath of $\mathbb{G}_1$ must have even size, and the size of all S cookies have opposite parity as m. (b) Otherwise, any S cookie in the middle subpath must have even size, and the sizes of the N cookies will have the opposite parity of m.*

Algorithm RECONFIGSUBRECT. If $\mathbb{G}_h$ is a middle sub-rectangle, we apply a *SweepDown* procedure, first placing the zipline on Row 1, and moving down two rows after each zip until we reach the S boundary. If m is odd, we get an $\mathcal{E} - W$ canonical path of $\mathbb{G}_h$ at this point. Otherwise, we will end up with unit size S cookies after the sweep; therefore, we *SweepLeft* to grow the S cookies all the way to the N boundary and obtain an $N - S$ canonical path of $\mathbb{G}_h$.

If $\mathbb{G}_1$ contains only the initial subpath of P, then we apply a *SweepLeft*, and then a *SweepDown* if we end up with unit size W cookies in Column 1 after the first sweep. Otherwise, depending on the W compatibility of η_1, we either *SweepDown* or *SweepUp*, and then we *SweepLeft* if we have unit size S or N cookies, respectively, after the first sweep. We can prove the correctness of this algorithm using the Structure Theorem (Theorem 1), and Lemmas 1–3.

Theorem 2. *Algorithm* RECONFIGSUBRECT *reconfigures a sub-rectangle $\mathbb{G}_h$ to a canonical form in $O(|\mathbb{G}_h|)$ switch operations.*

Merging the Canonical Sub-rectangles. For each $\mathbb{G}_h$, if the $\mathcal{E}$ boundary is a dummy column, and m is odd, we apply one vertical zip with the zipline on Column $n' - 2$ of $\mathbb{G}_h$ such that both Columns $n - 2$ and $n - 1$ become path segments after the zip (Fig. 5(b)). We remove the dummy edges; flip the sub-rectangle back, if it was flipped before; then add all the straight separators and edges on the N and S boundaries that were removed, to get an s, t Hamiltonian path P' of $\mathbb{G}$. If P' is not a canonical path, it must have "comb" shaped subpaths connected by straight separators as shown in Fig. 5(c). We then apply one more *SweepDown* to obtain an $\mathcal{E} - W$ canonical path of $\mathbb{G}$.

Theorem 3. *Algorithm* 1COMPLEXTOCANONICAL *reconfigures a 1-complex s, t Hamiltonian path to a canonical Hamiltonian path of $\mathbb{G}$ in $O(|\mathbb{G}|)$ time using $O(|\mathbb{G}|)$ switch operations.*

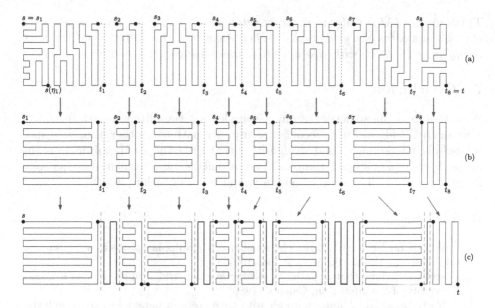

Fig. 5. (a) Extended sub-rectangles of the path in Fig. 2; (b) the canonical forms for the sub-rectangles; (c) the canonical sub-rectangles merged.

6   Reconfiguring Between 1-Complex Paths

In this section, we give an algorithm called 1COMPLEXTo1COMPLEX to reconfigure between any two 1-complex s, t Hamiltonian paths P_1 and P_2 in $O(|\mathbb{G}|)$ time. Our strategy is to use two *canonical Hamiltonian paths* $\mathbb{P}_1$ and $\mathbb{P}_2$ as intermediate paths, where the two canonical paths may or may not be the same based on the parity of m and n. The reconfiguration sequence is as follows: (a) P_1 to $\mathbb{P}_1$, (b) $\mathbb{P}_1$ to $\mathbb{P}_2$ if they are different, and finally (c) $\mathbb{P}_2$ to P_2. Algorithm 1COMPLEXToCANONICAL suffices for Steps (a) and (c), since reconfiguring $\mathbb{P}_2$ to P_2 is similar to reversing the steps of the reconfiguration of P_2 to $\mathbb{P}_2$. Now, we give an algorithm that we call CANONICALToCANONICAL to reconfigure one canonical Hamiltonian path to the other.

Let $\mathbb{P}_1$ and $\mathbb{P}_2$ be the two input canonical paths of $\mathbb{G}$ to Algorithm CANONICALToCANONICAL. We check the first edge on each path. If the edge is on the $\mathcal{W}$ boundary then the path is an $\mathcal{N}$-$\mathcal{S}$ canonical path; otherwise, it is an $\mathcal{E}$-$\mathcal{W}$ canonical path. If $\mathbb{P}_1$ and $\mathbb{P}_2$ are the same path, then $\mathbb{P}_1$ is returned by the algorithm. Otherwise, if $\mathbb{P}_1$ is $\mathcal{N}$-$\mathcal{S}$ and $\mathbb{P}_2$ is $\mathcal{E}$-$\mathcal{W}$, we apply a *SweepDown* procedure on $\mathbb{P}_1$ starting from Row 1 and ending on Row $m-2$. In the remaining case, when $\mathbb{P}_1$ is $\mathcal{E}$-$\mathcal{W}$ and $\mathbb{P}_2$ is $\mathcal{N}$-$\mathcal{S}$, we apply *SweepLeft* on $\mathbb{P}_1$ with the zipline sweeping from Column 1 to Column $n-2$. To conclude,

Theorem 4. *Let $\mathbb{P}_1$ and $\mathbb{P}_2$ be two canonical paths of $\mathbb{G}$. Then Algorithm CANONICALToCANONICAL reconfigures $\mathbb{P}_1$ to $\mathbb{P}_2$ in $O(|\mathbb{G}|)$ time using $O(|\mathbb{G}|)$ switches.*

Theorem 5. *(Main algorithmic result). Let P_1 and P_2 be two 1-complex s, t Hamiltonian paths of a grid graph $\mathbb{G}$. Then Algorithm* 1COMPLEXTO1COMPLEX *reconfigures P_1 to P_2 with zips in $O(|\mathbb{G}|)$ time, using $O(|\mathbb{G}|)$ switches.*

7   Conclusion

We established the structure of any 1-complex s, t Hamiltonian path in $\mathbb{G}$. We gave an $O(|\mathbb{G}|)$ algorithm to reconfigure any such path to another using *switches*. It would be interesting to find an algorithm that keeps the s, t Hamiltonian paths in the intermediate steps 1-complex. The reconfiguration problem remains open for grid graphs with arbitrary boundary, and in d-dimension, $d \geq 3$.

References

1. Collins, K.L., Krompart, L.B.: The number of Hamiltonian paths in a rectangular grid. Discrete Math. **169**(1–3), 29–38 (1997)
2. Everett, H.: Hamiltonian paths in nonrectangular grid graphs. Master's thesis, University of Saskatchewan, Canada (1986)
3. Fellows, M., et al.: Milling a graph with turn costs: a parameterized complexity perspective. In: Thilikos, D.M. (ed.) WG 2010. LNCS, vol. 6410, pp. 123–134. Springer, Heidelberg (2010). https://doi.org/10.1007/978-3-642-16926-7_13
4. Gorbenko, A., Popov, V., Sheka, A.: Localization on discrete grid graphs. In: He, X., Hua, E., Lin, Y., Liu, X. (eds.) Computer, Informatics, Cybernetics and Applications. LNEE, vol. 107, pp. 971–978. Springer, Dordrecht (2012). https://doi.org/10.1007/978-94-007-1839-5_105
5. Itai, A., Papadimitriou, C.H., Szwarcfiter, J.L.: Hamilton paths in grid graphs. SIAM J. Comput. **11**(4), 676–686 (1982)
6. Jacobsen, J.L.: Exact enumeration of Hamiltonian circuits, walks and chains in two and three dimensions. J. Phys. A Math. Theor. **40**, 14667–14678 (2007)
7. Keshavarz-Kohjerdi, F., Bagheri, A.: Hamiltonian paths in L-shaped grid graphs. Theor. Comput. Sci. **621**, 37–56 (2016)
8. Muller, P., Hascoet, J.Y., Mognol, P.: Toolpaths for additive manufacturing of functionally graded materials (FGM) parts. Rapid Prototyp. J. **20**(6), 511–522 (2014)
9. Nishat, R.I., Srinivasan, V., Whitesides, S.: Reconfiguring simple s, t Hamiltonian paths in rectangular grid graphs. In: Flocchini, P., Moura, L. (eds.) IWOCA 2021. LNCS, vol. 12757, pp. 501–515. Springer, Cham (2021). https://doi.org/10.1007/978-3-030-79987-8_35
10. Nishat, R.I., Whitesides, S.: Reconfiguring Hamiltonian cycles in L-shaped grid graphs. In: Sau, I., Thilikos, D.M. (eds.) WG 2019. LNCS, vol. 11789, pp. 325–337. Springer, Cham (2019). https://doi.org/10.1007/978-3-030-30786-8_25
11. Nishat, R.I., Whitesides, S.: Bend complexity and Hamiltonian cycles in grid graphs. In: Cao, Y., Chen, J. (eds.) COCOON 2017. LNCS, vol. 10392, pp. 445–456. Springer, Cham (2017). https://doi.org/10.1007/978-3-319-62389-4_37
12. Bodroža Pantić, O., Pantić, B., Pantić, I., Bodroža Solarov, M.: Enumeration of Hamiltonian cycles in some grid graphs. MATCH Commun. Math. Comput. Chem. **70**, 181–204 (2013)
13. Takaoka, A.: Complexity of Hamiltonian cycle reconfiguration. Algorithms **11**(9), 140 (2018)

Graph Drawing and Visualization

Aspect Ratio Universal Rectangular Layouts

Stefan Felsner[1], Andrew Nathenson[2], and Csaba D. Tóth[3(✉)]

[1] Institut für Mathematik, Technische Universität Berlin, Berlin, Germany
felsner@math.tu-berlin.de
[2] University of California, San Diego, CA, USA
anathenson@ucsd.edu
[3] California State University Northridge, Los Angeles, CA, USA
csaba.toth@csun.edu

Abstract. A *generic rectangular layout* (for short, *layout*) is a subdivision of an axis-aligned rectangle into axis-aligned rectangles, no four of which have a point in common. Such layouts are used in data visualization and in cartography. The contacts between the rectangles represent semantic or geographic relations. A layout is weakly (strongly) *aspect ratio universal* if any assignment of aspect ratios to rectangles can be realized by a weakly (strongly) equivalent layout. We give a combinatorial characterization for weakly and strongly aspect ratio universal layouts, respectively. Furthermore, we describe a quadratic-time algorithm that decides whether a given graph G is the dual graph of a strongly aspect ratio universal layout, and finds such a layout if one exists.

1 Introduction

A *rectangular layout* (a.k.a. *mosaic floorplan* or *rectangulation*) is a subdivision of an axis-aligned rectangle into axis-aligned rectangle faces, it is *generic* if no four faces have a point in common. In the *dual graph* $G(\mathcal{L})$ of a layout $\mathcal{L}$, the nodes correspond to rectangular faces, and an edge corresponds to a pair of rectangles whose common boundary contains a line segment [6,28,29].

Two generic layouts are *strongly equivalent* if they have isomorphic dual graphs, and the corresponding line segments between rectangles have the same orientation (horizontal or vertical); see Fig. 1 for examples. Two generic layouts are *weakly equivalent* if there is a bijection between their horizontal and vertical segments, resp., such that the contact graphs of the segments are isomorphic plane graphs. Strong equivalence implies weak equivalence [9]; however, for example the brick layouts in Figs. 4a and 4b are weakly equivalent, but not strongly equivalent. The closures of weak (resp., strong) equivalence classes under the uniform norm extend to nongeneric layouts, and a nongeneric layout may belong to the closures of multiple equivalence classes.

Rectangular layouts have been studied for more than 40 years, originally motivated by VLSI design [21,23,34] and cartography [26], and more recently by data

Research on this paper was partially supported by the NSF award DMS-1800734.

P. Mutzel et al. (Eds.): WALCOM 2022, LNCS 13174, pp. 73–84, 2022.
https://doi.org/10.1007/978-3-030-96731-4_7

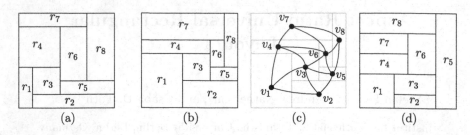

Fig. 1. (a–b) Two equivalent layouts. (c) Dual graph. (d) Another layout with the same dual graph. The layout in (d) is sliceable, none of them is one-sided.

visualization [17]. The weak equivalence classes of layouts are in bijection with Baxter permutations [1,27,35].

An (abstract) graph is called a *proper graph* if it is the dual of a generic layout. Every proper graph is a near-triangulation (a plane graph where every bounded face is a triangle, but the outer face need not be a triangle). But not every near-triangulation is a proper graph [28,29]. Ungar [33] gave a combinatorial characterization of proper graphs (see also [16,31]); and they can be recognized in linear time [12,22,24,25].

In data visualization and cartography [17,26], the rectangles correspond to entities (e.g., countries or geographic regions); adjacency between rectangles represents semantic or geographic relations, and the "shape" of a rectangle represent data associated with the entity. It is often desirable to use equivalent layouts to realize different statistics associated with the same entities. A generic layout $\mathcal{L}$ is *weakly (strongly) area universal* if any area assignment to the rectangles can be realized by a layout weakly (strongly) equivalent to $\mathcal{L}$. Wimer et al. [34] showed that every generic layout is weakly area universal (see also [9, Thm. 3]). Eppstein et al. [6] proved that a layout is strongly area universal if and only if it is one-sided (defined below). However, no polynomial-time algorithm is known for testing whether a given graph G is the dual of some area-universal layout.

In some applications, the aspect ratios (rather than the areas) of the rectangles are specified. For example, in word clouds adapted to multiple languages, the aspect ratio of (the bounding box of) each word depends on the particular language. The *aspect ratio* of an axis-aligned rectangle r is $\mathrm{height}(r)/\mathrm{width}(r)$. A generic layout $\mathcal{L}$ is *weakly (strongly) aspect ratio universal (ARU* for short) if any assignment of aspect ratios to the rectangles can be realized by a layout weakly (strongly) equivalent to $\mathcal{L}$.

Our Results. We characterize strongly and weakly aspect ratio universal layouts.

Theorem 1. *A generic layout is weakly aspect ratio universal if and only if it is sliceable.*

Theorem 2. *For a generic layout $\mathcal{L}$, the following properties are equivalent:*

(i) $\mathcal{L}$ is strongly aspect ratio universal;

(ii) $\mathcal{L}$ *is one-sided and sliceable;*

(iii) *the extended dual* $G^*(\mathcal{L})$ *of* $\mathcal{L}$, *admits a unique transversal structure.*

The terms in Theorems 1–2 are defined below. It is not difficult to show that one-sided sliceable layouts are strongly aspect ratio universal; and admit a unique transversal structure. Proving the converses, however, is more involved.

Algorithmic Results. In some applications, the rectangular layout is not specified, and we are only given the dual graph of a layout (i.e., a proper graph). This raises the following problem: Given a proper graph G with n vertices, find a strongly (resp., weakly) ARU layout $\mathcal{L}$ such that $G \simeq G(\mathcal{L})$ or report that none exists. Using structural properties of one-sided sliceable layouts that we develop here, we present an algorithm for recognizing the duals of strongly ARU layouts.

Theorem 3. *We can decide in* $O(n^2)$ *time whether a given graph* G *with* n *vertices is the dual of a one-sided sliceable layout.*

Thomassen [31] gave a linear-time algorithm to recognize proper graphs if the nodes corresponding to corner rectangles are specified, using combinatorial characterizations of layouts [33]. Kant and He [13, 15] described a linear-time algorithm to test whether a given graph G^* is the extended dual of a layout, using transversal structures. Later, Rahman et al. [12, 22, 24, 25] showed that proper graphs can be recognized in linear time (without specifying the corners). However, a proper graph may have exponentially many nonequivalent realizations, and prior algorithms may not find a one-sided sliceable realization even if one exists. Currently, no polynomial-time algorithm is known for recognizing the duals of sliceable layouts [5, 18, 36] (i.e., weakly ARU layouts); or one-sided layouts [6].

Background and Terminology. A *rectangular layout* (for short, *layout*) is a rectilinear graph in which each face is a rectangle, the outer face is also a rectangle, and the vertex degree is at most 3. A *sublayout* of a layout $\mathcal{L}$ is a subgraph of $\mathcal{L}$ which is a layout. A layout is *irreducible* if it does not contain any nontrivial sublayout. A *rectangular arrangement* is a 2-connected subgraph of a layout in which bounded faces are rectangles (the outer face need not be a rectangle).

One-Sided Layouts. A *segment* of a layout $\mathcal{L}$ is a path of collinear inner edges of $\mathcal{L}$. A segment of $\mathcal{L}$ that is not contained in any other segment is maximal. In a *one-sided* layout, every maximal line segment s must be a side of at least one rectangle R; in particular, any other segment orthogonal to s with an endpoint in the interior of s lies in a halfplane bounded by s, and points away from R.

Sliceable Layouts. A maximal line segment subdividing a rectangle or a rectangular union of rectangular faces is called a *slice*. A *sliceable layout* (a.k.a. *slicing floorplan* or *guillotine rectangulation*) is one that can be obtained through recursive subdivision with vertical or horizontal lines; see Fig. 1(d). The recursive subdivision can be represented by a *binary space partition tree* (*BSP-tree*), which is a binary tree where each vertex is associated with either a rectangle with a slice, or just a rectangle if it is a leaf [3]. For a nonleaf vertex, the two subrectangles on each side of the slice are associated with the two children. The number of (equivalence classes of) sliceable layouts with n rectangles is known to be the nth Schröder number [35]. One-sided sliceable layouts are in bijection with certain pattern-avoiding permutations, closed formulas for their number has been given by Asinowski and Mansour [2]; see also [20] and OEIS A078482 in the on-line encyclopedia of integer sequences (https://oeis.org/) for further references.

A *windmill* in a layout is a set of four pairwise noncrossing maximal line segments, called *arms*, which contain the sides of a central rectangle, and each arm has an endpoint on the interior of another (e.g., the maximal segments around r_3 or r_6 in Fig. 2 (a)). We orient each arm from the central rectangle to the other endpoint. A windmill is either *clockwise* or *counterclockwise*. It is well known that a layout is sliceable if and only if it does not contain a windmill [1].

Transversal Structure. The dual graph $G(\mathcal{L})$ of a layout $\mathcal{L}$ encodes adjacency between faces, but does not specify the relative positions between faces (above-below or left-right). The transversal structure (a.k.a. regular edge-labelling) was introduced by He [13,15] for the efficient recognition of proper graphs, and later used extensively for counting and enumerating (equivalence classes of) layouts [11]. The *extended dual graph* $G^*(\mathcal{L})$ is the contact graph of the rectangular faces *and* the four edges of the bounding box of $\mathcal{L}$; it is a triangulation in an outer 4-cycle without separating triangles; see Fig. 2.

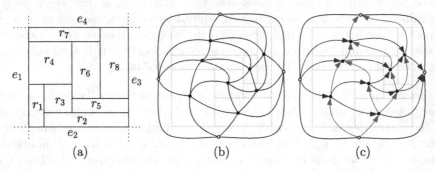

(a)   (b)   (c)

Fig. 2. (a) A layout $\mathcal{L}$ bounded by $e_1, \ldots, e_4$. (b) Extended dual graph $G^*(\mathcal{L})$ with an outer 4-cycle $(e_1, \ldots, e_4)$. (c) Transversal structure. (Color figure online)

A layout $\mathcal{L}$ is encoded by a *transversal structure* that comprises $G^*(\mathcal{L})$ and an orientation and bicoloring of the inner edges of $G^*(\mathcal{L})$, where red (resp., blue) edges correspond to above-below (resp., left-to-right) relation between two objects in contact. An (abstract) *transversal structure* is defined as a graph G^*, which is a 4-connected triangulation of an outer 4-cycle (S, W, N, E), together with a bicoloring and orientation of the inner edges of G^* such that all the inner edges incident to S, W, N, and E, respectively, are outgoing red, outgoing blue, incoming red, and incoming blue; and at each inner vertex the counterclockwise rotation of incident edges consists of four nonempty blocks of outgoing red, outgoing blue, incoming red, and incoming blue edges; see Fig. 2(c).

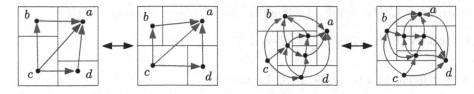

Fig. 3. A flip of an empty (left) and a nonempty (right) alternating cycle.

Flips and Alternating 4-Cycles. It is known that transversal structures are in bijection with the strong equivalence classes of generic layouts [8,11,15]. Furthermore, a sequence of flip operations can transform any transversal structure with n inner vertices into any other [7,11]. Each *flip* considers an *alternating 4-cycle* C, which comprises red and blue edges alternatingly, and changes the color of every edge in the interior of C; see Fig. 3. If, in particular, there is no vertex in the interior of C, then the flip changes the color of the inner diagonal of C. Furthermore, every flip operation yields a valid transversal structure on $G^*(\mathcal{L})$, hence a new generic layout $\mathcal{L}'$ that is strongly non-equivalent to $\mathcal{L}$. We can now establish a relation between geometric and combinatorial properties.

Lemma 1. *A layout $\mathcal{L}$ is one-sided and sliceable if and only if $G^*(\mathcal{L})$ admits a unique transversal structure.*

Proof. Assume that $\mathcal{L}$ is a layout where $G^*(\mathcal{L})$ admits two or more transversal structures. Consider a transversal structure of $G^*(\mathcal{L})$. Since any two transversal structures are connected by a sequence of flips, there exists an alternating 4-cycle. Any alternating 4-cycle with no interior vertex corresponds to a segment in $\mathcal{L}$ that is two-sided. Any alternating 4-cycle with interior vertices corresponds to a windmill in $\mathcal{L}$. Consequently, $\mathcal{L}$ is not one-sided or not sliceable.

Conversely, if $\mathcal{L}$ is not one-sided (resp., sliceable), then the transversal structure of $G^*(\mathcal{L})$ contains an alternating 4-cycle with no interior vertex (resp., with interior vertices). Consequently, we can perform a flip operation, and obtain another transversal structure for $G^*(\mathcal{L})$. □

2 Aspect Ratio Universality

An *aspect ratio assignment* to a layout $\mathcal{L}$ is a function that maps a positive real to each rectangle in $\mathcal{L}$. An aspect ratio assignment to $\mathcal{L}$ is *realizable* if there exists an equivalent layout $\mathcal{L}'$ with the required aspect ratios (a *realization*). A layout is *aspect ratio universal (ARU)* if every aspect ratio assignment is realizable. In this section, we characterize weakly and strongly ARU layouts (Theorems 1–2). We start with an easy observation. (Omitted proofs are in the full paper [10].)

Lemma 2. *Let $\mathcal{L}$ be a sliceable layout. If an aspect ratio assignment for $\mathcal{L}$ is realizable, then there is a unique realization up to scaling and translation. Furthermore, for every $\alpha > 0$ there exists a realizable aspect ratio assignment for which the bounding box of the realization has aspect ratio α.*

Corollary 1. *If $\mathcal{L}$ is one-sided and sliceable, then it is strongly ARU.*

Corollary 2. *If $\mathcal{L}$ is sliceable, then it is weakly ARU.*

2.1   Sliceable and One-Sided Layouts

Next we show that any sliceable layout that is strongly ARU must be one-sided. We present two types of simple layouts that are not aspect ratio universal, and then show that all other layouts that are not one-sided or not sliceable can be reduced to these prototypes.

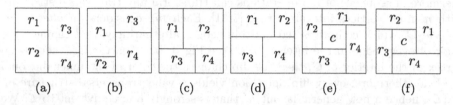

(a)   (b)   (c)   (d)   (e)   (f)

Fig. 4. Prototype layouts that are not aspect ratio universal: (a)–(d) brick layouts are sliceable but not one-sided; (e)–(f) windmills are one-sided but not sliceable.

Lemma 3. *The brick layouts in Figs. 4a–4d are not strongly ARU; the windmill layouts in Figs. 4e–4f are neither strongly nor weakly ARU.*

Proof. Suppose w.l.o.g. that a brick layout $\mathcal{L}_0$ in Fig. 4a is strongly ARU. Then there exists a strongly equivalent layout $\mathcal{L}$ for the aspect ratio assignment $\alpha(r_2) = \alpha(r_3) = 1$ and $\alpha(r_1) = \alpha(r_4) = 2$. Since width$(r_1) = $ width(r_2) and $\alpha(r_1) = 2\alpha(r_2)$, then height$(r_1) = 2\,$height$(r_2)$, and the left horizontal slice is below the median of $r_1 \cup r_2$. Similarly, width$(r_3) = $ width(r_4) and $\alpha(r_4) = 2\alpha(r_2)$ imply that the right horizontal slice is above the median of $r_3 \cup r_4$. Consequently, r_1 and r_4 are in contact, and $\mathcal{L}$ is not equivalent to $\mathcal{L}_0$, which is a contradiction.

Suppose w.l.o.g. that the windmill layout $\mathcal{L}_1$ in Fig. 4e is weakly ARU. Then there exists a weakly equivalent layout $\mathcal{L}$ for the aspect ratio assignment $\alpha(c) = \alpha(r_1) = \alpha(r_2) = \alpha(r_3) = \alpha(r_4) = 1$. In particular, $r_1, \ldots, r_4$ are squares; denote their side lengths by s_i, for $i = 1, \ldots, 4$. Note that one side of r_i strictly contains a side of r_{i-1} for $i = 1, \ldots, 4$ (with arithmetic modulo 4). Consequently, $s_1 < s_2 < s_3 < s_4 < s_1$, which is a contradiction.   □

Lemma 4. *If a layout is sliceable but not one-sided, then it is not strongly ARU.*

Proof. To show that a layout is not strongly ARU, it is sufficient to show that any of its sublayouts are not strongly ARU, because any nonrealizable aspect ratio assignment for a sublayout can be expanded arbitrarily to an aspect ratio assignment for the entire layout.

Let $\mathcal{L}$ be a sliceable but not one-sided layout. We claim that $\mathcal{L}$ contains a sublayout strongly equivalent to a layout in Figs. 4a–4d. Because $\mathcal{L}$ is not one-sided, it contains a maximal line segment ℓ which is not the side of any rectangle. Because $\mathcal{L}$ is sliceable, every maximal line segment in it subdivides a larger rectangle into two smaller rectangles. We may assume w.l.o.g. that ℓ is vertical. Because ℓ is not the side of any rectangle, the rectangles on the left and right of ℓ must be subdivided horizontally in the recursion. Let ℓ_{left} and ℓ_{right} be the first maximal horizontal line segments on the left and right of ℓ, respectively. Assume that they each subdivide a rectangle adjacent to ℓ into r_1 and r_2 (on the left) and r_3 and r_4 on the right. These rectangles comprise a layout equivalent to the one in Figs. 4a–4d; but they may be further subdivided recursively. By Lemma 3, there exists an aspect ratio assignment to $\mathcal{L}$ not realizable by a strongly equivalent layout.   □

In the remainder of this section, we prove that if a layout is not sliceable, then it contains a sublayout similar, in some sense, to a prototype in Figs. 4e–4f. In a nutshell, our proof goes as follows: Consider an arbitrary windmill in a nonsliceable layout $\mathcal{L}$. We subdivide the exterior of the windmill into four *quadrants*, by extending the arms of the windmill into rays $\ell_1, \ldots, \ell_4$ to the bounding box; see Fig. 5. Each rectangle of $\mathcal{L}$ lies in a quadrant or in the union of two consecutive quadrants. We assign aspect ratios to the rectangles based on which quadrant(s) it lies in. If these aspect ratios can be realized by a layout $\mathcal{L}'$ weakly equivalent to $\mathcal{L}$, then the rays $\ell_1, \ldots, \ell_4$ will be "deformed" into x- or y-monotone paths that subdivide $\mathcal{L}'$ into the center of the windmill and four arrangements of rectangles, each incident to a unique corner of the bounding box. We assign the aspect ratios for the rectangles in $\mathcal{L}'$ so that these arrangements can play the same role as rectangles $r_1, \ldots, r_4$ in the prototype in Figs. 4e–4f. We continue with the details.

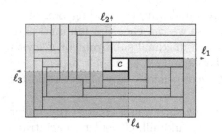 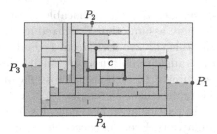

(a) A nonsliceable layout, a windmill, where rays $\ell_1, \ldots, \ell_4$ define quadrants.

(b) An equivalent layout, where four paths define rectangular arrangements.

Fig. 5. Rays $\ell_1, \ldots, \ell_4$ deform into monotone paths in an equivalent layout.

We clarify what we mean by a "deformation" of a (horizontal) ray ℓ.

Lemma 5. *Let a ray ℓ be the extension of a horizontal segment in a layout $\mathcal{L}$ such that ℓ does not contain any other segment and it intersects the rectangles*

$r_1, \ldots, r_k$ *in this order. Suppose that in a weakly equivalent layout* $\mathcal{L}'$, *the corresponding rectangles* $r'_1, \ldots, r'_k$ *are sliced by horizontal segments* $s_1, \ldots, s_k$. *Then there exists an x-monotone path comprised of horizontal edges* $s_1, \ldots, s_k$, *and vertical edges along vertical segment of the layout* $\mathcal{L}'$.

Proof. Assume w.l.o.g. that ℓ points to the right. Since ℓ does not contain any other segment and it intersects the rectangles $r_1, \ldots, r_k$ in this order, then r_i and r_{i+1} are on opposite sides of a vertical segment for $i = 1, \ldots, k-1$. The same holds for r'_i and r'_{i+1} as $\mathcal{L}'$ is weakly equivalent to $\mathcal{L}$. In particular, the right endpoint of s_i and the left endpoint of s_{i+1} are on the same vertical segment in $\mathcal{L}'$, for all $i = 1, \ldots k-1$. □

The next lemma allows us to bound the aspect ratio of the bounding box of a rectangular arrangement in terms of the aspect ratios of individual rectangles.

Lemma 6. *If every rectangle in a rectangular arrangement has aspect ratio* αm, *where m is the number of rectangles in the arrangement, then the aspect ratio of the bounding box of the arrangement is at least* α *and at most* αm^2.

Proof. Consider an arrangement A with m rectangles and a bounding box R. Let w be the maximum width of a rectangle in A. Then, width$(R) \leq mw$. A rectangle of width w has height $\alpha m w$, and so height$(R) \geq \alpha m w$. The aspect ratio of R is height$(R)/$width$(R) \geq (\alpha m w)/(mw) = \alpha$.

Similarly, let h be the maximum height of rectangle in A. Then height$(R) \leq mh$. A rectangle of height h has width $\frac{h}{\alpha m}$, and so width$(R) \geq \frac{h}{\alpha m}$. The aspect ratio of R is height$(R)/$width$(R) \leq mh/(\frac{h}{\alpha m}) = \alpha m^2$, as claimed. □

We can now complete the characterization of aspect ratio universal layouts.

Lemma 7. *If a layout* $\mathcal{L}$ *is not sliceable, it is not weakly ARU.*

Proof. Let R be a nonslicable layout of n rectangles in a bounding box of $\mathcal{L}$. We may assume that $\mathcal{L}$ is irreducible, otherwise we can choose a minimal nonsliceable sublayout $\mathcal{L}^*$ from $\mathcal{L}$, and replace each maximal sublayout of $\mathcal{L}^*$ with a rectangle to obtain an irreducible layout. By Lemma 2, a suitable aspect ratio assignment to each sliceable sublayout of $\mathcal{L}^*$ can generate any aspect ratio for the replacement rectangle.

In particular, $\mathcal{L}$ thus contains no slices, as any slice would create two smaller sublayouts. Every nonsliceable layout contains a windmill. Consider an arbitrary windmill in $\mathcal{L}$, assume w.l.o.g. that it is clockwise (cf. Fig. 4e). and let c be its central rectangle. By extending the arms of the windmill into rays, $\ell_1, \ldots, \ell_4$, we subdivide $R \setminus c$ into four *quadrants*, denoted by $Q_1, \ldots, Q_4$ in counterclockwise order starting with the top-right quadrant.

Note that at most one ray intersects the interior of a rectangle in $\mathcal{L}$. Indeed, any two points in two different rays, $p_i \in \ell_i$ and $p_j \in \ell_j$, span an axis-parallel rectangle that intersects the interior of c. Consequently, p_i and p_j cannot be in the same rectangle in $R \setminus c$. It follows that every rectangle of $\mathcal{L}$ in $R \setminus c$ lies in one quadrant or in the union of two consecutive quadrants.

We define an aspect ratio assignment α as follows: Let $\alpha(c) = 1$. If $r \subsetneq Q_1$ or $r \subsetneq Q_3$, let $\alpha(r) = 6n$; and if $r \subsetneq Q_2$ or $r \subsetneq Q_4$, let $\alpha(r) = (6n^2)^{-1}$. For a rectangle r split by a ray, we set $\alpha(r) = 6n + (6n^2)^{-1}$ if r is split by a horizontal ray ℓ_1 or ℓ_3; and $\alpha(r) = ((6n)^{-1} + (6n^2))^{-1}$ if split by a vertical ray ℓ_2 or ℓ_4.

Suppose that a layout $\mathcal{L}'$ weakly equivalent to $\mathcal{L}$ realizes α. Split every rectangle of aspect ratio $6n + (6n^2)^{-1}$ in $\mathcal{L}'$ horizontally into two rectangles of aspect ratios $6n$ and $(6n^2)^{-1}$. Similarly, split every rectangle of aspect ratio $((6n)^{-1} + (6n^2))^{-1}$ vertically into two rectangles of aspect ratios $6n$ and $(6n^2)^{-1}$; see Fig. 5b. By Lemma 5, there are four x- or y-monotone paths $P_1, \ldots, P_4$ from the four arms of the windwill to four distinct sides of the bounding box that pass through the slitting segments. The paths $P_1, \ldots, P_4$ subdivide the exterior of the windmill into four arrangements of rectangles, $A_1, \ldots, A_4$ that each contain a unique corner of the bounding box. By construction, every rectangle in A_1 and A_3 has aspect ratio $6n$, and every rectangle in A_2 and A_4 has aspect ratio $(6n^2)^{-1}$.

Let $R_1, \ldots, R_4$ be the bounding boxes of $A_1, \ldots, A_4$, respectively. By Lemma 6, both R_1 and R_3 have aspect ratios at least 6, and both R_2 and R_4 have aspect ratios at most $\frac{1}{6}$. By construction, the arrangements $A_1, \ldots, A_4$ each contain an arm of the windmill. This implies that width$(c) <$ min$\{$width(R_1), width$(R_3)\}$ and height$(c) <$ min$\{$height(R_2), height$(R_4)\}$. Consider the arrangement comprised of A_1, c, and A_3. It contains two opposite corners of R, and so its bounding box is R. Furthermore, height$(R) \geq$ max$\{$height(R_1), height$(R_3)\}$, and

$$\text{width}(R) \leq \text{width}(R_1) + \text{width}(c) + \text{width}(R_3) < 3 \max\{\text{width}(R_1), \text{width}(R_3)\}$$

$$\leq 3 \max\left\{\frac{\text{height}(R_1)}{6}, \frac{\text{height}(R_3)}{6}\right\} = \frac{\max\{\text{height}(R_1), \text{height}(R_3)\}}{2},$$

and so the aspect ratio of R is at least 2. Similarly, the bounding box of the arrangement comprised of A_2, c, and A_3 is also R, and an analogous argument implies that its aspect ratio must be at most $\frac{1}{2}$. We have shown that the aspect ratio of R is at least 2 and at most $\frac{1}{2}$, a contradiction. Thus the aspect ratio assignment α is not realizable, and so $\mathcal{L}$ is not weakly aspect ratio universal. □

This completes the proof of both Theorems 1 and 2. Specifically, Corollary 2 and Lemma 7 imply Theorem 1. For Theorem 2, we need to show that properties (i)–(iii) are equivalent: By Lemma 1, (ii) and (iii) are equivalent; Corollary 1 states that (ii) implies (i); and the converse follows from Lemmata 4 and 7.

2.2   Unique Transversal Structure

Subdividing a square into squares has fascinated humanity for ages [4,14,32]. For example, a *perfect square tiling* is a tiling with squares with distinct integer side lengths. Schramm [30] (see also [19, Chap. 6]) proved that every near triangulation with an outer 4-cycle is the extended dual of a (possibly degenerate or nongeneric) subdivision of a rectangle into squares. The result generalizes to rectangular faces of arbitrary aspect ratios (rather than squares):

Theorem 4. *(Schramm [30, Thm. 8.1]) Let $T = (V, E)$ be near triangulation with an outer 4-cycle, and $\alpha : V^* \to \mathbb{R}^+$ a function on the set V^* of the inner vertices of T. Then there exists a unique (but possibly degenerate or nongeneric) layout $\mathcal{L}$ such that $G^*(\mathcal{L}) = T$, and for every $v \in V^*$, the aspect ratio of the rectangle corresponding to v is $\alpha(v)$.*

The caveat in Schramm's result is that all rectangles in the interior of every separating 3-cycle must degenerate to a point, and rectangles in the interior of some of the separating 4-cycles may also degenerate to a point. We only use the *uniqueness* claim under the assumption that a nondegenerate and generic realization exists for a given aspect ratio assignment.

Lemma 8. *If a layout $\mathcal{L}$ is strongly ARU, then its extended dual $G^*(\mathcal{L})$ admits a unique transversal structure.*

Proof. Consider the extended dual graph $T = G^*(\mathcal{L})$ of a strongly ARU layout $\mathcal{L}$. As noted above, T is a 4-connected inner triangulation of a 4-cycle. If T admits two different transversal structures, then there are two strongly nonequivalent layouts, $\mathcal{L}$ and $\mathcal{L}'$, such that $T = G^*(\mathcal{L}) = G^*(\mathcal{L}')$, which in turn yield two aspect ratio assignments, α and α', on the inner vertices of T. By Theorem 4, the (non-degenerate) layouts $\mathcal{L}$ and $\mathcal{L}'$, that realize α and α', are unique. Consequently, neither of them can be strongly aspect ratio universal. □

Lemma 8 readily shows that Theorem 2(i) implies Theorem 2(iii), and provides an alternative proof for the geometric arguments in Lemmata 4 and 7.

3 Recognizing Duals of Aspect Ratio Universal Layouts

We describe an algorithm that, for a given graph G, either finds a one-sided sliceable layout $\mathcal{L}$ whose dual graph is G, or reports that no such layout exists. We can decide in $O(n)$ time whether a given graph is proper [12,22,24,25]. Every proper graph is a connected plane graph in which all bounded faces are triangles.

Problem Formulation. The input of our recursive algorithm will be an *instance* $I = (G, C, P)$, where $G = (V, E)$ is a near-triangulation, $C : V(G) \to \mathbb{N}_0$ is a *corner count*, and P is a set of ordered pairs (u, v) of vertices on the outer face of G. An instance $I = (G, C, P)$ is *realizable* if there exists a one-sided sliceable layout $\mathcal{L}$ such that G is the dual graph of $\mathcal{L}$, every vertex $v \in V$ corresponds to a rectangle in $\mathcal{L}$ incident to *at least* $C(v)$ corners of $\mathcal{L}$, and every pair $(a, b) \in P$ corresponds to a pair of rectangles in $\mathcal{L}$ incident to two ccw *consecutive* corners. When we have no information about corners, then $C(v) = 0$ for all $v \in V$, and $P = \emptyset$. In the full paper [10], we establish the following structural result.

Lemma 9. *Assume that (G, C, P) admits a realization $\mathcal{L}$ and $|V(G)| \geq 2$. Then G contains a vertex v with one of the following (mutually exclusive) properties.*

(I) Vertex v is a cut vertex in G. Then r_v is bounded by two parallel sides of R and by two parallel slices; and $C(v) = 0$.

(II) Rectangle r_v is bounded by three sides of R and a slice; and $0 \leq C(v) \leq 2$.

Based on property (II), a vertex v of G is a *pivot* if there exists a one-sided sliceable layout $\mathcal{L}$ with $G \simeq G(\mathcal{L})$ in which r_v is bounded three sides of R and a slice. If we find a cut vertex or a pivot v in G, then at least one side of r_v is a slice, so we can remove v and recurse on the connected components of $G - v$. We describe an analyze our algorithm for an instance I in the full paper [10].

4 Conclusions

We have shown that a layout $\mathcal{L}$ is weakly (strongly) ARU if and only if $\mathcal{L}$ is sliceable (one-sided and sliceable); and we can decide in $O(n^2)$-time whether a given graph G on n vertices is the dual of a one-sided sliceable layout. An immediate open problem is whether the runtime can be improved. Cut vertices and 2-cuts play a crucial role in our algorithm. We can show (in Sect. 4 of the full paper [10]) that the duals of one-sided sliceable layouts have vertex cuts of size at most 3. Perhaps 3-cuts can be utilized to speed up our algorithm. Recall that no polynomial-time algorithm is currently known for recognizing the duals of sliceable layouts [5,18,36] and one-sided layouts [6]. It remains open to settle the computational complexity of these problems.

References

1. Ackerman, E., Barequet, G., Pinter, R.Y.: A bijection between permutations and floorplans, and its applications. Discrete Appl. Math. **154**(12), 1674–1684 (2006)
2. Asinowski, A., Mansour, T.: Separable d-permutations and guillotine partitions. Ann. Comb. **14**, 17–43 (2010)
3. de Berg, M., Cheong, O., van Kreveld, M., Overmars, M.: Binary space partitions. In: Computational Geometry: Algorithms and Applications, Chap. 12, pp. 259–281. Springer, Heidelberg (2008). https://doi.org/10.1007/978-3-540-77974-2_12
4. Brooks, R.L., Smith, C.A.B., Stone, A.H., Tutte, W.T.: The dissection of rectangles into squares. Duke Math. J. **7**(1), 312–340 (1940)
5. Dasgupta, P., Sur-Kolay, S.: Slicible rectangular graphs and their optimal floorplans. ACM Trans. Design Autom. Electr. Syst. **6**(4), 447–470 (2001)
6. Eppstein, D., Mumford, E., Speckmann, B., Verbeek, K.: Area-universal and constrained rectangular layouts. SIAM J. Comput. **41**(3), 537–564 (2012)
7. Felsner, S.: Lattice structures from planar graphs. Electron. J. Comb. **11**(1) (2004)
8. Felsner, S.: Rectangle and square representations of planar graphs. In: Pach, J. (ed.) Thirty Essays on Geometric Graph Theory, pp. 213–248. Springer, New York (2013). https://doi.org/10.1007/978-1-4614-0110-0_12
9. Felsner, S.: Exploiting air-pressure to map floorplans on point sets. J. Graph Algorithms Appl. **18**(2), 233–252 (2014)
10. Felsner, S., Nathenson, A., Tóth, C.D.: Aspect ratio universal rectangular layouts. Preprint (2021). https://arxiv.org/abs/2112.03242
11. Fusy, É.: Transversal structures on triangulations: a combinatorial study and straight-line drawings. Discrete Math. **309**(7), 1870–1894 (2009)

12. Hasan, M.M., Rahman, M.S., Karim, M.R.: Box-rectangular drawings of planar graphs. J. Graph Algorithms Appl. **17**(6), 629–646 (2013)
13. He, X.: On finding the rectangular duals of planar triangular graphs. SIAM J. Comput. **22**(6), 1218–1226 (1993)
14. Henle, F.V., Henle, J.M.: Squaring the plane. Am. Math. Mon. **115**(1), 3–12 (2008)
15. Kant, G., He, X.: Regular edge labeling of 4-connected plane graphs and its applications in graph drawing problems. Theor. Comput. Sci. **172**(1–2), 175–193 (1997)
16. Koźmiński, K., Kinnen, E.: Rectangular duals of planar graphs. Networks **15**(2), 145–157 (1985)
17. van Kreveld, M.J., Speckmann, B.: On rectangular cartograms. Comput. Geom. **37**(3), 175–187 (2007)
18. Kusters, V., Speckmann, B.: Towards characterizing graphs with a sliceable rectangular dual. In: Di Giacomo, E., Lubiw, A. (eds.) GD 2015. LNCS, vol. 9411, pp. 460–471. Springer, Cham (2015). https://doi.org/10.1007/978-3-319-27261-0_38
19. Lovász, L.: Graphs and Geometry, Colloquium Publications, vol. 65. AMS, Providence, RI (2019)
20. Merino, A.I., Mütze, T.: Efficient generation of rectangulations via permutation languages. In: 37th International Symposium on Computational Geometry (SoCG). LIPIcs, vol. 189, pp. 54:1–54:18. Schloss Dagstuhl (2021)
21. Mitchell, W.J., Steadman, J.P., Liggett, R.S.: Synthesis and optimization of small rectangular floor plans. Environ. Plann. B Plann. Des. **3**(1), 37–70 (1976)
22. Nishizeki, T., Rahman, M.S.: Rectangular drawing algorithms. In: Handbook on Graph Drawing and Visualization, pp. 317–348. Chapman and Hall/CRC (2013)
23. Otten, R.H.J.M.: Automatic floorplan design. In: Proceedings of the 19th Design Automation Conference (DAC), pp. 261–267. ACM/IEEE (1982)
24. Rahman, M.S., Nakano, S., Nishizeki, T.: Rectangular grid drawings of plane graphs. Comput. Geom. **10**(3), 203–220 (1998)
25. Rahman, M.S., Nakano, S., Nishizeki, T.: Rectangular drawings of plane graphs without designated corners. Comput. Geom. **21**(3), 121–138 (2002)
26. Raisz, E.: The rectangular statistical cartogram. Geogr. Rev. **24**(2), 292–296 (1934)
27. Reading, N.: Generic rectangulations. Eur. J. Comb. **33**(4), 610–623 (2012)
28. Rinsma, I.: Existence theorems for floorplans. Ph.D. thesis, University of Canterbury, Christchurch, New Zealand (1987)
29. Rinsma, I.: Nonexistence of a certain rectangular floorplan with specified areas and adjacency. Environ. Plann. B Plann. Des. **14**(2), 163–166 (1987)
30. Schramm, O.: Square tilings with prescribed combinatorics. Isr. J. Math. **84**, 97–118 (1993)
31. Thomassen, C.: Plane representations of graphs. In: Progress in Graph Theory, pp. 43–69. Academic Press Canada (1984)
32. Tutte, W.T.: Squaring the square. Scientific American **199**, 136–142 (1958), in Gardner's 'Mathematical Games' column. Reprinted with addendum and bibliography in the US in M. Gardner. In: The 2nd Scientific American Book of Mathematical Puzzles & Diversions, Simon and Schuster, New York, pp. 186–209 (1961)
33. Ungar, P.: On diagrams representing maps. J. Lond. Math. Soc. **s1-28**(3), 336–342 (1953)
34. Wimer, S., Koren, I., Cederbaum, I.: Floorplans, planar graphs, and layouts. IEEE Trans. Circuits Syst. **35**(3), 267–278 (1988)
35. Yao, B., Chen, H., Cheng, C., Graham, R.L.: Floorplan representations: complexity and connections. ACM Trans. Des. Autom. Electr. Syst. **8**(1), 55–80 (2003)
36. Yeap, G.K.H., Sarrafzadeh, M.: Sliceable floorplanning by graph dualization. SIAM J. Discrete Math. **8**(2), 258–280 (1995)

Morphing Tree Drawings
in a Small 3D Grid

Aleksandra Istomina[✉], Elena Arseneva, and Rahul Gangopadhyay

Saint-Petersburg University, Saint Petersburg, Russia
st062510@student.spbu.ru, e.arseneva@spbu.ru, rahulg@iiitd.ac.in

Abstract. We study crossing-free grid morphs for planar tree drawings using the third dimension. A morph consists of morphing steps, where vertices move simultaneously along straight-line trajectories at constant speeds. There is a crossing-free morph between two drawings of an n-vertex planar graph G with $\mathcal{O}(n)$ morphing steps, and using the third dimension the number of steps can be reduced to $\mathcal{O}(\log n)$ for an n-vertex tree [Arseneva et al. 2019]. However, these morphs do not bound one practical parameter, the resolution. Can the number of steps be reduced substantially by using the third dimension while keeping the resolution bounded throughout the morph? We present a 3D crossing-free morph between two planar grid drawings of an n-vertex tree in $\mathcal{O}(\sqrt{n}\log n)$ morphing steps. Each intermediate drawing lies in a $3D$ grid of polynomial volume.

Keywords: morphing grid drawings · bounded resolution · 3D morphing

1 Introduction

Given an n-vertex graph G, a *morph* between two drawings (i.e., embeddings in $\mathbb{R}^d$) of G is a continuous transformation from one drawing to the other through a family of intermediate drawings. One is interested in well-behaved morphs, i.e., those that preserve essential properties of the drawing at any moment. Usually, this property is that the drawing is *crossing-free*; such morphs are called *crossing-free* morphs. This concept finds applications in multiple domains: animation, modeling, and computer graphics, etc. A drawing of G is a *straight-line drawing* if it maps each vertex of G to a point in $\mathbb{R}^d$ and each edge of G to the line segment whose endpoints correspond to the endpoints of this edge. In this work, we focus on the case of drawings in the Euclidean plane ($d = 2$) and $3D$ drawings ($d = 3$); a non-crossing drawing of a graph in $\mathbb{R}^2$ is called *planar*.

There is an interest in studying crossing-free morphs of straight-line drawings, where vertex trajectories are simple, in particular, *linear morphs*. A linear morph transforms one straight-line drawing Γ of a graph G to another such drawing Γ' through a sequence of straight-line drawings; each *morphing steps* or *step* is a linear interpolation between two consecutive drawings in that sequence.

© Springer Nature Switzerland AG 2022
P. Mutzel et al. (Eds.): WALCOM 2022, LNCS 13174, pp. 85–96, 2022.
https://doi.org/10.1007/978-3-030-96731-4_8

That is, during each morphing step each vertex of G moves along a straight-line segment at a constant speed. A linear morph is said to be *unidirectional* if all vertices move along parallel lines in the same direction. Alamdari et al. [1] showed that for any two topologically equivalent planar drawings of a graph G, there is a linear $2D$ morph that transforms one drawing to the other in $\Theta(n)$ steps. This bound is asymptotically optimal in the worst case even when the graph G is a path. A natural further question is how the situation changes when we involve the third dimension. For general 3D graph drawings the problem seems challenging since it is tightly connected to *unknot recognition* problem. If both the initial and the final drawing are planar and the given graph is a tree, then $\mathcal{O}(\log n)$ steps suffice [2]. In both algorithmic results [1,2], the intermediate steps use infinitesimal or very small distances, as compared to distances in the input drawings. This may blow up the space requirements and affect the aesthetical aspect. This raises a demand for morphing algorithms that operate on a small grid, i.e., of size that is polynomial in the size of the graph and parameters of the input drawings. All the intermediate drawings are then restricted to be *grid drawings*, where vertices map to vertices of the grid. Two crucial parameters of a straight-line grid drawing are: the area (or volume for the 3D case) of the required grid, and the *resolution*, that is the ratio between the maximum edge length and the minimum edge-edge distance. If the grid area (or volume) is polynomially bounded, then so is the resolution [3].

Very recently Barrera-Cruz et al. [3] gave an algorithm that linearly morphs between two planar straight-line grid drawings Γ and Γ' of an n-vertex rooted tree in $\mathcal{O}(n)$ steps while each intermediate drawing is also a planar straight-line drawing in a bounded grid. In particular, the maximum grid length and width are respectively $\mathcal{O}(D^3 n \cdot L)$ and $\mathcal{O}(D^3 n \cdot W)$, where $L = max\{l(\Gamma), l(\Gamma')\}$, $W = max\{w(\Gamma), w(\Gamma')\}$ and $D = max\{L, W\}$, $l(\Gamma)$ and $w(\Gamma)$ are the *length* and the *width* of the drawing Γ respectively. Note that D is $\Omega(\sqrt{n})$.

Let Γ and Γ' be two planar straight-line drawings of an n-vertex tree T. Throughout this paper, a morph $\mathcal{M} = \langle \Gamma_1, \Gamma_2, \ldots, \Gamma_k \rangle$ of T is a sequence of 3D straight-line drawings of T such that $\Gamma_1 = \Gamma, \Gamma_k = \Gamma'$ are the initial and the final drawings, and each $\langle \Gamma_i, \Gamma_{i+1} \rangle$ is a linear morph. Here we study the problem of morphing one straight-line grid drawing Γ to another such drawing Γ' in *sublinear* number of steps using the third dimension such that the resolutions of the intermediate drawings are bounded. We morph the initial planar drawing of tree T to its $3D$ canonical drawing $\mathcal{C}(T)$ and then analogously morph $\mathcal{C}(T)$ to the final planar drawing. Effectively we solve the same problem as in [2], but with the additional restriction that all drawings throughout the algorithm lie in a small grid. We give an algorithm that requires $\mathcal{O}(\sqrt{n} \log n)$ steps. All the intermediate drawings require a $3D$ grid of length $\mathcal{O}(d^3(\Gamma) \cdot \log n)$, width $\mathcal{O}(d^3(\Gamma) \cdot \log n)$ and height $\mathcal{O}(n)$, where $d(\Gamma) = max(d(\Gamma), d(\Gamma'))$. During the procedure, we use some known techniques, e.g., canonical drawing [2] and "Pinwheel" rotation [3] combined with several new ideas.

In Sect. 2, we introduce the definitions that are used in the paper. After introducing the necessary definitions and preliminaries in Sect. 2, we describe the tools that are the building blocks of our algorithm: stretching, mapping around the pole, rotating and shrinking subtrees (See Sect. 3). In Sect. 4, we introduce a technique of lifting paths such that the vertices on the path along with their subtrees go to the respective canonical positions and the drawing remains crossing-free. The morphing algorithm in Sect. 4 splits the given tree into disjoint paths that are lifted one by one in specific order. Since lifting each path takes constant number of steps, in the worst case this algorithm takes $\mathcal{O}(n)$ steps to lift a tree. In Sect. 5, we show how to lift a set of edges of the given tree simultaneously. This is used in the second morphing algorithm, that lifts the tree by lifting disjoint sets of its edges one after another. This algorithm takes $\mathcal{O}(h)$ steps to lift a tree of height h. We then combine two algorithms in Sect. 6 to produce the final algorithm that uses $o(n)$ morphing steps. It first lifts all paths of T of length at most $\sqrt{n}$ using the algorithm of Sect. 5. Since the total number of remaining paths is less than $\sqrt{n}$, we lift them one after another by using the algorithm of Sect. 4. The full version[1] of the paper contains detailed proofs and descriptions which are omitted here due to space constraints.

2   Preliminaries and Definitions

Tree Drawings. For a rooted tree T, let $r(T)$ be its root, and $T(v)$ be the subtree of T rooted at a vertex v of T. Let $E(T)$, $V(T)$ and $|T|$ denote respectively, the set of edges, the set of vertices, and the number of vertices of T. In a *straight-line drawing* of T, each vertex is mapped to a point in $\mathbb{R}^d$ and each edge is mapped to a straight-line segment connecting its end-points. A 3D- (respectively, a 2D-) *grid drawing* of T is a straight-line drawing where each vertex is mapped to a point with integer coordinates in $\mathbb{R}^3$ (respectively, $\mathbb{R}^2$). A drawing of T is said to be *crossing-free* if images of no two edges intersect except, possibly, at common end-points. A crossing-free 2D-grid drawing is called a *planar grid drawing*. For a crossing-free drawing Γ, let $B(\Gamma(v), r)$ denote the open disc of radius r in the horizontal plane centered at the image $\Gamma(v)$ of v. By the *projection*, denoted by $pr()$, we mean the vertical projection to the horizontal plane passing through the origin. Let $l(\Gamma)$, $w(\Gamma)$ and $h(\Gamma)$ respectively denote the *length*, *width* and *height* of the 3D drawing Γ of T, i.e., the maximum absolute difference between the x-, y- and z-coordinates of vertices in Γ. Let $d(\Gamma)$ denote the *diameter* of Γ, defined as the ceiling of the maximum pairwise (Euclidean) distance between its vertices. Note that $d(\Gamma)$ estimates the space required by Γ since $M \leq d(\Gamma) \leq \sqrt{3}M$, where $M = \max(l(\Gamma), w(\Gamma), h(\Gamma))$. Let $dist_\Gamma(v, e)$ (resp., $dist_\Gamma(v, u)$) be the distance between $\Gamma(v)$ and $\Gamma(e)$ (resp., between $\Gamma(v)$ and $\Gamma(u)$), where u, v are vertices of T and e is an edge of T. For a grid drawing Γ, we define the *resolution* of Γ as the ratio of the distances between the farthest and closest pairs of geometric objects of Γ (images of tree vertices and edges).

For any vertex v and edge e not incident to v in a crossing-free grid drawing Γ of T, $dist(v, e) \geq \frac{1}{d(\Gamma)}$. In a 3D grid drawing Γ of T, the distance $dist(e_1, e_2) \geq \frac{1}{2\sqrt{3}\,(d(\Gamma))^2}$ for a pair of non-adjacent edges e_1, e_2. This implies that 2D and

---

[1] arXiv:2106.04289.

$3D$ crossing-free grid drawings of T have polynomially bounded resolution. For a point $p = (p_x, p_y, p_z)$, we denote by YZ_p, XZ_p, XY_p planes $x = p_x, y = p_y, z = p_z$ respectively. Analogously, XZ_p^+ (resp., XZ_p^-) denotes the vertical half-plane $\{(x, y, z) : y = p_y, x \geq p_x (resp., x \leq p_x)\}$ and YZ_p^+ (resp., YZ_p^-) the half-plane $\{(x, y, z) : x = p_x, y \geq p_y (resp., y \leq p_y)\}$.

Path Decomposition. $\mathcal{P}$ of a tree T is a decomposition of its edges into a set of edge-disjoint paths as follows. Choose some root-to-leaf path in T and store it in the set $\mathcal{P}$ which is empty at the beginning. Remove the edges of this path from T. It may disconnect the tree; recurse on the remaining connected components while there are edges. In the end, $\mathcal{P}$ contains disjoint paths whose union is $E(T)$. The depth $dpt(v)$ of a vertex v in T is the length of the path from $r(T)$ to v. *Head* of a path P, denoted as $head(P)$, is the vertex $x \in P$ with the minimum depth in tree T. Let the *internal vertices* of path P be all vertices of P except $head(P)$. Any path decomposition $\mathcal{P}$ of T induces a linear order of the paths: path P' succeeds P, i.e., $P' \succ P$, if and only if P' is deleted before P during the construction of $\mathcal{P}$. Note that the subtree of each internal vertex of a path P is a subset of the union of the paths that precede P.

In the *long-path decomposition* [4] $\mathcal{L}(T)$, the path chosen in every iteration is the longest root-to-leaf path (ties are broken arbitrarily). Let $\mathcal{L} = \{L_1, \ldots, L_m\}$ be the ordered set of paths of a long-path decomposition of T. For $i < j$, $|L_i| \leq |L_j|$.

In the *heavy-rooted-pathwidth decomposition* $\mathcal{H}(T)$ (see, e.g., [2]), of a tree T, the root-to-leaf path chosen in every iteration maximizes the rooted pathwidth, $rpw(T).rpw(T)$ is defined recursively: for each leave v of T: $rpw(\{v\}) = 1$; for each internal vertex u and its children $v_1, \ldots, v_k$ we have $rpw(T(u)) = max(rpw(T(v_i))), 1 \leq i \leq k$ if $rpw(T(v_i))$ are not all equal, and $rpw(T(u)) = rpw(T(v_1)) + 1$ in the other case. It is known [5] that for a tree T with n vertices $rpw(T) = \mathcal{O}(\log n)$. Figure 1a and 1b show respectively the heavy-rooted-pathwidth and the long-path decomposition of a tree where heavy paths and long paths are shown in different colors.

Canonical 3D drawing $\mathcal{C}(T)$ of a tree T [2] is the crossing-free straight-line 3D drawing of T that maps each vertex v of T to its *canonical position* $\mathcal{C}(v)$ determined by the heavy-rooted pathwidth decomposition. We later use the fact that $\mathcal{C}(T)$ lies in XZ_0^+ inside a bounding box of height $|T|$ and width $rpw(T)$. For any vertex v of T, the *relative canonical drawing* $\mathcal{C}_{T_v}$ of $T(v)$ is the drawing of $T(v)$ obtained by cropping $\mathcal{C}(T)$ and translating the obtained drawing of $T(v)$ so that v is mapped to the origin. Since tree T never changes throughout our algorithm, we refer to $rpw(T)$ as to rpw.

3   Tools for Morphing Algorithms

We define stretching, mapping, rotation and shrinking of subtrees in this section. Each of these are fundamental tools used in our algorithm.

Stretching with a Constant $\mathcal{S}_1$. Let the drawing Γ lie in the XY_0 plane. During *stretching morph* $\langle \Gamma, \Gamma_1 \rangle$ each coordinate of each vertex in Γ is multiplied by a common positive integer constant $\mathcal{S}_1$ to obtain Γ_1. Thereby, it is a linear morph that "stretches" the vertices apart. Stretching morph is crossing-free.

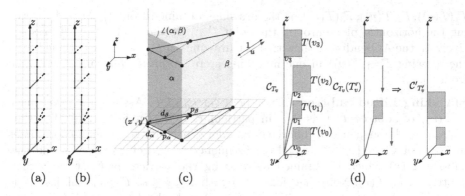

Fig. 1. Canonical drawing of a tree with (a) heavy and (b) long paths, where paths are colored with different colors, paths that consist of one edge are dashed. (c) The mapping morph, half-planes α, β sharing a common pole through point (x', y') and their vector of mapping. (d) The Shrinking morph when $l = 4$.

Lemma 1. *For any pair v_i, v_j of vertices disks $B(\Gamma_1(v_i), \frac{S_1}{2})$ and $B(\Gamma_1(v_j), \frac{S_1}{2})$ do not cross in the XY_0 plane. For a vertex v_i disk $B(\Gamma_1(v_i), \frac{S_1}{2 \cdot d(\Gamma)})$ does not enclose any other vertices or any part of edges non-incident to v_i in Γ_1. For every vertex v and every edge $e = (v, u)$ in Γ_1 there is lattice point z such that $z \in e$ and $z \in B(\Gamma_1(v_i), d(\Gamma))$.*

Mapping Around a Pole. Let the *pole through* (x', y') be the vertical line in $3D$ through a point $(x', y', 0)$. Let α, β be vertical half-planes containing the pole l through a point with integer coordinates. Suppose $\angle(\alpha, \beta) \notin \{0, \pi\}$ and α, β contain infinitely many points with integer coordinates. Mapping *around the pole* l is a morphing step to obtain a drawing Γ' which lies in β from Γ which lies in α. Each vertex moves along a horizontal vector between α and β. The direction of this vector is common for all vertices of Γ and is defined by α and β. Let us fix a horizontal plane h passing through the point $(0, 0, b)$ where b is an integer. Let p_α, p_β be points that lie on $h \cap \alpha$ and $h \cap \beta$, respectively; such that $dist(l, p_\alpha) = d_\alpha$ and $dist(l, p_\beta) = d_\beta$ be the minimum non-zero distances from the l to the integer points lying in $h \cap \alpha$ and $h \cap \beta$. The *vector of mapping* u is defined as $\frac{p_\beta - p_\alpha}{|p_\beta - p_\alpha|}$. Mapping is an unidirectional morph since all vertices of Γ move along the vectors parallel to the vector of mapping till they reach the half-plane β, see Fig. 1c. Since mapping is comprised of rotation and stretching in horizontal direction, it is a crossing-free morph that preserves grid drawings. Throughout the paper, we denote by *rotation* a mapping when α, β are half-planes of planes parallel to XZ_0, YZ_0 respectively or vice-versa. Similarly, we define mapping around horizontal pole, i.e., a pole parallel to the X-axis.

Rotating Horizontal Plane. Let $\Gamma_0(T(v))$ be the canonical drawing of a subtree $T(v)$ on the horizontal plane XY_v obtained by rotating the relative canonical drawing $\mathcal{C}_{T_v}$ around the horizontal pole through v. Let

$\Gamma_1(T(v)), \Gamma_2(T(v)), \Gamma_3(T(v))$ be the drawings obtained from $\Gamma_0(T(v))$ by rotating the horizontal plane around the point $\Gamma(v)$ by the angles $\frac{\pi}{2}, \pi, \frac{3\pi}{2}$, respectively. In the Appendix we show that the drawing $\Gamma_i(T(v))$ can be obtained from the drawing $\Gamma_{i-1}(T(v))$ in one morphing step—*rotating step*—using a lemma from [3].

Shrinking Lifted Subtrees. Let v be a vertex of T. Assume that the image $\Gamma(T(v))$ of subtree $T(v)$ is $\mathcal{C}_{T_v}$, in particular, it lies in $h = XZ_v^+$. Let $\mathcal{C} = \{v_1, \ldots, v_l\}$ be sequence of children of v, ordered according to their z-coordinates in $\mathcal{C}_{T_v}$. Let $\mathcal{C}' = \{v_{i_1}, \ldots, v_{i_k}\}$ be subsequence of $\mathcal{C}$. Let us consider the new subtree $T'(v)$ which is obtained by deleting the vertices in $\mathcal{C} \setminus \mathcal{C}'$ and their subtrees from $T(v)$. Note that, for each j with $1 \leq j \leq k$, $T'(v_{i_j})$ still lies inside a box of height $|T(v_{i_j})|$ and width $rpw(T(v_{i_j}))$ on h. We define the *shrink subtree* procedure on T'_v as follows. We move each vertex v_{i_j} along with their subtrees from $\mathcal{C}_{T_v}(v_{i_j})$ to $(\mathcal{C}_{T_v}(v_{i_j})_x, \mathcal{C}_{T_v}(v_{i_j})_y, \mathcal{C}_{T'_v}(v_j)_z)$. Let us denote the shrunk subtree by $\mathcal{C}'_{T'_v}$. The height of the shrunk subtree $\mathcal{C}'_{T'_v}$ is equal to the number of vertices in $T'(v)$. Also, note that shrinking is a crossing-free unidirectional morph.

4   Morphing Through Lifting Paths

Let T be an n-vertex tree and $\mathcal{P}$ be a path decomposition of T into k paths. In this section, we describe an algorithm that morphs a plane drawing $\Gamma = \Gamma_0$ in XY_0 plane of tree T to the canonical 3D drawing $\Gamma' = \mathcal{C}(T)$ of T in $\mathcal{O}(k)$ steps. It lifts the paths of $\mathcal{P}$ one by one applying procedure $Lift()$. Note that the final positions for the vertices in $\mathcal{C}(T)$ are independent of $\mathcal{P}$. Also, a morph from $\mathcal{C}(T)$ to Γ' can be obtained by applying the morph from Γ' to $\mathcal{C}(T)$ backwards. At all times during the algorithm, the following invariant holds: a path $P_i \in \mathcal{P}$ is lifted only after all the children of the internal vertices of P_i are lifted. After the execution of $Lift(P_i)$, path P_i moves to its canonical position with respect to $head(P_i)$, see Fig. 2 and 3.

Step 0: Preprocessing. This step is a single stretching morph $\langle \Gamma, \Gamma_1 \rangle$ with $S_1 = 2 \cdot (rpw + d(\Gamma))$. Note that stretching is a crossing-free morph.

$Lift(path)$ Procedure
Let $P_i = (v_0, v_1, \ldots, v_m)$ be the first path in $\mathcal{P}$ that has not been processed yet and Γ_t be the current drawing of T. We lift the path P_i. For any vertex v let *lifted subtree* $T'(v)$ be the portion of subtree $T(v)$ that has been lifted after execution of $Lift(P_j)$ for some $j < i$. Let *the processing vertices* be the internal vertices of P_i along with the vertices of their lifted subtrees. The subtrees of all internal vertices v_j in P_i are already lifted due to the ordering among the paths. Suppose the lifted subtrees are in the canonical position with respect to the roots, the maximum height of vertices in an intermediate drawing Γ_t is strictly less than n and the difference of width between a lifted vertex and its root is at most rpw. We provide a brief overview of the $Lift()$ procedure in the following.

The procedure $Lift(P_i)$ consists of 13 steps and results in moving vertices of path P_i along with their lifted subtrees to their canonical positions with respect

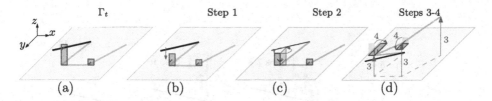

Fig. 2. (a) Drawing Γ_t in the beginning of the procedure $Lift(P_i)$, bounding boxes for lifted subtrees are violet, P_i consists of green edges. Directions of movement of the vertices are shown with red arrows. (b) **Step 1**, (c) **Step 2** and (d) **Steps 3–4** of $Lift()$. (Color figure online)

to the head of P_i, i.e.,vertex v_0. Since the height of any lifted vertex is strictly less than n and the difference of width between a lifted vertex and its root is at most rpw, preprocessing Step 0 and Lemma 1 guarantee that the already lifted subtrees lie in the disjoint right circular cylinders of radius rpw and height n.

Step 1: For every internal vertex v_j of the path P_i, its lifted subtree $T'(v_j)$ morphs into the shrunk lifted subtree, see Sect. 3. All subtrees are shrunk simultaneously in one morphing step. This step is needed to ensure that the maximum height of a vertex does not exceed $2n$ during the $Lift()$ procedure. It is a crossing-free morph since the subtrees move in mutually disjoint cylinders.

Step 2: It consists of steps $\langle \Gamma_t, \Gamma_{t+1} \rangle, \langle \Gamma_{t+1}, \Gamma_{t+2} \rangle$. For $0 \leq j < m - 1$, if projection $pr(T'(v_j))$ overlaps with $pr((v_j, v_{j+1}))$, we rotate twice the drawing of $T'(v_j)$ around the vertical pole through $\Gamma_t(v_j)$. Since every lifted subtree $T'(v_j)$ lies in $XZ_{v_j}^+$, after this step all lifted subtrees lie in $XZ_{v_j}^+$ or $XZ_{v_j}^-$. It is a crossing-free morph since the rotations of subtrees happen inside mutually disjoint cylinders.

Step 3: In the morphing step $\langle \Gamma_{t+2}, \Gamma_{t+3} \rangle$, each internal vertex $v_j, j \geq 1$ of path P_i moves vertically to the height defined recursively as follows: for v_1: $\Gamma_{t+3}(v_1)_z = n$; for $v_j, j > 1$: $\Gamma_{t+3}(v_j)_z = \Gamma_{t+3}(v_{j-1})_z + |T'(v_{j-1})|$. Note that $|T'(v_j)|$, a number of vertices in $T'(v_j)$, is equal to the height of this shrinked lifted subtree. This step is crossing free since the projections of different subtrees and the path edges to the XY_0 plane does not change during the morph. After this step the vertices of P_i are in the same vertical order as in the canonical drawing $\mathcal{C}(T)$.

Step 4: The lifted subtree of each internal vertex of P_i is rotated to lie in a horizontal plane passing through the corresponding vertex. This step places all $T'(v_j)$ in disjoint horizontal planes. The direction of rotation is chosen in such a way that $T'(v_j)$ does not cross with an edge (v_j, v_{j+1}).

Steps 5 and 6: In Step 5 ($\langle \Gamma_{t+4}, \Gamma_{t+5} \rangle$), each vertex $v_j (j \geq 2)$ of the path P_i moves together with its subtree $T'(v_j)$ along the vector $((v_{1_x} - v_{j_x}) + \mathcal{C}(v_j)_x - \mathcal{C}(v_1)_x, v_{1_y} - v_{j_y}, 0)$, where v_{1_x} denotes x-coordinate of vertex v_1 in drawing Γ_{t+4}. In Step 6 ($\langle \Gamma_{t+5}, \Gamma_{t+6} \rangle$), every vertex $v_j, j \geq 2$ of the path P_i moves together with

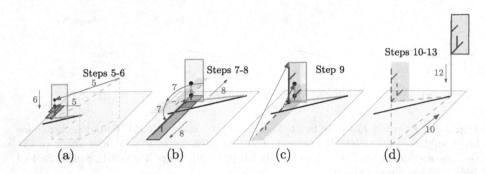

Fig. 3. Yellow plane is a vertical plane. (a) **Steps 5–6**, (b) **Steps 7–8**, (c) **Step 9** and (d) **Steps 10–13** (Steps 11, 13 do not make any changes in this example) of *Lift*(). (Color figure online)

its subtree $T'(v_j)$ along the same vertical vector $(0, 0, (v_{1_z} - v_{j_z}) + \mathcal{C}(v_j)_z - \mathcal{C}(v_1)_z)$, where v_{1_z} means z-coordinate of vertex v_1 in drawing Γ_{t+5}. Steps 5 and 6 move $v_2, \ldots, v_m$ to their canonical positions with respect to the vertex v_1

Steps 7, 8 and 9: Step 7, i.e., $\langle \Gamma_{t+6}, \Gamma_{t+7} \rangle, \langle \Gamma_{t+7}, \Gamma_{t+8} \rangle$, turns every lifted subtree $T''(v_j)$ of internal vertices of P_i to lie in positive x-direction with respect to v_j. Step 8, i.e., $\langle \Gamma_{t+8}, \Gamma_{t+9} \rangle$, morphs lifted subtrees of internal vertices of P_i in the horizontal planes from shrunk to the canonical size. In Step 9 ($\langle \Gamma_{t+9}, \Gamma_{t+10} \rangle$) all lifted subtrees $T'(v_j)$ of the internal vertices of P_i rotate around horizontal axes $(x, v_{j_y}, v_{j_z}), x \in \mathbb{R}$ to lie in vertical plane in positive direction such that the subtree $T(v_1)$ is in the canonical position with respect to v_1.

Step 10: In the morphing step $\langle \Gamma_{t+10}, \Gamma_{t+11} \rangle$, every internal vertex v_j of the path with its subtree $T'(v_j)$ moves horizontally in the direction $(v_{0_x} - v_{1_x}, v_{0_y} - v_{1_y}, 0)$. If in $\mathcal{C}(T)$ the edge (v_0, v_1) is vertical, vertex v_1 moves along this vector to get x, y-coordinates equal to (v_{0_x}, v_{0_y}). Otherwise, vertex v_1 moves along this vector as long as possible to get integer x and y coordinates not equal to (v_{0_x}, v_{0_y}). Step 10 ensures that Steps 11–13 move vertices only inside right circular cylinder of radius $rpw + d(\Gamma)$ and height $2n$ around v_0. During Steps 11–13 the processed part of the tree does not intersect with the unprocessed part since the above mentioned cylinders are disjoint for the vertices that are lying in XY_0.

Steps 11, 12 and 13: These steps differ depending on whether or not we have rotated $T'(v_0)$ during Step 2. In one case $T'(v_0)$ is in x-positive direction from v_0 and in the other $T(v_1)$ is in x-positive direction from v_0. Steps 11 and 13 make two rotations of the needed part of the tree to correct it's x, y-coordinates. Also, we need to move v_1 and its subtree to the canonical height with respect to v_0. In Step 12, we make the z-coordinate correction of $T(v_1)$. The steps are ordered in such a way that no intersections happen during their execution. Step 13 concludes the procedure $Lift(P_i)$ by placing all processing vertices into their canonical positions with respect to v_0.

In the end of these morphing steps, we observe that all the internal vertices of P_i along with their subtrees are placed in the canonical position with respect to v_0. The lifted subtrees that were in the relative canonical position at the beginning of $Lift(P_i)$, still maintain their positions. For any path P_k, such that $k > i$, its vertices still lie on the XY_0 plane and their positions do not change during these steps. We keep on lifting up paths until we obtain the canonical drawing of T. The following theorem summarises what we achieved in this section.

Theorem 1. *For every two planar straight-line grid drawings Γ, Γ' of tree T with n vertices there exists a crossing-free 3D-morph $\mathcal{M} = \langle \Gamma = \Gamma_0, \ldots, \Gamma_l = \Gamma' \rangle$ that takes $\mathcal{O}(k)$ steps where k is number of paths in some path decomposition of tree T. In this morph, every intermediate drawing $\Gamma_i, 1 \leq i \leq l$ is a straight-line 3D grid drawing lying in a grid of size $\mathcal{O}(d^2 \times d^2 \times n)$, where d is maximum of the diameters of the given drawings.*

5   Morphing Through Lifting Edges

In this section, we describe another algorithm that morphs a planar drawing Γ of tree T to the canonical drawing $\mathcal{C}(T)$ of T. This time one iteration of our algorithm lifts simultaneously a set of edges with at most one edge of each path of a selected path decomposition. Let $\Gamma = \Gamma_0$ be a planar drawing of T.

Step 0: Preprocessing. This step $\langle \Gamma, \Gamma_1 \rangle$ is a stretching morph with $\mathcal{S}_1 = 2 \cdot rpw \cdot d(\Gamma) \cdot (4 \cdot d(\Gamma) + 1)$. It is a crossing-free morph.

$\overline{Lift}(edges)$ ***procedure***
For edge e of T, let $st(e)$ (respectively, $end(e)$) be the vertex of e with smallest (respectively, largest) depth. Let $\mathcal{K} = \{K_1, \ldots, K_m\}$ be the partition of edges of T into disjoint sets such that $e \in K_i$ if and only if $dpt(st(e)) = m - i$, where m denotes the depth of T. We lift up sets K_i from $\mathcal{K}$ from $i = 1$ to $i = m$ by executing $\overline{Lift}(K_i)$ (Steps 1–5, see Fig. 4 and 5). Let Γ_t be the drawing of T before lifting set K_i. Let *lifted subtree* $T'(v_j)$ be the portion of subtree $T(v_j)$ lifted by the execution of $\overline{Lift}(K_j)$ where $j < i$. Suppose the drawing of $T'(v)$ in Γ_t is the canonical drawing of $T'(v)$ with respect to v; and the vertices that are incident to some non-processed edges lie in XY_0 plane.

Lemma 2. *For every edge $e = (v, u)$ with $st(e) = v$ in Γ_1 there is a lattice point $z_e \in e$ such that $B(\Gamma_1(z_e), rpw \cdot d(\Gamma)) \subset B(\Gamma_1(v), rpw \cdot d(\Gamma) \cdot (4 \cdot d(\Gamma) + 1))$. For distinct pair of edges $e_1, e_2 \in K_i \forall i = 1, \ldots, m$ disks $B(\Gamma_1(z_{e_1}), rpw)$ and $B(\Gamma_1(z_{e_2}), rpw)$ are disjoint. Also, for distinct pair of edges $e_1, e_2 \in K_i \forall i = 1, \ldots, m$ regions $\mathcal{F}_{e_1}, \mathcal{F}_{e_2}$ are disjoint, where $\mathcal{F}_e = \{x \in XY_0 : dist_{\Gamma_1}(x, (z_e, u)) \leq rpw\}$.*

Step 1: Shrink. In the step $\langle \Gamma_t, \Gamma_{t+1} \rangle$, for every edge $e \in K_i$ we move vertex $end(e)$ along with its lifted subtree towards $st(e)$ until $end(e)$ reaches point z_e.

Step 2: Go up. In morphing step $\langle \Gamma_{t+1}, \Gamma_{t+2} \rangle$, we move $end(e)$ with $T'(end(e))$ along the vector $(0, 0, \mathcal{C}(end(e))_z - \mathcal{C}(st(e))_z)$ for all $e \in K_i$.

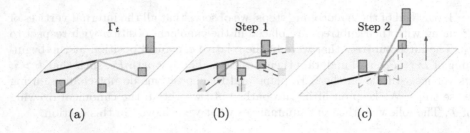

Fig. 4. (a) Drawing Γ_t in the beginning of the procedure $\overline{Lift}(K_i)$, bounding boxes for lifted subtrees are violet, K_i consists of green edges. (b) **Step 1** and (c) **Step 2** of $\overline{Lift}()$.

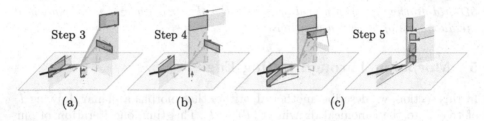

Fig. 5. (a) **Step 3**; (b) **Step 4**, (c) **Step 5**, consists of two morphing steps.

Step 3: Mapping. Morphing step $\langle \Gamma_{t+2}, \Gamma_{t+3} \rangle$ is a mapping morph, see Sect. 3. For every lifted subtree $T'(v_j)$, where $v_j = end(e), e \in K_i$, we define the half-planes of the mapping morph as follows: half-plane α is $XZ_{v_j}^+$, half-plane β is part of the vertical plane containing the edge e in such direction that $e \notin \beta$, the common vertical pole of α and β is a pole through v_j. All mapping steps are done simultaneously for all subtrees of end vertices of the edges of K_i.

Step 4: Shrink more. The morphing step $\langle \Gamma_{t+3}, \Gamma_{t+4} \rangle$ is a horizontal morph.

For each $v_j = end(e), e \in K_i$ we define a horizontal vector of movement as follows. If e is a vertical edges in canonical drawing then this vector is $(\Gamma_{t+3}(st(e))_x - \Gamma_{t+3}(end(e))_x, \Gamma_{t+3}(st(e))_y - \Gamma_{t+3}(end(e))_y, 0)$, in this case sub-tree $T'(end(e))$ is moving towards vertical pole through $st(e)$ until the image of the edge e becomes vertical. If e is not a vertical edge in canonical drawing, then $\mathcal{C}(end(e))_x - \mathcal{C}(st(e))_x = 1$ and we move the whole subtree $T'(end(e))$ towards the pole through $\Gamma_{t+3}(st(e))$ until $end(e)$ reaches the last point with integer coordinates before $(\Gamma_{t+3}(st(e))_x, \Gamma_{t+3}(st(e))_y, \Gamma_{t+3}(end(e))_z)$.

Step 5: Collide planes. During the following steps $\langle \Gamma_{t+4}, \Gamma_{t+5} \rangle, \dots,$ $\langle \Gamma_{t+5+\log k}, \Gamma_{t+5+\log k+1} \rangle$ we iteratively divide half-planes that contain $T'(end(e)), e \in K_i$ around each vertex $st(e), e \in K_i$ in pairs which are formed of neighboring half-planes in clockwise order around the pole through $st(e)$. If in some iteration there are odd number of planes around some pole, the plane without pair does not move in this iteration. In every iteration we map the drawing of one plane in the pair to another simultaneously in all pairs. As around each vertex we can have at most $k = \Delta(T)$ number of half-planes, we need at most $\mathcal{O}(\log k)$ number of mapping steps to collide all planes in one and to rotate the resulting image to $XZ^+_{st(e)}$

We perform $\overline{Lift}()$ for each $K_i \in \mathcal{K}$ till we obtain the canonical drawing of T. The following theorem summarises the result of this section.

Theorem 2. *For every two planar straight-line grid drawings Γ, Γ' of an n-vertex tree T, there exists a crossing-free 3D-morph $\mathcal{M} = \langle \Gamma = \Gamma_0, \ldots, \Gamma_k = \Gamma' \rangle$ that takes $\mathcal{O}(dpt(T) \cdot \log \Delta(T))$ steps and $\mathcal{O}(d^3 \cdot \log n \times d^3 \cdot \log n \times n)$ space such that every intermediate drawing $\Gamma_i, 0 \leq i \leq k$ is a straight-line 3D grid drawing, where d is maximum of the diameters of the given drawings. In the worst case the algorithm can take $\mathcal{O}(dpt(T) \cdot \log n)$ steps since the maximum degree of T can be $\mathcal{O}(n)$.*

6   Trade-off

Recall that $\mathcal{L}(T)$ is the set of paths induced by the long-path decomposition, see Sect. 2. Let $Long(T)$ be a set of paths from $\mathcal{L}(T)$, consisting of the paths whose length is at least $\sqrt{n}$, i.e. $Long(T) = \{L_i \in \mathcal{L}(T) : |L_i| \geq \sqrt{n}\}$, let the order in $Long(T)$ be induced from the order in $\mathcal{L}(T)$. We denote by $Short(T)$ a set of trees that are left after deleting from T edges of $Long(T)$.

Lemma 3. $|Long(T)| \leq \sqrt{n}$ *and for every tree T_i in $Short(T)$ depth of T_i is at most $\lfloor \sqrt{n} \rfloor$.*

We divide edges in $Short(T)$ into disjoint sets $Sh_1, \ldots Sh_{\lfloor \sqrt{n} \rfloor}$. An edge (v_i, v_j) in tree T_k lies in the set Sh_l if and only if $max(dpt(v_i), dpt(v_j)) = \lfloor \sqrt{n} \rfloor - l + 1$, where $dpt(v)$ is the depth of vertex v in the corresponding tree T_k. Since the maximum depth of any tree T_k is at most $\sqrt{n}$, $Sh_1, \ldots Sh_{\lfloor \sqrt{n} \rfloor}$ contain all the edges of these subtrees.

***Trade-off Algorithm*:** In the beginning we perform a stretching step with $S_1 = 2 \cdot rpw \cdot d(\Gamma) \cdot (4 \cdot d(\Gamma) + 1)$ as mentioned in Sect. 5. S_1 is big enough to perform $Lift()$ procedure mentioned in Sect. 4. Then, we lift edges from sets Sh_1 to $Sh_{\lfloor \sqrt{n} \rfloor}$ by $\overline{Lift}(Sh_i)$ procedure. It takes $\mathcal{O}(\sqrt{n} \cdot \log \Delta(T))$ steps in total by Theorem 2. After that, we lift paths in $Long(T)$ in the order induced by the path decomposition. As $|Long(T)| \leq \sqrt{n}$ and each $Lift()$ procedure consists of a constant number of morphing steps, this step takes $\mathcal{O}(\sqrt{n})$ steps.

Theorem 3. *For every two planar straight-line grid drawings Γ, Γ' of tree T with n vertices there exists a crossing-free 3D-morph $\mathcal{M} = \langle \Gamma = \Gamma_0, \ldots, \Gamma_l = \Gamma' \rangle$ that takes $\mathcal{O}(\sqrt{n} \cdot \log \Delta(T))$ steps ($\mathcal{O}(\sqrt{n} \cdot \log n)$ in the worst case) and $\mathcal{O}(d^3 \cdot \log n \times d^3 \cdot \log n \times n)$ space to perform, where d is maximum of the diameters of the given drawings. In this morph every intermediate drawing $\Gamma_i, 1 \leq i \leq l$ is a straight-line 3D grid drawing. It is possible to morph between Γ, Γ' using $\mathcal{O}(\sqrt{n})$ steps if maximum degree of T is a constant.*

7 Conclusion

In this paper, we presented an algorithm that morphs between two planar grid drawings of an n-vertex tree T in $\mathcal{O}(\sqrt{n}\log n)$ steps such that all intermediate drawings are crossing-free $3D$ grid drawings and lie inside a polynomially bounded $3D$-grid. Arseneva et al. [2] proved that $\mathcal{O}(\log n)$ steps are enough to morph between two planar grid drawings of an n-vertex tree T where intermediate drawings are allowed to lie in $\mathbb{R}^3$ but they did not guarantee that intermediate drawings have polynomially bounded resolution. Several problems are left open in this area of research. We mention some of them here. It is interesting to prove a lower bound on the number of morphing steps if intermediate drawings are allowed to lie in $\mathbb{R}^3$ (with or without the additional constraint of polynomially bounded resolution). Another intriguing question is if it possible to morph between two planar grid drawings in $o(n)$ number of steps for a richer class of graphs (e.g. outer-planar graphs) than trees if we are allowed to use the third dimension.

Acknowledgements. Elena Arseneva was partially supported by the Foundation for the Advancement of Theoretical Physics and Mathematics "BASIS". Elena Arseneva and Aleksandra Istomina were partially supported by RFBR, project 20-01-00488. Rahul Gangopadhyay was supported by Ministry of Science and Higher Education of the Russian Federation, agreement no. 075-15-2019-1619.

References

1. Alamdari, S., et al.: How to morph planar graph drawings. SIAM J. Comput. **46**(2), 824–852 (2017). https://doi.org/10.113716M1069171
2. Arseneva, E., et al.: Pole dancing: 3D morphs for tree drawings. J. Graph Algorithms Appl. **23**(3), 579–602 (2019)
3. Barrera-Cruz, F., et al.: How to morph a tree on a small grid. In: Friggstad, Z., Sack, J.-R., Salavatipour, M.R. (eds.) WADS 2019. LNCS, vol. 11646, pp. 57–70. Springer, Cham (2019). https://doi.org/10.1007/978-3-030-24766-9_5
4. Bender, M.A., Farach-Colton, M.: The level ancestor problem simplified. In: Rajsbaum, S. (ed.) LATIN 2002. LNCS, vol. 2286, pp. 508–515. Springer, Heidelberg (2002). https://doi.org/10.1007/3-540-45995-2_44
5. Biedl, T.: Optimum-width upward drawings of trees. arXiv preprint arXiv:1506.02096 (2015)

StreamTable: An Area Proportional Visualization for Tables with Flowing Streams

Jared Espenant and Debajyoti Mondal$^{(\boxtimes)}$

University of Saskatchewan, Saskatoon, Canada
{jae608,d.mondal}@usask.ca

Abstract. Let M be a two-dimensional table with each cell weighted by a nonzero positive number. A StreamTable visualization of M represents the columns as non-overlapping vertical streams and the rows as horizontal stripes such that the intersection between a stream and a stripe is a rectangle with area equal to the weight of the corresponding cell. To avoid large wiggle of the streams, it is desirable to keep the consecutive cells in a stream to be adjacent. Let B be the smallest axis-aligned bounding box containing the StreamTable. Then the difference between the area of B and the sum of the weights is referred to as the excess area. We attempt to optimize various StreamTable aesthetics (e.g., minimizing excess area, or maximizing cell adjacencies in streams).

- If the row permutation is fixed and the row heights are given, then we give an $O(rc)$-time algorithm to optimizes these aesthetics, where r and c are the number of rows and columns, respectively.
- If the row permutation is fixed but the row heights can be chosen, then we discuss a technique to compute an aesthetic (but not necessarily optimal) StreamTable by solving a quadratically-constrained quadratic program, followed by iterative improvements. If the row heights are restricted to be integers, then we prove the problem to be NP-hard.
- If the row permutations can be chosen, then we show that it is NP-hard to find a row permutation that optimizes the area or adjacency aesthetics.

Keywords: Geometric Algorithms · Table Cartogram · Streamgraphs

1 Introduction

Proportional area charts and cartographic visualizations commonly represent data values as geometric objects. Table cartogram [8] is a brilliant way to visualize tables as cartograms, where each table cell is mapped to a convex quadrilateral with area equal to the cell's weight. Furthermore, the visualization preserves

This research was undertaken thanks in part to funding from the Canada First Research Excellence Fund and to the Natural Sciences and Engineering Research Council of Canada (NSERC).

P. Mutzel et al. (Eds.): WALCOM 2022, LNCS 13174, pp. 97–108, 2022.
https://doi.org/10.1007/978-3-030-96731-4_9

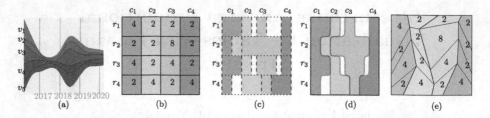

Fig. 1. (a) A streamgraph. (b) A table T. (c) A StreamTable for T. (d) A StreamTable visualization with smooth streams. (e) A table cartogram for T.

cell adjacencies and the quadrilaterals are packed together in a rectangle with no empty space in between (e.g., see Fig. 1(e)). However, since the cells in a table cartogram are represented with convex quadrilaterals, it may sometimes become difficult to follow the rows and columns [12]. This motivated us to look for a solution, where each row is represented with a horizontal *stripe* (i.e., a region bounded by two horizontal lines) and the cells in each row are represented with axis aligned rectangles inside the corresponding stripe.

Streamgraphs are examples where the columns can be thought of as vertical stripes. Given a set of variables, a *streamgraph* visualizes how their values change over time by representing each variable with a flowing river-like stream (e.g., an x-monotone polygon). The width of the stream at a timestamp is determined by the value of the variable at that time. Figure 1(a) illustrates a streamgraph with five variables. Streamgraphs are often used to create infographics of temporal data [4], e.g., box office revenues for movies [2], various statistics or demographics of a population over time [13], etc.

In this paper, we introduce StreamTable that extends this idea of a streamgraph to visualize tables or spreadsheets. We now formally define a StreamTable.

1.1   StreamTable

Let T be an $r \times c$ table with r rows and c columns, where each cell is weighted by a nonzero positive number. A *StreamTable* visualization of T is a partition of an axis-aligned rectangle R into r consecutive horizontal stripes that represent the rows of T, where each stripe is further divided into rectangles to represent the cells of its corresponding row. A column q of T is thus represented by a sequence of rectangles corresponding to the cells of q. By a *stream* we refer to such a sequence of rectangles that represents a column of T. Furthermore, a StreamTable must satisfy the following properties.

P_1. The left side of the leftmost stream (resp., the right side of the rightmost stream) must be aligned to the left side (resp., right side) of R.

P_2. For each cell of T, the area of its corresponding rectangle in the StreamTable must be equal to the cell's weight.

Property P_1 ensures an aesthetic alignment with the row labels and provides a sense of total visualization area. Property P_2 provides an area proportional

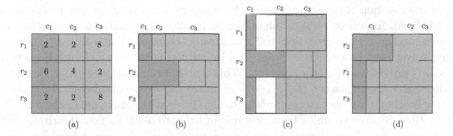

Fig. 2. (a) A table. (b) A StreamTable with no excess area and 2 splits. (c) A StreamTable with non-uniform row heights, non-zero excess area, but no split. (d) A StreamTable with no excess area and 1 split (obtained by reordering rows).

representation of the table cells. Figure 1(b) illustrates a table and Fig. 1(c) illustrates a corresponding StreamTable. The stripes (rows) are shown in dotted lines and the partition of the stripes are shown in dashed lines. Figure 1(d) illustrates an aesthetic visualization of the streams after smoothing the corners.

Note that a StreamTable may contain rectangular regions that do not correspond to any cell. We refer to such regions as *empty regions* and the sum of the area of all empty regions as the *excess area*. While computing a StreamTable, a natural optimization criterion is to minimize this excess area. However, minimizing excess area may sometimes result into disconnected streams. Figure 2(b) illustrates a StreamTable with zero excess area, where the consecutive rectangles for column c_2 are not adjacent (i.e., no two consecutive rectangles of c_2 share a common boundary point). If a pair of cells are consecutive in a column but the corresponding rectangles are nonadjacent in the stream, then they *split* the stream. To maintain the stream connectedness, it is desirable to minimize the number of such splits. As illustrated in Fig. 2(c)–(d), one may choose non-uniform row heights or reorder the rows to optimize the aesthetics. Such reordering operations also appear in matrix reordering problems [14] where the goal is to reveal clusters in matrix data. StreamTable computation also relates to floorplanning [5,16] and area-universal rectangular layout problems [3,6], where the horizontal adjacencies are not mandatory but vertical adjacencies must be preserved.

1.2 Our Contribution

We explore StreamTable from a theoretical perspective and consider the following two problems.

> **Problem 1 (StreamTable with no Split, Minimum Excess Area, and Fixed Row Ordering).** Given an $r \times c$ table T, can we compute a StreamTable for T in polynomial time with no split and minimum excess area? Note that in this problem, the StreamTable must respect the row ordering of T.

If the row heights are restricted to be integers, then we show the problem to be NP-hard. In general, the problem can be modeled leveraging a quadratically-constrained quadratic program, and a solution computed by non-linear programming solver may be iteratively improved by adjusting the row heights. However, this only provides a heuristic solution. While Problem 1 remains open, if the input additionally specifies a set $\{h_1, \ldots, h_r\}$ of nonzero positive numbers to be chosen as row heights, then we can compute a StreamTable with minimum excess area in $O(rc)$ time. Since choosing a fixed row height helps to obtain a fast algorithm and to compare the cell areas more accurately, we examined whether one can leverage the row ordering to further improve the StreamTable aesthetics.

Problem 2 (Row-Permutable StreamTable with Uniform Row Heights). Given a table T and a non-zero positive number $\delta > 0$, can we compute a StreamTable in polynomial time by setting δ as the row height, and minimizing the excess area (or, the number of splits)? Note that in this problem, the row ordering can be chosen.

We show that Problem 2 is NP-hard, i.e., we show that computing a StreamTable with no excess area and minimum number of split is NP-hard and similarly, computing a StreamTable with no split and minimum excess area is NP-hard.

2   No Split, Minimum Excess Area, Fixed Row Ordering

In this section we compute StreamTables by respecting the given row ordering of the input table. We first explore the case when the row heights are given, and then the case when the row heights can be chosen.

2.1   Fixed Row Heights

Let T be an $r \times c$ table and let $\{h_1, \ldots, h_r\}$ be a set of nonzero positive numbers to be chosen as row heights. We now introduce some notation for the rectangles and streams in the StreamTable. Let $w_{i,j}$ be the weight for the (i, j)th entry of T, where $1 \leq i \leq r$ and $1 \leq j \leq c$, and let $R_{i,j}$ be the rectangle with height h_i and width $(w_{i,j}/h_{i,j})$. Let $a_{i,j}$ and $b_{i,j}$ be the x-coordinates of the left and right side $R_{i,j}$. We now show that a StreamTable $\mathcal{R}$ for T with no split and minimum excess area can be constructed using a greedy algorithm $\mathcal{G}$, as follows:

Step 1. Draw the rectangles $R_{i,1}$ (first column) such that they are left aligned.
Step 2. For each $j < c$, draw the jth stream by minimizing the sum of x-coordinates $a_{i,j}$, but ensuring that the stream remains connected.
Step 3. Draw the rectangles $R_{i,c}$ of the last column by minimizing the maximum x-coordinate over $b_{i,c}$, but ensuring that the rectangles are right aligned.

For every column j, let $A(\mathcal{R}, j)$ be the orthogonal polygonal chain determined by the left side of $R_{i,j}$. Similarly, we define (resp., $B(\mathcal{R}, j)$) for the right side of $R_{i,j}$. We now have the following lemma.

Lemma 1. $\mathcal{G}$ *computes a no-split StreamTable* $\mathcal{R}$ *with minimum excess area.*

Proof. We employ an induction on the number of columns. For an $r \times c$ table T with $c = 2$, it is straightforward to verify the lemma. We now assume that the lemma holds for every table with j columns where $1 \leq j < c$. Consider now a table with c columns and let $\mathcal{R}^*$ be an optimal StreamTable with no split and minimum excess area.

We first show that the first two streams of $\mathcal{R}^*$ can be replaced with the corresponding streams of $\mathcal{R}$. To observe this first note that the stream for the first column must be drawn left-aligned, and since the rectangle heights are given, the right side of the streams $B(\mathcal{R}, 1)$ must coincide with $B(\mathcal{R}^*, 1)$. Consider now the left sides of the second streams. If $A(\mathcal{R}, 2)$ does not coincide with $A(\mathcal{R}^*, 2)$, then there must be non-zero area between them. Let A be an orthogonal polygonal chain constructed by taking the left envelope of these two chains. In other words, for each row, we choose the part of the chain that have the minimum x-coordinate. Since the streams for $\mathcal{R}$ and $\mathcal{R}^*$ are connected, the stream determined by A must be connected. Since the sum of x-coordinates is smaller for A, the polygonal chain $A(\mathcal{R}, 2)$ must coincide with A. Thus the right side of the stream, i.e., the polygonal chain $B(\mathcal{R}, 2)$, must remain to the left of $B(\mathcal{R}^*, 2)$.

We can now construct an $r \times (c-1)$ table T' by treating the polygonal chain $B(\mathcal{R}, 2)$ as $B(\mathcal{R}, 1)$. By induction, $\mathcal{G}$ provides a StreamTable $\mathcal{R}'$ with no split and minimum excess area. We now obtain the StreamTable $\mathcal{R}$ by replacing the first stream with the two streams that we constructed using the greedy approach. $\square$

We now have the following theorem whose proof is included in the full version [7].

Theorem 1. *Given an $r \times c$ table T and a height for each row, a StreamTable $\mathcal{R}$ for T with no split and minimum excess area can be computed in $O(rc)$ time such that $\mathcal{R}$ respects the row ordering of T.*

We now consider the case when a set $\{h_1, \ldots, h_r\}$ of row heights are given as an input. Here we show how to formulate a system of linear equations to compute a StreamTable for T with no split and minimum excess area such that the height of the ith row is set to h_i, where $1 \leq i \leq r$. This will be useful for the subsequent section. Let $d_{i,j}$ be a variable to model the adjacency between $R_{i,j}$ and $R_{i+1,j}$, where $1 \leq i \leq r-1$ and $1 \leq j \leq c$. We minimize the excess area:

$\sum\limits_{j=1}^{r} \sum\limits_{k=1}^{c-1} h_j(a_{j,k+1} - b_{j,k})$, subject to the following constraints.

1. $a_{j,1} = a_{j+1,1}$ and $b_{j,c} = b_{j+1,c}$, where $j = 1, \ldots, r-1$. This ensures StreamTable property P_1.
2. $b_{j,k} - a_{j,k} = (w_{j,k}/h_j)$, where $j = 1, \ldots, r$ and $k = 1, \ldots, c$. This ensures property P_2.
3. $a_{j,k} \leq d_{j,k} \leq b_{j,k}$ and $a_{j+1,k} \leq d_{j,k} \leq b_{j+1,k}$, where $1 \leq j \leq r-1$ and $1 \leq k \leq c$. This ensures that there is no split in the streams.

Since $h_1, \ldots, h_r$ are fixed, the above system with the constraint that the variables must be non-negative can be modeled as a linear program, e.g., see Fig. 3 (left), but that would take at least a quadratic time in the number of table cells.

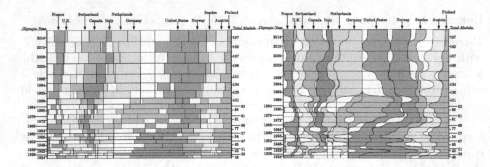

Fig. 3. StreamTables of a Winter Olympics dataset (left) using a linear program with row height proportional to the row sum, and (right) using Gurobi with a fixed total height and with corner smoothing.

2.2   Variable Row Heights

We model this case by treating $h_1, \ldots, h_j$ as variables. Hence the objective and constraint functions yield a quadratically-constrained quadratic program. Note that scaling down the height of a StreamTable by some $\delta \in (0, 1]$ and scaling up the width by $1/\delta$ do not change the excess area. Therefore, a non-linear program solver may end up generating a final StreamTable with bad aspect ratio. Hence we suggest to add another constraint: $h_1 + \ldots + h_k = H$, where H is the desired height of the visualization. Figure 3 (right) shows an example (not necessarily optimal) solution computed using a non-linear program solver Gurobi [11].

Local Improvement: We now show how a non-optimal StreamTable may be improved further by examining each empty cell individually, while deciding whether that cell can be removed by shrinking the height of the corresponding row. By $E_{i,j}$ we denote the empty rectangle between the rectangles $R_{i,j}$ and $R_{i,j+1}$. We first refer the reader to Fig. 4(a)–(b). Assume that we want to decide whether the empty cell $E_{i,j}(= E_{2,4})$ can be removed by scaling down the height of the second row. The idea is to grow the rectangles to the left (resp., right) of $E_{i,j}$ towards the right (resp., left) respecting the adjacencies and area.

Now consider a rectangle $R_{i,k}(= R_{2,2})$ before $E_{i,k}(= E_{2,4})$. Let $G_{i,k}$ be the rectangle determined by the ith row with left and right sides coinciding with the left and right sides of $R_{i,1}$ and $R_{i,k}$, respectively. Figure 4(a) shows $G_{2,2}$ in a falling pattern. Let $\ell_{i,k}$ be the length of $G_{i,k}$. Let $A_{i,k}$ is the initial area of $G_{i,k}$, and our goal is to keep this area fixed as we scale down the height of the ith row. The height of $G_{i,k}$ is defined by $f(\ell_{i,k}) = A_{i,k}/\ell_{i,k}$. Since the rectangles of the $(i-1)$th and $(i+1)$th rows do not move, $f(\ell_{i,k})$ does not split the $(k+1)$th stream as long as $\ell_{i,k}$ is upper bounded by the right sides of $R_{i-1,k+1}$ and $R_{i+1,k+1}$. Figure 4(c) plots these functions, where H_c is the current height of the second row. The height function for $G_{2,2}$ is drawn in thick purple in the interval $[\ell_{2,2}, \min\{q_{1,3}, q_{3,3}\}]$, where $q_{1,3}$ and $q_{3,3}$ are the right sides of $R_{1,3}$ and $R_{3,3}$, respectively.

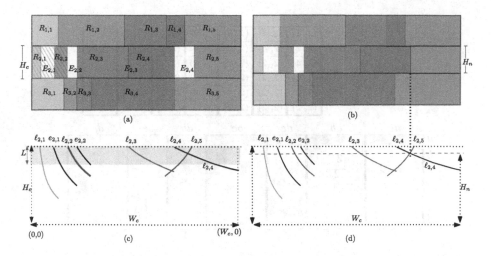

Fig. 4. (a) A StreamTable with width W_c and height H_c. (b) Removal of the empty rectangle $E_{2,4}$ (c)–(d) Illustration for computing the new height H_n of the second row.

We construct such functions also for all the empty rectangles $E_{i,k}$, where $1 \leq k < j$. These are labelled with $e_{i,k}$. Finally, we construct these functions symmetrically for the rectangles that appears after $E_{i,j}$. We then find a height H_n by determining the common interval L where all these functions are valid individually (Fig. 4(c)), and then determining the first intersection (if any) in this interval, as illustrated in Fig. 4(d). If no such intersection point exists, then we can shrink the row by an amount equal to the length of the interval L.

We iterate over the empty rectangles as long as we can find an empty rectangle to improve the solution, or to a maximum number of iterations. However, this only provides a heuristic algorithm, and thus Problem 1 remains open.

If the cells are allowed to have an area larger than their corresponding weights, then the problem can be modelled using a geometric programming (e.g., see the full version [7]). Furthermore, if the row heights are restricted to be positive integers, then we prove the problem to be NP-hard.

Theorem 2. *Given a table T and a positive integer H, it is NP-hard to compute a minimum-area no-split StreamTable of height H with row heights as integers respecting the row ordering of T.*

Proof. We reduce the NP-hard problem *clique* [10], where the input is a graph G and a positive integer k and the goal is to find a set of k vertices that are pairwise adjacent. The problem remains NP-hard even when $1 < k < n$. Given an instance G of the clique problem with n vertices and m edges, we construct a table T with n rows and m columns as follows.

1. For each edge $e \in E_G$, we create a column called an *edge column*, and label it by e (e.g., see Fig. 5(top)).

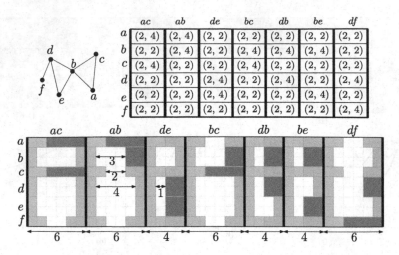

Fig. 5. Illustration for the proof of Theorem 2. Given a clique $\{b, e, d\}$, one can construct a StremTable with $\binom{3}{2}$ edge columns of width 4, where the remaining edge columns are of width 6. The line columns are shown in thick vertical lines.

2. We insert an additional column at the left and right sides of the table and also between every pair of adjacent columns. We refer to these columns as *line columns*. Each cell of a line column has a weight of $\epsilon = \frac{1}{n(m+1)}$.
3. For each vertex $v \in V_G$, we create a row and assign it the label v. For each edge column, we assign each cell a weight 6.
4. We now partition each cell $T_{v,e}$ into two cells (Fig. 5), as follows.
   (a) If vertex v is an end point of edge e, then the weight of the left and right cells are 2 and 4, respectively. We refer to these as a $(2, 4)$-*group*.
   (b) Otherwise, the weight of the left and right cells are 2 and 2, respectively. We refer to these as a $(2, 2)$-*group*.

It now suffices to show that G admits a clique of size k if and only if there exists a no-split StreamTable of height $H = (n + k)$ and width at most $(6m - 2\binom{k}{2}) + (m + 1)\epsilon$, where the row heights are integers (e.g., see Fig. 5(bottom)). We include the details in the full version [7].   □

3   Uniform Row Heights and Variable Row Ordering

In this section we consider the case when each row of the StreamTable must have the same height and the row permutations can be chosen. We prove that it is NP-hard to find a row permutation that minimizes the area (Sect. 3.1) or number of splits (Sect. 3.2).

3.1   No Split and Minimum Excess Area

We now show that computing StreamTables with no split while minimizing the excess area by reordering the rows is NP-hard. We reduce the NP-complete problem *betweenness* [15], where the input is a set of ordered triples of elements, and the problem is to decide whether there exists a total order σ of these elements, with the property that for each given triple, the middle element in the triple appears somewhere in σ between the other two elements.

Theorem 3. *Given a table T and a non-zero positive number $\delta > 0$, it is NP-hard to compute a StreamTable with no split and minimum excess area, where each row is of height δ and the ordering of the rows can be chosen.*

Proof. Let S be a set of c integer triples (an instance of betweenness) over r elements (integers), where $r, c \geq 5$. We now construct an $r \times (4c + 1)$ table T (Fig. 6(a)), as follows:

1. For every triple $t \in S$, we make a column (labelled with t). We refer to these columns as *triple columns*. Each of these columns will later be split into three more columns. For every element e, we create a row (labelled with e). Assume that each cell of a triple column has a weight of $w(= 15)$.
2. We insert an additional column at the left and right sides of the table and also between every pair of adjacent triple columns. We refer to these columns as *line columns*. Each cell of a line column has a weight of $\epsilon = \frac{1}{r(c+1)}$.
3. For every triple t and row i, we further partition the cell (t, i) into three cells and distribute the weight w among them, as follows:

   4a. If i is the left element of t, then the weight of the left, middle and right cells are $\frac{2w}{3}$, $\frac{w}{6}$ and $\frac{w}{6}$, respectively.

   4b. If i is the right element of t, then the weight of the left, middle and right cells are $\frac{w}{6}$, $\frac{w}{6}$ and $\frac{2w}{3}$, respectively.

   4c. If i is the centre element of t, then the weight of the left, middle and right cells are $\frac{w}{6}$, $\frac{2w}{3}$ and $\frac{w}{6}$, respectively.

   4d. Finally, if i does not belong to t, then the weight of the left, middle and right cells are $\frac{5w}{12}$, $\frac{w}{6}$ and $\frac{5w}{12}$, respectively.

We set δ to be 1. It now suffices to show that the betweenness instance S admits a total order σ, if and only if there exists a StreamTable with no split and at most $\frac{rcw}{12}$ excess area, where each row is of height δ.

First assume that S admits a total order σ. We draw the rectangles of each line column on top of each other (vertically aligned) and allocate a width of $(w + \frac{w}{12})$ for the triple columns. Since we order the rows by σ, a pair of rows that satisfy conditions (4a) and (4b) for a triple t must have a row k satisfying condition (4c) for t. Therefore, we can complete the drawing of the rectangles of the three streams within the allocated width without any split. Figure 6(b) illustrates a schematic representation of the rows for this scenario. Since $\delta = 1$, the excess area is at most $\frac{rcw}{12}$ in total. Figure 7 illustrates a StreamTable for the table from Fig. 6(a).

We now show that if there is a StreamTable for T with at most $\frac{rcw}{12}$ excess area, then the corresponding row ordering will yield the total order for the betweenness instance. Any three streams corresponding to a triple t must have exactly one (4a), one (4b) and one (4c) conditions. Suppose for a contradiction that for some triple t, the condition (4c) does not appear between conditions (4a) and (4b) (Fig. 6(c)). Then these streams would require a width

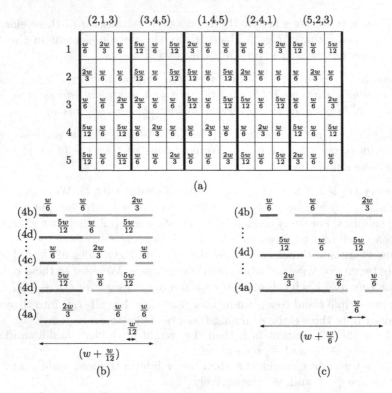

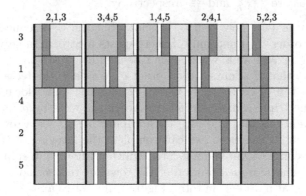

Fig. 6. (a) A table T obtained from a set of triples $\{(2,1,3), (3,4,5), (1,4,5), (2,4,1), (5,2,3)\}$. Here the thick black lines represent the line columns. (b)–(c) Illustration for the required width for different row orderings.

Fig. 7. A StreamTable for T, where $\sigma = \{3,1,4,2,5\}$.

of at least $(w + \frac{w}{6})$, i.e., the longest intervals in (4a) and (4b) cannot come any closer maintaining connectivity of the middle stream. Let t' be a triple that appears immediately after t. Since the stream (line column) between t

and t' is very narrow, they cannot share a width of more than $r\epsilon$. By construction, for each triple, we have three streams and they require a width of at least $(w + \frac{w}{12})$. Hence the total width of the visualization must be at least $\left(w + \frac{w}{6}\right) + (c-1)\left(w + \frac{w}{12}\right) - r(c+1)\epsilon = cw + \frac{cw}{12} + \frac{w}{12} - r(c+1)\epsilon = cw + \frac{cw}{12} + \frac{w}{12} - 1$.. Since $\epsilon = \frac{1}{r(c+1)}$, the sum of the weights of T is $(rcw + r(c+1)\epsilon) = (rcw+1)$. Since $r \geq 5$ and $w = 15$, we have an excess area of larger than $\frac{rcw}{12} + \frac{rw}{12} - r - 1 > \frac{rcw}{12}$.
□

3.2   No Excess Area and Minimum Number of Splits

We now show that computing StreamTables with no excess area while minimizing the number of splits by reordering the rows is NP-hard. We reduce the NP-complete problem *Hamiltonian path in a cubic graph* [9], where the input is a graph G with n vertices and m edges such that every vertex is of degree 3, and the problem is to decide whether there exists a total order of the vertices that determines a Hamiltonian path in G. A proof for the following theorem is included in the full version [7].

Theorem 4. *Given a table T and a non-zero positive number $\delta > 0$, it is NP-hard to compute a StreamTable with zero excess area and minimum number of splits, where each row is of height δ and the ordering of the rows can be chosen.*

4   Conclusion

In this paper we have introduced StreamTable, which is an area proportional visualization inspired by streamgraphs. We formulated algorithmic problems that need to be tackled to produce aesthetic StreamTables and examined two aesthetic criteria – excess area and number of splits.

We have showed that if row heights and row ordering are given, then a StreamTable with no split and minimum area can be computed via a linear program. However, the case when the row ordering is given but the row heights can be chosen needs further investigation. We only provided a quadratically-constrained quadratic program to model the problem and an NP-hardness proof when the row heights are constrained to be integers. However the original question remains open.

Open Problem 1: Given a table T and a positive integer H, does there exist a polynomial-time algorithm to compute a minimum-area no-split StreamTable of height H that respects the row ordering of T?

We also showed that if the row ordering can be chosen, then the problem of finding a minimum-area or a minimum-split StreamTable is NP-hard. In this setting, it would be interesting to find algorithms for computing zero-excess-area (resp., no split) StreamTables with good approximation on the number of splits (resp., excess area).

Open Problem 2: Design polynomial-time algorithms to find good approximation for StreamTable aesthetics (excess area or number of splits) in both the fixed and variable row ordering settings.

Recently a framework for ∃ℝ-completeness of packing problems has been proposed in [1]. It would be interesting to investigate ∃ℝ-completeness in this context, where the rows need to be packed inside a rectangle maintaining column adjacencies.

References

1. Abrahamsen, M., Miltzow, T., Seiferth, N.: Framework for ER-completeness of two-dimensional packing problems. In: Proceedings of the 61st IEEE Annual Symposium on Foundations of Computer Science (FOCS), pp. 1014–1021. IEEE (2020)
2. Bartolomeo, M.D., Hu, Y.: There is more to streamgraphs than movies: better aesthetics via ordering and lassoing. Comput. Graph. Forum **35**(3), 341–350 (2016)
3. Buchin, K., Eppstein, D., Löffler, M., Nöllenburg, M., Silveira, R.I.: Adjacency-preserving spatial treemaps. J. Comput. Geom. **7**(1), 100–122 (2016)
4. Byron, L., Wattenberg, M.: Stacked graphs - geometry & aesthetics. IEEE Trans. Vis. Comput. Graph. **14**(6), 1245–1252 (2008)
5. Chen, T., Fan, M.K.H.: On convex formulation of the floorplan area minimization problem. In: Sarrafzadeh, M. (ed.) Proceedings of the 1998 International Symposium on Physical Design (ISPD), pp. 124–128. ACM (1998)
6. Eppstein, D., Mumford, E., Speckmann, B., Verbeek, K.: Area-universal and constrained rectangular layouts. SIAM J. Comput. **41**(3), 537–564 (2012)
7. Espenant, J., Mondal, D.: StreamTable: an area proportional visualization for tables with flowing streams. arXiv:2103.15037 (2021)
8. Evans, W.S., et al.: Table cartogram. Comput. Geom. **68**, 174–185 (2018)
9. Garey, M.R., Johnson, D.S., Tarjan, R.E.: The planar Hamiltonian circuit problem is NP-complete. SIAM J. Comput. **5**(4), 704–714 (1976)
10. Garey, M.R., Johnson, D.S.: Computers and Intractability: A Guide to the Theory of NP-Completeness. W. H. Freeman, San Francisco (1979)
11. Gurobi Optimization, L.: Gurobi optimizer reference manual (2020). https://www.gurobi.com/wp-content/plugins/hd_documentations/documentation/9.1/refman.pdf
12. Hasan, M.R., Tasnim, D.M.J., Schneider, K.A.: Putting table cartograms into practice. In: Proceedings of the 16th International Symposium on Visual Computing (ISVC), vol. 13017, pp. 91–102. Springer, Cham (2021). https://doi.org/10.1007/978-3-030-90439-5_8
13. Havre, S., Hetzler, E.G., Whitney, P., Nowell, L.T.: ThemeRiver: visualizing thematic changes in large document collections. IEEE Trans. Vis. Comput. Graph. **8**(1), 9–20 (2002)
14. Mäkinen, E., Siirtola, H.: Reordering the reorderable matrix as an algorithmic problem. In: Anderson, M., Cheng, P., Haarslev, V. (eds.) Diagrams 2000. LNCS (LNAI), vol. 1889, pp. 453–468. Springer, Heidelberg (2000). https://doi.org/10.1007/3-540-44590-0_37
15. Opatrny, J.: Total ordering problem. SIAM J. Comput. **8**(1), 111–114 (1979)
16. Rosenberg, E.: Optimal module sizing in VLSI floorplanning by nonlinear programming. ZOR Meth. Model. Oper. Res. **33**(2), 131–143 (1989)

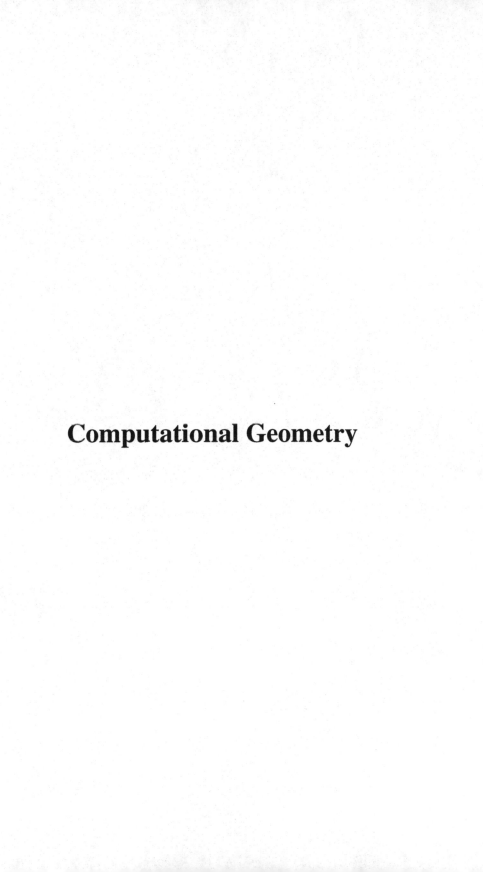

Computational Geometry

Vertex-to-Point Conflict-Free Chromatic Guarding is NP-Hard

Chuzo Iwamoto[(✉)][iD] and Tatsuaki Ibusuki

Hiroshima University, Higashi-Hiroshima 739-8521, Japan
chuzo@hiroshima-u.ac.jp
https://home.hiroshima-u.ac.jp/chuzo/

Abstract. The art gallery problem is to find a set of guards who together can observe every point of the interior of a polygon P. We study a *chromatic* variant of the problem, where each guard is assigned one of k distinct colors. A chromatic guarding is said to be *conflict-free* if at least one of the colors seen by every point in P is unique (i.e., each point in P is seen by some guard whose color appears exactly once among the guards visible to that point). In this paper, we consider *vertex-to-point* guarding, where the guards are placed on vertices of P, and they observe every point of the interior of P. The *vertex-to-point conflict-free chromatic art gallery problem* is to find a colored-guard set such that (i) guards are placed on P's vertices, and (ii) any point in P can see a guard of a unique color among all the visible guards. In this paper, it is shown that determining whether there exists a conflict-free chromatic vertex-guard set for a polygon with holes is NP-hard when the number of colors is $k = 2$.

1 Introduction

The art gallery problem is to determine the minimum number of guards who can observe the interior of a gallery. Chvátal [3] proved that $\lfloor n/3 \rfloor$ guards are always sufficient and sometimes necessary for observing the interior of an n-vertex simple polygon. This $\lfloor n/3 \rfloor$-bound is replaced by $\lfloor n/4 \rfloor$ if the instance is restricted to a simple orthogonal polygon [8].

Another perspective to the art gallery problem is to study the complexity of locating the minimum number of guards in a polygon. The NP-hardness and APX-hardness of this problem were shown by Lee and Lin [12] and by Eidenbenz et al. [4], respectively. Furthermore, Schuchardt and Hecker [16] proved that this problem remains NP-hard even if we restrict our attention to simple orthogonal polygons. Even guarding the vertices of a simple orthogonal polygon was shown to be NP-hard [11].

In this paper, we consider *vertex-to-point* guarding, where the guards are placed on vertices of a polygon P, and they observe every point inside P. We study a *chromatic* version of the art gallery problem, where each guard

---

This work was supported by JSPS KAKENHI Grant Number 16K00020.

P. Mutzel et al. (Eds.): WALCOM 2022, LNCS 13174, pp. 111–122, 2022.
https://doi.org/10.1007/978-3-030-96731-4_10

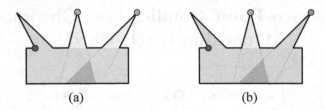

(a)   (b)

Fig. 1. (a) *Conflict-free* chromatic guarding. The dark gray area is seen by one red guard and two blue guards, where red is the unique color among the three guards. Each of the two light gray areas is seen by one red guard and one blue guard. (b) *Strong* chromatic guarding. The dark gray area is seen by three guards having three different colors. (Color figure online)

is assigned one of k distinct colors. There are two chromatic variants, which are called *conflict-free* chromatic guarding and *strong* chromatic guarding [9] (see Fig. 1). A chromatic guarding is said to be *conflict-free* if at least one of the guards seen by every point in P has a unique color [1]. It is *strong* if no two guards with the same color have overlapping visibility regions [5].

The *vertex-to-point conflict-free chromatic art gallery problem* is to find a colored-guard set such that (i) guards are placed on P's vertices, and (ii) any point inside P can see a guard of a unique color among all the visible guards. In this paper, it is shown that determining whether there exists a conflict-free chromatic vertex-guard set which together observe every point in a given polygon with holes is NP-hard when the number of colors is $k = 2$.

The chromatic art gallery problem was motivated by the following application [2,6]. Consider the problem of navigating a robot inside a polygon, where the robot communicates with radio beacons. The robot must be able to communicate with a radio beacon of a unique frequency in order to prevent interference. This motivates a *chromatic* version of the art gallery problem, where a guard corresponds to a radio beacon, and colors correspond to different frequencies.

The computational complexity of the chromatic art gallery problem was firstly investigated in [6]; the *point-to-point strong* chromatic art gallery problem was shown to be NP-hard for general polygons with holes. Recently, the current authors proved that the *point-to-point strong* chromatic art gallery problem with *r-visibility* is NP-hard for *orthogonal* polygons with holes [10]. Here, two points are said to be *r-visible* if the smallest axis-aligned rectangle containing them lies entirely within the polygon.

Çağırıcı et al. studied the *vertex-to-vertex conflict-free* chromatic guarding problem [2]; they proved the NP-hardness of the problem when the number of colors is $k = 2$. However, they mentioned that their proof does not imply the NP-hardness for the *vertex-to-point* case. Hence, the computational complexity of the vertex-to-point conflict-free chromatic guarding problem remained open. In the current paper, we solve this open problem.

Several results on the lower and upper bounds of the minimum number of colors can be found in [1,5,9] for general and orthogonal polygons under standard and orthogonal visibility conditions.

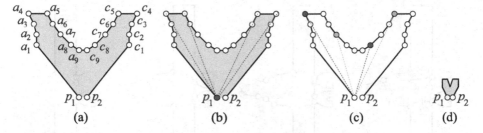

Fig. 2. (a) A *bowl*-shaped gadget [2]. (b) A guard on p_1 or p_2 can see all the vertices of $\{a_1, a_2, \ldots, a_9\} \cup \{c_1, c_2, \ldots, c_9\}$. (c) If no guards are placed on vertices of $\{p_1, p_2\}$, then there exists no conflict-free 2-color guard-set. Thus, a guard must be placed on at least one of p_1 and p_2 (see (b)). (d) is a simplified illustration of (a), where two vertices p_1 and p_2 are called *door vertices* of the bowl.

2   Definitions and Results

The definitions of a polygon and a polygon with holes are mostly from [13, 15]. A *polygon* is defined by a finite set of segments such that every segment endpoint is shared by exactly two segments and no subset of segments has the same property. The segments and their endpoints are called the *edges* and *vertices* of the polygon, respectively.

A *polygon with holes* is a polygonal domain defined by a polygon P enclosing several other polygons $H_1, H_2, \ldots, H_h$, the holes. None of the boundaries of P, $H_1, H_2, \ldots, H_h$ may intersect, and each of the holes is empty. P is said to bound a *multiply-connected* region with h holes: the region of the plane interior to or on the boundary of P, but exterior to or on the boundary of $H_1, H_2, \ldots, H_h$.

Two points v and u in a polygon P are said to be *visible* (or v *sees* u) if the line segment connecting them lies entirely within P. Here, the line segment may contain points on the boundary of P, but it must not across any hole of the polygon. An area is said to be *observed* by a point v if every point in the area is visible from v.

An instance of the *vertex-to-point conflict-free chromatic art gallery problem for polygons with holes* is $(P, H_1, H_2, \ldots, H_h; k)$, where P is a polygon with holes $H_1, H_2, \ldots, H_h$, and k is the number of colors. The problem asks whether there exists a conflict-free k-chromatic vertex-guard set which together observe every point in the polygonal domain defined by $(P, H_1, H_2, \ldots, H_h)$. (Color figure online)

Theorem 1. *The vertex-to-point conflict-free chromatic art gallery problem for polygons with holes is NP-hard when the number of colors is two.*

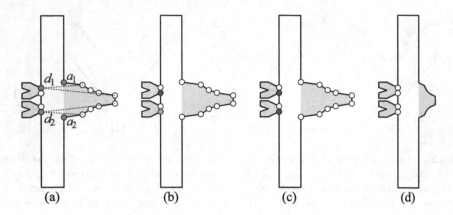

Fig. 3. (a) The green area surrounded by a 10-vertex chain from a_1 to a_2 is called a *pocket*-gadget [2]. (b) A pair of red and blue guards on door vertices of bowls can see the inside of the pocket in the conflict-free condition. (c) is an illegal 2-coloring. (d) is a simplified illustration of (a). (Color figure online)

3   NP-Completeness

3.1   3SAT Problem

The definition of 3SAT is mostly from [7,14]. Let $U = \{x_1, x_2, \ldots, x_n\}$ be a set of Boolean *variables*. Boolean variables take on values 0 (false) and 1 (true). If x is a variable in U, then x and $\overline{x}$ are *literals* over U. The value of $\overline{x}$ is 1 (true) if and only if x is 0 (false). A *clause* over U is a set of literals over U, such as $\{\overline{x_1}, x_3, x_4\}$. A clause is *satisfied* by a truth assignment if and only if at least one of its members is true under that assignment.

An instance of PLANAR 3SAT is a collection $C = \{c_1, c_2, \ldots, c_m\}$ of clauses over U such that (i) $|c_j| = 3$ for each $c_j \in C$ and (ii) the graph $G = (V, E)$, defined by $V = U \cup C$ and $E = \{(x_i, c_j) \mid x_i \in c_j \in C \text{ or } \overline{x_i} \in c_j \in C\}$, is planar. PLANAR 3SAT asks whether there exists some truth assignment for U that simultaneously satisfies all the clauses in C.

If E is replaced with

$$E_1 = E \cup \{(c_j, c_{j+1}) \mid 1 \le j \le m - 1\},$$

then the problem is called CLAUSE-LINKED PLANAR 3SAT. This problem is NP-complete, since VARIABLE-CLAUSE-LINKED PLANAR 3SAT was shown to be NP-complete in [14], where the edge set E_2 is defined as

$$E_2 = E \cup \{(x_i, x_{i+1}) \mid 1 \le i \le n - 1\} \cup \{(x_n, c_1)\}$$
$$\cup \{(c_j, c_{j+1}) \mid 1 \le j \le m - 1\} \cup \{(c_m, x_1)\}.$$

Note that CLAUSE-LINKED PLANAR 3SAT in this paper is defined by a *chain* connecting $c_1, c_2, \ldots, c_m$ of length $m-1$, while VARIABLE-CLAUSE-LINKED PLANAR 3SAT in [14] is defined by a *cycle* of length $m + n$.

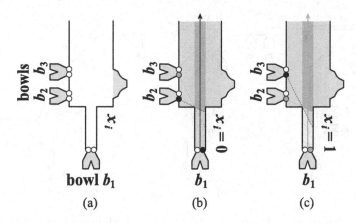

Fig. 4. (a) Variable gadget. (b) Conflict-free guarding when $x_i = 0$. (c) Conflict-free guarding when $x_i = 1$. (c) is obtained from (b) by switching red and blue guards. (Color figure online)

For example, $U = \{x_1, x_2, x_3, x_4\}$, $C = \{c_1, c_2, c_3\}$, and $c_1 = \{x_1, \overline{x_2}, x_3\}$, $c_2 = \{x_1, x_2, \overline{x_4}\}$, $c_3 = \{\overline{x_2}, \overline{x_3}, x_4\}$ provide an instance of CLAUSE-LINKED PLANAR 3SAT. For this instance, the answer is "yes," since there is a truth assignment $(x_1, x_2, x_3, x_4) = (1, 1, 0, 1)$ satisfying all clauses.

3.2 Guard-Fix Gadgets

In this section, we explain a *bowl*-shaped gadget (see Fig. 2(a)) and a *pocket* gadget (see Fig. 3(a)), introduced in [2].

A *bowl*-shaped gadget is a 20-vertex chain, which has the following two properties: (i) If a guard is placed on vertex p_1 or p_2 (see Fig. 2(b)), then it can see all the vertices of $\{a_1, a_2, \ldots, a_9\} \cup \{c_1, c_2, \ldots, c_9\}$. (ii) Suppose that no guard is placed on p_1 or p_2 (see Fig. 2(c)). Then, in order to observe vertices $a_1, a_2, \ldots, a_9$, both a red guard and a blue guard must be placed on two of the vertices in $\{a_1, a_2, \ldots, a_9\}$. Similarly, both a red guard and a blue guard must be placed on $\{c_1, c_2, \ldots, c_9\}$. Those four guards see vertices p_1 and p_2 simultaneously (see Fig. 2(c)). From the properties (i) and (ii), one can see that, *in any conflict-free 2-coloring of a bowl-shaped gadget, there is a guard placed on p_1 or p_2 (or both)* (see Fig. 2(b)). In the following, we use a simplified illustration shown in Fig. 2(d) as a bowl-shaped gadget. Vertices p_1 and p_2 are called *door vertices* of the bowl. The distance between p_1 and p_2 is assumed to be so tiny that there is no accidental visibility between a vertex inside the bowl and a vertex outside of the bowl.

In Fig. 3(a), the green area surrounded by a 10-vertex chain from a_1 to a_2 is called a *pocket* gadget. Vertices d_1 and d_2 can see all the 10 vertices of the pocket, but neither a_1 nor a_2 does so. If a pair of red and blue guards are placed on door vertices of bowls (see Fig. 3(b)), then they can see the inside of the pocket in the conflict-free condition. On the other hand, Fig. 3(c) is an illegal 2-coloring

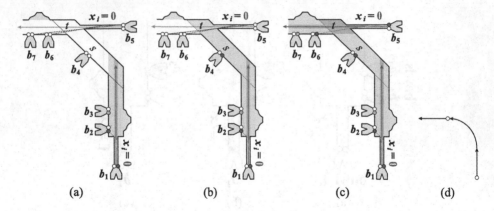

Fig. 5. (a) Left-turn gadget when $x_i = 0$. (b) Area s is observed by a red guard. (c) A red guard and a pair of red and blue guards are placed on b_5 and b_6, b_7, respectively. (d) is a simplified illustration of a left-turn gadget. A left-turn gadget when $x_i = 1$ is obtained from (c) by switching red and blue guards. (Color figure online)

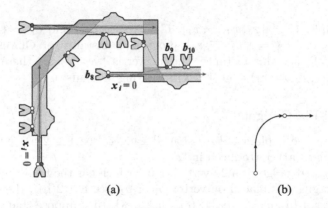

Fig. 6. (a) Right-turn gadget when $x_i = 0$. (b) Simplified illustration.

because of the following reason. In order to guard the inside of the pocket in the conflict-free condition, we need a *single* blue guard on the 10-vertex chain. However, the 10-vertex chain contains no single vertex which can see every point of the inside of the pocket. Figure 3(d) is a simplified illustration of Fig. 3(a).

3.3 Transformation from an Instance of CLAUSE-LINKED PLANAR 3SAT to a Polygon with Holes

We present a polynomial-time transformation from an arbitrary instance of clause-linked planar 3SAT C to a polygon with holes such that C is satisfiable if and only if there is a conflict-free 2-chromatic vertex-guard set which together observe every point in the polygon.

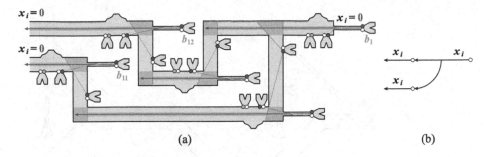

Fig. 7. (a) Branching gadget when $x_i = 0$. (b) is a simplified illustration.

Each variable $x_i \in \{x_1, x_2, \ldots, x_n\}$ is transformed into the variable gadget as illustrated in Fig. 4. In the gadget, there are three bowl-shaped gadgets b_1 and b_2, b_3 and one pocket gadget. From the reasons given in Sect. 3.2, a pair of red and blue guards must be placed on two of the door vertices of bowls b_2, b_3 (see Figs. 4(b) and 4(c)). A door vertex of bowl b_1 emits a beam of red or blue light upward. A red and blue beams in Fig. 4 correspond to the assignment $x_i = 0$ and $x_i = 1$, respectively.

In Fig. 4(b), a dark gray area is observed by two red guards and one blue guard, and a light gray area is observed by one red guard and one blue guard. Note that Fig. 4(c) is obtained from Fig. 4(b) by switching red and blue guards. In Figs. 5, 6, 7 and 8(a), we present figures only for $x_i = 0$.

Figure 5 is a left-turn gadget. (a) Suppose that bowl b_1 emits a beam of red light (see Fig. 5(a)). Since the dark gray area is observed by two red guards and one blue guard, area s must be observed by a red guard (see Fig. 5(b)). (b) Now, area t is observed by a red guard of b_4, and t is also seen by door vertices of b_5 and b_6, b_7. (c) In order to satisfy the conflict-free condition, we must place a red guard on b_5 and a pair of red and blue guards on b_6 and b_7, respectively. (d) is a simplified illustration of a left-turn gadget.

Figure 6 is a right-turn gadget. Bowl b_8 emitting a red beam is used so that a pair of bowls (see b_9, b_{10}) and a pocket are located on the left and right sides of the beam, respectively. (Bowls b_{11} and b_{12} in Fig. 7(a) are used for the same purpose.) Fig. 7 is a branching gadget. If bowl b_1 emits a red beam, then b_{11} and b_{12} also emit red beams.

Figure 8 is a NOR gadget. If $x_{i_1} = x_{i_2} = 0$ (see Fig. 8(a)), then the NOR gadget will emit a blue beam (= value 1) upward. By switching red and blue guards in Fig. 8(a), one can see that the NOR gadget outputs 0 if $x_{i_1} = x_{i_2} = 1$. (Fig. 8(b) is explained later; the case $x_{i_1} \neq x_{i_2}$ will have to be treated carefully.)

A NOT gadget (see Fig. 9(a)) is obtained by connecting a branching gadget and a NOR gadget. In the NOT gadget, the input is $x_i = 0$ if and only if the output is $\overline{x_i} = 1$. An OR gadget (see Fig. 9(b)) is obtained by connecting a NOR gadget and a NOT gadget. Note that, if $x_{i_1} = x_{i_2} = 0$, the OR gadget outputs 0. By switching red and blue guards, one can see that the OR gadget outputs 1 if $x_{i_1} = x_{i_2} = 1$. (The case $x_{i_1} \neq x_{i_2}$ is explained later.)

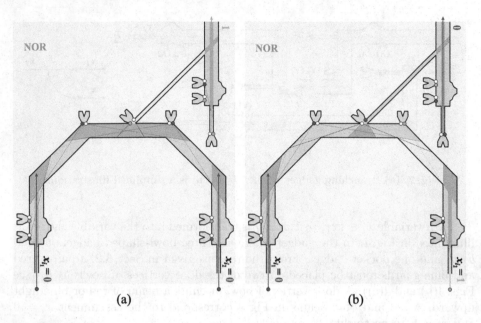

Fig. 8. NOR gadget. (a) If $x_{i_1} = x_{i_2} = 0$, this gadget outputs value 1 upward. By switching red and blue guards, one can see that the gadget outputs 0 if $x_{i_1} = x_{i_2} = 1$. (b) When $x_{i_1} \neq x_{i_2}$, the gadget *can* output 0 (see the body text for details). (Color figure online)

A clause gadget $c_j = \{x_{i_1}, x_{i_2}, x_{i_3}\}$ (see Fig. 10) contains three OR gadgets. If $x_{i_1} = x_{i_2} = x_{i_3} = 0$, the clause gadget outputs value $c_j = 0$. By switching red and blue guards, one can see that the clause gadget outputs $c_j = 1$ if $x_{i_1} = x_{i_2} = x_{i_3} = 1$. (The remaining cases are explained in the next paragraph.)

Consider a NOR gadget when $x_{i_1} \neq x_{i_2}$ (see Fig. 8(b)). In this case, there exists a conflict-free 2-chromatic guard set (see red and blue guards in Fig. 8(b)) so that the clause gadget emits a red beam (=value 0) upward. Namely, the NOR gadget *can* output 0 when $x_{i_1} \neq x_{i_2}$. Thus, in Fig. 9(b), the OR gadget *can* output 1 when $x_{i_1} \neq x_{i_2}$. Hence, in Fig. 10, a clause gadget *can* output $c_j = 1$ if at least one of x_{i_1}, x_{i_2}, and x_{i_3} is 1. In summary, if $(x_{i_1}, x_{i_2}, x_{i_3}) = (0, 0, 0)$ (resp. $(1, 1, 1)$) then the clause gadget *must* output $c_j = 0$ (resp. $c_j = 1$) (see the previous paragraph), and if $(x_{i_1}, x_{i_2}, x_{i_3}) \notin \{(0, 0, 0), (1, 1, 1)\}$ then the clause gadget *can* output $c_j = 1$. (In Fig. 12, if $(x_1, x_2, x_3, x_4) = (1, 1, 0, 1)$, there exists a conflict-free 2-chromatic guard set so that all of c_1, c_2, and c_3 output 1. Here, $(1, 1, 0, 1)$ satisfies the 3SAT instance given in the caption.)

Figure 11 is an XNOR gadget, which connects clause gadgets c_j and c_{j+1} for every $j \in \{0, 1, ..., m - 1\}$ (see also Fig. 12). Figure 11(a) is an invalid placement of red and blue guards, since we cannot place neither a red guard nor a blue guard on a door vertex of bowl b_{13} in order to observe area u_j. On the other hand, if both clauses c_j and c_{j+1} have value 1 (resp. value 0), then area u_j can be observed by a blue guard (resp. red guard) (see Figs. 11(b,c)).

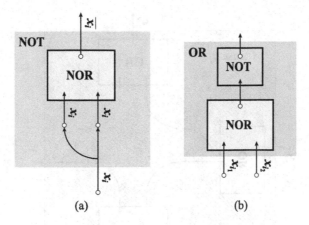

(a)   (b)

Fig. 9. (a) NOT gadget. The input is $x_i = 0$ if and only if the output is $\overline{x_i} = 1$. (b) OR gadget. If $x_{i_1} = x_{i_2} = 0$, the OR gadget outputs 0. By switching red and blue guards, one can see that the OR gadget outputs 1 if $x_{i_1} = x_{i_2} = 1$. When $x_{i_1} \neq x_{i_2}$, the OR gadget *can* output 1 (see the body text for details). (Color figure online)

Figure 12 is a sketch of a polygon P with holes transformed from $U = \{x_1, x_2, x_3, x_4\}$ and $C = \{c_1, c_2, c_3\}$, where $c_1 = \{x_1, \overline{x_2}, x_3\}$, $c_2 = \{x_1, x_2, \overline{x_4}\}$, and $c_3 = \{\overline{x_2}, \overline{x_3}, x_4\}$. In Fig. 12, $c_0 = \{x_0, x_0, x_0\}$ is a dummy clause, where x_0 is a dummy variable.

Lemma 1. *The instance C of 3SAT is satisfiable if and only if there exists a conflict-free 2-chromatic vertex-guard set for the polygon P with holes.*

Proof. ($\Rightarrow$) Suppose that the instance C of 3SAT is satisfiable. In Fig. 12, clause gadget c_0 can emit a blue beam upward, since the dummy clause $c_0 = \{x_0, x_0, x_0\}$ is satisfied if the dummy variable $x_0 = 1$. Then, area u_0 can be observed by a blue guard if c_1 is satisfied. Suppose that c_0 and c_1 are satisfied. Then, area u_1 can be observed by a blue guard if c_2 is satisfied. By continuing this observation, one can see that all areas $u_0, u_1, \ldots, u_{m-1}$ can be observed by blue guards if all of $c_1, c_2, \ldots, c_m$ are satisfied.

($\Leftarrow$) Suppose that the instance C of 3SAT is not satisfiable. Consider an arbitrary assignment $(b_1, b_2, \ldots, b_n) \in \{0,1\}^n$ for $(x_1, x_2, \ldots, x_n)$. Since C is not satisfiable, there exists at least one clause $c_j = \{x_{h_1}, x_{h_2}, x_{h_3}\}$ such that $x_{h_1} = x_{h_2} = x_{h_3} = 0$ when the assignment is $(b_1, b_2, \ldots, b_n)$. Here, each of x_{h_1}, x_{h_2}, and x_{h_3} is a positive or negative literal. Furthermore, for the same assignment $(b_1, b_2, \ldots, b_n)$, there exists at least one clause $c_k = \{x_{l_1}, x_{l_2}, x_{l_3}\}$ such that $x_{l_1} = x_{l_2} = x_{l_3} = 1$ because of the following reason: Assume for contradiction that there is no $c_k = \{x_{l_1}, x_{l_2}, x_{l_3}\}$ such that $x_{l_1} = x_{l_2} = x_{l_3} = 1$. Then, every clause contains at least one literal x whose value is 0. Now, consider the "inverted" assignment $(\overline{b_1}, \overline{b_2}, \ldots, \overline{b_n})$. For the inverted assignment, every clause contains at least one literal of value $\overline{x} = 1$. This implies that C is satisfiable, a contradiction.

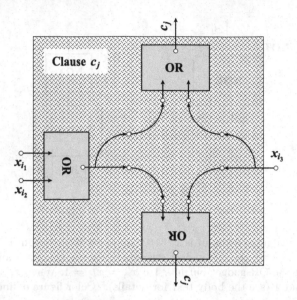

Fig. 10. Clause gadget $c_j = \{x_{i_1}, x_{i_2}, x_{i_3}\}$. If $x_{i_1} = x_{i_2} = x_{i_3} = 0$, then the clause gadget outputs $c_j = 0$. By switching red and blue guards, one can see that the clause gadget outputs $c_j = 1$ when $x_{i_1} = x_{i_2} = x_{i_3} = 1$. On the other hand, if at least one of x_{i_1}, x_{i_2}, and x_{i_3} is 1, then the clause gadget *can* output $c_j = 1$. (Color figure online)

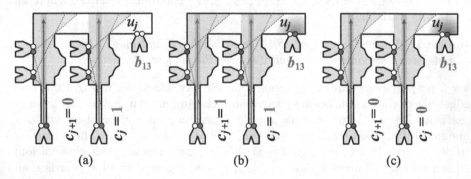

Fig. 11. XNOR gadget. This gadget connects clause gadgets c_j and c_{j+1}. (a) is an invalid placement of red and blue guards. (b,c) If clauses c_j and c_{j+1} have value 1 (resp. value 0), then area u_j can be observed by a blue guard (resp. red guard). (Color figure online)

Therefore, in any unsatisfiable instance C of 3SAT, there are two clauses $c_j = \{x_{h_1}, x_{h_2}, x_{h_3}\}$ and $c_k = \{x_{l_1}, x_{l_2}, x_{l_3}\}$ such that $x_{h_1} = x_{h_2} = x_{h_3} = 0$ and $x_{l_1} = x_{l_2} = x_{l_3} = 1$ for every assignment $(b_1, b_2, \ldots, b_n) \in \{0, 1\}^n$. If $j < k$, there exists an integer $j' \in \{j, j+1, \ldots, k-1\}$ such that the area $u_{j'}$ is observed by neither a blue guard nor a red guard (see Fig. 11(a)). The case $k < j$ is similar. This completes the proof of Lemma 1.

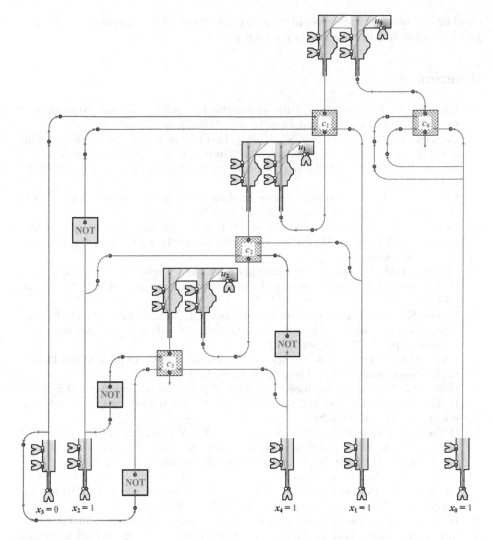

Fig. 12. Sketch of a polygon with holes transformed from $U = \{x_1, x_2, x_3, x_4\}$ and $C = \{c_1, c_2, c_3\}$, where $c_1 = \{x_1, \overline{x_2}, x_3\}$, $c_2 = \{x_1, x_2, \overline{x_4}\}$, and $c_3 = \{\overline{x_2}, \overline{x_3}, x_4\}$. In this figure, $c_0 = \{x_0, x_0, x_0\}$ is a dummy clause, where x_0 is a dummy variable. The polygon with holes constructed according to this figure can be guarded by red and blue guards in the conflict-free condition. From the positions of red and blue guards, one can see that $(x_1, x_2, x_3, x_4) = (1, 1, 0, 1)$ satisfies all the clauses. (Color figure online)

4   Conclusion

In this paper, we studied a chromatic variant of the art gallery problem under the conflict-free conditions. It was shown that the *vertex-to-point* conflict-free chromatic art gallery problem for polygons with holes is NP-hard when the number of colors is two. Proving the NP-hardness of the problem under the *point-to-point* condition remains an open problem.

References

1. Bärtschi, A., Suri, S.: Conflict-free chromatic art gallery coverage. Algorithmica **68**, 265–283 (2014). https://doi.org/10.1007/s00453-012-9732-5
2. Çağırıcı, O., Ghosh, S.K., Hliněný, P., Roy, B.: On conflict-free chromatic guarding of simple polygons. In: Li, Y., Cardei, M., Huang, Y. (eds.) COCOA 2019. LNCS, vol. 11949, pp. 601–612. Springer, Cham (2019). https://doi.org/10.1007/978-3-030-36412-0_49
3. Chvátal, V.: A combinatorial theorem in plane geometry. J. Comb. Theory B. **18**, 39–41 (1975). https://doi.org/10.1016/0095-8956(75)90061-1
4. Eidenbenz, S.J., Stamm, C., Widmayer, P.: Inapproximability results for guarding polygons and terrains. Algorithmica **31**, 79–113 (2001). https://doi.org/10.1007/s00453-001-0040-8
5. Erickson, L.H., LaValle, S.M.: A chromatic art gallery problem. Technical report, Department of Computer Science, University of Illinois at Urbana-Champaign (2011)
6. Fekete, S.P., Friedrichs, S., Hemmer, M., Mitchell, J.B.M., Schmidt, C.: On the chromatic art gallery problem. In. 30th Canadian Conference on Computational Geometry, pp. 73–79. Halifax (2014)
7. Garey, M.R., Johnson, D.S.: Computers and Intractability: A Guide to the Theory of NP-Completeness. W.H. Freeman, New York (1979)
8. Hoffmann, F.: On the rectilinear art gallery problem. In: Paterson, M.S. (ed.) ICALP 1990. LNCS, vol. 443, pp. 717–728. Springer, Heidelberg (1990). https://doi.org/10.1007/BFb0032069
9. Hoffman, F., Kriegel, K., Suri, S., Verbeek, K., Willert, M.: Tight bounds for conflict-free chromatic guarding of orthogonal art galleries. Comput. Geom. Theor. Appl. **73**, 24–34 (2018). https://doi.org/10.1016/j.comgeo.2018.01.003
10. Iwamoto, C., Ibusuki, T.: Computational complexity of the chromatic art gallery problem for orthogonal polygons. IEICE Trans. Fundam. Electron. **E104-A**(9), 1108–1115 (2021). https://doi.org/10.1007/978-3-030-39881-1_13
11. Katz, M.J., Roisman, G.S.: On guarding the vertices of rectilinear domains. Comput. Geom. Theor. Appl. **39**(3), 219–228 (2008). https://doi.org/10.1016/j.comgeo.2007.02.002
12. Lee, D.T., Lin, A.K.: Computational complexity of art gallery problems. IEEE Trans. Inf. Theory **32**(2), 276–282 (1986). https://doi.org/10.1109/TIT.1986.1057165
13. O'Rourke, J.: Art Gallery Theorems and Algorithms. Oxford University Press, New York (1987)
14. Pilz, A.: Planar 3-SAT with a clause/variable cycle. In: Eppstein, D. (eds.) SWAT 2018, no. 31, pp. 31:1–31:13 (2018). https://doi.org/10.4230/LIPIcs.SWAT.2018.31
15. Preparata, F.P., Shamos, M.I.: Computational Geometry: An Introduction. Springer, New York (1985). https://doi.org/10.1007/978-1-4612-1098-6
16. Schuchardt, D., Hecker, H.-D.: Two NP-hard art-gallery problems for ortho-polygons. Math. Log. Q. **41**(2), 261–267 (1995)

The Polygon Burning Problem

William Evans and Rebecca Lin[(✉)]

University of British Columbia, Vancouver, Canada
will@cs.ubc.ca, ryelin@student.ubc.ca

Abstract. Motivated by the k-center problem in location analysis, we consider the *polygon burning* (PB) problem: Given a polygonal domain P with h holes and n vertices, find a set S of k vertices of P that minimizes the maximum geodesic distance from any point in P to its nearest vertex in S. Alternatively, viewing each vertex in S as a site to start a fire, the goal is to select S such that fires burning simultaneously and uniformly from S, restricted to P, consume P entirely as quickly as possible. We prove that PB is NP-hard when k is arbitrary. We show that the discrete k-center of the vertices of P under the geodesic metric on P provides a 2-approximation for PB, resulting in an $O(n^2 \log n + hkn \log n)$-time 3-approximation algorithm for PB. Lastly, we define and characterize a new type of polygon, the sliceable polygon. A sliceable polygon is a convex polygon that contains no Voronoi vertex from the Voronoi diagram of its vertices. We give a dynamic programming algorithm to solve PB exactly on a sliceable polygon in $O(kn^2)$ time.

Keywords: k-center · Polygon covering · Voronoi diagram

1 Introduction

Given a set S of n points representing clients or demands, the *k-center* problem asks to determine a collection C of k center points for placing facilities so as to minimize the maximum distance from any demand to its nearest facility. Geometrically speaking, the goal is to find the centers of k equal-radius balls whose union covers S and whose common radius, the *radius* of the k-center, is as small as possible. This paper assumes the discrete version of the k-center problem where centers are selected from S.

The k-center problem is NP-hard when k is an arbitrary input parameter [11] and NP-hard to approximate within a factor of $2 - \epsilon$ for any $\epsilon > 0$ [9]. However, there exist several 2-approximation algorithms that hold in any metric space [6, 10]. Gonzalez, for one, gave a greedy approach: Select the first center from S arbitrarily, and while $|C| < k$, repeatedly find the point in S whose minimum distance to the chosen centers is maximized and add it to C.

In many real-world applications, demands are not restricted to a discrete set but may be distributed throughout an area. Consider, for example, installing charging stations in a warehouse so that the worst-case travel time of robots to their nearest stations is minimal. In practice, regions of demand are often

© Springer Nature Switzerland AG 2022
P. Mutzel et al. (Eds.): WALCOM 2022, LNCS 13174, pp. 123–134, 2022.
https://doi.org/10.1007/978-3-030-96731-4_11

modelled using polygonal domains. A *polygonal domain* P with h holes and n vertices is a connected region whose boundary ∂P comprises n line segments that form $h + 1$ simple closed polygonal chains. If P is without holes, then it is a *simple polygon*. We define the *geodesic distance* $d(s, t)$ between any two points $s, t \in P$ to be the Euclidean length of the shortest path connecting s and t that is contained in P.

Given a polygonal domain P, the geodesic k-center problem on P asks to find a set C of k points in P that minimizes the maximum geodesic distance from any point in P to its closest point in C. We call C the *k-center* of P. Asano and Toussaint [2] gave the first algorithm for computing the 1-center of a simple polygon with n vertices; it runs in $O(n^4 \log n)$ time. This result was later improved by Pollack et al. [14] to $O(n \log n)$, and recently, Ahn et al. [1] presented an optimal linear-time algorithm. Following these explorations, Oh et al. [12] gave an $O(n^2 \log^2 n)$-time algorithm for computing the 2-center of a simple polygon. However, it appears that no results are known for $k > 2$ in the case of simple polygons. Likewise, for polygons with one or more holes, results are limited: only the 1-center problem has been solved with a running time of $O(n^{11} \log n)$ [16].

In practice, facilities are often restricted to feasible locations. Hence, there has been some interest in constrained versions of the geodesic k-center problem on polygonal domains. Oh et al. [13] considered the problem of computing the 1-center of a simple polygon constrained to a set of line segments or simple polygonal regions in the polygon. Du and Xu [4] proposed a 1.8841-approximation algorithm for computing the k-center of a convex polygon P with centers restricted to the boundary of P.

In this paper, we consider a new variant of the geodesic k-center problem that restricts facilities to the vertices of the given polygonal domain. Unlike the original problem and the constrained versions above, our problem is a combinatorial optimization problem: We draw centers from a finite set of points rather than a region in the plane. Viewing each vertex as a potential site to start a fire, we arrive at the following problem formulation we adopt in this paper.

Definition 1 (Polygon Burning). *Given a polygonal domain P with h holes and n vertices and an integer $k \in [1, n]$, find a set S of k vertices of P such that P is consumed as quickly as possible when burned simultaneously and uniformly from S.*

Section 2 is devoted to the background required for our study. In Sect. 3, we prove that PB is NP-hard when k is part of the input. In Sect. 4, we show that the k-center of the vertices of P under the geodesic metric on P provides a 2-approximation for PB on P. This result applying to Gonzalez's greedy algorithm leads to an $O(n^2 \log n + hkn \log n)$-time 3-approximation algorithm for PB. Finally, given the NP-hardness of PB in general, we shift our focus to restricted instances. In Sect. 5, we consider convex polygons that contain no Voronoi vertex from the Voronoi diagram of their vertices. We call such instances sliceable. Their structure admits a natural ordering of separable subproblems, permitting an exact $O(kn^2)$ algorithm using the dynamic programming technique.

2   Preliminaries

Unless stated otherwise, the distance metric d we use on a polygonal domain P is the geodesic metric on P. The *diameter* of P, $diam(P)$, is the largest distance between any two points in P.

Let $S = \{s_1, s_2, \ldots, s_k\}$ be a set of k points, called *sites* or *burn sites*, in a region R. The Voronoi diagram $\mathrm{VD}_R(S)$ of S is the subdivision of R into k Voronoi regions, one per site $s_i \in S$, such that any point in the Voronoi region of s_i is closer to s_i (using the geodesic metric on R) than to any other site in S. We refer to $\mathrm{VD}_{\mathbb{R}^2}(S)$ as $\mathrm{VD}(S)$.

Consider a polygonal domain P with vertices $V = \{v_1, v_2, \ldots, v_n\}$. Let $S \subseteq V$ be a selection of k burn sites. Each Voronoi region P_i in the Voronoi diagram $\mathrm{VD}_P(S)$ of S is the set of points in P burned by the fire from site $s_i \in S$. We associate with each point p in P_i the time it burns, which is the distance travelled by the fire from s_i to p. It follows that P burns in time $t_S(P) = \max_{s_i \in S} \max_{p \in P_i} d(s_i, p)$. As described in Definition 1, PB asks to find a set $S \subseteq V$, $|S| = k$, that minimizes $t_S(P)$. We let $S_k(P)$ denote such an optimizing set and let $\mathrm{OPT}_k(P)$ be the minimum burning time of P.

A *geodesic disk* of radius r centered at a point $p \in P$ is the set of points in P at most geodesic distance r from p. By definition, the union of k geodesic disks of radius $\mathrm{OPT}_k(P)$ centered at the sites in $S_k(P)$ contains P. Observe that $diam(P) \leq 2k \cdot \mathrm{OPT}_k(P)$ since P cannot be covered by k geodesic disks of radius $\mathrm{OPT}_k(P)$ otherwise. The time to burn P given any non-empty selection of burn sites is at most $diam(P)$. Hence any non-empty selection of burn sites from V gives a $2k$-approximation for PB with k sites on P.

3   Hardness

In this section, we show that PB is NP-hard on polygonal domains. We reduce from 4-Planar Vertex Cover (4VPC): Given a planar graph G with max-degree four and an integer κ, does G contain a vertex cover (i.e., a set of vertices $C \subseteq V(G)$ such that every edge in G contains at least one vertex in C) of size at most κ? This problem is known to be NP-hard [5].

Given an instance G, κ of 4PVC, we construct an equivalent instance of PB. First we compute an orthogonal drawing Γ of G with $O(n)$ bends on an integer grid of $O(n^2)$ area (Fig. 1a) using an $O(n)$-time algorithm due to Tomassia and Tollis [15]. Every edge $uv \in E(G)$ is represented as a sequence of connected line segments $\overline{p_1 p_2}, \overline{p_2 p_3}, \ldots, \overline{p_{i-1} p_i}$ in Γ, denoted $\Gamma(uv)$, where $p_1 = \Gamma(u)$ and $p_i = \Gamma(v)$ correspond to the endpoints of uv and $p_2, \ldots, p_{i-1}$ are *bends* in $\Gamma(uv)$. The length $|\Gamma(uv)|$ of $\Gamma(uv)$ is the sum of the lengths of its line segments.

Next we transform Γ into a constrained straight-line drawing Π of a subdivision H of G in two steps. First we add a vertex at every bend in Γ (Fig. 1b). Then we replace each segment $\overline{p_j p_{j+1}}$ ($1 \leq j < i$) along $\Gamma(uv)$ with either $3|\overline{p_j p_{j+1}}|$ or $3|\overline{p_j p_{j+1}}| + 1$ equal-length edges depending on the parity required to ensure that the overall number ℓ_{uv} of segments along $\Gamma(uv)$ is odd (Fig. 1c). Property 1 and 2

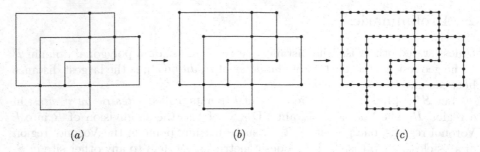

Fig. 1. (a) A planar orthogonal grid drawing Γ of G, (b) a straight-line grid drawing (step 1), and (c) the drawing Π of the subdivision H of G satisfying Property 1 and 2 (step 2).

follow from these steps. Property 2 is due to the fact that a double subdivision of an edge in G increases the size of any vertex cover of G by one.

Property 1. For every $uv \in E(H)$, $\frac{1}{4} \leq |\Pi(uv)| \leq \frac{1}{3}$.

Property 2. G has a vertex cover of size κ if and only if H has a vertex cover of size $K(G) := \kappa + \frac{1}{2} \sum_{uv \in E} (\ell_{uv} - 1)$.

Finally, we convert Π into a polygonal domain $P(G)$ by thickening each line segment in Π as follows. For every vertex $v \in V(H)$, we replace $\Pi(v)$ with a set $S(v)$ of four vertices at $\Pi(v) + (-\epsilon, \epsilon)$, $\Pi(v) + (\epsilon, \epsilon)$, $\Pi(v) + (\epsilon, -\epsilon)$, and $\Pi(v) + (-\epsilon, -\epsilon)$, where $\epsilon < \frac{1}{120}$ is a fixed constant. Let $R(uv)$ denote the convex hull of $S(u) \cup S(v)$. We define $P(G)$ to be the union of the collection of regions $R(uv)$ for all $uv \in E(H)$.

It is straightforward to verify that the above transformation of an instance G of 4PVC to an instance $P(G)$ of PB runs in $O(n)$ time. Furthermore, $P(G)$ has $O(n)$ vertices, and the number of bits required in the binary representation of each vertex coordinate is bounded by a polynomial in n. It remains to demonstrate that:

Lemma 1. *G has a vertex cover of size at most κ if and only if $P(G)$ can be burned in time $\frac{1}{3} + 3\epsilon$ using $K(G)$ sites.*

Proof. It suffices to show that for any $uv \in E(H)$, $R(uv)$ can be burned in time $\frac{1}{3} + 3\epsilon$ if and only if at least one vertex in $S(u) \cup S(v)$ is a burn site. The forward direction follows from observing that $\frac{1}{3} + 3\epsilon$ is a loose upper bound on the burning time of $R(uv)$ given that a site is located in either $S(u)$ or $S(v)$ (Property 1). For the reverse direction, suppose no vertices in $S(u)$ or $S(v)$ are selected. We obtain a lower bound on the burning time of $R(uv)$ by considering the scenario where $R(uv)$ is burned the quickest: First, for each vertex $w \in H$ adjacent to either u or v, let all vertices in $S(w)$ be burn sites. Second, assume u and v have as many adjacent edges as possible in $E(H)$ to assist in burning $R(uv)$. At most one of these two adjacent vertices can have degree greater than two since at

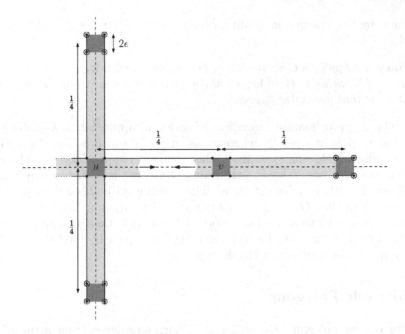

Fig. 2. A scenario where $R(uv)$ is burned the quickest assuming that no sites (circled) are selected from either $S(u)$ or $S(v)$. The two dashed lines are the only integer grid lines in the figure.

most one is on the integer grid, and this vertex, say u, can have degree at most four. The other vertex v can have degree two, but its adjacent edges must be colinear in the drawing. Finally, suppose all these edges are as short as possible in the drawing Π ($\frac{1}{4}$ by Property 1). We find that the burning time of $R(uv)$, if no vertex in $S(u)$ or $S(v)$ is a site, is bounded below by $\frac{3}{8} - 2\epsilon > \frac{1}{3} + 3\epsilon$ (see Fig. 2). The lemma then follows from Property 2. □

As a result, we obtain:

Theorem 1. *PB is NP-hard on polygonal domains.*

4 Approximation by a k-Center

We present a straightforward 3-approximation algorithm for PB by considering the k-center problem described in the introduction.

Theorem 2. *The radius of a k-center of the vertices V of P, using the geodesic metric on P, provides a 2-approximation of $\mathsf{OPT}_k(P)$.*

Proof. Let $C \subseteq V$ denote a k-center of V and let r denote its radius. Observe two facts: First, $\mathsf{OPT}_k(P) \geq r$ since $P \supseteq V$. Second, each point $p \in P$ is within $\mathsf{OPT}_k(P)$ of a vertex v of P, and v is at most r from some center c in C.

Therefore, by the triangle inequality, $d(p, c) \leq \mathsf{OPT}_k(P) + r \leq 2\mathsf{OPT}_k(P)$ as desired. $\quad\square$

Corollary 1. *Applying Gonzalez's greedy 2-approximation algorithm for finding a k-center of V yields an $O(n^2 \log n + hkn \log n)$-time 3-approximation algorithm for PB on P that uses $O(n^2)$ space.*

Proof. The 2-approximation algorithm provides an approximate k-center of V whose radius r' is at most $2r$ where r, as in the above proof, is the optimal k-center radius. Following that proof, this yields a 3-approximation. The time and space complexity are due to performing $O(kn)$ geodesic distance queries on P using an algorithm by Guo et al. [8]. Their algorithm builds a data structure of size $O(n^2)$ in time $O(n^2 \log n)$ to support $O(h \log n)$-time geodesic distance queries between any two points in P. Note, if P is simple, then a 3-approximation for PB can be found in $O(kn \log n)$ time using $O(n)$ space by the faster geodesic distance queries of Guibas and Hershberger [7]. $\quad\square$

5   Sliceable Polygons

Even for convex polygons, the choice of a burn site depends on many of the choices of other burn sites since a site may have many Voronoi neighbors. To obtain an efficient algorithm, we consider a family of polygons where the number of such interactions between burn sites is small.

Definition 2. *A sliceable polygon P is convex and contains no Voronoi vertex from the Voronoi diagram VD(V) of its vertices V.*

Every Voronoi edge in VD(V) that intersects P slices through P (Fig. 3). We can solve PB on P using dynamic programming, as P admits a total ordering of vertices with the property that if burn sites u, v, and w satisfy $u < v < w$, then the region of P burned by u does not share a boundary with the region of P burned by w (Lemma 2). We start with a simple example that indicates the use of this property.

5.1   Polygons in One Dimension

Let P be a 1-dimensional polygon with n vertices $v_1, v_2, \ldots, v_n$ ordered by x-coordinate. Let $P[i, j]$ be the segment of P from v_i to v_j. The minimum time to burn P using k sites is

$$\mathsf{OPT}_k(P) = \begin{cases} \min_{i \in [n]} \max\{d(v_1, v_i), \mathrm{LNR}(i, k-1)\} & \text{if } k > 0 \\ \infty & \text{otherwise,} \end{cases}$$

where $d(v_1, v_i)$ is the time to burn $P[1, i]$ from site v_i and $\mathrm{LNR}(i, k)$ denotes the minimum time to burn $P[i, n]$ using k sites in addition to v_i. If $k > 0$, then $\mathrm{LNR}(i, k)$ is achieved by choosing the next site v_j ($i < j \leq n$) to minimize the

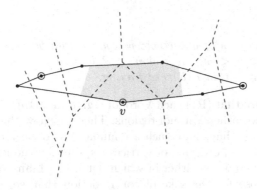

Fig. 3. A sliceable polygon P overlaid with the Voronoi diagram (dashed) of its vertices. By Lemma 2, the region of P (shaded) burned by a site v separates the regions burned by sites (circled) before v in the ordering from regions burned by sites after v, no matter what those sites are. This holds for every v.

larger of two values: (i) the time $d(v_i, v_j)/2$ to burn $P[i, j]$ and (ii) the minimum time to burn $P[j, n]$ knowing v_j is a burn site with $k - 1$ burn sites remaining. If $k = 0$, no sites are allowed beyond v_i, in which case the minimum time to burn $P[i, n]$, with v_i as a burn site, is $d(v_i, v_n)$.

$$\text{LNR}(i, k) = \begin{cases} \min_{i < j \leq n} \max\{d(v_i, v_j)/2, \text{LNR}(j, k - 1)\} & \text{if } k > 0, \\ d(v_i, v_n) & \text{otherwise.} \end{cases}$$

This recurrence relation relies only on the property that any burn site preceding the burn site v_i is farther from every point in $P[i, j]$ than v_i for $j > i$. We will prove a similar property for sliceable polygons.

A dynamic programming algorithm follows directly from the recurrence.

Theorem 3. *PB can be solved in $O(kn^2)$ time on a 1-dimension polygon with n vertices.*

Proof. (Sketch) Use dynamic programming. Two observations hold on each iteration of the algorithm: (i) The choice of the following site v_j is unaffected by the sites selected before the current site v_i, and (ii) we evaluate every possible choice v_j and take the best amongst them. The natural ordering of subproblems implied by (i) combined with the virtue of an exhaustive search as noted in (ii) allows us to successfully compute the solution to the original problem from the solutions to the recursive subproblems.

The algorithm populates a table of size $O(kn)$. To fill each entry, it computes the minimum of $O(n)$ previous entries. Therefore, the total running time is $O(kn^2)$. □

5.2   Ordering

Lemma 2. *The vertices of a sliceable polygon P can be ordered such that for any burn sites $u < v < w$, the region of P burned from u does not share a boundary with the region in P burned from w.*

Proof. We first prove that (P1) each Voronoi region in $VD_P(V)$ shares a boundary with at most two other Voronoi regions. Then we show that (P2) the graph joining two vertices if they share such a boundary is connected and thus forms a path, which defines an ordering of vertices required by the lemma. (The path can be directed in two ways, either of which defines such an ordering.)

For (P1), suppose for the sake of contradiction that vertex u of P forms Voronoi edges in $VD(V)$ that cross P with three other vertices, say r, s, and t. Since P is sliceable, the endpoints (Voronoi vertices) of these Voronoi edges lie outside P.

Let P' be the convex hull of $\{u, r, s, t\}$. The Voronoi edge between u and r in $VD(\{u, r, s, t\})$ contains the corresponding Voronoi edge in $VD(V)$ since every point that is closest to u and r among all vertices of V is still closest to u and r among a subset of V. The same is true for the Voronoi edges between u and s and between u and t. Thus, since all three of these Voronoi edges cross P in $VD(V)$ the corresponding edges in $VD(\{u, r, s, t\})$ cross P and hence cross $P' \subsetneq P$ as well. It follows that a sliceable polygon P with a vertex u that creates Voronoi edges crossing P with three different vertices r, s, and t implies the existence of a sliceable quadrilateral P' with the same property. To obtain a contradiction and establish (P1), we will argue that no such quadrilateral exists.

Assume r, s, and t are labelled so that the circumcenters c_1 of $\triangle urs$ and c_2 of $\triangle ust$ are the two Voronoi vertices shared by these three Voronoi edges. Since the boundary of the Voronoi region of u intersects P' in three segments that do not contain c_1 or c_2, c_1 lies on the side of the line through rs opposite u and c_2 lies on the side of the line through st opposite of u. It follows that $\angle rus$ and $\angle sut$ are obtuse. Thus the interior angle of P' at u is greater than π, contradicting the convexity of P' (inset). This result establishes (P1).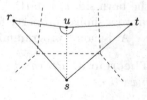

For (P2), assume for a contradiction that the graph has more than one connected component. Then no inter-component vertices form Voronoi boundaries with each other in $VD_P(V)$. It follows that the fires burning from separate connected components never meet, and hence P cannot be burned entirely. This contradiction establishes (P2).   □

5.3   Sliceability of Subsets

In this section, we study the sliceability of subsets of sliceable polygons. In particular, we show that a sliceable polygon P contains no Voronoi vertex from $VD(S)$ for any subset $S \subseteq V$. While the existence of a dynamic programming algorithm does not require this result, it adds to our characterization of sliceable polygons and allows us to define a simpler recurrence for PB on P that yields a faster dynamic programming algorithm.

The *Delaunay triangulation* of a set S of sites, denoted DT(S), is the dual graph of VD(S). It is a triangulation of S such that no circumcircle of any triangle in DT(S) contains a site. The circumcenters of the triangles are the vertices of VD(S).

Lemma 3. *Let T be a triangulation of a convex polygon P. Suppose there exist adjacent triangles pqr and prs in T that form a convex quadrilateral. If P contains the circumcenter of $\triangle pqr$ and s is interior to the circumcircle of $\triangle pqr$, then P contains the circumcenter of $\triangle pqs$ or the circumcenter of $\triangle qrs$, or both.*

Proof. Assume the vertices of quadrilateral $pqrs$ are labelled in counter-clockwise order. By the conditions of the lemma, triangles pqs and qrs form the Delaunay triangulation of quadrilateral $pqrs$. Orient P so that $\overline{pq}$ is aligned with the x-axis with r and s lying above it (Fig. 4). Let f, g, and h denote the circumcenters of $\triangle pqr$, $\triangle pqs$, and $\triangle qrs$, respectively. Since r lies outside C_{pqs} above $\overline{qs}$, C_{pqs} lies below C_{pqr}, implying that g is below f. Similarly, since p lies outside C_{qrs} left of $\overline{qs}$, C_{qrs} lies right of C_{pqr}, which implies that h is right of f. To prove that either g or h lies in P given that f is in P, consider two cases:

Case 1: Suppose h lies on or left of $\overline{qr}$. Let m be the midpoint of $\overline{qr}$. Since h is right of f and both f and h lie on the bisector of q and r, h lies along $\overline{fm}$. Hence, by the convexity of P, h lies in P.

Case 2: Otherwise, h lies right of $\overline{qr}$. Then $\angle qsr > \frac{\pi}{2}$. We show that g must lie on or above $\overline{pq}$ in this scenario. Assume for a contradiction that g lies below $\overline{pq}$. Then $\angle psq > \frac{\pi}{2}$. This yields $\angle psr = \angle psq + \angle qsr > \pi$, which implies that P is not convex. This contradiction establishes that g lies above $\overline{pq}$. By the same analysis provided in the previous case, we conclude that P contains g.   □

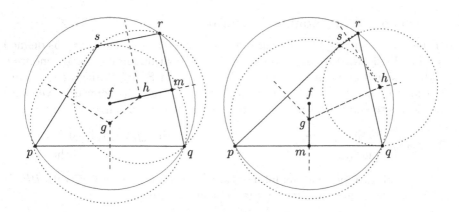

Fig. 4. Illustration of Case 1 (left) and Case 2 (right) of Lemma 3 with C_{pqr} (solid), C_{pqs} and C_{qrs} (dotted), and VD$(\{p,q,r,s\})$ (dashed).

Lemma 4. *Consider a triangulation T of a convex polygon P. If P contains the circumcenter of a triangle in T, then it contains the circumcenter of a triangle in the Delaunay triangulation $DT(V)$ of V.*

Proof. Let $\overline{pr}$ be an edge in T incident to two triangles pqr and prs that form a convex quadrilateral. We say $\overline{pr}$ is an *illegal edge* if s lies in C_{pqr}. A new triangulation T' of P can be obtained from T by replacing $\overline{pr}$ with $\overline{qs}$. This *edge flip* operation creates $\triangle pqs$ and $\triangle qrs$ in place of $\triangle pqr$ and $\triangle prs$. If $\overline{pr}$ is illegal, then, by Lemma 3, P contains the circumcenter of $\triangle pqs$ or $\triangle qrs$ (or both) if it contains the circumcenter of $\triangle pqr$ or $\triangle prs$. More generally, assuming that T' is obtained by flipping an illegal edge in T, P contains the circumcenter of some triangle in T' if it contains the circumcenter of some triangle in T. We can compute $DT(V)$ by flipping illegal edges in T until none exist [3]. Therefore, by repeated application of Lemma 3, P contains the circumcenter of some triangle in $DT(V)$ if it contains the circumcenter of some triangle in T. $\square$

Theorem 4. *If a convex polygon P does not contain the circumcenter of any triangle in $DT(V)$, then P does not contain the circumcenter of any triangle in $DT(S)$ for any $S \subseteq V$.*

Proof. For completeness, we restate the theorem in terms of Voronoi diagrams. If a convex polygon P does not contain any Voronoi vertex of $VD(V)$, then P does not contain any Voronoi vertex of $VD(S)$ for any $S \subseteq V$.

We provide a contrapositive proof. Suppose P contains the circumcenter of $\triangle pqr$ in $DT(S)$. Let T be any triangulation of P containing $\triangle pqr$. Of course, P contains the circumcenter of a triangle in T, implying that P contains the circumcenter of a triangle in $DT(V)$ by Lemma 4. $\square$

Corollary 2. *If P is sliceable, then the convex hull of S is sliceable for any $S \subseteq V$.*

Proof. Since P contains no Voronoi vertex of $VD(S)$ for any $S \subseteq V$ by Theorem 4, neither does any subset of P, including the convex hull of S. $\square$

5.4   Dynamic Programming Algorithm

Let P be a sliceable polygon. The ordering of its vertices $v_1, v_2, \ldots, v_n$ as defined in Lemma 2 permits a dynamic programming algorithm similar to the one used for 1-dimensional polygons that solves PB on P. Let $\mathrm{SLB}(i, k)$ denote the minimum time to burn the subset of P from the bisector of $\overline{v_{i-1}v_i}$ onward given that v_i is a burn site and k sites remain to be chosen. It can be defined recursively as

$$\mathrm{SLB}(i, k) = \begin{cases} \min_{i < j \leq n} \max\{d(v_j, p_{ij}), d(v_j, q_{ij}), \mathrm{SLB}(j, k-1)\} & \text{if } k > 0, \\ d(v_i, v_n) & \text{otherwise,} \end{cases}$$

where p_{ij} and q_{ij} represent the intersections of the bisector of $\overline{v_i v_j}$ with ∂P. It follows that the minimum time to burn P using k sites is

$$\mathsf{OPT}_k(P) = \begin{cases} \min_{i \in [n]} \max\{d(v_1, v_i), \mathrm{SLB}(i, k-1)\} & \text{if } k > 0 \\ \infty & \text{otherwise.} \end{cases}$$

Theorem 5. *Using a dynamic programming algorithm, PB can be solved in $O(kn^2)$ time on an n-vertex sliceable polygon.*

Proof. (Sketch) We prove that the recurrence for $\text{SLB}(i, k)$ is correct by showing that the maximum distance from burn site v_j to a point in the region P_j that is burnt by v_j is correctly calculated in $\text{SLB}(i, k)$. Let v_i be the burn site preceding v_j, and v_ℓ be the burn site following v_j in the vertex ordering. The region P_j is bounded by the perpendicular bisectors of segments $\overline{v_i v_j}$ and $\overline{v_j v_\ell}$ which intersect P in segments $\overline{p_{ij} q_{ij}}$ and $\overline{p_{j\ell} q_{j\ell}}$ respectively (Fig. 5). It suffices to show that the time to burn P_j from v_j is the larger of $\max\{d(v_j, p_{ij}), d(v_j, q_{ij})\}$, considered in $\text{SLB}(i, k)$, and $\max\{d(v_j, p_{j\ell}), d(v_j, q_{j\ell})\}$, considered in $\text{SLB}(j, k - 1)$. If no site precedes v_j then the recurrence correctly, by Lemma 2, uses $d(v_1, v_j)$ instead of $\max\{d(v_j, p_{ij}), d(v_j, q_{ij})\}$. Likewise, if no site follows v_j then the recurrence correctly uses $d(v_j, v_n)$ instead of $\max\{d(v_j, p_{j\ell}), d(v_j, q_{j\ell})\}$.

Since P_j is convex, the point in P_j farthest from site v_j is some vertex of P_j. We show that this vertex is in $\{p_{ij}, q_{ij}, p_{j\ell}, q_{j\ell}\}$. Suppose, for the sake of contradiction, it is not and that there exists a vertex u of P in P_j such that the circle C centered at v_j through u contains P_j.

First, $\angle v_j u v_i$ is acute since for P_j to lie inside C, both edges of P_j incident to u must form acute angles with its radius $v_j u$. Hence, since P is convex, both the edge uv_i and the edge uv_ℓ form an acute angle with $v_j u$. Second, both v_i and v_ℓ lie outside the circle with diameter $v_j u$, otherwise u would be closer to v_i or v_ℓ than to v_j and hence not be burned by v_j. This implies that $\angle u v_i v_j$ is acute. Finally, (i) if $v_i < u < v_j$ in the vertex ordering then $\angle v_i v_j u$ is acute, otherwise the perpendicular bisector of uv_j would not separate v_i from v_j which violates the properties of the ordering. Similarly, (ii) if $v_j < u < v_\ell$ then $\angle v_j v_\ell u$ is acute.

Combining these three observations, we have in case (i) that $\triangle v_i u v_j$ is acute and in case (ii) that $\triangle v_j u v_\ell$ is acute, both of which contradict Corollary 2.   □

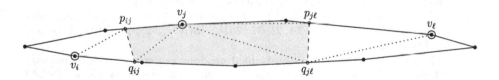

Fig. 5. Region P_j (shaded) induced by sites v_i, v_j, and v_ℓ (circled), overlaid with the distances considered by algorithm (dotted).

6   Conclusion

In this paper, we proved PB to be NP-hard on general polygonal domains. Nevertheless, the hardness for simple and convex polygons remains open. In addition, we gave an $O(n^2 \log n + hkn \log n)$-time 3-approximation algorithm for PB. Finally, we considered sliceable polygons on which we can obtain a dynamic programming solution for PB. Avenues for future research are to improve the approximation algorithm, to expand the class of polygons solvable using dynamic programming, and to resolve the complexity of PB on simple polygons.

References

1. Ahn, H.K., Barba, L., Bose, P., De Carufel, J.L., Korman, M., Oh, E.: A linear-time algorithm for the geodesic center of a simple polygon. Discrete Comput. Geom. **56**(4), 836–859 (2016)
2. Asano, T., Toussaint, G.: Computing the geodesic center of a simple polygon. In: Johnson, D.S., Nishizeki, T., Nozaki, A., Wilf, H.S. (eds.) Discrete Algorithms and Complexity, pp. 65–79. Academic Press (1987)
3. Berg, M.D., Cheong, O., Kreveld, M.V., Overmars, M.: Computational Geometry: Algorithms and Applications, 3rd edn. Springer, Heidelberg (2008). https://doi.org/10.1007/978-3-540-77974-2
4. Du, H., Xu, Y.: An approximation algorithm for k-center problem on a convex polygon. J. Combin. Optim. **27**(3), 504–518 (2014)
5. Garey, M.R., Johnson, D.S.: Computers and Intractability: A Guide to the Theory of NP-Completeness. W. H. Freeman & Co., New York (1979)
6. Gonzalez, T.F.: Clustering to minimize the maximum intercluster distance. Theor. Comput. Sci. **38**, 293–306 (1985)
7. Guibas, L.J., Hershberger, J.: Optimal shortest path queries in a simple polygon. J. Comput. Syst. Sci. **39**(2), 126–152 (1989)
8. Guo, H., Maheshwari, A., Sack, J.-R.: Shortest path queries in polygonal domains. In: Fleischer, R., Xu, J. (eds.) AAIM 2008. LNCS, vol. 5034, pp. 200–211. Springer, Heidelberg (2008). https://doi.org/10.1007/978-3-540-68880-8_20
9. Hochbaum, D.: Approximation Algorithms for NP-Hard Problems. PWS Publishing Company, Boston (1995)
10. Hochbaum, D.S., Shmoys, D.B.: A best possible heuristic for the k-center problem. Math. Oper. Res. **10**(2), 180–184 (1985)
11. Kariv, O., Hakimi, S.L.: An algorithmic approach to network location problems. I: the p-centers. SIAM J. Appl. Math. **37**(3), 513–538 (1979)
12. Oh, E., De Carufel, J.L., Ahn, H.K.: The geodesic 2-center problem in a simple polygon. Comput. Geom. **74**, 21–37 (2018). https://doi.org/10.1016/j.comgeo.2018.02.008
13. Oh, E., Son, W., Ahn, H.K.: Constrained geodesic centers of a simple polygon. In: 15th Scandinavian Symposium and Workshops on Algorithm Theory (SWAT 2016), vol. 53, pp. 29:1–29:13 (2016). https://doi.org/10.4230/LIPIcs.SWAT.2016.29
14. Pollack, R., Sharir, M., Rote, G.: Computing the geodesic center of a simple polygon. Discrete Comput. Geom. **4**(6), 611–626 (1989). https://doi.org/10.1007/BF02187751
15. Tamassia, R., Tollis, I.G.: Planar grid embedding in linear time. IEEE Trans. Circ. Syst. **36**(9), 1230–1234 (1989)
16. Wang, H.: On the geodesic centers of polygonal domains. In: 24th Annual European Symposium on Algorithms (ESA 2016), vol. 57, pp. 77:1–77:17 (2016). https://doi.org/10.4230/LIPIcs.ESA.2016.77

Reverse Shortest Path Problem in Weighted Unit-Disk Graphs

Haitao Wang and Yiming Zhao$^{(\boxtimes)}$

Department of Computer Science, Utah State University, Logan, UT 84322, USA
{haitao.wang,yiming.zhao}@usu.edu

Abstract. Given a set P of n points in the plane, a unit-disk graph $G_r(P)$ with respect to a parameter r is an undirected graph whose vertex set is P such that an edge connects two points $p, q \in P$ if the (Euclidean) distance between p and q is at most r (the weight of the edge is 1 in the unweighted case and is the distance between p and q in the weighted case). Given a value $\lambda > 0$ and two points s and t of P, we consider the following *reverse shortest path problem*: Compute the smallest r such that the shortest path length between s and t in $G_r(P)$ is at most λ. In this paper, we study the weighted case and present an $O(n^{5/4} \log^{5/2} n)$ time algorithm. We also consider the L_1 version of the problem where the distance of two points is measured by the L_1 metric; we solve the problem in $O(n \log^3 n)$ time for both the unweighted and weighted cases.

1 Introduction

Given a set P of n points in the plane and a parameter r, the *unit-disk graph* $G_r(P)$ is an undirected graph whose vertex set is P such that an edge connects two points $p, q \in P$ if the (Euclidean) distance between p and q is at most r. The weight of each edge of $G_r(P)$ is defined to be one in the *unweighted* case and is defined to the distance between the two vertices of the edge in the *weighted* case. Alternatively, $G_r(P)$ can be viewed as the intersection graph of the set of congruous disks centered at the points of P with radii equal to $r/2$, i.e., two vertices are connected if their disks intersect. The *length* of a path in $G_r(P)$ is the sum of the weights of the edges of the path.

Computing shortest paths in unit-disk graphs with different distance metrics and different weights assigning methods has been extensively studied, e.g., [5–7,12,13,17,19,20]. Although a unit-disk graph may have $\Omega(n^2)$ edges, geometric properties allow to solve the single-source-shortest-path problem (SSSP) in subquadratic time. Roditty and Segal [17] first proposed an algorithm of $O(n^{4/3+\epsilon})$ time for unit-disk graphs for both unweighted and weighted cases, for any $\epsilon > 0$. Cabello and Jejčič [5] gave an algorithm of $O(n \log n)$ time for the unweighted case. Using a dynamic data structure for bichromatic closest pairs [1], they also solved the weighted case in $O(n^{1+\epsilon})$ time [5]. Chan and Skrepetos [6] gave an

---

This research was supported in part by NSF under Grant CCF-2005323.

P. Mutzel et al. (Eds.): WALCOM 2022, LNCS 13174, pp. 135–146, 2022.
https://doi.org/10.1007/978-3-030-96731-4_12

$O(n)$ time algorithm for the unweighted case, assuming that all points of P are presorted. Kaplan et al. [13] developed a new randomized result for the dynamic bichromatic closest pair problem; applying the new result to the algorithm of [5] leads to an $O(n \log^{12+o(1)} n)$ expected time randomized algorithm for the weighted case. Recently, Wang and Xue [19] proposed a new algorithm that solves the weighted case in $O(n \log^2 n)$ time.

The L_1 version of the SSSP problem has also been studied, where the distance of two points in the plane is measured under the L_1 metric when defining $G_r(P)$. Note that in the L_1 version a "disk" is a diamond. The SSSP algorithms of [5,6] for the L_2 unweighted version can be easily adapted to the L_1 unweighted version. Wang and Zhao [20] recently solved the L_1 weighted case in $O(n \log n)$ time. It is known that $\Omega(n \log n)$ is a lower bound for the SSSP problem in both L_1 and L_2 versions [5,20]. Hence, the SSSP problem in the L_1 weighted/unweighted case as well as in the L_2 unweighted case has been solved optimally.

In this paper, we consider the following *reverse shortest path* (RSP) problem. In addition to P, given a value $\lambda > 0$ and two points $s, t \in P$, the problem is to find the smallest value r such that the distance between s and t in $G_r(P)$ is at most λ. Throughout the paper, we let r^* denote the optimal value r for the problem. The goal is therefore to compute r^*.

Observe that r^* must be equal to the distance of two points in P in any case (i.e., L_1, L_2, weighted, unweighted). For the L_2 unweighted case, Cabello and Jejčič [5] mentioned a straightforward solution that can solve it in $O(n^{4/3} \log^3 n)$ time, by using the distance selection algorithm of Katz and Sharir [14] to perform binary search on all interpoint distances of P; Wang and Zhao [21] later gave two algorithms with time complexities $O(\lfloor \lambda \rfloor \cdot n \log n)$ and $O(n^{5/4} \log^{7/4} n)$,[1] respectively, using the parametric search technique. The first algorithm is interesting for small λ and the second algorithm uses the first one as a subroutine.

In this paper, we study the L_2 weighted case of the RSP problem and present an algorithm of $O(n^{5/4} \log^{5/2} n)$ time. We also consider the L_1 version of the RSP problem and solve it in $O(n \log^3 n)$ time for both unweighted and weighted cases.

Recently, Katz and Sharir [15] proposed randomized algorithms of $O(n^{6/5+\epsilon})$ expected time for the L_2 RSP problem for both the unweighted and weighted cases, for arbitrary small $\epsilon > 0$.[2]

The RSP problem has been studied in the literature under various problem settings. Intuitively, the problem is to modify the graph (e.g., modify edge weights) so that certain desired constraints related to shortest paths can be satisfied, e.g., [4,22]. As a motivation of our problem, consider the following scenario. Suppose $G_r(P)$ represents a wireless sensor network in which each sensor is represented by a disk centered at a point in P and two sensors can communicate

---

[1] The time complexity given in [21] is $O(n^{5/4} \log^2 n)$, but can be easily improved to $O(n^{5/4} \log^{7/4} n)$ by changing the threshold for defining large cells from $n^{3/4}$ to $(n/\log n)^{3/4}$ in Sect. 4 [21].

[2] It is not explicitly stated in [15] that the algorithm is randomized. A key subroutine used in the algorithm is Theorem 1 [15], which is from [2] and is a randomized algorithm (see Sect. 4 in [2]).

with each other (e.g., directly transmit a message) if they are connected by an edge in $G_r(P)$. The disk radius is proportional to the energy of the sensor. The latency of transmitting a message between two neighboring sensors is proportional to their distance. For two sensors s and t, we want to know the minimum energy for all sensors so that the total latency of transmitting messages between s and t is no more than a target value λ. It is not difficult to see that this is equivalent to our RSP problem.

1.1 Our Approach

Our algorithm for the L_2 weighted RSP problem follows the parametric search scheme. Let $d_r(s,t)$ denote the distance from s to t in $G_r(P)$. Given any r, the *decision problem* is to decide whether $r^* \leq r$. Observe that $r^* \leq r$ holds if and only if $d_r(s,t) \leq \lambda$. Hence, the shortest path algorithm of Wang and Xue [19] (referred to the WX algorithm) can be used to solve the decision problem in $O(n \log^2 n)$ time. To compute r^*, since r^* is equal to the distance of two points of P, one could first compute all interpoint distances of points of P and then use the WX algorithm to perform binary search among these distances to compute r^*. Clearly, the algorithm takes $\Omega(n^2)$ time. Alternatively, as mentioned in [5], one can perform binary search by using the distance selection algorithm of Katz and Sharir [14] (i.e., given any k with $1 \leq k \leq \binom{n}{2}$, the algorithm finds the k-th smallest distance among all interpoint distances of P) without explicitly computing all these $\Omega(n^2)$ distances. As the algorithm of Katz and Sharir [14] runs in $O(n^{4/3} \log^2 n)$, this approach can compute r^* in $O(n^{4/3} \log^3 n)$ time.

We propose a more efficient parametric search algorithm, by "parameterizing" the decision algorithm, i.e., the WX algorithm. Like typical parametric search, we run the decision algorithm with a parameter $r \in (r_1, r_2]$ by simulating the decision algorithm on the unknown r^*. At each step, we call the decision algorithm on certain "critical values" r to compare r and r^*, and the algorithm will proceed accordingly based on the result of the comparison. The interval $(r_1, r_2]$ will also be shrunk after these comparisons but is guaranteed to contain r^* throughout the algorithm. The algorithm terminates once t is reached, at which moment we can prove that r^* is equal to r_2 of the current interval $(r_1, r_2]$.

For the L_1 RSP problem, we use an approach similar to the distance selection algorithm in [14]. As in the L_2 case, the decision problem can be solved in $O(n \log n)$ time by applying the SSSP algorithms for both the unweighted case and the weighted case [5,6,20,21] (more precisely, for the unweighted case, the decision problem can be solved in $O(n)$ time after $O(n \log n)$ time preprocessing for sorting the points of P [6]). Let Π denote the set of all pairwise distances of all points of P. In light of the observation that r^* is in Π, each iteration of our algorithm computes an interval $(a_j, b_j]$ (initially, $a_0 = -\infty$ and $b_0 = \infty$) such that $r^* \in (a_j, b_j]$ and the number of values of Π in $(a_j, b_j]$ is a constant fraction of the number of values of Π in $(a_{j-1}, b_{j-1}]$. In this way, r^* can be found within $O(\log n)$ iterations. Each iteration will call the decision algorithm to perform binary search on certain values. The total time of the algorithm is $O(n \log^3 n)$.

A by-product of our technique is an $O(n \log^3 n)$ time algorithm that can compute the k-th smallest L_1 distance among all pairs of points of P, for any given k with $1 \leq k \leq \binom{n}{2}$. As mentioned before, the L_2 version of the problem can be solved in $O(n^{4/3} \log^2 n)$ time [14].

Outline. In the following, we tackle the L_2 problem in Sects. 2. Due to the space limit, many proofs and the discussion about the L_1 problem are omitted but can be found in the full paper of [21] (the two papers are merged).

2   The L_2 RSP Problem

We follow the notation introduced in Sect. 1, e.g., P, $G_r(P)$, $d_r(s,t)$, r^*. Our goal is to compute r^*. As we will parameterize the WX algorithm, we first review the WX algorithm. For any two points p and q in the plane, let $\|p - q\|$ denote the Euclidean distance between them.

2.1   A Review of the WX Algorithm

Given P, r, and a source point $s \in P$, we consider the SSSP problem to compute shortest paths from s to all points of P in the unit-disk graph $G_r(P)$. The WX algorithm can solve the problem in $O(n \log^2 n)$ time.

For any point p, denote by $\bigodot_p$ the disk centered at p with radius r.

The first step is to implicitly build a grid $\Psi_r(P)$ of square cells whose side lengths are $r/2$. For simplicity of discussion, we assume that every point of P lies in the interior of a cell of $\Psi_r(P)$. A *patch* of $\Psi_r(P)$ refers to a square area consisting of 5×5 cells. For a point $p \in P$, we use $\square_p$ to denote the cell of $\Psi_r(P)$ containing p and use $\boxplus_p$ to denote the patch whose central cell is $\square_p$ (e.g., see Fig. 1). We refer to cells of $\boxplus_p \setminus \square_p$ as the *neighboring cells* of $\square_p$. As the side length of each cell of $\Psi_r(P)$ is $r/2$, any two points of P in a single cell of $\Psi_r(P)$ must be connected by an edge in $G_r(P)$. Moreover, if an edge connects two points p and q in $G_r(P)$, then q must lie in $\boxplus_p$ and vice versa. For any subset $Q \subseteq P$ and a cell $\square$ (resp.,a patch $\boxplus$) of $\Psi_r(P)$, define $Q_\square = Q \cap \square$ (resp., $Q_\boxplus = Q \cap \boxplus$). The step of implicitly building the grid actually computes the subset $P_\square$ for each cell $\square$ of $\Psi_r(P)$ that contains at least one point of P as well as associate pointers to each point $p \in P$ so that given any $p \in P$, the list of points of $P_{\square_p}$ (resp., $P_{\boxplus_p}$) can be accessed immediately. Building $\Psi_r(P)$ implicitly as above can be done in $O(n \log n)$ time and $O(n)$ space [19].

The WX algorithm follows the basic idea of Dijkstra's algorithm and computes an array $dist[\cdot]$ for each point $p \in P$, where $dist[p]$ will be equal to $d_r(s,p)$ when the algorithm terminates. Different from Dijkstra's shortest path algorithm, which picks a single vertex in each iteration to update the shortest path information of other adjacent vertices, the WX algorithm aims to update in each iteration the shortest path information for all points within one single cell of $\Psi_r(P)$ and pass on the shortest path information to vertices lying in the neighboring cells.

A key subroutine used in the WX algorithm is UPDATE(U, V), which updates the shortest path information for a subset $V \subseteq P$ of points by using the shortest path information of another subset $U \subseteq P$ of points. Specifically, the subroutine finds, for each $v \in V$, $q_v = \arg\min_{u \in U \cap \odot_v}\{dist[u] + \|u - v\|\}$ and update $dist[v] = \min\{dist[v], dist[q_v] + \|q_v - v\|\}$.

With the subroutine UPDATE(U, V), the WX algorithm works as follows.

Initially, we set $dist[s] = 0$, $dist[p] = \infty$ for all other points $p \in P \setminus \{s\}$, and $Q = P$. Then we enter the main (while) loop. In each iteration, we find a point z with minimum $dist$-value from Q, and then execute two update subroutines UPDATE($Q_{\boxplus_z}$, $Q_{\square_z}$) and UPDATE($Q_{\square_z}$, $Q_{\boxplus_z}$). Next, points of $Q_{\square_z}$ are removed from Q, because it can be shown that $dist[p]$ for all points $p \in Q_{\square_z}$ have been correctly computed [19]. The algorithm stops once Q becomes $\emptyset$. The efficiency of the algorithm hinges on the implementation of the two update subroutines. We give some details below, which are needed in our RSP algorithm as well.

The First Update. For the first update UPDATE($Q_{\boxplus_z}$, $Q_{\square_z}$), the crucial step is finding a point $q_v \in Q_{\boxplus_z} \cap \odot_v$ for each point $v \in Q_{\square_z}$ such that $dist[q_v] + \|q_v - v\|$ is minimized. If we assign $dist[q]$ as a weight to each point $q \in Q_{\boxplus_z}$, then the problem is equivalent to finding the additively-weighted nearest neighbor q_v from $Q_{\boxplus_z} \cap \odot_v$ for each $v \in Q_{\square_z}$. To this end, Wang and Xue [19] proved a *key observation* that any point $q \in Q_{\boxplus_z}$ that minimizes $dist[q] + \|q - v\|$ must lie in $\odot_v$. This implies that for each point $v \in Q_{\square_z}$, its additively-weighted nearest neighbor in $Q_{\boxplus_z}$ is also its additively-weighted nearest neighbor in $Q_{\boxplus_z} \cap \odot_v$. As such, q_v for all $v \in Q_{\square_z}$ can be found by first building an additively-weighted Voronoi Diagram on points of $Q_{\boxplus_z}$ [9] and then performing point locations for all $v \in Q_{\square_z}$ [8,16,18]. In this way, since $\sum_{z_i} |P_{\boxplus_{z_i}}| = O(n)$, where z_i refers to the point z in the i-th iteration of the main loop, the first updates for all iterations of the main loop can be done in $O(n \log n)$ time in total [19].

The Second Update. The second update UPDATE($Q_{\square_z}$, $Q_{\boxplus_z}$) is more challenging because the above key observation no longer holds. Since $Q_{\boxplus_z}$ has $O(1)$ cells of $\Psi_r(P)$, it suffices to perform UPDATE($Q_{\square_z}$, $Q_{\square}$) for all cells $\square \in \boxplus_z$.

If $\square$ is $\square_z$, then $Q_{\square_z} = Q_{\square}$. Since the distance between any two points in $\square_z$ is at most r, we can easily implement UPDATE($Q_{\square_z}$, $Q_{\square}$) in $O(|Q_{\square_z}| \log |Q_{\square_z}|)$ time, by first building a additively-weighted Voronoi diagram on points of $Q_{\square_z}$ (each point $q \in Q_{\square_z}$ is assigned a weight equal to $dist[q]$), and then using it to find the additively-weighted nearest neighbor q_v for each point $v \in Q_{\square_z}$.

If $\square$ is not $\square_z$, a useful property is that $\square$ and $\square_z$ are separated by an axis-parallel line. The WX algorithm implements UPDATE($Q_{\square_z}$, $Q_{\square}$) with the following three steps. Let $U = Q_{\square_z}$ and $V = Q_{\square}$.

1. Sort points of U as $\{u_1, u_2, ..., u_{|U|}\}$ such that $dist[u_1] \leq dist[u_2] \leq ... \leq dist[u_{|U|}]$.
2. Compute $|U|$ disjoint subsets $\{V_1, ..., V_{|U|}\}$ with $V_i = \{v \in V \mid v \in \odot_{u_i}$ and $v \notin \odot_{u_j}$ for all $1 \leq j < i\}$. Equivalently, for each point $v \in V$, v is in V_{i_v}, where i_v is the smallest index i (if exists) such that $\odot_{u_i}$ contains v.

3. Initialize $U' = \emptyset$. Proceed with $|U|$ iterations for $i = |U|, |U| - 1, ..., 1$ sequentially and do the following in each iteration for i: (1) Add u_i to U'; (2) for each point $v \in V_i$, compute $q_v = \arg\min_{u \in U'}\{dist[u] + \|u - v\|\}$; (3) update $dist[v] = \min\{dist[v], dist[q_v] + \|q_v - v\|\}$.

By the definition of V_i, $U \cap \bigodot_v \subseteq U' = \{u_{|U|}, u_{|U|-1}, ..., u_i\}$ for each $v \in V_i$ in the iteration for i of Step 3. Wang and Xue [19] proved that q_v found for each $v \in V_i$ in Step 3 must lie in $\bigodot_v$. They gave a method to implement Step 2 in $O(k \log k)$ time by making use of the property that U and V are separated by an axis-parallel line, where $k = |U| + |V|$. Step 3 can be considered as an offline insertion-only additively-weighted nearest neighbor searching problem and the WX algorithm solves the problem in $O(k \log^2 k)$ time using the standard logarithmic method [3], with $k = |U| + |V|$.

As such, the second updates for all iterations in the WX algorithm takes $O(n \log^2 n)$ time in total [19], which dominates the entire algorithm (other parts of the algorithm together takes $O(n \log n)$ time).

2.2 The RSP Algorithm

We now tackle the RSP problem, i.e., computing r^* for two points $s, t \in P$ and a value λ, by "parameterizing" the WX algorithm.

Recall that the decision problem is to decide whether $r^* \leq r$ for a given r. Notice that $r^* \leq r$ holds if and only if $d_r(s, t) \leq \lambda$. The decision problem can be solved in $O(n \log^2 n)$ time by running the WX algorithm on r. In the following, we refer to the WX algorithm as the *decision algorithm*. We say that r is a *feasible value* if $r^* \leq r$ and an *infeasible value* otherwise.

As discussed in Sect. 1, to find r^*, we run the decision algorithm with a parameter r in an interval $(r_1, r_2]$ by simulating the algorithm on the unknown r^*. The interval always contains r^* but will be shrunk during course of the algorithm (for simplicity, when we say $(r_1, r_2]$ is shrunk, this also include the case that $(r_1, r_2]$ does not change). Initially, we set $r_1 = 0$ and $r_2 = \infty$.

The first step is to build a grid for P. The goal is to shrink $(r_1, r_2]$ so that it contains r^* and if $r^* \neq r_2$ (and thus $r^* \in (r_1, r_2)$), for any $r \in (r_1, r_2)$, the grid $\Psi_r(P)$ has the same combinatorial structure as $\Psi_{r^*}(P)$ in the following sense: (1) Both grids have the same number of rows and columns; (2) for any point $p \in P$, p lies in the i-th row and j-th column of $\Psi_r(P)$ if and only if p lies in the i-th row and j-th column of $\Psi_{r^*}(P)$. This step is also needed in the algorithm of [21] for solving the unweighted case of the RSP problem and an $O(n \log n)$ time algorithm was given in [21] to achieve this by using the sorted matrix searching technique [10,11] along with the linear-time decision algorithm for the unweighted case [6] (more specifically, the decision algorithm is called $O(\log n)$ times). Here in our weighted problem, we can apply exactly the same algorithm except that we use our $O(n \log^2 n)$ time decision algorithm instead and the total time thus becomes $O(n \log^3 n)$.

Let $(r_1, r_2]$ denote the interval after building the grid. We pick any $r \in (r_1, r_2)$ and call the WX algorithm on r to compute a grid $\Psi_r(P)$. Recall from Sect. 2.1 that by "computing $\Psi_r(P)$", we mean to compute the following *grid information*: $P_\square$ for each cell $\square$ of $\Psi_r(P)$ that contains at least one point of P as well as the associated information (e.g., for finding cells of $P_{\boxplus_p}$). These information is the same as that of $\Psi_{r^*}(P)$ if $r^* \neq r_2$. Below, we will simply use $\Psi(P)$ to refer to the grid information computed above, meaning that it does not change with respect to $r \in (r_1, r_2)$.

We use $dist_r[\cdot]$, $Q(r)$, $z(r)$ respectively to refer to $dist[\cdot]$, Q, z in the WX algorithm running on a parameter r. We start with setting $dist_r[s] = 0$, $dist_r[p] = \infty$ for all $p \in P \setminus \{s\}$, and $Q(r) = P$.

Next we enter the main loop. As long as $Q(r) \neq \emptyset$, each iteration finds a point $z(r)$ with the minimum $dist_r$-value from $Q(r)$ and update $dist_r$-values for points in $Q(r)_{\square_{z(r)}} \cup Q(r)_{\boxplus_{z(r)}}$. Points in $Q(r)_{\square_{z(r)}}$ are then removed from $Q(r)$. Each iteration will shrink $(r_1, r_2]$ such that the following algorithm invariant is maintained: $(r_1, r_2]$ contains r^* and if $r^* \neq r_2$, the following holds for all $r \in (r_1, r_2)$: $z(r) = z(r^*)$, $Q(r) = Q(r^*)$, and $dist_r[p] = dist_{r^*}[p]$ for all $p \in P$.

Consider an iteration of the main loop. We assume that the invariant holds before the iteration on the interval $(r_1, r_2]$, which is true before the first iteration. In the following, we describe our algorithm for the iteration and we will show that the invariant holds after the iteration. We assume that $r^* \neq r_2$. According to our invariant, for any $r \in (r_1, r_2)$, we have $z(r) = z(r^*)$, $Q(r) = Q(r^*)$, and $dist_r[p] = dist_{r^*}[p]$ for all $p \in P$.

We first find a point $z(r) \in Q(r)$ with the minimum $dist_r$-value. Since the invariant holds before the iteration, we have $z(r) = \arg\min_{p \in Q(r)} dist_r[p] = \arg\min_{p \in Q(r^*)} dist_{r^*}[p] = z(r^*)$. If ties happen, we follow the same way as the WX algorithm to break ties and ensure $z(r) = z(r^*)$. Hence, no "parameterization" is needed in this step, i.e., all involved values in the computation of this step are independent of r.

Next, we perform the first update $\text{UPDATE}(Q(r)_{\boxplus_{z(r)}}, Q(r)_{\square_{z(r)}})$. This step also does not need parameterization. Indeed, for each point $p \in Q(r)_{\boxplus_{z(r)}}$, we assign $dist_r[p]$ to p as a weight, and then construct the additively-weighted Voronoi diagram on $Q(r)_{\boxplus_{z(r)}}$. For each point $v \in Q(r)_{\square_{z(r)}}$, we use the diagram to find its additively-weighted nearest neighbor $q_v(r) \in Q(r)_{\boxplus_{z(r)}}$ and update $dist_r[v] = \min\{dist_r[v], dist_r[q_v(r)] + \|q_v(r) - v\|\}$. Since $z(r) = z(r^*)$, and $Q(r) = Q(r^*)$, we have $Q(r)_{\boxplus_{z(r)}} = Q(r^*)_{\boxplus_{z(r^*)}}$ and $Q(r)_{\square_{z(r)}} = Q(r^*)_{\square_{z(r^*)}}$. Further, since $dist_r[p] = dist_{r^*}[p]$ for all $p \in P$, for each point $v \in Q(r)_{\square_{z(r)}}$, $q_v(r) = q_v(r^*)$ and each updated $dist_r[v]$ in our algorithm is equal to the corresponding updated $dist_{r^*}[v]$ in the same iteration of the WX algorithm running on r^*. As such, the invariant still holds after the first update.

Implementing the second update $\text{UPDATE}(Q(r)_{\square_{z(r)}}, Q(r)_{\boxplus_{z(r)}})$ is more challenging and parameterization is necessary. It suffices to implement the updates $\text{UPDATE}(Q(r)_{\square_{z(r)}}, Q(r)_{\square})$ for all cells $\square \in \boxplus_{z(r)}$.

If $\square$ is $\square_{z(r)}$, then $Q(r)_{\square_{z(r)}} = Q(r)_{\square}$. In this case, again no parameterization is needed. Since the distance between any two points in $\square_{z(r)}$ is at most r, we can easily implement $\text{UPDATE}(Q(r)_{\square_{z(r)}}, Q(r)_{\square})$ in $O(|Q(r)_{\square_{z(r)}}| \log |Q(r)_{\square_{z(r)}}|)$ time, by first building an additively-weighted Voronoi diagram on points of $Q(r)_{\square_{z(r)}}$ (each point $p \in Q(r)_{\square_{z(r)}}$ is assigned a weight equal to $dist_r[p]$), and then using it to find the additively-weighted nearest neighbor $q_v(r)$ for each point $v \in Q(r)_{\square_z}$. By an analysis similar to the above first update, the invariant still holds.

We now consider the case where $\square$ is not $\square_{z(r)}$. In this case, $\square$ and $\square_{z(r)}$ are separated by an axis-parallel line ℓ. Without loss of generality, we assume that ℓ is horizontal and $\square_{z(r)}$ is below ℓ. Since $z(r) = z(r^*)$ and $Q(r) = Q(r^*)$ for all $r \in (r_1, r_2)$, we let $U = Q(r)_{\square_{z(r)}}$ and $V = Q(r)_{\square}$, meaning that both U and V are independent of $r \in (r_1, r_2)$. Recall that there are three steps in the second update of the decision algorithm. Our algorithm needs to simulate all three steps. As will be seen later, only the second step needs parameterization.

The first step is to sort points in U by their $dist_r$-values. Since $dist_r[p] = dist_{r^*}[p]$ for all $p \in P$, the sorted list $\{u_1, u_2, ..., u_{|U|}\}$ of U obtained in our algorithm is the same as that obtained in the decision algorithm running on r^*.

For any r, denote by $\bigodot_p(r)$ the disk centered at a point p with radius r.

The second step is to compute $|U|$ disjoint subsets $\{V_1(r), V_2(r), ..., V_{|U|}(r)\}$ of V such that $V_i(r) = \{v \mid i_v(r) = i, v \in V\}$, where $i_v(r)$ is the smallest index such that $\bigodot_{u_{i_v(r)}}(r)$ contains point v. This step needs parameterization. We will shrink the interval $(r_1, r_2]$ so that it still contains r^* and if $r^* \neq r_2$, then for any $r \in (r_1, r_2)$, $V_i(r) = V_i(r^*)$ holds for all $1 \leq i \leq |U|$ (it suffices to ensure $i_v(r) = i_v(r^*)$ for all $v \in V$). Our algorithm relies on the following observation, which is based on the definition of $i_v(r)$.

Observation 1. *For any point $v \in V$, if $\bigodot_{u_j}(r)$ contains v with $1 \leq j \leq |U|$, then $i_v(r) \leq j$.*

For a subset $P' \subseteq P$, let $\mathcal{F}_r(P')$ denote the union of the disks centered at points of P' with radius r. We first solve a subproblem in the following lemma.

Lemma 1. *Suppose $(r_1, r_2]$ contains r^* such that if $r^* \neq r_2$, then for all $r \in (r_1, r_2)$, $dist_r[p] = dist_{r^*}[p]$ for all points $p \in P$. For a subset $U' \subseteq U$ and a subset $V' \subseteq V$, in $O(n \log^2 n \cdot \log(|U'| + |V'|))$ time we can shrink $(r_1, r_2]$ so that it still contains r^* and if $r^* \neq r_2$, then for all $r \in (r_1, r_2)$, for any $v \in V'$, v is contained in $\mathcal{F}_r(U')$ if and only if v is contained in $\mathcal{F}_{r^*}(U')$.*

Recall that we have an interval $(r_1, r_2]$. Our goal is to shrink it so that it still contains r^* and if $r^* \neq r_2$, then for any $r \in (r_1, r_2)$, $V_i(r) = V_i(r^*)$ holds for all $1 \leq i \leq |U|$. With Observation 1 and Lemma 1, we have the following lemma.

Lemma 2. *We can shrink the interval $(r_1, r_2]$ in $O(n \log^4 n)$ time so that it still contains r^* and if $r^* \neq r_2$, then for any $r \in (r_1, r_2)$, $V_i(r) = V_i(r^*)$ holds for all $1 \leq i \leq |U|$.*

Proof. To have $V_i(r) = V_i(r^*)$ for all $1 \leq i \leq |U|$, it suffices to ensure $i_v(r) = i_v(r^*)$ for all points $v \in V$. Let $M = |U|$ and $N = |V|$. Note that $M \leq n$ and $N \leq n$.

As defined in the proof of Lemma 1, for any subset $U' \subseteq U$ and any r, denote by $\mathcal{U}_r(U')$ the upper envelope of the portions of $\bigodot_u(r)$ above ℓ for all $u \in U'$.

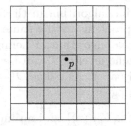

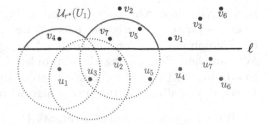

Fig. 1. The red cell that contains the point p is $\Box_p$ and the square area bounded by blue segments is the patch $\boxplus_p$. All adjacent vertices of p in $G_r(P)$ must lie in the grey region. (Color figure online)

Fig. 2. Illustrating U_1 and V_1, where $U_1 = \{u_1, u_2, u_3\}$ and $V_1 = \{v_4, v_5, v_7\}$. The solid arcs are on $\mathcal{U}_{r^*}(U_1)$.

In light of Observation 1, we use the divide and conquer approach. Recall that $U = \{u_1, u_2, \ldots, u_M\}$. Consider the following subproblem on (U, V): shrink $(r_1, r_2]$ so that it still contains r^* and if $r^* \neq r_2$, then for any $r \in (r_1, r_2)$, for any $v \in V$, v is below $\mathcal{U}_r(U_1)$ if and only if v is below $\mathcal{U}_{r^*}(U_1)$, where U_1 is the first half of U, i.e., $U_1 = \{u_1, u_2, \ldots, u_{\lfloor \frac{M}{2} \rfloor}\}$. The subproblem can be solved in $O(n \log^3 n)$ time by applying Lemma 1. Next, we pick any $r \in (r_1, r_2)$ and compute $\mathcal{U}_r(U_1)$ and find the subset V_1 of the points of V that are below $\mathcal{U}_r(U_1)$ (e.g., see Fig. 2). By Observation 1, for each point $v \in V$, $i_v(r) \leq \lfloor \frac{M}{2} \rfloor$ if $v \in V_1$ and $i_v(r) > \lfloor \frac{M}{2} \rfloor$ otherwise. By the above property of $(r_1, r_2]$, for each point $v \in V$, we also have $i_v(r^*) \leq \lfloor \frac{M}{2} \rfloor$ if $v \in V_1$ and $i_v(r^*) > \lfloor \frac{M}{2} \rfloor$ otherwise.

Next, we solve two subproblems recursively: one on (U_1, V_1) and the other on $(U \setminus U_1, V \setminus V_1)$. Both subproblems use $(r_1, r_2]$ as their "input intervals" and solving each subproblem will produce a shrunk "output interval" $(r_1, r_2]$. Consider a subproblem on (U', V') with $U' \subseteq U$ and $V' \subseteq V$. If $|U'| = 1$, then we solve the problem "directly" (i.e., this is the base case) as follows. Assume that $r^* \neq r_2$ and let r be any value in (r_1, r_2). Let u_j be the only point of U'. If $j < M$, according to our algorithm and based on Observation 1, $i_v(r) = i_v(r^*) = j$ holds for all points $v \in V'$. If $j = M$, however, for each point $v \in V'$, it is possible that v is not contained in $\bigodot_u(r^*)$ for any point $u \in U$, in which case v is not below $\mathcal{U}_{r^*}(U)$ and thus is not below $\mathcal{U}_{r^*}(U')$. On the other hand, if v is below $\mathcal{U}_{r^*}(U')$, then $i_v(r^*) = M$. To solve the problem, we can simply apply Lemma 1 on U' and V', after which we obtain an interval $(r_1, r_2]$. Then, we pick any $r \in (r_1, r_2)$ and for any $v \in V'$ with v contained in $\bigodot_{u_M}(r)$, $i_v(r) = i_v(r^*) = M$ holds if $r^* \neq r_2$.

The above divide-and-conquer algorithm can be viewed as a binary tree structure T in which each node represents a subproblem. Clearly, the height of T is $O(\log M)$ and T has $\Theta(M)$ nodes. If we solve each subproblem individually by Lemma 1 as described above, then the algorithm would take $\Omega(Mn)$ time because there are $\Omega(M)$ subproblems and solving each subproblem by Lemma 1 takes $\Omega(n)$ time, which would result in an $\Omega(n^2)$ time algorithm in the worst case. To reduce the runtime, instead, we solve subproblems at the same level of T simultaneously (or "in parallel") by applying the algorithm of Lemma 1. We can show that solving all subproblems in the same level of T can be done in $O(n \log^3 n)$ time. The details are given in our full paper. As T has $O(\log M)$ levels, the total time of the overall algorithm is $O(n \log^4 n)$. □

With Lemma 2, we obtain subsets $\{V_1(r), V_2(r), ..., V_{|U|}(r)\}$ and an interval $(r_1, r_2]$ containing r^* such that if $r^* \neq r_2$, for any $r \in (r_1, r_2)$, $V_i(r) = V_i(r^*)$ holds for all $1 \leq i \leq |U|$. Note that neither the array $dist_r[\cdot]$ nor $Q(r)$ is modified during the algorithm of Lemma 2. Hence, if $r^* \neq r_2$, for all $r \in (r_1, r_2]$, we still have $Q(r) = Q(r^*)$ and $dist_r[p] = dist_{r^*}[p]$ for all points $p \in P$. Thus, our algorithm invariant still holds. This finishes the second step of the second update.

The third step of the second update is to solve the offline insertion-only additively-weighted nearest neighbor searching problem. This step does not need parameterization. Similar to the first update, we pick any $r \in (r_1, r_2)$ and apply the WX algorithm directly. Indeed, the algorithm on r^* only relies on the following information: U and its sorted list by $dist_{r^*}[\cdot]$ values and the subsets $V_1(r^*), \ldots, V_{|U|}(r^*)$. Recall that if $r^* \neq r_2$, then for all $r \in (r_1, r_2)$, $dist_r[p] = dist_{r^*}[p]$ for all $p \in P$, and $V_i(r) = V_i(r^*)$ for all $1 \leq i \leq |U|$. As such, if we pick any $r \in (r_1, r_2)$ and apply the WX algorithm directly, $dist_r[v] = dist_{r^*}[v]$ holds for all points $v \in V$ after this step. Therefore, as in the WX algorithm, this step can be done in $O(k \log^2 k)$ time, where $k = |U| + |V|$.

This finishes the second update of the algorithm. As discussed above, the algorithm invariant holds for the interval $(r_1, r_2]$.

The final step of the iteration is to remove points in $Q(r)_{\square_{z(r)}}$ from $Q(r)$. Since if $r^* \neq r_2$, for all $r \in (r_1, r_2)$, $Q(r) = Q(r^*)$, $z(r) = z(r^*)$, and $Q(r)_{\square_{z(r)}} = Q(r^*)_{\square_{z(r^*)}}$, $Q(r) = Q(r^*)$ still holds after this point removal operation. Therefore, our algorithm invariant holds after the iteration.

In summary, each iteration of our algorithm takes $O(n \log^4 n)$ time. If the point t is contained in $\square_{z(r)}$ (i.e., t is reached) in the current iteration, then we terminate the algorithm. The following lemma shows that we can simply return r_2 as r^*.

Lemma 3. *Suppose that t is contained in $\square_{z(r)}$ in an iteration of our algorithm and $(r_1, r_2]$ is the interval after the iteration. Then $r^* = r_2$.*

The algorithm may take $\Omega(n^2)$ time because t may be reached in $\Omega(n)$ iterations. A further improvement is discussed in the next subsection.

2.3   A Further Improvement

To further reduce the runtime of the algorithm, we borrow a technique from [21] to partition the cells of the grid into large and small cells.

As before, we first compute the grid information $\Psi(P)$ and obtain an interval $(r_1, r_2]$. Let C denote the set of all non-empty cells of $\Psi(P)$ (i.e., cells that contain at least one point of P). For each cell $C \in \mathcal{C}$, let $N(C)$ denote the set of non-empty neighboring cells of C in $\mathcal{C}$ and $P(C)$ the set of points of P contained in cell C. We have $|N(C)| = O(1)$ and $|\mathcal{C}| = O(n)$. A cell C of $\mathcal{C}$ is a *large cell* if it contains at least $n^{3/4} \log^{3/2} n$ points of P, i.e., $|P(C)| \geq n^{3/4} \log^{3/2} n$, and a *small cell* otherwise. Clearly, $\mathcal{C}$ has at most $n^{1/4} / \log^{3/2} n$ large cells. For all pairs of non-empty neighboring cells (C, C'), with $C \in \mathcal{C}$ and $C' \in N(C)$, (C, C') is a *small-cell pair* if both C and C' are small cells, and a *large-cell pair* otherwise, i.e., at least one cell is a large cell. Since $N(C) = O(1)$ for each cell $C \in \mathcal{C}$, there are $O(n^{1/4} / \log^{3/2} n)$ large-cell pairs.

We first provide some intuition about our approach and then fresh out the details. Notice that in each iteration of the main loop in our previous algorithm, only the second step of the second update parameterizes the WX algorithm (i.e., the decision algorithm is called on certain critical values); in that step, we need to process $O(1)$ pairs of cells (C, C') with $C \in \mathcal{C}$ and $C' \in N(C)$. No matter how many points of P contained in the two cells, we need $O(n \log^4 n)$ time to perform the parametric search due to Lemma 2. To reduce the time, we preprocess all small-cell pairs so that the algorithm only needs to perform the parametric search for large-cell pairs. Since there are only $O(n^{1/4} / \log^{3/2} n)$ large-cell pairs, the total time we spend on parametric search can be reduced to $O(n^{5/4} \log^{5/2} n)$. For those small-cell pairs, the preprocessing provides sufficient information to allow us to simply run the original WX algorithm without resorting to parametric search. Specifically, before we enter the main loop of the algorithm (and after the grid information $\Psi(P)$ is computed, along with an interval $(r_1, r_2]$), we preprocess all small-cell pairs using the following lemma which is similar to [21].

Lemma 4. *In $O(n^{5/4} \log^{5/2} n)$ time we can shrink the interval $(r_1, r_2]$ so that it still contains r^* and if $r^* \neq r_2$, then for any $r \in (r_1, r_2)$, for any small-cell pair (C, C') with $C \in \mathcal{C}$ and $C' \in N(C)$, an edge connects a point $p \in P(C)$ and a point $p' \in P(C')$ in $G_r(P)$ if and only if an edge connects p and p' in $G_{r^*}(P)$.*

Let $(r_1, r_2]$ denote the interval obtained after the preprocessing for all small-cell pairs in Lemma 4. Lemma 4 essentially guarantees that if $r^* \neq r_2$, then for any $r \in (r_1, r_2)$, the adjacency relation of points in any small-cell pair in $G_r(P)$ is the same as that in $G_{r^*}(P)$. Note that if $(r_1, r_2]$ is shrunk so that it still contains r^*, then the above property still holds for the shrunk interval. Based on this property, combining with our previous algorithm, we have the following theorem.

Theorem 1. *The reverse shortest path problem for unit-disk graphs in the L_2 weighted case can be solved in $O(n^{5/4} \log^{5/2} n)$ time.*

References

1. Agarwal, P., Efrat, A., Sharir, M.: Vertical decomposition of shallow levels in 3-dimensional arrangements and its applications. SIAM J. Comput. **29**, 912–953 (1999)

2. Avraham, R.B., Filtser, O., Kaplan, H., Katz, M.J., Sharir, M.: The discrete and semicontinuous Fréchet distance with shortcuts via approximate distance counting and selection. ACM Trans. Algorithms **11**(4), 1–29 (2015). Article No. 29

3. Bentley, J.: Decomposable searching problems. Inf. Process. Lett. **8**, 244–251 (1979)

4. Burton, D., Toint, P.: On an instance of the inverse shortest paths problem. Math. Program. **53**, 45–61 (1992)

5. Cabello, S., Jejčič, M.: Shortest paths in intersection graphs of unit disks. Comput. Geom. Theory Appl. **48**(4), 360–367 (2015)

6. Chan, T., Skrepetos, D.: All-pairs shortest paths in unit-disk graphs in slightly subquadratic time. In: Proceedings of the 27th International Symposium on Algorithms and Computation (ISAAC), pp. 24:1–24:13 (2016)

7. Chan, T., Skrepetos, D.: Approximate shortest paths and distance oracles in weighted unit-disk graphs. In: Proceedings of the 34th International Symposium on Computational Geometry (SoCG), pp. 24:1–24:13 (2018)

8. Edelsbrunner, H., Guibas, L., Stolfi, J.: Optimal point location in a monotone subdivision. SIAM J. Comput. **15**(2), 317–340 (1986)

9. Fortune, S.: A sweepline algorithm for Voronoi diagrams. Algorithmica **2**, 153–174 (1987). https://doi.org/10.1007/BF01840357

10. Frederickson, G., Johnson, D.: Generalized selection and ranking: sorted matrices. SIAM J. Comput. **13**(1), 14–30 (1984)

11. Frederickson, G., Johnson, D.: Finding kth paths and p-centers by generating and searching good data structures. J. Algorithms **4**(1), 61–80 (1983)

12. Gao, J., Zhang, L.: Well-separated pair decomposition for the unit-disk graph metric and its applications. SIAM J. Comput. **35**(1), 151–169 (2005)

13. Kaplan, H., Mulzer, W., Roditty, L., Seiferth, P., Sharir, M.: Dynamic planar Voronoi diagrams for general distance functions and their algorithmic applications. In: Proceedings of the 28th Annual ACM-SIAM Symposium on Discrete Algorithms (SODA), pp. 2495–2504 (2017)

14. Katz, M., Sharir, M.: An expander-based approach to geometric optimization. SIAM J. Comput. **26**(5), 1384–1408 (1997)

15. Katz, M.J., Sharir, M.: Efficient algorithms for optimization problems involving distances in a point set. arXiv:2111.02052 (2021)

16. Kirkpatrick, D.: Optimal search in planar subdivisions. SIAM J. Comput. **12**(1), 28–35 (1983)

17. Roditty, L., Segal, M.: On bounded leg shortest paths problems. Algorithmica **59**(4), 583–600 (2011)

18. Sarnak, N., Tarjan, R.: Planar point location using persistent search trees. Commun. ACM **29**, 669–679 (1986)

19. Wang, H., Xue, J.: Near-optimal algorithms for shortest paths in weighted unit-disk graphs. Discret. Comput. Geom. **64**, 1141–1166 (2020)

20. Wang, H., Zhao, Y.: An optimal algorithm for L_1 shortest paths in unit-disk graphs. In: Proceedings of the 33rd Canadian Conference on Computational Geometry (CCCG), pp. 211–218 (2021)

21. Wang, H., Zhao, Y.: Reverse shortest path problem for unit-disk graphs. In: Proceedings of the 17th International Symposium of Algorithms and Data Structures (WADS), pp. 655–668 (2021). https://arxiv.org/abs/2104.14476

22. Zhang, J., Lin, Y.: Computation of the reverse shortest-path problem. J. Global Optim. **25**(3), 243–261 (2003)

Computational Complexity

Happy Set Problem on Subclasses of Co-comparability Graphs

Hiroshi Eto[1], Takehiro Ito[1], Eiji Miyano[2], Akira Suzuki[1],
and Yuma Tamura[1(✉)]

[1] Graduate School of Information Sciences, Tohoku University, Sendai, Japan
{hiroshi.eto.b4,takehiro,akira,tamura}@tohoku.ac.jp
[2] School of Computer Science and Systems Engineering, Kyushu Institute
of Technology, Iizuka, Japan
miyano@ai.kyutech.ac.jp

Abstract. In this paper, we investigate the complexity of the MAXIMUM HAPPY SET problem on subclasses of co-comparability graphs. For a graph G and its vertex subset S, a vertex $v \in S$ is happy if all v's neighbors in G are contained in S. Given a graph G and a non-negative integer k, MAXIMUM HAPPY SET is the problem of finding a vertex subset S of G such that $|S| = k$ and the number of happy vertices in S is maximized. In this paper, we first show that MAXIMUM HAPPY SET is NP-hard even for co-bipartite graphs. We then give an algorithm for n-vertex interval graphs whose running time is $O(k^2 n^2)$; this improves the best known running time $O(kn^8)$ for interval graphs. We also design an algorithm for n-vertex permutation graphs whose running time is $O(k^3 n^2)$. These two algorithmic results provide a nice contrast to the fact that MAXIMUM HAPPY SET remains NP-hard for chordal graphs, comparability graphs, and co-comparability graphs.

1 Introduction

Easley and Kleinberg [7] said that *homophily* is one of the most basic notions governing the structure of social networks. Homophily is the principle that we are likely to associate with people who are similar in characteristics, such as their ages, their occupations and their interests. Motivated from homophily of social networks, Zhang and Li [11] formulated two graph coloring problems, and recently Asahiro et al. [1] introduced another formulation on graphs. In this paper, we study the latter formulation, defined as follows.

For a graph $G = (V, E)$ and a subset $S \subseteq V$, a vertex $v \in S$ is *happy* if all its neighbors in G are contained in S. Given an undirected graph $G = (V, E)$ and a non-negative integer k, MAXIMUM HAPPY SET is the problem of finding a subset $S \subseteq V$ such that $|S| = k$ and the number of happy vertices in S is maximized. For example, the set $S = \{v_1, v_2, v_6, v_7\}$ in Fig. 1(b) is an optimal solution to MAXIMUM HAPPY SET for the graph G in Fig. 1(a) and $k = 4$, where only two vertices v_1 and v_7 are happy.

© Springer Nature Switzerland AG 2022
P. Mutzel et al. (Eds.): WALCOM 2022, LNCS 13174, pp. 149–160, 2022.
https://doi.org/10.1007/978-3-030-96731-4_13

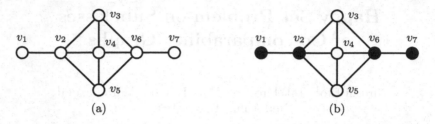

Fig. 1. (a) A graph G, and (b) an optimal solution $S = \{v_1, v_2, v_6, v_7\}$ for G and $k = 4$, where only v_1 and v_7 are happy vertices.

1.1 Known Results

Although MAXIMUM HAPPY SET was proposed recently,[1] it has been already studied from various viewpoints such as polynomial-time solvability, approximability, and fixed-parameter tractability.

Polynomial-time solvability: MAXIMUM HAPPY SET is NP-hard even for bipartite graphs [2], cubic graphs [2], and split graphs [1]. On the other hand, the problem is solvable in $O(k^2 n)$ time for block graphs [2], and solvable in $O(kn^8)$ time for interval graphs [2], where n is the number of vertices in a graph.

Approximability: MAXIMUM HAPPY SET admits a polynomial-time approximation algorithm whose approximation ratio depends on the maximum degree of a graph [2].

Fixed-parameter tractability: MAXIMUM HAPPY SET is W [1]-hard when parameterized by k even on split graphs [1], and hence it is very unlikely that the problem admits a fixed-parameter algorithm even when restricted to split graphs and parameterized by k. On the other hand, the problem admits fixed-parameter algorithms when parameterized by graph structural parameters such as tree-width, clique-width, neighborhood diversity, and twin-cover number of a graph [1].

1.2 Our Contributions

In this paper, we further investigate the polynomial-time solvability of MAXIMUM HAPPY SET, by focusing on subclasses of co-comparability graphs. In particular, we consider co-bipartite graphs, interval graphs, and permutation graphs. See the relationship of graph classes illustrated in Fig. 2.

We first show that MAXIMUM HAPPY SET is NP-hard even for co-bipartite graphs. As far as we know, this is the first intractability result of MAXIMUM HAPPY SET on subclasses of co-comparability graphs. We thus need to focus on other subclasses of co-comparability graphs, in order to seek polynomial-time solvable cases, as below.

---

[1] We note that the graph coloring problem introduced by Zhang and Li [11] is called a similar name, MAXIMUM HAPPY VERTICES, but it is a different problem from ours.

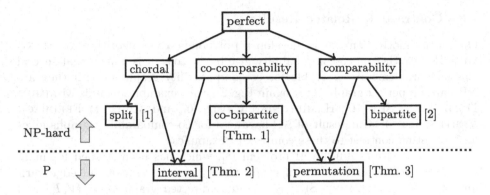

Fig. 2. Our results together with known ones for MAXIMUM HAPPY SET on subclasses of perfect graphs. Each arrow represents the inclusion relationship between graph classes: $A \rightarrow B$ means that the graph class B is a subclass of the graph class A.

We then give a polynomial-time algorithm for interval graphs. Recall that the polynomial-time solvability for interval graphs is already known [2]. However, our algorithm runs in $O(k^2 n^2)$ time for n-vertex interval graphs, which improves the best known running time $O(k n^8)$ [2].

We finally give an algorithm for n-vertex permutation graphs which runs in $O(k^3 n^2)$ time. This is a new polynomial-time solvable case, and gives a nice contrast to the known fact that MAXIMUM HAPPY SET is NP-hard for comparability graphs. (See also Fig. 2.) We note that if k is a constant, then both algorithms for interval graphs and permutation graphs run in $O(n^2)$ time. Proofs for the claims marked with (∗) are omitted from this extended abstract.

Technical highlight: Both our polynomial-time algorithms for interval graphs and permutation graphs employ basically the same technique, that is, a dynamic programming approach based on graph representation models. Details and formal definitions will be given later, but we here explain the key point. Given an n-vertex graph $G = (V, E)$, we define a subgraph $G_i = (V_i, E_i)$ for each integer $i = 1, 2, \ldots, n$, depending on a representation model for G. Then, we wish to compute a partial solution $S_i = S^* \cap V_i$ for each G_i, where S^* is an optimal solution of G. Note that S_i is not always optimal for G_i, and hence it is not enough to compute an optimal solution of G_i. The key of our algorithms is that partial solutions S_i of G_i can be characterized by only two vertices that are *not* contained in S^*, when G is an interval graph or a permutation graph. In this paper, we will prove only the correctness for interval graphs in Lemma 2 due to page limitation. This efficient characterization of partial solutions leads to improving the running time for interval graphs.

1.3   Contrasts to Related Results

Our initial motivation was to develop a polynomial-time algorithm for MAXIMUM HAPPY SET on co-comparability graphs, because it is known that several classical problems are tractable for co-comparability graphs even if they are NP-hard on perfect graphs. (See again Fig. 2.) Such examples include MINIMUM DOMINATING SET [10], HAMILTONIAN CYCLE [6], and MINIMUM FEEDBACK VERTEX SET [4]. Our result of NP-hardness for co-comparability graphs gives an interesting contrast to these complexity examples.

The DENSEST k-SUBGRAPH problem [8], which has been studied for more than two decades in the field of graph theory, can be seen as an edge variant of MAXIMUM HAPPY SET: given an undirected graph $G = (V, E)$ and a non-negative integer k, the task of the problem is to find a vertex subset $S \subseteq V$ of size exactly k such that the number of edges whose both endpoints are contained in S is maximized. Interestingly, the complexity of DENSEST k-SUBGRAPH remains open for interval graphs, permutation graphs, and planar graphs. Although results on MAXIMUM HAPPY SET cannot be converted directly to DENSEST k-SUBGRAPH, our complexity results in this paper may give new insights to DENSEST k-SUBGRAPH.

2   Preliminaries

Let $G = (V, E)$ be a graph; we denote by $V(G)$ and $E(G)$ the vertex set and the edge set of G, respectively. We assume that all graphs in this paper are simple, undirected, and unweighted. For a vertex v of G, we denote by $N_G(v)$ and $N_G[v]$ the open and closed neighborhood of v in G, respectively, that is, $N_G(v) = \{w \in V(G) : vw \in E(G)\}$ and $N_G[v] = N_G(v) \cup \{v\}$. For a vertex subset $V' \subseteq V$, we denote by $G - V'$ the subgraph of G obtained by deleting all the vertices in V' and their incident edges. We shall often write $G - v$ instead of $G - \{v\}$ for a vertex $v \in V$.

For a graph $G = (V, E)$ and its vertex subset $S \subseteq V$, we say that a vertex $v \in V$ is *happy with respect to S on G* if $N_G[v] \subseteq S$; otherwise v is *unhappy with respect to S on G*. We denote by $H(G; S)$ the set of happy vertices with respect to S on G. We note that $H(G; \emptyset) = \emptyset$. Given a graph $G = (V, E)$ and a non-negative integer k, MAXIMUM HAPPY SET is the problem of finding a vertex subset $S \subseteq V$ such that $|S| = k$ and the size of $H(G; S)$ is maximized. For simplicity, our algorithms in this paper only compute the maximum value of $|H(G; S)|$. However, one can easily modify the algorithms so that they find an actual subset S in the same time complexity.

3   NP-hardness for Co-bipartite Graphs

A graph is *co-bipartite* if it is the complement of a bipartite graph. In other words, a co-bipartite graph is a graph whose vertex set can be partitioned into two cliques. In this section, we give the following hardness result.

Theorem 1. MAXIMUM HAPPY SET *is NP-hard for co-bipartite graphs.*

Proof. We give a polynomial-time reduction from MAXIMUM HAPPY SET on general graphs to MAXIMUM HAPPY SET on co-bipartite graphs. Let (G, k) be an instance of MAXIMUM HAPPY SET on general graphs. We construct an instance (G', k') of MAXIMUM HAPPY SET on co-bipartite graphs from (G, k) as follows. Let $n = |V(G)|$ and $V(G) = \{v_1, v_2, \ldots, v_n\}$. The graph G' consists of disjoint two cliques whose vertex subsets are $A = \{a_1, a_2, \ldots, a_n\}$ and $B = \{b_1, b_2, \ldots, b_{n+k+1}\}$. In addition, G' has an edge $a_i b_j$ if and only if $i = j$ or $v_i v_j \in E(G)$ for each $i, j \in \{1, 2, \ldots, n\}$. This completes the construction of G'. Finally, we let $k' = n + k$. Clearly, this reduction can be done in polynomial time.

In the remainder of this proof, we show that for any integer $\ell \geq 0$, there exists a subset $S \subseteq V(G)$ such that $|S| = k$ and $|H(G; S)| \geq \ell$ if and only if there exists a subset $S' \subseteq V(G')$ such that $|S'| = k'$ and $|H(G'; S')| \geq \ell$.

We first prove the necessity. We assume that there exists a subset $S \subseteq V(G)$ such that $|S| = k$ and $|H(G; S)| \geq \ell$. Let $S' = \{a_i \mid i = 1, 2, \ldots, n\} \cup \{b_j \mid v_j \in S\}$. Clearly, $|S'| = n + k = k'$. In order to show that $|H(G'; S')| \geq \ell$, we will show that for any happy vertex $v_i \in H(G; S)$, the corresponding vertex $a_i \in A$ of G' is also happy with respect to S'. Obviously, $N_{G'}[a_i] \cap A \subseteq S'$. Moreover, since v_i is happy with respect to S on G, any vertex $v_j \in N_G[v_i]$ is in S. By the construction of S', we have $b_j \in S'$. Therefore, we have $N_{G'}[a_i] \subseteq S'$, that is, $a_i \in H(G'; S')$. This implies that $|H(G'; S')| \geq |H(G; S)| \geq \ell$.

Conversely, we prove the sufficiency. We assume that there exists a subset $S' \subseteq V(G')$ such that $|S'| = k' = n + k$ and $|H(G'; S')| \geq \ell$. Since $|S'| = n + k$ and B forms a clique of $n + k + 1$ vertices, at least one vertex $b_j \in B$ is not in S' and hence every vertex in B is unhappy with respect to S'. This means that any happy vertex with respect to S' on G' is in A. In the remainder of this proof, we assume that $A \subseteq S'$; otherwise $|H(G'; S')| = 0$. In addition, if there exists a vertex $b_j \in S'$ for an integer j with $n+1 \leq j \leq n+k+1$, then we remove b_j from S' and add a vertex $b_{j'} \notin S'$ into S' for an integer j' with $0 \leq j' \leq n$. We denote by S'' a set obtained by repeating this operation until $S'' \cap \{b_{n+1}, \ldots, b_{n+k+1}\} = \emptyset$. Clearly, $S'' \cap B \subseteq \{b_1, b_2, \ldots, b_n\}$. Moreover, since no vertex in A is adjacent to b_j for an integer j with $n + 1 \leq j \leq n + k + 1$ from the construction of G', $H(G'; S') \subseteq H(G'; S'')$ holds. Let $S = \{v_j \mid j = 1, 2, \ldots, n$ and $b_j \in S''\}$. Note that, since S'' contains all vertices in A, S'' contains exactly k vertices in B and hence $|S| = k$. We now show that for any happy vertex $a_i \in H(G'; S'')$, the corresponding vertex $v_i \in V(G)$ is also happy with respect to S. Since a_i is happy, any vertex $b_j \in N_{G'}[a_i] \cap B$ is in S'' and hence $v_j \in S$. This implies that $N_G[v_i] \subseteq S$. Therefore, $|H(G; S)| \geq |H(G'; S'')| \geq |H(G'; S')| \geq \ell$. This completes the proof. □

4 Polynomial-Time Algorithm for Interval Graphs

A graph $G = (V, E)$ with vertices $v_1, v_2, \ldots, v_n$ is called an *interval graph* if, for some family $\mathcal{I} = \{I_1, I_2, \ldots, I_n\}$ of intervals on the real line, there is a one-to-one correspondence between V and $\mathcal{I}$ such that $v_i v_j \in E$ if and only if I_i intersects I_j for each $i, j \in \{1, 2, \ldots, n\}$. Such a family $\mathcal{I}$ of intervals is called an *interval representation* of G. In this section, we give a polynomial-time algorithm for MAXIMUM HAPPY SET on interval graphs.

Theorem 2. *Given an n-vertex interval graph G and a non-negative integer k, MAXIMUM HAPPY SET is solvable in $O(k^2 n^2)$ time.*

Before the detailed description of our algorithm, we give the following simple but useful lemma.

Lemma 1. (∗). *Let $G = (V, E)$ be a graph and let V', S be subsets of V. Then, it holds that $H(G; S) \setminus V' \subseteq H(G - V'; S \setminus V')$. Moreover, if $V' \subseteq S$, then it holds that $H(G; S) \setminus V' = H(G - V'; S \setminus V')$.*

Notice that Lemma 1 is applicable to general graphs as well as interval graphs.

To explain our algorithm, we need several assumptions and notations. Given an interval graph G, an interval representation $\mathcal{I}$ of G can be constructed in linear time [3,5,9]. Therefore, we may assume without loss of generality that an interval graph G and its interval representation $\mathcal{I}$ are both given. In the remainder of this section, we do not distinguish between vertices of G and intervals of $\mathcal{I}$, that is, we regard I_i as not only an interval of $\mathcal{I}$ but also a vertex of G. We denote by $\mathsf{left}(I_i)$ and $\mathsf{right}(I_i)$ the left endpoint and the right endpoint of an interval $I_i \in \mathcal{I}$, respectively. It is easy to see that an interval representation $\mathcal{I}$ can be transformed into another one without changing G so that distinct integers between 1 and $2n$ are assigned to the endpoints $\mathsf{left}(I_i)$ and $\mathsf{right}(I_i)$ of every interval I_i. Moreover, we assume that intervals of $\mathcal{I}$ are sorted in increasing order of the right endpoints, that is, $\mathsf{right}(I_i) < \mathsf{right}(I_j)$ for any integers i, j such that $1 \le i < j \le n$. We then add dummy intervals I_0 and I_{n+1} with $\mathsf{left}(I_0) = -1, \mathsf{right}(I_0) = 0, \mathsf{left}(I_{n+1}) = 2n + 1$ and $\mathsf{right}(I_{n+1}) = 2n + 2$ into $\mathcal{I}$. Note that, the dummy intervals I_0 and I_{n+1} correspond to the isolated vertices of G. The addition of I_0 and I_{n+1} is not essential for proving Theorem 2, but this simplifies the description of our algorithm. In the remainder of this section, we assume that G has I_0 and I_{n+1}. Let G_i be the subgraph of G induced by a vertex set $\mathcal{I}_i = \{I_0, I_1, \ldots, I_i\}$. We also define $\mathcal{I}_i^+ = N_{G_i}[I_i]$ and $\mathcal{I}_i^- = \mathcal{I}_i \setminus \mathcal{I}_i^+$.

We describe the idea of our algorithm. Let S^* be a subset of $V(G) \setminus \{I_0, I_{n+1}\}$ such that S^* maximizes $|H(G; S^*)|$ among all subsets of $V(G) \setminus \{I_0, I_{n+1}\}$ of size k. Since I_0 and I_{n+1} are the isolated vertices on G, S^* is also the optimal solution of the original graph that has no dummy vertices I_0 and I_{n+1}. In order to find S^*, we wish to compute a partial solution $S_i = S^* \cap V(G_i)$ for each $i = 0, 1, \ldots, n+1$ by means of dynamic programming. Since $G = G_{n+1}$, we have $S^* = S_{n+1}$. Notice that a partial solution S_i is not always optimal for G_i, because a happy vertex $I_{i'} \in \mathcal{I}_i$ with respect to S_i on G_i may be unhappy with respect to S^* on G. This implies that it is not enough to find only an optimal solution of G_i. To correctly compute S_i, we guess integers r, u, k' that satisfy the following three conditions for S^*:

- the interval I_r has the smallest left endpoint among all intervals in $V(G) \setminus (V(G_i) \cup S^*)$;

- the interval I_u has the largest right endpoint among all intervals in $V(G_i)\setminus S^*$, that is, $I_u \notin S^*$ and $I_{i'} \in S^*$ for every i' with $u < i' \leq i$; and
- $|S_i| = k'$.

We say that a quadruple (i, r, u, k') is *compatible* with S^* if i, r, u, k' satisfy the above three conditions. Clearly, if (i, r, u, k') is compatible with S^*, then $0 \leq u \leq i < r \leq n + 1$ holds. For integers i and r, we denote by $G_{i,r}$ the subgraph of G induced by $V(G_i) \cup \{I_r\}$. We then obtain the following lemma.

Lemma 2. *Let S^* be a subset of $V(G) \setminus \{I_0, I_{n+1}\}$ such that S^* maximizes $|H(G; S^*)|$ among all subsets of $V(G) \setminus \{I_0, I_{n+1}\}$ of size k, and let $S_i = S^* \cap V(G_i)$ for an integer i with $0 \leq i \leq n$. For integers r, u, k' with $0 \leq u \leq i < r \leq n + 1$ and $k' \leq k$, suppose that a quadruple (i, r, u, k') is compatible with S^*. Then, S_i maximizes $|H(G_{i,r}; S_i)|$ among all subsets $S_i \subseteq V(G_{i,r}) \setminus \{I_0, I_r, I_u\}$ of size k' such that $I_{i'} \in S_i$ for every i' with $u < i' \leq i$.*

Proof. Assume for a contradiction that there exists a subset $S_i' \subseteq V(G_{i,r}) \setminus \{I_0, I_r, I_u\}$ such that $|H(G_{i,r}; S_i)| < |H(G_{i,r}; S_i')|$, $|S'| = k'$ and $I_{i'} \in S_i'$ for every i' with $u < i' \leq i$. We let $S^* = (S^* \setminus S_i) \cup S_i'$ and show that the following three inequalities: (I) $|H(G; S^*) \cap V(G_{i,r})| \leq |H(G_{i,r}; S_i)|$; (II) $|H(G_{i,r}; S_i')| \leq |H(G; S^*) \cap V(G_{i,r})|$; and (III) $|H(G; S^*) \setminus V(G_{i,r})| \leq |H(G; S^*) \setminus V(G_{i,r})|$. Since $|H(G_{i,r}; S_i)| < |H(G_{i,r}; S_i')|$, we have $|H(G; S^*)|$ from the inequalities (I)–(III) as follows:

$$
\begin{aligned}
|H(G; S^*)| &= |H(G; S^*) \cap V(G_{i,r})| + |H(G; S^*) \setminus V(G_{i,r})| \\
&\leq |H(G_{i,r}; S_i)| + |H(G; S^*) \setminus V(G_{i,r})| \\
&< |H(G_{i,r}; S_i')| + |H(G; S^*) \setminus V(G_{i,r})| \\
&\leq |H(G; S^*) \cap V(G_{i,r})| + |H(G; S^*) \setminus V(G_{i,r})| \\
&= |H(G; S^*)|.
\end{aligned}
$$

This contradicts the maximality of $|H(G; S^*)|$.

We first show the inequality (I). By setting $V' = V(G) \setminus V(G_{i,r})$ and $S = S^*$ on Lemma 1, we have

$$
\begin{aligned}
H(G; S^*) \cap V(G_{i,r}) &= H(G; S^*) \setminus V' \\
&\subseteq H(G - V'; S^* \setminus V') \\
&= H(G_{i,r}; S^* \cap V(G_{i,r})).
\end{aligned}
$$

Since $S_i = S^* \cap V(G_i)$ and $I_r \notin S^*$, we have $|H(G; S^*) \cap V(G_{i,r})| \leq |H(G_{i,r}; S^* \cap V(G_{i,r}))| = |H(G_{i,r}; S_i)|$.

We next show the inequality (II). Suppose that $I_\ell \in H(G_{i,r}; S_i')$ for an integer ℓ with $0 \leq \ell \leq i$. Then, $I_\ell \in S_i'$ holds and I_ℓ is adjacent to none of vertices in $V(G_{i,r}) \setminus S_i' = V(G_{i,r}) \setminus (S^* \cap V(G_i))$. In particular, since $\ell \leq i < r$ holds and intervals of $\mathcal{I}$ are sorted in increasing order of the right endpoints, we have $\text{right}(I_\ell) < \text{left}(I_r)$. For any integer ℓ' such that $i < \ell' \leq n + 1$ and $I_{\ell'} \notin S^*$, we also have $\text{left}(I_r) \leq \text{left}(I_{\ell'})$ from the definition of I_r. Thus, $\text{right}(I_\ell) < \text{left}(I_{\ell'})$

holds and hence I_ℓ is also not adjacent to $I_{\ell'}$. As a conclusion, I_ℓ is adjacent to none of vertices in $V(G) \setminus S^\star$, that is, $I_\ell \in H(G; S^\star) \cap V(G_{i,r})$. This means that $|H(G_{i,r}; S_i')| \leq |H(G; S^\star) \cap V(G_{i,r})|$.

Finally, we show the inequality (III). Suppose that $I_\ell \in H(G; S^\star)$ for an integer ℓ with $i < \ell \leq n + 1$. Then, $I_\ell \in S^\star$ holds and I_ℓ is adjacent to none of vertices in $V(G) \setminus S^\star$. In particular, I_ℓ is not adjacent to I_u and hence $\mathrm{right}(I_u) < \mathrm{left}(I_\ell)$. Recall that $S^\star = (S^\star \setminus S_i) \cup S_i'$ and $S_i, S_i' \subseteq V(G_i)$. We thus have $S^\star \setminus V(G_i) = S^\star \setminus V(G_i)$. This means that $I_\ell \in S^\star$ and I_ℓ is adjacent to none of vertices in $V(G) \setminus (V(G_i) \cup S^\star)$. Moreover, for any integer ℓ' such that $I_{\ell'} \in V(G_i) \setminus S^\star$, it follows from the definitions of I_u that $\mathrm{right}(I_{\ell'}) \leq \mathrm{right}(I_u)$ because $I_u \notin S_i'$ and hence $I_u \notin S^\star$. Combined with $\mathrm{right}(I_u) < \mathrm{left}(I_\ell)$, we have $\mathrm{right}(I_{\ell'}) < \mathrm{left}(I_\ell)$, that is, I_ℓ is not adjacent to $I_{\ell'}$ because $\ell' \leq i < \ell$. Therefore, I_ℓ is adjacent to none of vertices in $V(G) \setminus S^\star$ and thus $I_\ell \in H(G; S^\star)$. Since $i < \ell \leq n + 1$, we conclude that $|H(G; S^\star) \setminus V(G_{i,r})| \leq |H(G; S^\star) \setminus V(G_{i,r})|$. This completes the proof of Lemma 2. □

Lemma 2 suggests that, for the sake of computing S_i for each i, it suffices to guess integers r, u, k' and compute S maximizes $|H(G_{i,r}; S)|$ among all subsets $S \subseteq V(G_{i,r}) \setminus \{I_0, I_r, I_u\}$ of size k' such that $I_{i'} \in S$ for every i' with $u < i' \leq i$. In fact, to compute the size of $S^\star$, our algorithm uses the following two main functions $f_{\mathrm{in}}(G_{i,r}; k')$, $f_{\mathrm{out}}(G_{i,r}; k')$ and the subfunction $f_{\mathrm{in}}'(G_{i,r}; j, k')$, where i, r, j, k' are integers such that $0 \leq i < r \leq n + 1$, $0 \leq j \leq \min\{k', i\} - 1$ and $0 \leq k' \leq k$;

- $f_{\mathrm{in}}(G_{i,r}; k')$ returns the maximum of $|H(G_{i,r}; S)|$ among all subsets $S \subseteq V(G_{i,r}) \setminus \{I_0, I_r\}$ such that $I_i \in S$ and $|S| = k'$;
- $f_{\mathrm{out}}(G_{i,r}; k')$ returns the maximum of $|H(G_{i,r}; S)|$ among all subsets $S \subseteq V(G_{i,r}) \setminus \{I_0, I_i, I_r\}$ such that $|S| = k'$; and
- $f_{\mathrm{in}}'(G_{i,r}; j, k')$ returns the maximum of $|H(G_{i,r}; S)|$ among all subsets $S \subseteq V(G_{i,r}) \setminus \{I_0, I_{i-j-1}, I_r\}$ such that $\{I_i \ldots, I_{i-j}\} \subseteq S$ and $|S| = k'$.

We let $f_{\mathrm{in}}(G_{i,r}; k') = -\infty$, $f_{\mathrm{out}}(G_{i,r}; k') = -\infty$ and $f_{\mathrm{in}}'(G_{i,r}; j, k') = -\infty$ if there exists no subset S that satisfies all the prescribed conditions for f_{in}, f_{out} and f_{in}', respectively. We remark that $f_{\mathrm{out}}(G_{i,r}; k')$ corresponds to the case where $u = i$ and $f_{\mathrm{in}}'(G_{i,r}; j, k')$ corresponds to the case where $u = i - j - 1$ on Lemma 2. The main function $f_{\mathrm{in}}(G_{i,r}; k')$ is used to improve the running time of our algorithm. We also remark that the integer j must be less than k' and i because $j \geq k'$ violates $|S| = k'$ and $j \geq i$ violates $I_0 \notin S$. We will compute values $f_{\mathrm{in}}(G_{i,r}; k')$, $f_{\mathrm{out}}(G_{i,r}; k')$ and $f_{\mathrm{in}}'(G_{i,r}; j, k')$ by means of dynamic programming. By taking the maximum of $f_{\mathrm{in}}(G_{n,n+1}; k)$ and $f_{\mathrm{out}}(G_{n,n+1}; k)$, we obtain the maximum size of $H(G; S)$ such that $S \subseteq V(G) \setminus \{I_0, I_{n+1}\}$ and $|S| = k$.

The Computation of $f_{\mathrm{in}}(G_{i,r}; k')$

If $i = 0$, then we have $f_{\mathrm{in}}(G_{i,r}; k') = -\infty$ for any r and k' because there is no subset $S \subseteq V(G_{i,r}) \setminus \{I_0, I_r\}$ such that $I_i \in S$. Similarly, if $k' = 0$, then we have $f_{\mathrm{in}}(G_{i,r}; k') = -\infty$ for any i and r. Suppose that $i > 0$ and $k' > 0$.

We then compute $f_{\text{in}}(G_{i,r}; k')$ from $f''_{\text{in}}(G_{i,r}; j, k')$ under the assumption that $f'_{\text{in}}(G_{i,r}; j, k')$ has already been computed for each j with $0 \le j \le \min\{k', i\} - 1$. Obviously, we have

$$f_{\text{in}}(G_{i,r}; k') = \max_{0 \le j \le \min\{k', i\} - 1} f'_{\text{in}}(G_{i,r}; j, k').$$

We explain how to compute $f'_{\text{in}}(G_{i,r}; j, k')$ for each quadruple (i, r, j, k'). We assume that the main function f_{out} and the subfunction f'_{in} have been already computed in accordance with the lexicographical order of (i, r, j, k'). We consider the two subcases: (I) $j = 0$ and (II) $j > 0$.

Case (I): $j = 0$

In this case, $f'_{\text{in}}(G_{i,r}; j, k')$ returns the maximum of $|H(G_{i,r}; S)|$ such that $S \subseteq V(G_{i,r}) \setminus \{I_0, I_{i-1}, I_r\}$, $I_i \in S$ and $|S| = k'$. From Lemma 1, it holds that $H(G_{i,r}; S) \setminus \{I_i\} = H(G_{i-1,r}; S \setminus \{I_i\})$. We thus compute $f'_{\text{in}}(G_{i,r}; j, k')$ from $f_{\text{out}}(G_{i-1,r}; k' - 1)$ by deciding whether the vertex I_i is happy with respect to S on $G_{i,r}$. Clearly, if I_i is adjacent to the vertex I_{i-1} or I_r, then I_i is unhappy. Conversely, if I_i is adjacent to neither I_{i-1} nor I_r, then I_i is the isolated vertex on $G_{i,r}$ from the assumption that the intervals of $\mathcal{I}$ are sorted in increasing order of the right endpoints. Thus, I_i is happy and we have $f'_{\text{in}}(G_{i,r}; j, k')$ in this case as follows:

$$f'_{\text{in}}(G_{i,r}; j, k') = \begin{cases} f_{\text{out}}(G_{i-1,r}; k' - 1) & \text{if } I_i I_{i-1} \in E(G_{i,r}) \text{ or } I_i I_r \in E(G_{i,r}), \\ f_{\text{out}}(G_{i-1,r}; k' - 1) + 1 & \text{otherwise.} \end{cases}$$

Case (II): $j > 0$

This case means that $f'_{\text{in}}(G_{i,r}; j, k')$ returns the maximum of $|H(G_{i,r}; S)|$ such that $S \subseteq V(G_{i,r}) \setminus \{I_0, I_{i-j-1}, I_r\}$, $\{I_i, \ldots, I_{i-j}\} \subseteq S$ and $|S| = k'$. In particular, $I_{i-1} \in S$ holds. We thus take a value $f'_{\text{in}}(G_{i-1,r}; j - 1, k' - 1)$ to compute $f'_{\text{in}}(G_{i,r}; j, k')$. We then determine whether the vertex I_i is happy with respect to S on $G_{i,r}$. If I_i is adjacent to I_{i-j-1} or I_r on $G_{i,r}$, then I_i is unhappy because $I_{i-j-1}, I_r \notin S$. If I_i is adjacent to neither I_{i-j-1} nor I_r on $G_{i,r}$, then $I_{i-j-1} \in \mathcal{I}_i^-$. This implies that $\mathcal{I}_i^+ \subseteq \{I_i, \ldots, I_{i-j}\} \subseteq S$ and hence I_i is happy. Therefore, it suffices to check whether I_i is adjacent to I_{i-j-1} or I_r on $G_{i,r}$, and we have $f'_{\text{in}}(G_{i,r}; j, k')$ in this case as follows:

$$f'_{\text{in}}(G_{i,r}; j, k') = \begin{cases} f'_{\text{in}}(G_{i-1,r}; j - 1, k' - 1) & \text{if } I_i I_{i-j-1} \in E(G_{i,r}) \text{ or } I_i I_r \in E(G_{i,r}), \\ f'_{\text{in}}(G_{i-1,r}; j - 1, k' - 1) + 1 & \text{otherwise.} \end{cases}$$

The Computation of $f_{\text{out}}(G_{i,r}; k')$

Let S be a subset of $V(G_{i,r}) \setminus \{I_0, I_i, I_r\}$ such that S maximizes $|H(G_{i,r}; S)|$ among all subsets $S \subseteq V(G_{i,r}) \setminus \{I_0, I_i, I_r\}$ with $|S| = k'$. Then, all vertices in $\mathcal{I}_i^+$ and I_r are unhappy with respect to S on $G_{i,r}$. However, some vertices in $\mathcal{I}_i^+ \setminus \{I_i\}$ may be contained in S because they can be used to make vertices in $\mathcal{I}_i^-$ happy. We thus consider which vertices in $\mathcal{I}_i^+ \setminus \{I_i\}$ are contained in S. In the naive way, we enumerate all subsets of $\mathcal{I}_i^+ \setminus \{I_i\}$ of size at most k'; it takes superpolynomial time in general. The following lemma provides us that the number of subsets of $\mathcal{I}_i^+ \setminus \{I_i\}$ to be enumerated is at most k'.

Lemma 3. *Let $G_{i,r}$ be an interval graph and suppose that there exist intervals $I_x, I_y \in \mathcal{I}_i^+ \setminus \{I_i\}$ such that $\mathsf{left}(I_x) < \mathsf{left}(I_y)$ for integers x, y. Then, for any subset $S \subseteq V(G_{i,r})$ such that $I_i, I_x, I_r \notin S$ and $I_y \in S$, it holds that $H(G_{i,r}; S) \subseteq H(G_{i,r}; S \cup \{I_x\} \setminus \{I_y\})$.*

Proof. Let I_z be an interval of $\mathcal{I}_i$ such that $I_z \in H(G_{i,r}; S)$. To prove Lemma 3, it suffices to show that $I_z \in H(G_{i,r}; S \cup \{I_x\} \setminus \{I_y\})$. Clearly, I_z intersects no interval in $\mathcal{I}_i \cup \{I_r\} \setminus S$. Thus, $I_z \in \mathcal{I}_i^-$, and $\mathsf{right}(I_z) < \mathsf{left}(I_{z'})$ holds for any interval $I_{z'} \in \mathcal{I}_i^+ \cup \{I_r\} \setminus S$ from the assumption that the intervals of $\mathcal{I}_i \cup \{I_r\}$ are sorted in increasing order of the right endpoints. In particular, we have $\mathsf{right}(I_z) < \mathsf{left}(I_x)$ and hence $\mathsf{right}(I_z) < \mathsf{left}(I_y)$ from the assumption in Lemma 3. This implies that the vertex I_z is adjacent to none of the vertex I_y and vertices in $V(G_i) \setminus S$, that is, $I_z \in H(G_{i,r}; S \cup \{I_x\} \setminus \{I_y\})$. □

Suppose that $|S \cap \mathcal{I}_i^+| = p$ for an integer p. We note that p is not greater than k' and $|\mathcal{I}_i^+| - 1$ because $p > k'$ violates $|S| = k'$ and $p > |\mathcal{I}_i^+| - 1$ violates $I_i \notin S$. We denote by $\mathcal{I}_i^p$ the set produced by picking the first p intervals in increasing order of the left endpoints of intervals in $\mathcal{I}_i^+ \setminus \{I_i\}$, that is, $\mathsf{left}(I_x) < \mathsf{left}(I_y)$ for any $I_x \in \mathcal{I}_i^p$ and any $I_y \in \mathcal{I}_i^+ \setminus (\mathcal{I}_i^p \cup \{I_i\})$. By applying Lemma 3 to S iteratively, we can obtain a subset $S' \subseteq V(G_{i,r}) \setminus \{I_0, I_i, I_r\}$ with $|S'| = k'$ such that $S' \cap \mathcal{I}_i^+ = \mathcal{I}_i^p$ and $H(G_{i,r}; S) \subseteq H(G_{i,r}; S')$. From the maximality of $|H(G_{i,r}; S)|$, S' also maximizes $|H(G_{i,r}; S')|$ among all subsets of $V(G_{i,r}) \setminus \{I_0, I_i, I_r\}$ of size k'. Thus, without enumerating all subsets of $\mathcal{I}_i^+ \setminus \{I_i\}$ of size at most k', it suffices to guess exactly p vertices in $\mathcal{I}_i^+ \setminus \{I_i\}$ are contained in S and assume that $S \cap \mathcal{I}_i^+ = \mathcal{I}_i^p$.

We next give another lemma that plays a central role in the computation of $f_{\text{out}}(G_{i,r}; k')$. Let i' be an integer such that the interval $I_{i'}$ has the largest right endpoint among all intervals in $\mathcal{I}_i^-$.

Lemma 4. *Let S be a subset of $V(G_{i,r}) \setminus \{I_i, I_r\}$ and let $S' = S \cap \mathcal{I}_i^+$. In addition, let r' be an integer such that the interval $I_{r'}$ has the smallest left endpoint among all intervals in $\mathcal{I}_i^+ \cup \{I_r\} \setminus S'$. Then, $H(G_{i,r}; S) = H(G_{i',r'}; S \setminus S')$.*

Proof. We first show that $H(G_{i,r}; S) \subseteq H(G_{i',r'}; S \setminus S')$. We set $V' = (\mathcal{I}_i^+ \cup \{I_r\}) \setminus \{I_{r'}\}$. Since $I_i, I_r \notin S$, we have $H(G_{i,r}; S) \cap V' = \emptyset$ and hence $H(G_{i,r}; S) \setminus V' = H(G_{i,r}; S)$. We also note that $V(G_{i,r}) \setminus V' = V(G_{i',r'})$ and $S \setminus V' = S \setminus S'$ because $S' \subseteq \mathcal{I}_i^+$ and $I_{r'} \notin S$. From Lemma 1, we then have

$$H(G_{i,r}; S) = H(G_{i,r}; S) \setminus V' \subseteq H(G_{i,r} - V'; S \setminus V') = H(G_{i',r'}; S \setminus S').$$

We next show that $H(G_{i',r'}; S \setminus S') \subseteq H(G_{i,r}; S)$. Since the vertex $I_{r'}$ of $G_{i',r'}$ is not contained in $S \setminus S'$, we have $\mathsf{right}(I_x) < \mathsf{left}(I_{r'})$ for any integer

x such that $I_x \in H(G_{i',r'}; S \setminus S')$. In addition, for any integer y such that $I_y \in \mathcal{I}_i^+ \cup \{I_r\} \setminus S'$, $\text{left}(I_{r'}) \leq \text{left}(I_y)$ holds from the definition of $I_{r'}$. We thus have $\text{right}(I_x) < \text{left}(I_y)$, that is, I_x does not intersect I_y. This implies that $I_x \in H(G_{i,r}; S)$ if $I_x \in H(G_{i',r'}; S \setminus S')$, and hence $H(G_{i',r'}; S \setminus S') \subseteq H(G_{i,r}; S)$. This completes the proof. $\qquad\square$

We have prepared for computing $f_{\text{out}}(G_{i,r}; k')$ for a triple (i, r, k') of integers such that $0 \leq i < r \leq n+1$ and $0 \leq k' \leq k$. If $i = 0$, the graph $G_{i,r}$ consists of the two isolated vertices I_0 and I_r. Only $S = \emptyset$ satisfies the prescribed conditions for $f_{\text{out}}(G_{i,r}; k')$. Thus, for any integer $r > 0$, we have $f_{\text{out}}(G_{i,r}; k') = 0$ if $k' = 0$; otherwise $f_{\text{out}}(G_{i,r}; k') = -\infty$.

Suppose that $i > 0$. Let S be a subset of $V(G_{i,r}) \setminus \{I_0, I_i, I_r\}$ such that S maximizes $|H(G_{i,r}; S)|$ among all subsets $S \subseteq V(G_{i,r}) \setminus \{I_0, I_i, I_r\}$ with $|S| = k'$. As mentioned above, if $|S \cap \mathcal{I}_i^+| = p$ for an integer p with $0 \leq p \leq \min\{k', |\mathcal{I}_i^+| - 1\}$, we can assume that $S \cap \mathcal{I}_i^+ = \mathcal{I}_i^p$ from Lemma 3. Let r' be an integer such that the interval $I_{r'}$ has the smallest left endpoint among intervals in $\mathcal{I}_i^+ \cup \{I_r\} \setminus \mathcal{I}_i^p$. For an optimal solution S^* of G, if the quadruple (i, r, i, k') is compatible with S^*, then the quadruple $(i', r', u, k' - p)$ is also compatible with S^* for some integer u with $0 \leq u \leq i'$. Therefore, by setting $S' = \mathcal{I}_i^p$ on Lemma 4, we compute $f_{\text{out}}(G_{i,r}; k')$ as follows:

$$f_{\text{out}}(G_{i,r}; k') = \max_{0 \leq p \leq \min\{k', |\mathcal{I}_i^+| - 1\}} \{f_{\text{in}}(G_{i',r'}; k' - p), f_{\text{out}}(G_{i',r'}; k' - p)\}.$$

The total running time of our algorithm is $O(k^2 n^2)$, as claimed in Theorem 2. The details are omitted due to page limitation.

5 Polynomial-Time Algorithm for Permutation Graphs

Let $\pi = (\pi(1), \pi(2), \ldots, \pi(n))$ be a permutation of the integers from 1 to n. We denote by $\pi^{-1}(i)$ the position of the number i in π. A graph $G = (V, E)$ with vertices $v_1, v_2, \ldots, v_n$ is called a *permutation graph* if there exists a permutation π between 1 and n such that for any two integers i and j, $v_i v_j \in E$ if and only if $(i - j)(\pi^{-1}(i) - \pi^{-1}(j)) < 0$. We give a polynomial-time algorithm for MAXIMUM HAPPY SET on permutation graphs by the same algorithmic approach as interval graphs.

Theorem 3 (*). *Given an n-vertex permutation graph G and a non-negative integer k, MAXIMUM HAPPY SET is solvable in $O(k^3 n^2)$ time.*

6 Conclusion

In this paper, we studied the complexity of MAXIMUM HAPPY SET on subclasses of co-comparability graphs; co-bipartite graphs, interval graphs and permutation graphs. Especially, our algorithm for interval graphs improved the best known running time $O(kn^8)$. Our polynomial-time algorithms employ basically the same technique. We believe that the technique is applicable to MAXIMUM HAPPY SET on other graph classes.

The complexity of MAXIMUM HAPPY SET has been studied for various graph classes. However, the (in)tractability of MAXIMUM HAPPY SET on planar graphs remains open. We note that the complexity of the edge variant of MAXIMUM HAPPY SET is also unknown for planar graphs.

Acknowledgments. This work is partially supported by JSPS KAKENHI Grant Numbers JP18H04091, JP19K11814, JP20H05793, JP20H05794, JP20K11666, JP21K 11755 and JP21K21302, Japan.

References

1. Asahiro, Y., Eto, H., Hanaka, T., Lin, G., Miyano, E., Terabaru, I.: Parameterized algorithms for the happy set problem. In: Rahman, M.S., Sadakane, K., Sung, W.-K. (eds.) WALCOM: Algorithms and Computation, pp. 323–328. Springer, Cham (2020)
2. Asahiro, Y., Eto, H., Hanaka, T., Lin, G., Miyano, E., Terabaru, I.: Complexity and approximability of the happy set problem. Theoret. Comput. Sci. **866**, 123–144 (2021)
3. Booth, K.S., Lueker, G.S.: Testing for the consecutive ones property, interval graphs, and graph planarity using PQ-tree algorithms. J. Comput. Syst. Sci. **13**(3), 335–379 (1976)
4. Coorg, S.R., Rangan, C.P.: Feedback vertex set on cocomparability graphs. Networks **26**(2), 101–111 (1995)
5. Corneil, D.G., Olariu, S., Stewart, L.: The LBFS structure and recognition of interval graphs. SIAM J. Discret. Math. **23**(4), 1905–1953 (2010)
6. Deogun, J.S., Steiner, G.: Polynomial algorithms for Hamiltonian cycle in cocomparability graphs. SIAM J. Comput. **23**(3), 520–552 (1994)
7. Easley, D., Kleinberg, J.: Networks, Crowds, and Markets: Reasoning About a Highly Connected World. Cambridge University Press, Cambridge (2010)
8. Feige, U., Kortsarz, G., Peleg, D.: The dense k-subgraph problem. Algorithmica **29**(3), 410–421 (2001)
9. Hsu, W.-L., Ma, T.-H.: Fast and simple algorithms for recognizing chordal comparability graphs and interval graphs. SIAM J. Comput. **28**(3), 1004–1020 (1998)
10. Kratsch, D., Stewart, L.: Domination on cocomparability graphs. SIAM J. Discret. Math. **6**(3), 400–417 (1993)
11. Zhang, P., Li, A.: Algorithmic aspects of homophyly of networks. Theoret. Comput. Sci. **593**, 117–131 (2015)

Finding Geometric Representations
of Apex Graphs is NP-Hard

Dibyayan Chakraborty[1](✉) and Kshitij Gajjar[2]

[1] Univ Lyon, LABEX, ENS de Lyon, Université Claude Bernard Lyon 1,
LIP UMR5668, Lyon, France
`dibyayan.chakraborty@ens-lyon.fr`
[2] National University of Singapore, Singapore, Singapore

Abstract. Planar graphs can be represented as intersection graphs of different types of geometric objects in the plane, *e.g.*, circles (Koebe, 1936), line segments (Chalopin & Gonçalves, SODA 2009), L-shapes (Gonçalves *et al.*, SODA 2018). For general graphs, however, even deciding whether such representations exist is often NP-hard. We consider apex graphs, *i.e.*, graphs that can be made planar by removing one vertex from them. We show, somewhat surprisingly, that deciding whether geometric representations exist for apex graphs is NP-hard.

More precisely, we show that for every positive integer g, recognizing every graph class $\mathscr{G}$ such that PURE-2-DIR $\subseteq \mathscr{G} \subseteq$ 1-STRING is NP-hard, even if the inputs are apex graphs of girth at least g. Here, PURE-2-DIR is the class of intersection graphs of axis-parallel line segments (where intersections are allowed only between horizontal and vertical segments), and 1-STRING is the class of intersection graphs of simple curves (where two curves cross at most once) in the plane. This partially answers an open question raised by Kratochvíl & Pergel (COCOON, 2007).

Most known NP-hardness reductions for these problems are from variants of 3-SAT. We reduce from the PLANAR HAMILTONIAN PATH COMPLETION problem, which uses the more intuitive notion of planarity. As a result, our proof is much simpler and encapsulates several classes of geometric graphs.

Keywords: Hamiltonian path · planar graph · apex graph · NP-hard · recognition problem · geometric intersection graph · VLSI design · 1-STRING · PURE-2-DIR

1 Introduction

The recognition a graph class is the decision problem of determining whether a given simple, undirected, unweighted graph belongs to the graph class.

D. Chakraborty—This work was done when the author was a postdoctoral fellow at the Indian Institute of Science, Bangalore.

K. Gajjar—This project has received funding from the European Union's Horizon 2020 research and innovation programme under grant agreement No. 682203-ERC-[Inf-Speed-Tradeoff].

P. Mutzel et al. (Eds.): WALCOM 2022, LNCS 13174, pp. 161–174, 2022.
https://doi.org/10.1007/978-3-030-96731-4_14

Recognition of graph classes is a fundamental problem in graph theory with a wide range of applications. In particular, when the graph class relates to intersection patterns of geometric objects, the corresponding recognition problem finds usage in disparate areas like VLSI design [13,14,41], map labelling [1], wireless networks [32], and computational biology [48].

The study of graphs that arise out of intersection patterns of geometric objects began with the celebrated circle packing theorem in 1936 [26] (also see [2,45]), which states that all *planar graphs* can be expressed as intersection graphs[1] of touching disks[2]. Since then, there has been a long line of research on finding representations of planar graphs using other types of geometric objects. In his PhD thesis, Scheinerman [40] conjectured that all planar graphs can be expressed as intersection graphs of line segments. After two decades of active research, Scheinerman's conjecture was finally proved in 2009 by Chalopin & Gonçalves [9]. In fact, Schinerman's conjecture has motivated researchers to study representations of planar graphs using many different types of geometric objects, mostly of them culminating in elegant results [10,15,19–22,24].

A similar line of research began when Benzer [4] initiated the study of *string graphs* (intersection graphs of simple curves[3] in the plane) over half a century ago. Benzer's motivation was to study the topology of genetic structures, and the patterns that arise therein. Since then, many fascinating results have been shown for string graphs, surveyed in the 2009 invited talk by János Pach [36] titled, *"Why are string graphs so beautiful?"*

Another application of geometric intersection graphs is in the construction of telecommunication networks, where the range of a broadcasting station is modelled by a circular disk and two stations can communicate if their corresponding disks overlap [17,33,37]. If the underlying network topology is known beforehand and the objective is to determine the existence of a placement of the broadcasting stations realizing the given topology, then the problem is boils down to deciding whether a particular graph can be expressed as an intersection graph of (unit) disks in the plane [7,8,25].

Geometric intersection graphs are also central to the field of VLSI design, where each electronic component on the VLSI circuit-board can be modelled as a geometric object in the plane [37]. The underlying circuitry (*i.e.*, which components are connected to and disjoint from which other components) [13] is usually known beforehand. The challenge lies in placing the components on the circuit-board in a way that respects this circuitry [14]. From a technical standpoint, the circuitry can be modelled as a graph, and the circuit-board can

---

[1] For a set of geometric objects $\mathscr{C}$, its intersection graph, $I(\mathscr{C})$, has $\mathscr{C}$ as the vertex set and two vertices are adjacent if and only if the corresponding geometric objects intersect.

[2] Formally, two closed disks are said to touch each other if they share exactly one point.

[3] Formally, a simple curve is a subset of the plane which is homeomorphic to the interval $[0, 1]$.

be thought of as a plane surface. Again, this is a problem of checking if a graph can be represented by certain types of geometric objects in the plane.

In the 1960s, Sinden [42] asked whether the recognition of string graphs is decidable. Kratochvíl [27] took the first steps towards answering Sinden's question by showing that recognizing string graphs is NP-hard. In 2003, Schaefer, Sedgwick & Štefankovič [39] showed that recognizing string graphs also lies in NP, and is thus NP-complete.

For his NP-hardness proof, Kratochvíl [27] introduced a variant of 3-SAT called PLANAR 3-CONNECTED 3-SAT. In subsequent years, other researchers also used variants of PLANAR 3-CONNECTED 3-SAT to show NP-hardness of recognizing various geometric intersection graph classes [11,12,28,30,31,34]. Unfortunately, the construction of the "variable gadgets" and "clause gadgets" used in these reductions is often quite involved, which ends up making the proofs rather complicated.

In this paper, we simplify and unify these earlier NP-hardness proofs by giving a single proof which holds for several graph classes. Furthermore, when the input graphs are restricted to be planar, most of these graph classes can be recognized in polynomial time. We ask, *what if the input graphs are "almost planar"?* In particular, we study the computational complexity of finding geometric representations of *apex graphs*.

Definition 1 [43,46]. *A graph is an apex graph if it contains a vertex whose removal makes it planar.*

As our main contribution, we show that recognizing various classes of geometric intersection graphs remains NP-hard even when the input graphs are both bipartite and apex (Theorem 1). This is slightly surprising, given the fact that an apex graph is simply a planar graph with one additional vertex.

Our proof technique deviates significantly from that of Kratochvíl [27] and other similar NP-hardness proofs that reduce from PLANAR 3-CONNECTED 3-SAT. We reduce from a different NP-hard problem called PLANAR HAMILTONIAN PATH COMPLETION, which uses the more intuitive notion of planarity, making our proof easier to understand.

Organisation of the Paper: In Sect. 2, we state our main result and its significance. We describe our proof techniques in Sect. 3, and prove our main result in Sect. 4.

2   Main Result

For our main result, we are particularly interested in two natural and well-studied classes of geometric intersection graphs called PURE-2-DIR and 1-STRING.

Definition 2. PURE-2-DIR *is the class of all graphs G, such that G is the intersection graph of axis-parallel line segments in the plane, where intersections are allowed only between horizontal and vertical segments (Fig. 1 (Left)).*

Definition 3. 1-STRING *is the class of all graphs G, such that G is the inter-section graph of simple curves in the plane, where two intersecting curves share exactly one point, at which they must cross each other (Fig. 1 (Right)).*

Theorem 1 (Main Result). *Let g be a fixed positive integer and $\mathscr{G}$ be a graph class such that*

$$\text{PURE-2-DIR} \subseteq \mathscr{G} \subseteq 1\text{-STRING}.$$

Then it is NP-*hard to decide whether an input graph belongs to $\mathscr{G}$, even when the input graphs are restricted to bipartite apex graphs of girth at least g.*

We reduce from the NP-complete PLANAR HAMILTONIAN PATH COMPLETION problem [3], which in turn was inspired by another NP-complete problem known as the PLANAR HAMILTONIAN CYCLE COMPLETION problem [47]. We explain these decision problems and the main ideas behind our reduction in Sect. 3.

2.1 Significance of the Main Result

An unfortunate consequence of our main result is that several known polynomial-time algorithms for planar graphs cannot be extended to even mildly non-planar graphs.

PURE-2-DIR $\mathscr{G}$ 1-STRING

Fig. 1. A visual depiction of Theorem 1. Note that this figure is for representational purposes only.

Let us state this more precisely. STRING is the class of intersection graphs of simple curves in the plane. Kratochvíl & Pergel [31] asked if STRING can be recognized in polynomial time when the inputs are restricted to graphs with large girth. The above question was answered by Mustață & Pergel [34], where they showed that recognizing STRING is NP-hard, even when the inputs are restricted to graphs of arbitrarily large girth. However, the graphs they constructed were far from planar. Since 1-STRING $\subsetneq$ STRING, Theorem 1 partially answers Kratochvíl & Pergel's [31] question for recognizing 1-STRING graphs when the inputs are restricted to apex graphs of arbitrarily large girth.

Chalopin & Gonçalves [9] showed that every planar graph can be represented as an intersection graph of line segments in polynomial time. By putting this graph class as $\mathscr{G}$ in Theorem 1, we obtain that a similar result cannot hold for apex graphs unless P = NP.

Corollary 1. *For every fixed positive integer g, recognizing intersection graphs of line segments is* NP-*hard, even for bipartite apex graphs with girth at least g.*

Gonçalves, Isenmann & Pennarun [21] showed that every planar graph can be represented as an intersection graph of L-shapes in polynomial time. The following corollary shows that a similar result cannot hold for apex graphs unless P = NP.

Corollary 2. *For every fixed positive integer g, recognizing intersection graphs of* L-*shapes is* NP-*hard, even for bipartite apex graphs with girth at least g.*

Our main result also has a connection to a graph invariant called *boxicity*. The boxicity of a graph is the minimum integer d such that the graph can be represented as an intersection graph of d-dimensional axis-parallel boxes. Thomassen showed that the boxicity of every planar graph is at most three [44]. It is easy to check if the boxicity of a planar graph is one [6], but the complexity of determining whether a planar graph has boxicity two or three is not yet known. A result of Hartman, Newman & Ziv [24] states that the class of bipartite graphs with boxicity two is precisely PURE-2-DIR. Combined with our main result, this implies that deciding the boxicity of apex graphs is NP-hard.

Corollary 3. *For every fixed positive integer g, recognizing graphs with boxicity 2 is* NP-*hard, even for bipartite apex graphs with girth at least g.*

A graph is c-apex if it contains a set of c vertices whose removal makes it planar. This is a natural generalization of apex graphs. Our main result implies that no graph class $\mathscr{G}$ satisfying PURE-2-DIR $\subseteq \mathscr{G} \subseteq$ 1-STRING can be recognized in $n^{f(c)}$ time, where f is a computable function depending only on c. This means recognizing $\mathscr{G}$ is XP-hard, and thus not fixed-parameter tractable [16,35] for c-apex graphs when parameterized by c.

Corollary 4. *Let g be a fixed positive integer and $\mathscr{G}$ be a graph class such that*

$$\text{PURE-2-DIR} \subseteq \mathscr{G} \subseteq \text{1-STRING}.$$

Then assuming P $\neq$ NP, *there is no deterministic $f(c) \cdot n^{O(1)}$-time algorithm that recognizes $\mathscr{G}$ (where f is a computable function depending only on c), even for bipartite c-apex graphs with girth at least g.*

Planar graphs are precisely the $(K_5, K_{3,3})$-minor free graphs [38]. Interestingly, the set of forbidden minors for apex graphs is not known, although it is known that the set is finite [23]. It is easy to see that apex graphs are K_6-minor free, implying the following.

Corollary 5. *Let g be a fixed positive integer and $\mathscr{G}$ be a graph class such that*

$$\text{PURE-2-DIR} \subseteq \mathscr{G} \subseteq \text{1-STRING}.$$

Then it is NP-*hard to decide whether an input graph belongs to $\mathscr{G}$, even for bipartite K_6-minor free graphs with girth at least g.*

Finally, using techniques different from ours, Kratochvíl & Matoušek [29] had shown that recognizing PURE-2-DIR is NP-hard, and so is the recognition of line segment intersection graphs. Theorem 1 and Corollary 1 show that these recognition problems remain NP-hard even if the inputs are restricted to bipartite apex graphs of large girth.

3   Proof Techniques

Let us now describe the PLANAR HAMILTONIAN PATH COMPLETION problem [3]. A Hamiltonian path in a graph is a path that visits each vertex of the graph exactly once.

Definition 4. PLANAR HAMILTONIAN PATH COMPLETION *is the following decision problem.*
Input: *A planar graph G.*
Output: *Yes, if G is a subgraph of a planar graph with a Hamiltonian path; no, otherwise.*

Theorem 2 (Auer & Gleißner [3]). PLANAR HAMILTONIAN PATH COMPLETION *is* NP-*hard.*

We will use Theorem 2 to show Theorem 1. Similar to Mustaţă & Pergel [34], we show NP-hardness for graph classes "sandwiched" between two classes of geometric intersection graphs. A more technical formulation of Theorem 1 is as follows.

Theorem 3. *For every planar graph G and positive integer g, there exists a bipartite apex graph G_{apex} of girth at least g which can be obtained in polynomial time from G, satisfying the following properties.*

(a) *If G_{apex} is in* 1-STRING, *then G is a yes-instance of* PLANAR HAMILTONIAN PATH COMPLETION.
(b) *If G is a yes-instance of* PLANAR HAMILTONIAN PATH COMPLETION, *then G_{apex} is in* PURE-2-DIR.

Proof of Theorem 1 assuming Theorem 3. Let $\mathscr{G}$ be a graph class satisfying the condition PURE-2-DIR $\subseteq \mathscr{G} \subseteq$ 1-STRING, and let G be a planar graph. If G is a yes-instance of PLANAR HAMILTONIAN PATH COMPLETION, then using Theorem 3 (b), we obtain that $G_{\text{apex}} \in$ PURE-2-DIR $\subseteq \mathscr{G}$. And if $G_{\text{apex}} \in \mathscr{G} \subseteq$ 1-STRING, then by Theorem 3 (a), G is a yes-instance of PLANAR HAMILTONIAN PATH COMPLETION.

Thus, $G_{\mathsf{apex}} \in \mathscr{G}$ if and only if G is a yes-instance of PLANAR HAMILTONIAN PATH COMPLETION. Since PLANAR HAMILTONIAN PATH COMPLETION is NP-hard (Theorem 2) and G_{apex} can be obtained in polynomial time from G, this implies that deciding whether the bipartite apex graph G_{apex} belongs to $\mathscr{G}$ is NP-hard.   □

Therefore, as Theorem 3 implies our main result (Theorem 1), the rest of this paper is devoted to the proof of Theorem 3.

4   Proof of the Main Result

4.1   Construction of the Apex Graph

We begin our proof of Theorem 3 by describing the construction of G_{apex}. Let G be a planar graph. G_{apex} is constructed in two steps.

$$G \to G_{k\text{-div}} \to G_{\mathsf{apex}}.$$

Let $g \geq 6$ be a positive integer constant, and $k \geq 3$ be the minimum odd integer greater or equal to $g - 3$. Let $G_{k\text{-div}}$ be the full k-subdivision of G, i.e., $G_{k\text{-div}}$ is the graph obtained by replacing each edge of G by a path with $k + 1$ edges. Figure 2 (a) denotes a graph G, and Fig. 2 (b) denotes the full 3-subdivision of G. Formally, we replace each $e = (x, y) \in E(G)$ by the path $(x, u_e^1, u_e^2, u_e^3, \ldots, u_e^k, y)$.

$$V(G_{k\text{-div}}) = V(G) \cup \{u_e^1, u_e^2, u_e^3, \ldots, u_e^k \mid e \in E(G)\};$$
$$E(G_{k\text{-div}}) = \{xu_e^1, u_e^1u_e^2, u_e^2u_e^3, \ldots, u_e^{k-1}u_e^k, u_e^ky \mid e = xy \in E(G)\}.$$

We call the vertices of $V(G) \subseteq V(G_{k\text{-div}})$ as the *original vertices* of $G_{k\text{-div}}$ and the remaining vertices as the *subdivision vertices* of $G_{k\text{-div}}$. Finally, we construct G_{apex} by adding a new vertex a to $G_{k\text{-div}}$ and making it adjacent to all the original vertices of $G_{k\text{-div}}$ (Fig. 2 (c)). Formally, G_{apex} is defined as follows.

$$V(G_{\mathsf{apex}}) = V(G_{k\text{-div}}) \cup \{a\};$$
$$E(G_{\mathsf{apex}}) = E(G_{k\text{-div}}) \cup \{av \mid v \in V(G)\}.$$

Observation A. *If G is planar, then G_{apex} is a bipartite apex graph of girth at least g.*

Proof. G is a planar graph and subdivision does not affect planarity, so $G_{k\text{-div}}$ is also planar, implying that G_{apex} is an apex graph. The vertex set of G_{apex} can be expressed as the disjoint union of two sets A and B, where

$$A = \{x \mid x \in V(G)\} \cup \{u_e^i \mid e \in E(G), i \text{ is even}\};$$
$$B = \{a\} \cup \{u_e^i \mid e \in E(G), i \text{ is odd}\}.$$

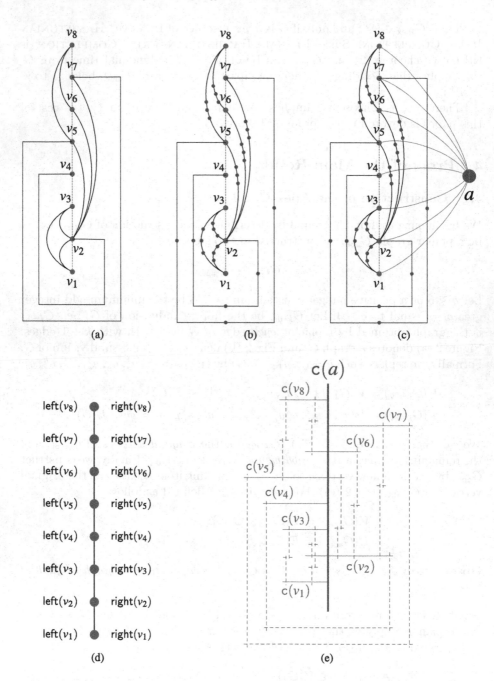

Fig. 2. (a) G, a yes-instance of PLANAR HAMILTONIAN PATH COMPLETION; (b) $G_{k\text{-div}}$ for $k = 3$; (c) G_{apex}; (d) left and right semi-disks representing the vertices of G; (e) C, a PURE-2-DIR representation of G_{apex}. (See Subsect. 4.1 for detailed explanations of (a), (b) and (c).)

Note that A induces an independent set in G_{apex}, and so does B. Thus, G_{apex} is a bipartite apex graph. As for the girth, note that every cycle in G_{apex} contains at least $k+2$ vertices $x, u_e^1, u_e^2, u_e^3, \ldots, u_e^k, y$, for some $e = (x, y) \in E(G)$. At least one more vertex is needed to complete the cycle, implying that the girth of G_{apex} is at least $k + 3 \geq g$. $\qquad\qquad\qquad\qquad\qquad\qquad\qquad\qquad\qquad\qquad\qquad\quad\square$

It is easy to see that this entire construction of G_{apex} from G can be carried out in polynomial time. In Subsect. 4.2, we will prove Theorem 3 (a). In Subsect. 4.3, we will provide a sketch of the proof of Theorem 3 (b).

4.2 Proof of Theorem 3 (a)

In this section, we will show that if G_{apex} is in 1-STRING, then G is a yes-instance of PLANAR HAMILTONIAN PATH COMPLETION. In other words, if G_{apex} has a 1-STRING representation, then G is a subgraph of a planar graph with a Hamiltonian path.

In our proofs, we will demonstrate the planarity of our graphs by embedding them in the plane. Typically, a planar graph is defined as a graph whose vertices are *points* in the plane and edges are *strings* connecting pairs of points such that no two strings intersect (except possibly at their end points). The same definition holds in more generality, *i.e.*, if the vertices are also allowed to be strings (see Fig. 3). Let us now state this formally.

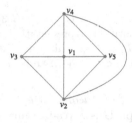

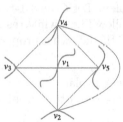

Definition 5 (Planarizable representation of a graph). A graph G on n vertices and m edges is said to admit a planarizable representation if there are two mutually disjoint sets of strings V and E (with $|V|= n$ and $|E|= m$) in the plane such that

- the strings of V correspond to the vertices of G, and those of E correspond to the edges of G;
- no two strings of V intersect;
- no two strings of E intersect, except possibly at their end points;
- apart from its two end points, a string of E does not intersect any string of V;
- for every vertex v and every edge $e = (x, y)$ of G, an end point of the string corresponding to e intersects the string corresponding to v if and only if $v = x$ or $v = y$.

Fig. 3. (Above) A standard representation of a planar graph with $n = 5$ vertices and $m = 9$ edges, where the vertices are points and the edges are strings. (Below) A planarizable representation (Definition 5) of the same graph, where the vertices as well as the edges are strings.

Figure 3 illustrates a planar graph and a planarizable representation of it.

Lemma 4. *A graph admits a planarizable representation if and only if it is planar.*

Lemma 4 may seem obvious. For completeness, we shall provide a formal proof of it in the full version of this paper. We now use this lemma to prove Theorem 3 (a).

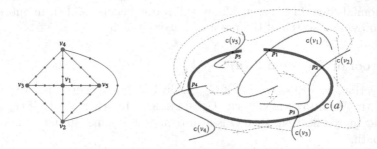

Fig. 4. (Left) $G_{k\text{-div}}$ ($k = 3$) for a planar graph G on $n = 5$ vertices. (Right) C, a planarizable representation of G_{apex}. The thickest string denotes $\mathsf{c}(a)$, the apex vertex of G_{apex}. The bold strings denote the original vertices of G. The thin dashed strings denote $\mathsf{c}(u_e^1)$, $\mathsf{c}(u_e^2)$ and $\mathsf{c}(u_e^3)$.

Proof of Theorem 3(a). Given $G_{\mathsf{apex}} \in 1\text{-}\mathrm{STRING}$, we will show that the planar graph G is a yes-instance of the PLANAR HAMILTONIAN PATH COMPLETION problem. Let C be a $1\text{-}\mathrm{STRING}$ representation of G_{apex} in the plane. It is helpful to follow Fig. 4 while reading this proof. We will use C to construct a graph G_{pl} with the following properties.

(a) G_{pl} is a supergraph of G on the same vertex set as G.
(b) G_{pl} is planar.
(c) G_{pl} has a Hamiltonian path.

Note that (a), (b), (c) together imply that G is a subgraph of a planar graph with a Hamiltonian path (*i.e.*, G is a yes-instance of PLANAR HAMILTONIAN PATH COMPLETION). Let $n = |V(G)|$ and assume that $n \geq 4$. Along with our construction of G_{pl}, we will also describe $\mathrm{DRAW}(G_{\mathsf{pl}})$, a planarizable representation (Definition 5) of G_{pl} in the plane.

In C, consider the strings corresponding to the n original vertices (the large vertices in Fig. 4 (Left)) of G. Since the original vertices form an independent set in G_{apex}, the bold strings are pairwise disjoint. We add these n strings to $\mathrm{DRAW}(G_{\mathsf{pl}})$, which correspond to the n vertices of G_{pl}.

Proof of (c): So far, G_{pl} has no edge. We will now add $n - 1$ edges to G_{pl} to connect these vertices via a Hamiltonian path. Recall that all n original vertices are adjacent to the apex vertex a in G_{apex}, implying that each of the n bold strings intersects $\mathsf{c}(a)$ at exactly one point (as C is a $1\text{-}\mathrm{STRING}$ representation). Starting from one end point of $\mathsf{c}(a)$ and travelling along the curve $\mathsf{c}(a)$ until we reach its other end point, we encounter these n points one-by-one. Let $(v_1, v_2, \ldots, v_n)$ be the order in which they are encountered.

For each $i \in [n]$, let p_i be the point at which $c(v_i)$ intersects $c(a)$. For each $i \in [n-1]$, let s_i be the substring of $c(a)$ between p_i and p_{i+1}. Add the strings $s_1, s_2, \ldots, s_{n-1}$ as edges to $\text{DRAW}(G_{\text{pl}})$, where s_i represents the edge between v_i and v_{i+1}. Thus the edges corresponding to the $n-1$ strings $s_1, s_2, \ldots, s_{n-1}$ constitute a Hamiltonian path $(v_1, v_2, \ldots, v_n)$ in G_{pl}. This shows (c).

Proof of (a): To show (a), we need to add all the edges of G to G_{pl} (other than those already added by the previous step), so that G_{pl} becomes a supergraph of G. For each edge $e = v_i v_j \in E(G)$, there are k strings $c(u_e^1), c(u_e^2), \ldots, c(u_e^k)$ (corresponding to the subdivision vertices $u_e^1, u_e^2, \ldots, u_e^k$ in G_{apex}) in C. Note that for each $t \in \{1, 2, \ldots, k\}$, the string $c(u_e^t)$ intersects exactly two other strings. Let $s(u_e^t)$ be the substring of $c(u_e^t)$ between those two intersection points. Let s_e be the string obtained by concatenating the k substrings thus obtained.

$$s_e \triangleq \bigcup_{t=1}^{k} s(u_e^t). \tag{1}$$

If the edge $e = v_i v_j$ is not already present in G_{pl}, then add the string s_e to $\text{DRAW}(G_{\text{pl}})$, where s_e represents the edge between v_i and v_j (one end point of s_e lies on $c(v_i)$ and the other on $c(v_j)$). This completes the construction of $\text{DRAW}(G_{\text{pl}})$, and shows (a).

Proof of (b): To show (b), it is enough to show that $\text{DRAW}(G_{\text{pl}})$ is a planarizable representation of G_{pl} (Lemma 4). Note that there are three types of strings in $\text{DRAW}(G_{\text{pl}})$: (i) substrings of $c(a)$, (ii) strings of the type s_e, for some $e = v_i v_j \in E(G)$, and (iii) n strings corresponding to the original vertices of G.

Two strings of type (i) are either disjoint or intersect at their end points, since $c(a)$ is non-self-intersecting. More precisely, for each $i \in [n-1]$, the point p_{i+1} (the unique intersection point of s_i and s_{i+1}) lies on $c(v_{i+1})$, which denotes a vertex in $\text{DRAW}(G_{\text{pl}})$. A string of type (ii) intersects exactly two strings, $c(v_i)$ and $c(v_j)$, which denote vertices in $\text{DRAW}(G_{\text{pl}})$. Finally, strings of type (iii) are mutually disjoint. This shows (b).

4.3 Proof of Theorem 3 (b)

Let us elucidate the main idea behind our proof. We are given a plane drawing of G in which its vertices are placed in a collinear fashion on a vertical line, respecting their ordering on the Hamiltonian path (Fig. 2 (a)). Our construction, in essence, modifies this plane drawing to obtain a PURE-2-DIR representation of G_{apex} (Fig. 2 (e)).

The apex segment $c(a)$ is placed on the vertical line. The vertices of the original graph within G_{apex} are represented using horizontal segments (of appropriate length). The edges of G, which were strings in the plane drawing, are now replaced by rectilinear piecewise linear curves where each individual orthogonal segment represents a subdivided vertex of G_{apex}. If we were allowed a large (unbounded) number of rectilinear pieces for each edge, then this construction is trivial, since every curve can be viewed as a series of infinitesimally small vertical and horizontal segments. Our proof formally justifies that this can always be done even when the number of allowed rectilinear pieces is a fixed odd integer greater than or equal to three.

We achieve this through a slightly modified version of a folklore observation concerning *book embeddings* of graphs [5]: if a graph is embedded in a book and a, b, c, d are four vertices on the spine of the book arranged in the order $a < b < c < d$, then (a, c) and (b, d) cannot both be edges on the same page of the book. Also note that Fig. 2 (a) is for representational purposes only. Owing to the observation above, our construction does not rely on the topology of the strings representing the edges of G in Fig. 2 (a). We provide a detailed proof of Theorem 3 (b) in the full version of this paper.

Acknowledgements. This work was done when Kshitij Gajjar was a postdoctoral researcher at Technion, Israel. Kshitij Gajjar's work is partially supported by NUS ODPRT Grant, WBS No. R-252-000-A94-133. Both authors thank the organisers of GRAPHMASTERS 2020 [18] for providing the virtual environment that initiated this research.

References

1. Agarwal, P.K., Van Kreveld, M.J.: Label placement by maximum independent set in rectangles. Comput. Geom. **11**(3–4), 209–218 (1998)
2. Andreev, E.M.: On convex polyhedra in Lobacevskii spaces. Math. USSR-Sbornik **10**(3), 413 (1970)
3. Auer, C., Gleißner, A.: Characterizations of deque and queue graphs. In: Kolman, P., Kratochvíl, J. (eds.) WG 2011. LNCS, vol. 6986, pp. 35–46. Springer, Heidelberg (2011). https://doi.org/10.1007/978-3-642-25870-1_5
4. Benzer, S.: On the topology of the genetic fine structure. Proc. Natl. Acad. Sci. U.S.A. **45**(11), 1607 (1959)
5. Bernhart, F., Kainen, P.C.: The book thickness of a graph. J. Comb. Theory Ser. B **27**(3), 320–331 (1979)
6. Booth, K.S., Lueker, G.S.: Testing for the consecutive ones property, interval graphs, and graph planarity using PQ-tree algorithms. J. Comput. Syst. Sci. **13**(3), 335–379 (1976)
7. Bowen, C., Durocher, S., Löffler, M., Rounds, A., Schulz, A., Tóth, C.D.: Realization of simply connected polygonal linkages and recognition of unit disk contact trees. In: Di Giacomo, E., Lubiw, A. (eds.) GD 2015. LNCS, vol. 9411, pp. 447–459. Springer, Cham (2015). https://doi.org/10.1007/978-3-319-27261-0_37
8. Breu, H., Kirkpatrick, D.G.: Unit disk graph recognition is NP-hard. Comput. Geom. **9**(1–2), 3–24 (1998)
9. Chalopin, J., Gonçalves, D.: Every planar graph is the intersection graph of segments in the plane. In: STOC, pp. 631–638 (2009)
10. Chalopin, J., Gonçalves, D., Ochem, P.: Planar graphs have 1-string representations. Discrete Comput. Geom. **43**(3), 626–647 (2010)
11. Chaplick, S., Jelínek, V., Kratochvíl, J., Vyskočil, T.: Bend-bounded path intersection graphs: sausages, noodles, and waffles on a grill. In: Golumbic, M.C., Stern,

M., Levy, A., Morgenstern, G. (eds.) WG 2012. LNCS, vol. 7551, pp. 274–285. Springer, Heidelberg (2012). https://doi.org/10.1007/978-3-642-34611-8_28

12. Chmel, P.: Algorithmic aspects of intersection representations. Bachelor's thesis (2020)

13. Chung, F., Leighton, F.T., Rosenberg, A.: Diogenes: a methodology for designing fault-tolerant VLSI processor arrays. Department of Electrical Engineering and Computer Science, Massachusetts Institute of Technology, Microsystems Program Office (1983)

14. Chung, F.R.K., Leighton, F.T., Rosenberg, A.L.: A graph layout problem with applications to VLSI design (1984)

15. de Castro, N., Cobos, F.J., Dana, J.C., Márquez, A., Noy, M.: Triangle-free planar graphs as segments intersection graphs. In: Kratochvíyl, J. (ed.) GD 1999. LNCS, vol. 1731, pp. 341–350. Springer, Heidelberg (1999). https://doi.org/10.1007/3-540-46648-7_35

16. Downey, R.G., Fellows, M.R.: Parameterized Complexity. Springer, New York (2012). https://doi.org/10.1007/978-1-4612-0515-9

17. Gao, J., Guibas, L.J., Hershberger, J., Zhang, L., Zhu, A.: Geometric spanners for routing in mobile networks. IEEE J. Sel. Areas Commun. **23**(1), 174–185 (2005)

18. Gasieniec, L., Klasing, R., Radzik, T.: IWOCA 2020, vol. 12126. Springer, Cham (2020)

19. Gonçalves, D.: 3-colorable planar graphs have an intersection segment representation using 3 slopes. In: Sau, I., Thilikos, D.M. (eds.) WG 2019. LNCS, vol. 11789, pp. 351–363. Springer, Cham (2019). https://doi.org/10.1007/978-3-030-30786-8_27

20. Gonçalves, D.: Not all planar graphs are in PURE-4-DIR. J. Graph Algorithms Appl. **24**(3), 293–301 (2020)

21. Gonçalves, D., Isenmann, L., Pennarun, C.: Planar graphs as L-intersection or L-contact graphs. In: SODA, pp. 172–184. SIAM (2018)

22. Gonçalves, D., Lévêque, B., Pinlou, A.: Homothetic triangle representations of planar graphs. J. Graph Algorithms Appl. **23**(4), 745–753 (2019)

23. Gupta, A., Impagliazzo, R.: Computing planar intertwines. In: FOCS, pp. 802–811. Citeseer (1991)

24. Hartman, I.B., Newman, I., Ziv, R.: On grid intersection graphs. Discret. Math. **87**(1), 41–52 (1991)

25. Klemz, B., Nöllenburg, M., Prutkin, R.: Recognizing weighted disk contact graphs. In: Di Giacomo, E., Lubiw, A. (eds.) GD 2015. LNCS, vol. 9411, pp. 433–446. Springer, Cham (2015). https://doi.org/10.1007/978-3-319-27261-0_36

26. Koebe, P.: Kontaktprobleme der konformen Abbildung. Hirzel (1936)

27. Kratochvíl, J.: String graphs. II. Recognizing string graphs is NP-hard. J. Comb. Theory Series B **52**(1), 67–78 (1991)

28. Kratochvíl, J.: A special planar satisfiability problem and a consequence of its NP-completeness. Discret. Appl. Math. **52**(3), 233–252 (1994)

29. Kratochvíl, J., Matoušek, J.: NP-hardness results for intersection graphs. Comment. Math. Univ. Carol. **30**(4), 761–773 (1989)

30. Kratochvíl, J., Matousek, J.: Intersection graphs of segments. J. Comb. Theory Ser. B **62**(2), 289–315 (1994)

31. Kratochvíl, J., Pergel, M.: Geometric intersection graphs: do short cycles help? In: Lin, G. (ed.) COCOON 2007. LNCS, vol. 4598, pp. 118–128. Springer, Heidelberg (2007). https://doi.org/10.1007/978-3-540-73545-8_14

32. Kuhn, F., Wattenhofer, R., Zollinger, A.: Ad hoc networks beyond unit disk graphs. Wireless Netw. **14**(5), 715–729 (2008)

33. Li, X.-Y.: Algorithmic, geometric and graphs issues in wireless networks. Wirel. Commun. Mob. Comput. **3**(2), 119–140 (2003)
34. Mustaţă, I., Pergel, M.: On unit grid intersection graphs and several other intersection graph classes. Acta Math. Univ. Comenian. **88**(3), 967–972 (2019)
35. Niedermeier, R.: Invitation to Fixed-Parameter Algorithms. OUP, Oxford (2006)
36. Pach, J.: Why are string graphs so beautiful? In: Eppstein, D., Gansner, E.R. (eds.) GD 2009. LNCS, vol. 5849, pp. 1–1. Springer, Heidelberg (2010). https://doi.org/10.1007/978-3-642-11805-0_1
37. Rajaraman, R.: Topology control and routing in ad hoc networks: a survey. ACM SIGACT News **33**(2), 60–73 (2002)
38. Robertson, N., Seymour, P.D.: Graph minors XX Wagner's conjecture. J. Comb. Theory Ser. B **92**(2), 325–357 (2004)
39. Schaefer, M., Sedgwick, E., Štefankovič, D.: Recognizing string graphs in NP. J. Comput. Syst. Sci. **67**(2), 365–380 (2003)
40. Scheinerman, E.R.: Intersection Classes and Multiple Intersection Parameters of Graphs. Princeton University (1984)
41. Sherwani, N.A.: Algorithms for VLSI Physical Design Automation. Springer, New York (2007)
42. Sinden, F.W.: Topology of thin film RC circuits. Bell Syst. Tech. J. **45**(9), 1639–1662 (1966)
43. Thilikos, D.M., Bodlaender, H.L.: Fast partitioning l-apex graphs with applications to approximating maximum induced-subgraph problems. Inf. Process. Lett. **61**(5), 227–232 (1997)
44. Thomassen, C.: Interval representations of planar graphs. J. Comb. Theory Ser. B **40**(1), 9–20 (1986)
45. Thurston, W.: Hyperbolic geometry and 3-manifolds. Low-dimensional topology (Bangor, 1979) **48**, 9–25 (1982)
46. Welsh, D.J.A.: Knots and braids: some algorithmic questions. Contemp. Math. **147** (1993)
47. Wigderson, A.: The complexity of the Hamiltonian circuit problem for maximal planar graphs. Technical report, EECS 198, Princeton University, USA (1982)
48. Xu, J., Berger, B.: Fast and accurate algorithms for protein side-chain packing. J. ACM (JACM) **53**(4), 533–557 (2006)

The Complexity of $L(p, q)$-Edge-Labelling

Gaétan Berthe[1], Barnaby Martin[2(✉)], Daniël Paulusma[2], and Siani Smith[2]

[1] ENS de Lyon, Lyon, France
[2] Department of Computer Science, Durham University, Durham, UK
barnaby.d.martin@durham.ac.uk

Abstract. The $L(p, q)$-EDGE-LABELLING problem is the edge variant of the well-known $L(p, q)$-LABELLING problem. It is equivalent to the $L(p, q)$-LABELLING problem itself if we restrict the input of the latter problem to line graphs. So far, the complexity of $L(p, q)$-EDGE-LABELLING was only partially classified in the literature. We complete this study for all $p, q \geq 0$ by showing that whenever $(p, q) \neq (0, 0)$, the $L(p, q)$-EDGE-LABELLING problem is NP-complete. We do this by proving that for all $p, q \geq 0$ except $p = q = 0$, there is an integer k so that $L(p, q)$-EDGE-k-LABELLING is NP-complete.

1 Introduction

This paper studies a problem that falls under the distance-constrained labelling framework. Given any fixed nonnegative integer values p and q, an $L(p, q)$-k-*labelling* is an assignment of *labels* from $\{0, \ldots, k-1\}$ to the vertices of a graph such that adjacent vertices receive labels that differ by at least p, and vertices connected by a path of length 2 receive labels that differ by at least q [5]. Some authors instead define the latter condition as being vertices at distance 2 receive labels which differ by at least q (e.g. [7]). These definitions are the same so long as $p \geq q$ and much of the literature considers only this case (e.g. [11]). If $q > p$, the definitions diverge. For example, in an $L(1, 2)$-labelling, the vertices of a triangle K_3 need labels $\{0, 1, 2\}$ in the second definition but $\{0, 2, 4\}$ in the first. We use the *first* definition, in line with [5]. The decision problem of testing if for a given integer k, a given graph G admits an $L(p, q)$-k-labelling is known as $L(p, q)$-LABELLING. If k is *fixed*, that is, not part of the input, we denote the problem as $L(p, q)$-k-LABELLING.

The $L(p, q)$-LABELLING problem has been heavily studied, both from the combinatorial and computational complexity perspectives. For a starting point, we refer the reader to the comprehensive survey of Calamoneri [5].[1] The $L(1, 0)$-LABELLING is the traditional GRAPH COLOURING problem (COL), whereas $L(1, 1)$-LABELLING is known as (PROPER) INJECTIVE COLOURING [2,3,9] and DISTANCE 2 COLOURING [13,16]. The latter problem is studied explicitly in many papers (see [5]), just as is $L(2, 1)$-LABELLING [8,11,12] (see also [5]). The $L(p, q)$-LABELLING problem is also studied for special graph classes, see in particular [6]

---

[1] See http://wwwusers.di.uniroma1.it/~calamo/survey.html for later results.

© Springer Nature Switzerland AG 2022
P. Mutzel et al. (Eds.): WALCOM 2022, LNCS 13174, pp. 175–186, 2022.
https://doi.org/10.1007/978-3-030-96731-4_15

Table 1. Table of results.The fourth row follows from [14] (which proves the case $p = q = 1$) and applying Lemma 1. The eighth row is obtained from a straightforward generalization of the result in [12] for the case where $p = 2$ and $q = 1$. The fourth column gives the minimal k for which we prove NP-completeness. In the second row choose minimal $n \geq 4$ so that $(n - 3)p \geq q$.

Regime	Reduction from	Place in article	k at least
$p = 0$ and $q > 0$	3-COL	Sect. 3	$3q$
$2 \leq q/p$	NAE-3-SAT	Appendix A	$(n - 1)p + q + 1$
$1 < q/p \leq 2$	NAE-3-SAT	Sect. 4	$5p + 1$
$q/p = 1$	3-COL	[14]	$4p$
$2/3 < q/p \leq 1$	3-COL	Sect. 5	$3p + q + 1$
$q/p = 2/3$	1-in-3-SAT	Appendix B	$4p$
$1/2 < q/p < 2/3$	2-in-4-SAT	Appendix C	$p + 4q + 1$
$0 < q/p \leq 1/2$	NAE-3-SAT	Appendix D [12]	$3p + 1$
$p > 0$ and $q = 0$	3-COL	Sect. 2	$3p$

for a complexity dichotomy for trees. Janczewski et al. [11] proved that if $p > q$, then $L(p,q)$-LABELLING is NP-complete for planar bipartite graphs.

We consider the edge version of the problem. The *distance* between two edges e_1 and e_2 is the length of a shortest path that has e_1 as its first edge and e_2 as its last edge minus 1 (we say that e_1 and e_2 are *adjacent* if they share an end-vertex or equivalently, are of distance 1 from each other). The $L(p,q)$-EDGE-LABELLING problem considers an assignment of the labels to the edges instead of the vertices, and now the corresponding distance constraints are placed instead on the edges. Owing to space constraints some proofs and cases are omitted. Please see the full version of this article at [1]. In particular, references to the appendix are intended for that version.

In [12], the complexity of $L(2,1)$-EDGE-k-LABELLING is classified. It is in P for $k < 6$ and is NP-complete for $k \geq 6$. In [14], the complexity of $L(1,1)$-EDGE-k-LABELLING is classified. It is in P for $k < 4$ and is NP-complete for $k \geq 4$. In this paper we complete the classification of the complexity of $L(p,q)$-EDGE-k-LABELLING in the sense that, for all $p, q \geq 0$ except $p = q = 0$, we exhibit k so we can show $L(p,q)$-EDGE-k-LABELLING is NP-complete. That is, we do not exhibit the border for k where the problem transitions from P to NP-complete (indeed, we do not even prove the existence of such a border). The authors of [12] were looking for a more general result, similar to ours, but found the case $(p,q) = (2,1)$ laborious enough to fill one paper [15]. In fact, their proof settles for us all cases where $p \geq 2q$. We now give our main result.

Theorem 1. *For all $p, q \geq 0$ except if $p = q = 0$, there exists an integer k so that $L(p,q)$-EDGE-k-LABELLING is NP-complete.*

The proof follows by case analysis as per Table 1, where the corresponding section for each of the subresults is specified. We are able to reduce to the case that

$\gcd(p, q) = 1$, due to the forthcoming Lemma 1. We prove NP-hardness by reduction from graph 3-colouring and several satisfiability variants. These latter are known to be NP-hard from Schaefer's classification [17]. Each section begins with a theorem detailing the relevant NP-completeness. The case $p = q = 0$ is trivial (never use more than one colour) and is therefore omitted. Our hardness proofs involve gadgets that have certain common features, for example, the vertex-variable gadgets are generally star-like. For one case, we have a computer-assisted proof (as we will explain in detail).

By Theorem 1 we obtain a complete classification of $L(p, q)$-EDGE-LABELLING.

Corollary 1. *For all $p, q \geq 0$ except $p = q = 0$, $L(p, q)$-EDGE-LABELLING is* NP-*complete.*

Note that $L(p, q)$-EDGE-LABELLING is equivalent to $L(p, q)$-LABELLING for line graphs (the line graph of a graph G has vertex set $E(G)$ and two vertices e and f in it are adjacent if and only if e and f are adjacent edges in G). Hence, we obtain another dichotomy for $L(p, q)$-LABELLING under input restrictions, besides the ones for trees [6] and if $p > q$, (planar) bipartite graphs [11].

Corollary 2. *For all $p, q \geq 0$ except $p = q = 0$, $L(p, q)$-LABELLING is* NP-*complete for the class of line graphs.*

2 Preliminaries

We use the terms colouring and labelling interchangeably. A special role will be played by the *extended n-star* (especially for $n = 4$). This is a graph built from an n-star $K_{1,n}$ by subdividing each edge (so it becomes a path of length 2). Instead of referring to the problem as $L(p, q)$-LABELLING (or $L(h, k)$-LABELLING) we will use $L(a, b)$-LABELLING to free these other letters for alternative uses.

The following lemma is folklore and applies equally to the vertex- or edge-labelling problem. Note that $\gcd(0, b) = b$.

Lemma 1. *Let $\gcd(a, b) = d > 1$. Then the identity is a polynomial time reduction from $L(a/d, b/d)$-(EDGE)-k-LABELLING to $L(a, b)$-(EDGE)-kd-LABELLING.*

This result and the known NP-completeness of Edge-3-COLOURING [10] imply:

Corollary 3. *For all $a > 0$, $L(a, 0)$-EDGE-$3a$-LABELLING is* NP-*complete.*

3 Case $a = 0$ and $b > 0$

By Lemma 1 we only have to consider $a = 0$ and $b = 1$.

Theorem 2. *The problem $L(0, 1)$-EDGE-3-LABELLING is* NP-*complete.*

Let us use colours $\{0, 1, 2\}$. Our NP-hardness proof involves a reduction from 3-COL but we retain the nomenclature of variable gadget and clause gadget (instead of vertex gadget and edge gadget) in deference to the majority of our other sections. Our variable gadget consists of a triangle attached on one of its vertices to a leaf vertex of a star. Our clause gadget consists of a bull, each of whose pendant edges (vertices of degree 1) has an additional pendant edge added (that is, they are subdivided). This is equivalent to a triangle with a path of length 2 added to each of two of the three vertices. We draw our variable gadget in Fig. 1 and our clause gadget in Fig. 2.

Lemma 2. *In any valid $L(0, 1)$-edge-3-labelling of the variable gadget, each of the pendant edges must be coloured the same.*

Proof. Each of the edges in the triangle must be coloured distinctly as there is a path of length two from each to any other (by this we mean with a single edge in between, though they are also adjacent). Suppose the triangle edge that has two nodes of degree 2 in the variable gadget is coloured i. It is this colour that must be used for all of the pendant edges. The remaining edge may be coloured by anything from $\{0, 1, 2\} \backslash \{i\}$. However, we will always choose the option $i - 1 \bmod 3$.   □

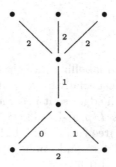

Fig. 1. The variable gadget for Theorem 2.

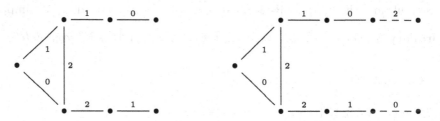

Fig. 2. The clause gadget for Theorem 2 (left) drawn also together with its interface with a variable gadget (right). The dashed line is an inner edge of the variable gadget.

Lemma 3. *In any valid $L(0,1)$-edge-3-labelling of the clause gadget, the two pendant edges must be coloured distinctly.*

Proof. Each of the edges in the triangle must be coloured distinctly as there is a path of length two from each to any other. Suppose the triangle edge that has two nodes of degree 3 in the clause gadget is coloured (w.l.o.g.) 2. The remaining edges in the triangle must be given 0 and 1, in some order. This then determines the colours of the remaining edges and enforces that the two pendant edges must be coloured distinctly. However, suppose we had started first by colouring distinctly the pendant edges. We could then choose a colouring of the remaining edges of the clause gadget so as to enforce the property that, if a pendant edge is coloured i, then its neighbour (in the clause gadget) is coloured $i + 1 \bmod 3$. This is the colouring we will always choose. □

We are now ready to prove Theorem 2.

Proof (Proof of Theorem 2.). We reduce from 3-COL. Let G be an instance of 3-COL involving n vertices and m edges. Let us explain how to build an instance G' for $L(0,1)$-EDGE-3-LABELLING. Each particular vertex may only appear in at most m edges (its degree), so for each vertex we take a copy of the variable gadget which has m pendant edges. For each edge of G we use a clause gadget to unite an instance of these pendant edges from the corresponding two variable gadgets. We use each pendant edge from a variable gadget in at most one clause gadget. We identify the pendant edge of a variable gadget with a pendant edge from a clause gadget so as to form a path from one to the other. We claim that G is a yes-instance of 3-COL iff G' is a yes-instance of $L(0,1)$-EDGE-3-LABELLING.

(Forwards.) Take a proper 3-colouring of G and induce these colours on the pendant edges of the corresponding variable gadgets. Distinct colours on pendant edges can be consistently united in a clause gadget since we choose, for a pendant edge coloured i: $i - 1 \bmod 3$ for its neighbour in the variable gadget, and $i + 1 \bmod 3$ for its neighbour in the clause gadget.

(Backwards.) From a valid $L(0,1)$-edge-3-labelling of G', we infer a 3-colouring of G by reading the pendant edge labels from the variable gadget of the corresponding vertex. The consistent labelling of each vertex follows from Lemma 2 and the fact that it is proper follows from Lemma 3. □

4   Case $1 < \frac{b}{a} \le 2$

In this section we prove the following result.

Theorem 3. *If $1 < \frac{b}{a} \le 2$, the problem $L(a,b)$-EDGE-$(5a + 1)$-LABELLING is NP-complete.*

We proceed by a reduction from (monotone) NAE-3-SAT. This case is relatively simple as the variable gadget is built from a series of extended 4-stars chained together, where each has a pendant 5-star to enforce some benign property. We will use colours from the set $\{0, \ldots, 5a\}$.

Lemma 4. *Let* $1 < \frac{b}{a} \le 2$. *In any valid* $L(a,b)$*-edge-*$(5a+1)$*-labelling of the extended 4-star, if one pendant edge is coloured* 0 *then all pendant edges are coloured in the interval* $\{0, \ldots, a\}$; *and if one pendant edge is coloured* $5a$ *then all pendant edge are coloured in the interval* $\{4a, \ldots, 5a\}$.

Proof. Suppose some pendant edge is coloured by 0 and another pendant is coloured by $l' \notin \{0, \ldots, a\}$. There are four inner edges of the star that are at distance 1 or 2 from these, and one another. If $l' < 2a$, then at least $2a$ labels are ruled out, which does not leave enough possibilities for the inner edges to be labelled in (at best) $\{2a+1, \ldots, 5a\}$. If $l' \ge 2a$, then it is not possible to use labels for the inner edges that are all strictly above l'. It is also not possible to use labels for the inner edges that are all strictly below l. In both cases, at least $2a$ labels are ruled out. Thus the labels, read in ascending order, must start no lower than a and have a jump of $2a$ at some point. It follows they are one of: $a, 3a, 4a, 5a$; or $a, 2a, 4a, 5a$; or $a, 2a, 3a, 5a$. This implies that l' is itself a multiple of a (whichever one was omitted in the given sequence). But now, since $b > a$, there must be a violation of a distance 2 constraint from l'.   □

We would like to chain extended 4-stars together to build our variable gadgets, where the pendant edges represent variables (and enter into clause gadgets) and we interpret one of the regimes $\{0, \ldots, a\}$ and $\{4a, \ldots, 5a\}$ as true, and the other as false. However, the extended 4-star can be validly $L(a, b)$-edge-$(5a+1)$-labelled in other ways that we did not yet consider. We can only use Lemma 4 if we can force one pendant edge in each extended 4-star to be either 0 or $5a$. Fortunately, this is straightforward: take a 5-star and add a new edge to one of the edges of the 5-star creating a path of length 2 from the centre of the star to the furthest leaf. This new edge can only be coloured 0 or $5a$. In Fig. 3 we show how to chain together copies of the extended 4-star, together with pendant 5-star gadgets at the bottom, to produce many copies of exactly one of the regimes $\{0, \ldots, a\}$ and $\{4a, \ldots, 5a\}$. Note that the manner in which we attach the pendant 5-star only produces a valid $L(a, b)$-edge-$(5a + 1)$-labelling because $2a \ge b$ (otherwise some distance 2 constraints would fail). So long as precisely one pendant edge per extended 4-star is used to encode a variable, then each encoding can realise all labels within each of these regimes, and again this can be seen by considering the pendant edges drawn top-most in Fig. 3, which can all be coloured anywhere in $\{4a, \ldots, 5a\}$. Let us recap, a *variable gadget* (to be used for a variable that appears in an instance of NAE-3-SAT m times) is built from chaining together m extended 4-stars, each with a pendant 5-star, exactly as is depicted in Fig. 3 for $m = 3$. The following is clear from our construction. The designation *top* is with reference to the drawing in Fig. 3. In Fig. 3, the case drawn corresponds to $\{4a, \ldots, 5a\}$, where the case $\{0, \ldots, a\}$ is symmetric.

Lemma 5. *Any valid* $L(a, b)$*-edge-*$(5a + 1)$*-labelling of a variable gadget is such that the top pendant edges are all coloured from precisely one of the sets* $\{0, \ldots, a\}$ *and* $\{4a, \ldots, 5a\}$. *Moreover, any colouring of the top pendant edges from one of these sets is valid.*

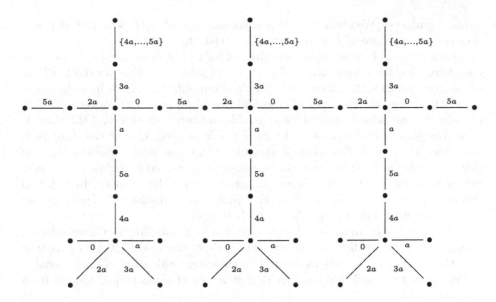

Fig. 3. Three extended 4-stars chained together, each with a pendant 5-star below, to form a variable gadget for Theorem 3. The pendant edges drawn on the top will be involved in clauses gadget and each of these three edges can be coloured with anything from $\{4a, \ldots, 5a\}$. If the top pendant edge is coloured $5a$ it may be necessary that the inner star edge below it is coloured not $3a$ but $2a$ (cf. Fig. 4). This is fine, the chaining construction works when swapping $2a$ and $3a$.

The clause gadget will be nothing more than a 3-star (a claw) which is formed from a new vertex uniting three (top) pendant edges from their respective variable gadgets. The following is clear.

Lemma 6. *A clause gadget is in a valid $L(a,b)$-edge-$(5a+1)$-labelling in the case where two of its edges are coloured $0, a$ and the third $5a$; or two of its edges are coloured $4a, 5a$ and the third 0. If all three edges come from only one of the regimes $\{0, \ldots, a\}$ and $\{4a, \ldots, 5a\}$, it can not be in a valid $L(a,b)$-edge-$(5a+1)$-labelling.*

We are now ready to prove Theorem 3.

Proof (Proof of Theorem 3.). We reduce from (monotone) NAE-3-SAT. Let Φ be an instance of NAE-3-SAT involving n occurrences of (not necessarily distinct) variables and m clauses. Let us explain how to build an instance G for $L(a,b)$-Edge-$(5a+1)$-Labelling. Each particular variable may only appear at most n times, so for each variable we take a copy of the variable gadget which is n extended 4-stars, each with a pendant 5-star, chained together. Each particular instance of the variable belongs to one of the free (top) pendant edges of the variable gadget. For each clause of Φ we use a 3-star to unite an instance of these free (top) pendant edges from the corresponding variable gadgets. Thus, we add a single vertex for each clause, but no new edges (they already existed in the variable gadgets). We claim that Φ is a yes-instance of NAE-3-SAT if and only if G is a yes-instance of $L(a,b)$-EDGE-$(5a+1)$-LABELLING.

(Forwards.) Take a satisfying assignment for Φ. Let the range $\{0,\ldots,a\}$ represent true and the range $\{4a,\ldots,5a\}$ represent false. This gives a valid labelling of the inner edges in the extended 4-stars, as exemplified in Fig. 3. In each clause, either there are two instances of true and one of false; or the converse. Let us explain the case where the first two variable instances are true and the third is false (the general case can easily be garnered from this). Colour the (top) pendant edge associated with the first variable as 0, the second variable a and the third variable $5a$. Plainly these can be consistently united in a claw by the new vertex that appeared in the clause gadget. We draw the situation in Fig. 4 to demonstrate that this will not introduce problems at distance 2. Thus, we can see this is a valid $L(a,b)$-edge-$(5a+1)$-labelling of G.

(Backwards.) From a valid $L(a,b)$-edge-$(5a+1)$-labelling of G, we infer an assignment Φ by reading, in the variable gadget, the range $\{0,\ldots,a\}$ as true and the range $\{4a,\ldots,5a\}$ as false. The consistent valuation of each variable follows from Lemma 5 and the fact that it is in fact not-all-equal follows from Lemma 6. □

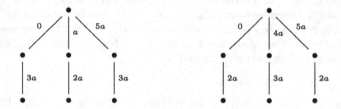

Fig. 4. The clause gadget and its interface with the variable gadgets (where we must consider distance 2 constraints). Both possible evaluations for not-all-equal are depicted.

5   Case $\frac{2}{3} < \frac{b}{a} < 1$

In this section we prove the following result.

Theorem 4. *If $\frac{2}{3} < \frac{b}{a} < 1$, then the problem $L(a,b)$-EDGE-$(3a+b+1)$-LABELLING is NP-complete.*

The regimes of the following lemma are drawn in Fig. 5.

Lemma 7. *Let $1 < \frac{a}{b} < \frac{3}{2}$. In an $L(a,b)$-edge-$(3a+b+1)$-labelling c of the extended 4-star, there are three regimes for the pendant edges. The first is $\{b,\ldots,a\}$, the second is $\{2a+b,\ldots,3a\}$, and the third is $\{a+b,\ldots,2a\}$.*

Proof. In a valid $L(a,b)$-edge-$(3a+b+1)$-labelling, we note $c_1 < c_2 < c_3 < c_4$ the colours of the 4 edges in the middle of the extended 4-star, and l_1, l_2, l_3, l_4 the colours of the pendant edges such that l_i is the colour of the pendant edge connected to the edge of colour c_i.

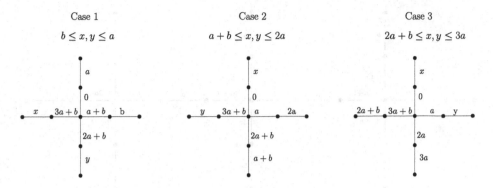

Fig. 5. The regimes of Theorem 4.

Claim 1. For all i, $c_1 < l_i < c_4$.

We only have to prove one inequality, as the other one is obtained by symmetry. If $l_i \leq c_1$ (bearing in mind also $b < a$), we have:

$$3a + b \geq c_4 - l_i = (c_1 - l_i) + (c_2 - c_1) + (c_3 - c_2) + (c_4 - c_3) \geq 3a + b.$$

So $(c_1, c_2, c_3, c_4) = (b, a+b, 2a+b, 3a+b)$, but $a > b$ so there is no possible value for l_1, which is not possible. So $c_1 < l_i$, and by symmetry $l_i < c_4$.

Claim 2. There exists $i \in \{1, 2, 3\}$ such that $c_{i+1} - c_i \geq a + b$.

We suppose the contrary. We have proved $c_1 < l_2, l_3 < c_4$. If $l_2 < c_2$, then $c_2 - c_1 = c_2 - l_2 + l_2 - c_1 \geq a + b$, impossible. If $c_2 < l_2 < c_3$, then $c_3 - c_2 = c_3 - l_2 + l_2 - c_2 \geq a + b$, impossible. So $c_3 < l_2 < c_4$. Symmetrically, we obtain $c_1 < l_3 < c_2$. So $c_1 < l_3 < c_2 < c_3 < l_2 < c_4$, and we get: $c_4 - c_1 \geq (l_3 - c_1) + (c_2 - l_3) + (c_3 - c_2) + (l_2 - c_3) + (c_4 - l_2) \geq 4b + a > 3a + b$, which is not possible.

Now we are in a position to derive the lemma, with the three regimes coming from the three possibilities of Claim 2. If $i = 1$, then the inner edges of the star are $0, a + b, 2a + b, 3a + b$ and the pendant edges come from $\{b, \ldots, a\}$. If $i = 2$, then the inner edges of the star are $0, a, 2a + b, 3a + b$ and the pendant edges come from $\{a + b, \ldots, 2a\}$. If $i = 3$, then the inner edges of the star are $0, a, a + b, 3a + b$ and the pendant edges come from $\{2a + b, \ldots, 3a\}$. □

The *variable gadget* may be taken as a series of extended 4-stars chained together. In the following, the "top" pendant edges refer to one of the two free pendant edges in each extended 4-star (not involved in the chaining together). The following is a simple consequence of Lemma 7 and is depicted in Fig. 6.

Lemma 8. *Any valid $L(a, b)$-edge-$(3a + b + 1)$-labelling of a variable gadget is such that the top pendant edges are all coloured from precisely one of the sets $\{b, \ldots, a\}$, $\{a + b, \ldots, 2a\}$ or $\{2a + b, \ldots, 3a\}$. Moreover, any colouring of the top pendant edges from one of these sets is valid.*

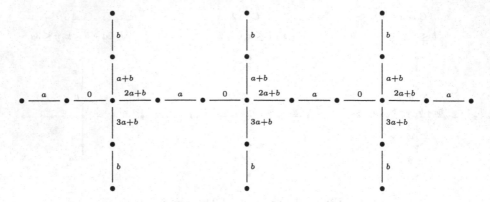

Fig. 6. Three extended 4-stars chained together, to form a variable gadget for Theorem 4. The pendant edges drawn on the top will be involved in clauses gadget. Suppose the top pendant edges are coloured b (as is drawn). In order to fulfill distance 2 constraints in the clause gadget, we may need the inner star vertices adjacent to them to be coloured not always $a + b$ (for example, if that pendant edge b is adjacent in a clause gadget to another edge coloured $a + b$). This is fine, the chaining construction works when swapping inner edges $a + b$ and $3a + b$ wherever necessary.

The clause gadget will be nothing more than a 2-star (a path) which is formed from a new vertex uniting two (top) pendant edges from their respective variable gadgets. The following is clear.

Lemma 9. *A clause gadget is in a valid $L(a, b)$-edge-$(3a + b + 1)$-labelling in the case where its edges are coloured distinctly. If they are coloured the same, then it can not be in a valid $L(a, b)$-edge-$(3a + b + 1)$-labelling.*

We are now ready to prove Theorem 4.

Proof (Proof of Theorem 4.). We reduce from 3-COL. Let G be an instance of 3-COL involving n vertices and m edges. Let us explain how to build an instance G' for $L(a, b)$-Edge-$(3a + b + 1)$-LABELLING. Each particular vertex may only appear in at most m edges (m is an upper ground on its degree), so for each vertex we take a copy of the variable gadget which is m extended 4-stars chained together. Each particular instance of the vertex belongs to one of the free (top) pendant edges of the variable gadget. For each edge of G we use a 2-star to unite an instance of these free (top) pendant edges from the corresponding two variable gadgets. Thus, we add a single vertex for each edge of G, but no new edges in G' (they already existed in the variable gadgets). We claim that G is a yes-instance of 3-COL if and only if G' is a yes-instance of $L(a, b)$-EDGE-$(3a + b + 1)$-LABELLING.

(Forwards.) Take a proper 3-colouring of G and induce these pendant edge labels on the corresponding variable gadgets according to the three regimes of Lemma 7. For example, map colours 1, 2, 3 to $b, a + b, 2a + b$. Plainly distinct pendant edge labels can be consistently united in a 2-claw by the new vertex that appeared in the clause gadget. Thus, we can see this is a valid $L(a,b)$-edge-$(3a+b+1)$-labelling of G'.

(Backwards.) From a valid $L(a,b)$-edge-$(3a+b+1)$-labelling of G', we infer a 3-colouring of G by reading the pendant edge labels from the variable gadget of the corresponding vertex and mapping these to their corresponding regime. The consistent valuation of each variable follows from Lemma 8 and the fact that it is proper (not-all-equal) follows from Lemma 9.   □

6   Final Remarks

We give several directions for future work. First, determining the boundary for k between P and NP-complete, in $L(p,q)$-EDGE-k-LABELLING, for all p,q is still open except if $(p,q)=(1,1)$ and $(p,q)=(2,1)$. For $(p,q)=(1,1)$ it is known to be 4 (it is in P for $k<4$ and is NP-complete for $k\geq 4$) [14]; and for $(p,q)=(2,1)$ it is known to be 6 (it is in P for $k<6$ and is NP-complete for $k\geq 6$) [12].

A second open line of research concerns $L(p,q)$-LABELLING for classes of graphs that omit a single graph H as an induced subgraph (such graphs are called H-free). A rich line of work in this vein includes [3], where it is noted, for $k\geq 4$, that $L(1,1)$-k-LABELLING is in P over H-free graphs, when H is a linear forest; for all other H the problem remains NP-complete. If k is part of the input and $p=q=1$, the only remaining case is $H=P_1+P_4$ [2]. Corollary 2 covers, for every $(p,q)\neq(0,0)$, the case where H contains an induced claw (as every line graph is claw-free). For bipartite graphs, and thus for H-free graphs for all H with an odd cycle, the result for $L(p,q)$-k-LABELLING is known from [11], at least in the case $p>q$.

As our final open problem, for $d\geq 1$, the complexity of $L(p,q)$-LABELLING on graphs of diameter at most d has, so far, only been determined for $a,b\in\{1,2\}$ [4].

References

1. Berthe, G., Martin, B., Paulusma, D., Smith, S.: The complexity of l(p, q)-edge-labelling. CoRR abs/2008.12226 (2020). https://arxiv.org/abs/2008.12226
2. Bok, J., Jedličková, N., Martin, B., Paulusma, D., Smith, S.: Injective colouring for H-free graphs. In: Santhanam, R., Musatov, D. (eds.) CSR 2021. LNCS, vol. 12730, pp. 18–30. Springer, Cham (2021). https://doi.org/10.1007/978-3-030-79416-3_2
3. Bok, J., Jedličková, N., Martin, B., Paulusma, D., Smith, S.: Acyclic, star and injective colouring: a complexity picture for H-free graphs. In: Proceedings of ESA 2020, LIPIcs, vol. 173, pp. 22:1–22:22 (2020)
4. Brause, C., Golovach, P., Martin, B., Paulusma, D., Smith, S.: Acyclic, star, and injective colouring: bounding the diameter. In: Kowalik, L, et al. (eds.) WG 2021. LNCS, vol. 12911, pp. 336–348. Springer, Cham (2021). https://doi.org/10.1007/978-3-030-86838-3_26
5. Calamoneri, T.: The $L(h,k)$-labelling problem: an updated survey and annotated bibliography. Comput. J. **54**, 1344–1371 (2011)

6. Fiala, J., Golovach, P.A., Kratochvíl, J.: Computational complexity of the distance constrained labeling problem for trees (Extended abstract). In: Aceto, L., Damgård, I., Goldberg, L.A., Halldórsson, M.M., Ingólfsdóttir, A., Walukiewicz, I. (eds.) ICALP 2008. LNCS, vol. 5125, pp. 294–305. Springer, Heidelberg (2008). https://doi.org/10.1007/978-3-540-70575-8_25

7. Fiala, J., Kloks, T., Kratochvíl, J.: Fixed-parameter complexity of lambda-labelings. Discret. Appl. Math. **113**, 59–72 (2001)

8. Griggs, J.R., Yeh, R.K.: Labelling graphs with a condition at distance 2. SIAM J. Discret. Math. **5**, 586–595 (1992)

9. Hahn, G., Kratochvíl, J., Širáň, J., Sotteau, D.: On the injective chromatic number of graphs. Discret. Math. **256**, 179–192 (2002)

10. Holyer, I.: The NP-completeness of edge-coloring. SIAM J. Comput. **10**, 718–720 (1981)

11. Janczewski, R., Kosowski, A., Małafiejski, M.: The complexity of the $L(p, q)$-labeling problem for bipartite planar graphs of small degree. Discret. Math. **309**, 3270–3279 (2009)

12. Knop, D., Masařík, T.: Computational complexity of distance edge labeling. Discret. Appl. Math. **246**, 80–98 (2018)

13. Lloyd, E.L., Ramanathan, S.: On the complexity of distance-2 coloring. Proc. ICCI **1992**, 71–74 (1992)

14. Mahdian, M.: On the computational complexity of strong edge coloring. Discret. Appl. Math. **118**, 239–248 (2002)

15. Masařík, T.: Private communication (2020)

16. McCormick, S.: Optimal approximation of sparse hessians and its equivalence to a graph coloring problem. Math. Program. **26**, 153–171 (1983)

17. Schaefer, T.J.: The complexity of satisfiability problems. In: STOC 1978, pp. 216–226 (1978)

Trains, Games, and Complexity: 0/1/2-Player Motion Planning Through Input/Output Gadgets

Joshua Ani[1], Erik D. Demaine[1] ⓘ, Dylan Hendrickson[1] ⓘ,
and Jayson Lynch[2](✉) ⓘ

[1] MIT Computer Science and Artificial Intelligence Laboratory, 32 Vassar Street,
Cambridge, MA 02139, USA
{joshuaa,edemaine,dylanhen}@mit.edu
[2] University of Waterloo, Waterloo, ON, Canada
jayson.lynch@uwaterloo.ca

Abstract. We analyze the computational complexity of motion planning through local "input/output" gadgets with separate entrances and exits, and a subset of allowed traversals from entrances to exits, each of which changes the state of the gadget and thereby the allowed traversals. We study such gadgets in the zero-, one-, and two-player settings, in particular extending past motion-planning-through-gadgets work [3,4] to zero-player games for the first time, by considering "branchless" connections between gadgets that route every gadget's exit to a unique gadget's entrance. Our complexity results include containment in L, NL, P, NP, and PSPACE; as well as hardness for NL, P, NP, and PSPACE. We apply these results to show PSPACE-completeness for certain mechanics in Factorio, [the Sequence], and a restricted version of Trainyard, improving the result of [1]. This work strengthens prior results on switching graphs, ARRIVAL [5], and reachability switching games [6].

Keywords: gadgets · motion planning · hardness of games

1 Introduction

Imagine a train proceeding along a track within a railroad network. Tracks are connected together by "switches": upon reaching one, the switch chooses the train's next track deterministically based on the state of the switch and where the train entered the switch; furthermore, the traversal changes the switch's state, affecting the next traversal. ARRIVAL [5] is one game of this type, where every switch has a single input and two outputs, and alternates between sending the train along the two outputs; the goal is to determine whether the train ever reaches a specified destination. Even this seemingly simple game has unknown complexity, but is known to be in NP ∩ coNP [5], so cannot be NP-hard unless NP = coNP. More recent work shows a stronger result of containment in UP ∩ coUP as well as CLS [9], PLS [11], and UEOPL [7]. But what about other types of switches?

© Springer Nature Switzerland AG 2022
P. Mutzel et al. (Eds.): WALCOM 2022, LNCS 13174, pp. 187–198, 2022.
https://doi.org/10.1007/978-3-030-96731-4_16

In this paper, we introduce a very general notion of "input/output gadgets" that models the possible behaviors of a switch, and analyze the resulting complexity of motion planning/prediction (does the train reach a desired destination?) while navigating a network of switches/gadgets. This framework gives us an expressive set of problems with different complexity classes to use as the basis for reductions for other problems of interest. For example, it is related to the generalization of ARRIVAL in [6] which define Reachability Switching Games. The paper further also describes how these Reachability Switching Games are related to switching systems and Propp machines, both of independent interest. In addition to ARRIVAL, our model captures other toy-train models, including those in the video game Factorio or the puzzle game Trainyard. In some cases, we obtain PSPACE-hardness, enabling building of a (polynomial-space) computer out of a railway system with a single train. Intuitively, our model is similar to a circuit model of computation, but where the state is stored in the gates (gadgets) instead of the wires, and gates update only according to visits by a single deterministically controlled agent (the train).

This work builds off of prior work on the computational complexity of agent-based motion planning [3,4], extending it zero-player situations. An analogous generalization of computational problems based on the number of players and boundedness of moves can be found in Constraint Logic [10] which has served as a basis for a large number of hardness proofs for reconfiguration problems, as well as games and puzzles. However this line of work differs from Constraint Logic because it involves the changes to the system being localized in a single agent, whereas all edges in a constraint logic puzzle are available for any given move. This is helpful in constructing hardness proofs where action is geographically constrained. Further Constraint Logic is an inherently reversible system and generalizing beyond that constraint can be helpful in hardness reductions.

Motion Planning Through Gadgets. Our model is a natural zero-player adaptation of the ***motion-planning-through-gadgets*** framework developed in [4] (after its introduction at FUN 2018 [3]), so we begin with a summary of that framework. A ***gadget*** consists of a finite set L of ***locations*** (entrances/exits), a finite set S of ***states***, and for each state $s \in S$, a labeled directed graph $G_s = (L, E_s)$ on the locations, where a directed edge (a, b) with label s' means that an agent can ***traverse*** the gadget by entering the gadget at location a and exiting at location b while changing the state of the gadget to s'. In general, a location might serve as the entrance for one traversal and the exit for another traversal; however, we consider in this paper the special case where each location serves exclusively as an entrance or an exit, but not both. Equivalently, a gadget is specified by its ***transition graph***, a directed graph whose vertices are state/location pairs $\in S \times L$, where a directed edge from (s, a) to (s', b) represents that an agent can traverse the gadget from a to b if it is in state s, and that such traversal changes the gadget's state to s'. We sometimes also consider the ***state-transition graph*** of a gadget, which is the directed graph with a vertex for each state $\in S$ and a directed edge (s, s') for each transition from (s, a) to (s', b) for any $a, b \in L$.

A **system of gadgets** consists of a set of gadgets, their initial states, and a **connection graph** on the gadgets' locations. If two locations a, b of two gadgets (possibly the same gadget) are connected by a path in the connection graph, then an agent can traverse freely between a and b (outside the gadgets). (Equivalently, we can think of locations a and b as being identified.) Gadgets are **local** in the sense that traversing a gadget does not change the state of any other gadgets.

In **one-player motion planning**, we are given the initial location of a single agent in a system of gadgets, and the problem asks whether there is a sequence of traversals that brings that agent to its goal location.

Past work [4] analyzed (and in many cases, characterized) the complexity of these motion-planning problems for gadgets satisfying a few additional properties, specifically, gadgets that are "reversible deterministic k-tunnel" or that are "DAG k-tunnel", defined as follows:

- A gadget is **k-tunnel** if it has $2k$ locations and there is a perfect matching, whose matching edges are called **tunnels**, such that the gadget only allows traversals between endpoints of a tunnel.
- A gadget is **deterministic** if its transition graph has maximum out-degree ≤ 1, i.e., an agent entering the gadget at some location a in some state s can exit at only one location b and in only one state s'.
- A gadget is **reversible** if its transition graph has the reverse of every edge, i.e., every traversal could be immediately undone.
- A gadget is a **DAG** if it has an acyclic state-transition graph. Such gadgets can necessarily be traversed only a bounded number of times (at most the number of states).

Input/Output Gadgets and Zero-Player Motion Planning. We define a gadget to be **input/output** if its locations can be partitioned into **input** locations (entrances) and **output** locations (exits) such that every traversal brings an agent from an input location to an output location, and in every state, there is at least one traversal from each input location. In particular, deterministic input/output gadgets have exactly one traversal from each input location in each state. Note that input/output gadgets cannot be reversible nor DAGs, so prior characterizations [4] do not apply to this setting.

An input/output gadget is **output-disjoint** if, for each output location, all of the transitions to it (including those from different states) are from the same input location. This notion is still more general than k-tunnel: it allows a one-to-many relation from a single input to multiple outputs.

With deterministic input/output gadgets, we can define a natural **zero-player motion-planning game** as follows. A system of gadgets is **branchless** if each connected component of the connection graph contains at most one input location.[1] Intuitively, if an agent finds itself in such a connected component, then there is only one gadget location it can enter, uniquely defining how it should

---

[1] Originally in [3] the gadget model was inherently branchless and non-deterministic, 1-state 'branching hallway' gadgets were used to connect multiple locations.

Table 1. Five subunits for 2-state, output-disjoint, input/output gadgets. We consider the 2-state gadgets to have the states *Up* and *Down*. Some subunits will set the state to a specific value such as Up, while some others always change the state when they are traversed.

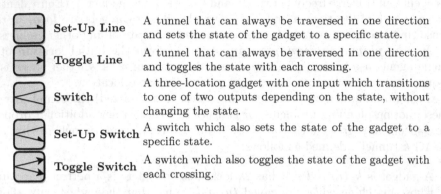

	Set-Up Line	A tunnel that can always be traversed in one direction and sets the state of the gadget to a specific state.
	Toggle Line	A tunnel that can always be traversed in one direction and toggles the state with each crossing.
	Switch	A three-location gadget with one input which transitions to one of two outputs depending on the state, without changing the state.
	Set-Up Switch	A switch which also sets the state of the gadget to a specific state.
	Toggle Switch	A switch which also toggles the state of the gadget with each crossing.

proceed. (If an agent finds itself in a connected component with no input locations, it is stuck in a dead-end and the game ends.) We can think of edges in the connection graph as directed wires that point from output locations to the input location in the same connected component. Note branchless systems can still have multiple output locations in a connected component which functions as a fan-in.

In a branchless system of deterministic input/output gadgets, there are never any choices to make: in the connection graph, there is at most one reachable input location, and when the agent enters an input location there is exactly one transition it can make. Thus we define **zero-player motion planning** with a set of deterministic input/output gadgets to be the one-player motion-planning game restricted to branchless systems of gadgets. Lacking any agency, the decision problem is equivalent to whether the agent ever reaches the goal location while following the unique path available to it.

Classifying Output-Disjoint Deterministic 2-State Input/Output Gadgets. In this paper, we are primarily interested in output-disjoint deterministic 2-state input/output gadgets. In this section, we omit the adjectives and refer to them simply as "gadgets", and give a categorization of these gadgets, into 'trivial,' 'bounded,' and 'unbounded' gadgets. For each category, we will show that every gadget in the category can simulate at least one of a finite set of gadgets. The behavior of an input location to a gadget is described by how it changes the state and which output location it sends the agent to in each state. If the input location doesn't change the state and always uses the same output location, it can be ignored (the path can be 'shortcut' to skip that transition). Otherwise, the input location corresponds to one of the following five nontrivial subunits, and the gadget is a disjoint union of some of these subunits (which interact by sharing state). These subunits are given in Table 1.

The ARRIVAL problem [5] is equivalent to zero-player motion planning with the toggle switch: we replace each vertex in their switch graph with a toggle switch, or vice versa. We will use their terminology when referring to switch graphs in the ARRIVAL paper; however, when referring to gadgets in our model, a switch is a gadget (or part of a gadget) which does not change state when crossed. More generally, zero-player motion planning with an arbitrary set of deterministic single-input input/output gadgets (with gadgets specified as part of the instance) is equivalent to explicit zero-player reachability switching games, as defined in [6].

We call the states of any such two state gadget *up* and *down*, and assume that each switch transitions to the top output in the up state and the bottom output in the down state; because we are not concerned with planarity, this assumption is fully general. There are two versions of the set line and set switch: one to set the gadget to each state. For example, a gadget with a set-up line and set-down switch is meaningfully different from a set-up line and set-up switch. We draw the set-down line and switch as the reflections of the set-up version. To represent the current state of a gadget, we make one of the lines in each switch dashed, so that the next transition would be made along a solid line. We categorize gadgets into three families:

1. *Trivial* gadgets have either no state change or no state-dependent behavior; they are composed entirely of either switches or toggle and set lines. They are equivalent to collections of simple tunnels, and zero-player motion planning with them is in L by straightforwardly simulating the agent for a number of steps equal to the number of locations.
2. *Bounded* gadgets have state-dependent behavior (i.e., some kind of switch) and one-way state change, either only to the up state or only to the down state. They naturally give rise to bounded games (a game in which the maximum number of turns is polynomially bounded before ending or repeating), because each gadget can change its state at most once.
3. *Unbounded* gadgets have state-dependent behavior and can change state in both directions. They naturally give rise to unbounded games.

We will find that the complexity of motion planning with a given gadget also depends on whether the gadget is *single-input*, meaning it has only one input location, or multiple nontrivial inputs. A non-trivial input must contain at least one transition from it, and that transition must either change the state of the gadget or must not exist in all states of the gadget. The only nontrivial single-input gadgets are the set switch and the toggle switch, which are bounded and unbounded, respectively. Recall Table 1 gives definitions for the pieces of 2-state input-output gadgets. The full version of the paper proves for all 2-state input-output gadgets with multiple inputs, there is a system of those gadgets with the same behavior as one of eight gadgets made of pairs of the subunits.

Lemma 1. *Let G be an output-disjoint deterministic 2-state input/output gadget with multiple nontrivial inputs.*

Table 2. Summary of results for output-disjoint deterministic 2-state input/output gadgets.

	Trivial (No state change or on tunnels)	Bounded, multiple nontrivial inputs	Unbounded, multiple nontrivial inputs
Zero-player (Fully Deterministic) (Sect. 2)	L	P-complete	PSPACE-complete
One-player (Sect. 1)	NL-complete	NP-complete	PSPACE-complete

Table 3. Summary of results for single-input input/output gadgets. These results can be found in the full version of the paper [2].

	Contained in	Hard for
Zero-player (Fully Deterministic)	UP ∩ coUP [9]	NL (cf. [6])
One-player	NP (cf. [6])	NP (cf. [6])
Two-Player	EXPTIME (cf. [6])	PSPACE (cf. [6])

- *If G is bounded, then it simulates either a switch/set-up line or a set-up switch/set-up line.*
- *If G is unbounded, then it simulates one of the following gadgets:*
  1. *switch/toggle line,*
  2. *switch/set-up line/set-down line,*
  3. *set-up switch/toggle line,*
  4. *set-up switch/set-down line,*
  5. *toggle switch/toggle line, or*
  6. *toggle switch/set-up line.*

Our Results. Table 2 summarizes our results on output-disjoint deterministic 2-state input/output gadgets. While our main motivation was to analyze zero-player motion planning, we also characterize the complexity of one-player motion planning for contrast. A full version of this paper is available [2].

We also consider motion planning with single-input input/output gadgets summarized in Table 3. This is a more immediate generalization of ARRIVAL [5], and is equivalent to the reachability switching games studied in [6]. We strengthen the results of [6] by showing that the containments in NP and EXP-TIME still hold when we allow nondeterministic gadgets, and by showing hardness with specific gadgets—the toggle switch for zero-player, and each of the toggle switch and set switch for one- and two-player—instead of having gadgets specified as part of the instance.

In the full version of the paper, we apply this framework to prove PSPACE-completeness of the mechanics in several video games: one-train colorless Train-yard, the game [the Sequence], trains in Factorio, and transport belts in Factorio are all PSPACE-complete. The first result improves a previous PSPACE-completeness result for two-color Trainyard [1] by using a strict subset of game features. Factorio in general is trivially PSPACE-complete, as players have explicitly built computers using the circuit network; here we prove hardness for the restricted problems with only train-related objects and only transport-belt-related objects.

2   Zero Players

In this section, we consider unbounded gadgets with multiple inputs, which are naturally PSPACE-complete. The full version of the paper considers unbounded gadgets with only a single input and bounded gadgets with multiple inputs, which are naturally P-complete.

We show that zero-player motion planning with any unbounded output-disjoint deterministic 2-state input/output gadget which has multiple nontrivial inputs is PSPACE-complete through a reduction from Quantified Boolean Formula (QBF), which is PSPACE-complete, to zero-player motion planning with the switch/set-up line/set-down line, and by showing that every such gadget simulates the switch/set-up line/set-down line.

Edge Duplicators. Many of our simulations involve building an ***edge duplicator*** An edge duplicator is a construction which allows us to effectively make a copy of a line from X to X' in a gadget. For example, we might want to take a switch/toggle-line and build a three input gadget made of a switch and two separate toggle-lines. Edge duplication is achieved by routing two inputs A and B to X, and then sending the agent from X' to one of two exits A' or B' corresponding to the input used. The details of the construction of an edge duplicator depend on the gadget used; see Fig. 1 for an example.

2.1   PSPACE-Hardness of the Switch/Set-Up Line/Set-Down Line

In this section, we show that zero-player motion planning with the switch/set-up line/set-down line is PSPACE-hard through a reduction from QBF. The switch/set-up line/set-down line is a 2-state input/output gadget with three inputs: one sets the state to up, one sets it to down, and one sends the agent to one of two outputs based on the current state.

Theorem 2. *Zero-player motion planning with the switch/set-up line/set-down line is PSPACE-hard.*

A full proof can be found in the full version of the paper but a sketch is provided here. We first build an edge duplicator, shown in Fig. 1. This allows us to use gadgets with multiple set-up or set-down lines. Each quantifier gadget

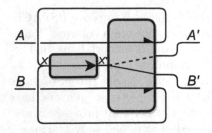

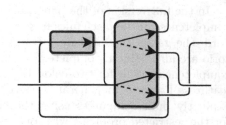

Fig. 1. An edge duplicator for the switch/set-up line/set-down line. A robot entering on the left sets the state of the switch, goes across the duplicated tunnel, and exits based on the state it set the switch to.

Fig. 2. An edge duplicator for the toggle switch/toggle switch. The tunnel on the left is duplicated.

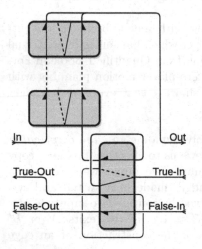

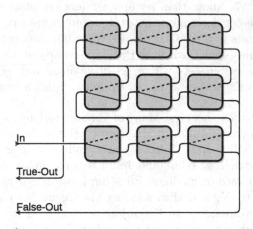

Fig. 3. The universal quantifier for the switch/set-up line/set-down line. An edge duplicator (Fig. 1) is used to give the bottom gadget two set-down lines.

Fig. 4. Three clauses of CNF evaluation for the switch/set-up line/set-down line; each clause is a row of three switches. The switches are part of gadgets in the quantifiers. We assume the top exit of each switch corresponds to that literal being true.

has three inputs, called In, True-In, and False-In, and three outputs, called Out, True-Out, and False-Out. The agent will always first arrive at In. This sets the variable controlled by that quantifier to true, and the agent leaves at Out, which sends it to the next quantifier gadget. The universal quantifier gadget is shown in Fig. 3. The existential quantifier is identical except that True-Out and False-Out are swapped, and True-In and False-In are swapped. Figure 4 shows how the variables can be incorporated into a CNF formula.

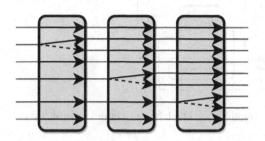 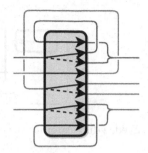

Fig. 5. A simulation of three toggle lines and three toggle switches from gadgets with one toggle switch and 5, 6, and 7 toggle lines. The red tunnels are toggle lines and the blue tunnels are toggle switches. (Color figure online)

Fig. 6. A simulation of a switch/set-up line/set-down line from the gadget built in Fig. 5. The switch, set-up line, and set-down line are red, green, and blue, respectively. (Color figure online)

2.2 Other Gadgets Simulate the Switch/Set-Up Line/Set-Down Line

In this section, we show that every unbounded output-disjoint deterministic 2-state input/output gadget with multiple nontrivial inputs can simulate the switch/set-up/set-down. We only need to show that the five other gadgets from Lemma 1 simulate the switch/set-up/set-down. It follows that zero-player motion planning with any such gadget is PSPACE-complete, since we can replace each gadget in a system of switch/set-up/set-down with a simulation of it. Some cases are presented here, see the full version of the paper for the remaining cases.

Toggle Switch/Toggle Switch. We begin with the toggle switch/toggle switch, which is not part of our basis of gadgets but will be a useful intermediate gadget. It builds an edge duplicator, shown in Fig. 2. We can merge the two outputs of one of the toggle switches to simulate a toggle switch/toggle line, and then duplicate the toggle line to make a gadget with one toggle switch and any number of toggle lines. By putting such gadgets in series, we can simulate a gadget with any number of toggle lines and any number of toggle switches. Figure 5 shows this for three toggle lines and three toggle switches, which is as large as we need. This simulated gadget can finally simulate the switch/set-up line/set-down line, shown in Fig. 6.

Switch/Toggle Line. We first build an edge duplicator, shown in Fig. 7. We can then duplicate the toggle line and put one copy in series with the switch, constructing a toggle switch/toggle line.

Set-Up Switch/Toggle Line. We first build an edge duplicator, shown in Fig. 8. We then simulate the switch/toggle line, shown in Fig. 9.

Toggle Switch/Set-Up Line. We simulate a set-up line/set-down switch using the toggle switch/ set-up line, as shown in Fig. 10; this is equivalent to a set-up switch/set-down line.

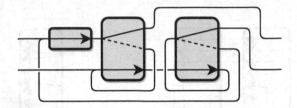

Fig. 7. An edge duplicator for the switch/toggle line. The leftmost tunnel is duplicated.

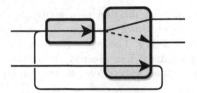

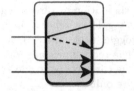

Fig. 8. An edge duplicator for the set-up switch/toggle line. The leftmost tunnel is duplicated.

Fig. 9. A simulation of the switch/toggle line using the set-up switch/toggle line. Red is the switch and blue is the toggle line. (Color figure online)

These simulations, together with Lemma 1, give the following theorem. The details of these cases are given in the full version of the paper.

Theorem 3. *Every unbounded output-disjoint deterministic 2-state input/output gadget with multiple nontrivial inputs simulates the switch/set-up line/set-down line.*

Corollary 4. *Let G be an unbounded output-disjoint deterministic 2-state input/output gadget with multiple nontrivial inputs. Zero-player motion planning with G is PSPACE-complete.*

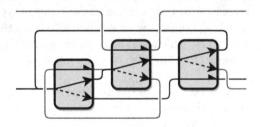

Fig. 10. A simulation of a set-up line/set-down switch from the set-up line/toggle switch. The state of the simulated gadget is the same as the state of the center gadget. The red path corresponds to the set-up line. When it enters the set-down switch, the robot goes along the blue lines if the state is down, the green lines if the state is up, and the black lines in both cases. (Color figure online)

3   Open Problems

One interesting problem left open by our paper and several before it [5,6,9] is the complexity of zero-player motion planning with deterministic single-input input/output gadgets, or equivalently ARRIVAL and zero-player reachability switching games; this is known to be between NL-hard and NP ∩ coNP, which is a large gap. We conjecture that many of these single input gadgets are P-hard and we would be interested to see such a result. We also leave open the complexity of two-player one-agent motion planning, or two-player reachability switching games, which is between PSPACE-hard and EXPTIME.

Since input/output gadgets seem to be a natural and rich class of gadgets, one could expand our characterization of zero-player motion planning to include input/output gadgets beyond the output-disjoint deterministic 2-state ones. Another question is whether these gadgets remain hard in the planar case.

Finally, although we have only defined zero-player motion planning with input/output gadgets (and the Trainyard gadget), many other classes of gadgets could be explored in the zero-player model. This model begins to look a lot more like a typical circuit or computing model with the unusual constraint that only a single signal is ever propagating through the system. In particular, a reasonable zero-player motion planning problem with reversible, deterministic gadgets (like those studied in [3] and [4]) is similar to asynchronous ballistic reversible logic [8] introduced to explore potential low-power computing architectures.

Acknowledgments. We thank Jeffrey Bosboom for suggesting applying the gadget framework to railroad switches (specifically, a switch/tripwire gadget) in 2017, and Mikhail Rudoy for pointing us to the subsequent analysis of ARRIVAL [5]. We also thank Jeffrey Bosboom for providing simplified constructions for the set-up switch/set-down line and toggle switch/set-up line, and for general discussion on topics in and related to this paper. Some of this work was done during open problem solving in the MIT class on Algorithmic Lower Bounds: Fun with Hardness Proofs (6.892) taught by Erik Demaine in Spring 2019. We thank the other participants of that class for related discussions and providing an inspiring atmosphere.

References

1. Almanza, M., Leucci, S., Panconesi, A.: Tracks from hell - when finding a proof may be easier than checking it. In: Ito, H., Leonardi, S., Pagli, L., Prencipe, G. (eds.) Proceedings of the 9th International Conference on Fun with Algorithms (FUN 2018). LIPIcs, vol. 100, pp. 4:1–4:13. La Maddalena, Italy, June 2018. https://doi.org/10.4230/LIPIcs.FUN.2018.4
2. Ani, J., Demaine, E.D., Hendrickson, D.H., Lynch, J.: Trains, games, and complexity: 0/1/2-player motion planning through input/output gadgets. arXiv preprint arXiv:2005.03192 (2020)
3. Demaine, E.D., Grosof, I., Lynch, J., Rudoy, M.: Computational complexity of motion planning of a robot through simple gadgets. In: Proceedings of the 9th International Conference on Fun with Algorithms (FUN 2018). LIPIcs, vol. 100, pp. 18:1–18:21, La Maddalena, Italy, June 2018

4. Demaine, E.D., Hendrickson, D., Lynch, J.: Toward a general theory of motion planning complexity: Characterizing which gadgets make games hard. In: Proceedings of the 11th Conference on Innovations in Theoretical Computer Science (ITCS 2020), pp. 62:1–62:42, Seattle, Washington, January 2020
5. Dohrau, J., Gärtner, B., Kohler, M., Matoušek, J., Welzl, E.: ARRIVAL: a zero-player graph game in NP ∩ coNP. In: Loebl, M., Nešetřil, J., Thomas, R. (eds.) A Journey Through Discrete Mathematics, pp. 367–374. Springer, Cham (2017). https://doi.org/10.1007/978-3-319-44479-6_14
6. Fearnley, J., Gairing, M., Mnich, M., Savani, R.: Reachability switching games. In: Chatzigiannakis, I., Kaklamanis, C., Marx, D., Sannella, D. (eds.) Proceedings of the 45th International Colloquium on Automata, Languages, and Programming (ICALP 2018). LIPIcs, vol. 107, pp. 124:1–124:14, Prague, Czech Republic, July 2018. https://doi.org/10.4230/LIPIcs.ICALP.2018.124
7. Fearnley, J., Gordon, S., Mehta, R., Savani, R.: Unique end of potential line. J. Comput. Syst. Sci. **114**, 1–35 (2020)
8. Frank, M.P.: Asynchronous ballistic reversible computing. In: Proceedings of the IEEE International Conference on Rebooting Computing (ICRC), pp. 1–8. Washington, DC, November 2017
9. Gärtner, B., Hansen, T.D., Hubáček, P., Král, K., Mosaad, H., Slívová, V.: ARRIVAL: next stop in CLS. In: Chatzigiannakis, I., Kaklamanis, C., Marx, D., Sannella, D. (eds.) Proceedings of the 45th International Colloquium on Automata, Languages, and Programming (ICALP 2018). LIPIcs, vol. 107, pp. 60:1–60:13. Prague, Czech Republic, July 2018. https://doi.org/10.4230/LIPIcs.ICALP.2018.60
10. Hearn, R.A., Demaine, E.D.: Games, Puzzles, and Computation. CRC Press, Cambridge (2009)
11. Karthik, C.: Did the train reach its destination: the complexity of finding a witness. Inf. Process. Lett. **121**, 17–21 (2017)

Online and Property Testing

An Optimal Tester for k-Linear

Nader H. Bshouty[✉]

Technion, Haifa, Israel
bshouty@cs.technion.ac.il

Abstract. A Boolean function $f : \{0,1\}^n \to \{0,1\}$ is k-linear if it returns the sum (over the binary field F_2) of k coordinates of the input. In this paper, we study property testing of the classes k-Linear, the class of all k-linear functions, and k-Linear*, the class $\cup_{j=0}^k j$-Linear. We give a non-adaptive distribution-free two-sided ϵ-tester for k-Linear that makes

$$O\left(k \log k + \frac{1}{\epsilon}\right)$$

queries. This matches the lower bound known from the literature.

We then give a non-adaptive distribution-free one-sided ϵ-tester for k-Linear* that makes the same number of queries and show that any non-adaptive uniform-distribution one-sided ϵ-tester for k-Linear must make at least $\tilde{\Omega}(k) \log n + \Omega(1/\epsilon)$ queries. The latter bound, almost matches the upper bound $O(k \log n + 1/\epsilon)$ known from the literature. We then show that any adaptive uniform-distribution one-sided ϵ-tester for k-Linear must make at least $\tilde{\Omega}(\sqrt{k}) \log n + \Omega(1/\epsilon)$ queries.

1 Inroduction

Property testing of Boolean function was first considered in the seminal works of Blum, Luby and Rubinfeld [5] and Rubinfeld and Sudan [19] and has recently become a very active research area. See for example works referenced in the surveys and books [11,12,16,17].

A Boolean function $f : \{0,1\}^n \to \{0,1\}$ is said to be linear if it returns the sum (over the binary field F_2) of some coordinates of the input, k-linear if it returns the sum of k coordinates of the input, and, k-linear* if it returns the sum of at most k coordinates. The class Linear (resp. k-Linear and k-Linear*) is the classes of all linear functions (resp. all k-linear functions and $\cup_{i=0}^k k$-Linear). Those classes has been of particular interest to the property testing community [3–6,8–10,12,14,16–18,20].

1.1 The Model

Let f and g be two Boolean functions $\{0,1\}^n \to \{0,1\}$ and let $\mathcal{D}$ be a distribution on $\{0,1\}^n$. We say that f is ϵ-*far* from g with respect to (w.r.t.) $\mathcal{D}$ if $\mathbf{Pr}_{\mathcal{D}}[f(x) \neq g(x)] \geq \epsilon$ and ϵ-*close* to g w.r.t. $\mathcal{D}$ if $\mathbf{Pr}_{\mathcal{D}}[f(x) \neq g(x)] \leq \epsilon$.

In the uniform-distribution and distribution-free property testing model, we consider the problem of testing a class of Boolean function C. In the distribution-free testing model (resp. uniform-distribution testing model), the *tester* is a

© Springer Nature Switzerland AG 2022
P. Mutzel et al. (Eds.): WALCOM 2022, LNCS 13174, pp. 201–212, 2022.
https://doi.org/10.1007/978-3-030-96731-4_17

randomized algorithm that has access to a Boolean function $f : \{0,1\}^n \to \{0,1\}$ via a black-box oracle that returns $f(x)$ when a string x is queried. The tester also has access to unknown distribution $\mathcal{D}$ (resp. uniform distribution) via an oracle that returns $x \in \{0,1\}^n$ chosen randomly according to the distribution $\mathcal{D}$ (resp. according to the uniform distribution). A *distribution-free tester*, [13], (resp. *uniform-distribution tester*) $\mathcal{A}$ for C is a tester that, given as input a distance parameter ϵ and the above two oracles to a Boolean function f,

1. if $f \in C$ then $\mathcal{A}$ accepts with probability at least $2/3$.
2. if f is ϵ-far from every $g \in C$ w.r.t. $\mathcal{D}$ (resp. uniform distribution) then $\mathcal{A}$ rejects with probability at least $2/3$.

We will also call $\mathcal{A}$ an *ϵ-tester for the class C* or an algorithm for *ϵ-testing C*. We say that $\mathcal{A}$ is *one-sided* if it always accepts when $f \in C$; otherwise, it is called *two-sided* tester. The *query complexity of $\mathcal{A}$* is the maximum number of queries $\mathcal{A}$ makes on any Boolean function f. If the query complexity is q then we call the tester a *q-query tester* or a tester with *query complexity q*.

In the *adaptive testing* (uniform-distribution or distribution-free) the queries can depend on the answers of the previous queries where in the *non-adaptive testing* all the queries are fixed in advance by the tester.

In this paper we study testers for the classes k-Linear and k-Linear*.

1.2   Prior Results

Blum et al. [5] gave an $O(1/\epsilon)$-query non-adaptive uniform-distribution one-sided ϵ-tester (called BLR tester) for Linear. Halevy and Kushilevitz, [14], used a self-corrector (an algorithm that computes $g(x)$ from a black box query to f that is ϵ-close to g) to reduce distribution-free testability to uniform-distribution testability. This reduction gives an $O(1/\epsilon)$-query non-adaptive distribution-free one-sided ϵ-tester for Linear. The reduction can be applied to any subclass of Linear. In particular, any q-query uniform-distribution ϵ-tester for k-Linear (k-Linear*) gives a $O(q)$-query distribution-free ϵ-tester.

It is well known that if there is a q_1-query uniform-distribution ϵ-tester for Linear and a q_2-query uniform-distribution ϵ-tester for the class k-Junta[1] then there is an $O(q_1 + q_2)$-query uniform-distribution $O(\epsilon)$-tester for k-Linear*. Since k-Linear $= k$-Linear*$\setminus(k-1)$-Linear*, if there is a q-query uniform-distribution ϵ-tester for k-Linear* then there is an $O(q)$-query uniform-distribution two-sided ϵ-tester for k-Linear. Therefore, all the results for testing k-Junta are also true for k-Linear* and k-Linear in the uniform-distribution model.

For non-adaptive testers Fisher, et al. [9] gave the lower bound $\Omega(\sqrt{k})$. Goldreich [10], gave the lower bound $\Omega(k)$. In [4], Blais and Kane gave the lower bound $2k - o(k)$. Then in [3], Blais et al. gave the lower bound $\Omega(k \log k)$. For adaptive testers, Goldreich [10], gave the lower bound $\Omega(\sqrt{k})$. Then Blais et al. [3] gave the lower bound $\Omega(k)$ and in [4], Blais and Kane gave the lower bound $k - o(k)$. Then in [20], Saglam gave the lower bound $\Omega(k \log k)$. This bound with

---

[1] The class of boolean functions that depends on at most k coordinates.

the trivial $\Omega(1/\epsilon)$ lower bound gives the lower bound $\Omega(k \log k + 1/\epsilon)$ for the query complexity of any adaptive uniform-distribution (and distribution-free) two-sided testers.

For upper bounds for uniform-distribution two-sided ϵ-testing k-Linear, Fisher, et al. [9] gave the first adaptive tester that makes $O(k^2/\epsilon)$ queries. In [8], Buhrman et al. gave a non-adaptive tester that makes $O(k \log k)$ queries for any constant ϵ. As is mentioned above, testing k-Linear can be done by first testing if the function is k-Junta and then testing if it is Linear. Therefore, using Blais [1,2] adaptive and non-adaptive testers for k-Junta we get adaptive and non-adaptive uniform-distribution testers for k-Linear that makes $O(k \log k + k/\epsilon)$ and $\tilde{O}(k^{1.5}/\epsilon)$ queries, respectively.

For upper bounds for two-sided distribution-free testing k-Linear, as is mentioned above, from Halevy et al. reduction in [14], an adaptive and non-adaptive distribution-free ϵ-tester can be constructed from adaptive and non-adaptive uniform-distribution ϵ-testers. This gives an adaptive and non-adaptive distribution-free two-sided testers for k-Linear that makes $O(k \log k + k/\epsilon)$ and $\tilde{O}(k^{1.5}/\epsilon)$ queries, respectively.

1.3   Our Results

In this paper we prove

Theorem 1. *For any $\epsilon > 0$, there is a polynomial time non-adaptive distribution-free one-sided ϵ-tester for k-Linear* that makes $O(k \log k + 1/\epsilon)$ queries.*

By the 2-sided reduction from k-Linear to k-Linear*, we get

Theorem 2. *For any $\epsilon > 0$, there is a polynomial time non-adaptive distribution-free two-sided ϵ-tester for k-Linear that makes $O(k \log k + 1/\epsilon)$ queries.*

This improves [8] by allowing ϵ to be general instead of a fixed constant, improves [2] by making the tester non-adaptive and improving the k/ϵ dependence to $1/\epsilon$.

For one-sided testers for k-Linear we prove

Theorem 3. *Any non-adaptive uniform-distribution one-sided ϵ-tester for k-Linear must make at least $\tilde{\Omega}(k) \log n + \Omega(1/\epsilon)$ queries.*

This almost matches the upper bound $O(k \log n + 1/\epsilon)$ that follows from the reduction of Goldreich et al. [13] and the non-adaptive deterministic exact learning algorithm of Hofmeister [15] that learns k-Linear with $O(k \log n)$ queries.

For adaptive testers we prove

Theorem 4. *Any adaptive uniform-distribution one-sided ϵ-tester for k-Linear must make at least $\tilde{\Omega}(\sqrt{k}) \log n + \Omega(1/\epsilon)$ queries.*

2   Overview of the Testers and Lower Bounds

In this section we give overview of the techniques used for proving the results in this paper.

2.1   One-Sided Tester for k-Linear*

The tester for k-Linear* first runs the tester BLR of Blum et al. [5] to test if the function f is ϵ'-close to Linear w.r.t. the uniform distribution, where $\epsilon' = \Theta(1/(k \log k))$. BLR is one-sided tester and therefore, if f is k-linear then BLR accepts with probability 1. If f is ϵ'-far from Linear w.r.t. the uniform distribution then, with probability at least $2/3$, BLR rejects. Therefore, if the tester BLR accepts, we may assume that f is ϵ'-close to Linear w.r.t. the uniform distribution. Let $g \in$ Linear be the function that is ϵ'-close to f. If f is k-linear* then $f = g$. This is because $\epsilon' < 1/8$ and the distance[2] between every two linear functions is $1/2$. BLR makes $O(1/\epsilon') = O(k \log k)$ queries.

In the second stage, the tester tests if g (not f) is k-linear*. Let us assume for now that we can query g in every string. Since $g \in$ Linear, we need to distinguish between functions in k-Linear* and functions in Linear$\setminus k$-Linear*. We do that with two tests. We first test if $g \in 8k$-Linear* and then test if it is in k-Linear* assuming that it is in $8k$-Linear*. In the first test, the tester "throws", uniformly at random, the variables of g into $16k$ bins and tests if there is more than k non-empty bins. If g is k-linear* then the number of non-empty bins is always less than k. If it is k'-linear for some $k' > 8k$ then with high probability (w.h.p.) the number of non-empty bins is greater than k. Notice that if f is k-linear* then the test always accepts and therefore it is one-sided. This tests makes $O(k)$ queries to g.

The second test is testing if g is in k-Linear* assuming that it is in $8k$-Linear*. This is done by projecting the n coordinates of g into $r = O(k^2)$ coordinates[3] uniformly at random[4] and learning (finding exactly) the projected function using the non-adaptive deterministic Hofmeister's algorithm, [15], that makes $O(k \log r) = O(k \log k)$ queries. Since $g \in 8k$-Linear*, w.h.p., the relevant coordinates[5] of the function are projected to different coordinates, and therefore, w.h.p., the learning gives a linear function that has exactly the same number of relevant coordinates as g. The tester accepts if the number of relevant coordinates in the projected function is at most k. If $g \in k$-Linear*, then the projected function is in k-Linear* with probability 1 and therefore this test is one-sided. This test makes $O(k \log k)$ queries.

We assumed that we can query g. We now show how to query g in $O(k \log k)$ strings so we can apply the above two tests. For this, the tester uses self-corrector, [5]. To compute $g(z)$, the self-corrector chooses a uniform random string $a \in \{0,1\}^n$ and computes $f(z + a) + f(a)$. Since f is $O(1/(k \log k))$-close to g w.r.t. the uniform distribution, we have that for any string $z \in \{0,1\}^n$ and an $a \in \{0,1\}^n$ chosen uniformly at random, with probability at least $1 - O(1/(k \log k))$, $f(z+a) + f(a) = g(z+a) + g(a) = g(z)$. Therefore, w.h.p., the

---

[2] w.r.t. the uniform distribution, i.e., $\mathbf{Pr}[f_1 \neq f_2]$ where f_1 and f_2 are distinct linear functions.

[3] That is, defining a function $f(x_{\phi(1)}, \ldots, x_{\phi(n)})$ where $\phi : [n] \to [r]$ is random uniform.

[4] We may assume that $k \leqslant \sqrt{n}$. This is because the one-sided non-adaptive testing in [13] asks $O(k \log n + 1/\epsilon)$ queries which is $O(k \log k + 1/\epsilon)$ queries for $k > \sqrt{n}$.

[5] Coordinates that the function depends on.

self-corrector computes correctly the values of g in $O(k \log k)$ strings. If $f \in k$-Linear then $g = f$ and $f(z + a) + f(z) = f(z) = g(z)$, i.e., the self-corrector gives the value of g with probability 1. This shows that the above two tests are one-sided.

Now, if f is k-linear* then $f = g$. If f is ϵ-far from every function in k-Linear* w.r.t. $\mathcal{D}$ then it is ϵ-far from g w.r.t. $\mathcal{D}$.

In the final stage the tester tests whether f is equal to g or ϵ-far from g w.r.t. $\mathcal{D}$. Here again the tester uses self-corrector. It asks for a sample $\{(z^{(i)}, f(z_i)) | i \in [t]\}$ according to the distribution $\mathcal{D}$ of size $t = O(1/\epsilon)$ and tests if $f(z^{(i)}) = f(z^{(i)} + a^{(i)}) + f(a^{(i)})$ for every $i \in [t]$, where $a^{(i)}$ are i.i.d. uniform random strings. If $f(z^{(i)}) = f(z^{(i)} + a^{(i)}) + f(a^{(i)})$ for all i then it accepts, otherwise, it rejects. If f is k-linear then $f(z^{(i)}) = f(z^{(i)} + a^{(i)}) + f(a^{(i)})$ for all i and the tester accepts with probability 1. Now suppose f is ϵ-far from g w.r.t. $\mathcal{D}$. Since f is ϵ'-close to g w.r.t. the uniform distribution and $\epsilon' \leq 1/8$ we have that, with probability at least $7/8$, $f(z^{(i)} + a^{(i)}) + f(a^{(i)}) = g(z^{(i)} + a^{(i)}) + g(a^{(i)}) = g(z^{(i)})$. Therefore, assuming the latter happens, then, with probability at least $1 - \epsilon$ we have $f(z^{(i)}) \neq g(z^{(i)}) = f(z^{(i)} + a^{(i)}) + f(a^{(i)})$. Thus, w.h.p, there is i such that $f(z^{(i)}) \neq f(z^{(i)} + a^{(i)}) + f(a^{(i)})$ and the tester rejects. This stage is one-sided and makes $O(1/\epsilon)$ queries.

2.2   Two-Sided Testers for k-Linear

As we mentioned in the introduction, the one-sided q-query uniform-distribution ϵ-tester for k-Linear* gives a two-sided uniform-distribution $O(q)$-query ϵ-tester for k-Linear. This is because, in the uniform distribution, the linear functions are $1/2$-far from each other and therefore, for any $\epsilon < 1/4$, if f is ϵ-close to a k-linear function g then it is $(1/2 - \epsilon)$-far from $(k-1)$-Linear*. This is not true for any distribution $\mathcal{D}$, and therefore, cannot be applied here.

The algorithm in the previous subsection can be changed to a two-sided tester for k-Linear as follows. The only part that should be changed is the test that g is in k-Linear* assuming that it is in $8k$-Linear*. We replace it with a test that g is in k-Linear assuming that it is in $8k$-Linear*. The tester rejects if the number of relevant coordinates in the function that is learned is not *equal* to k. This time the test is two-sided. The reason is that the projection to $O(k^2)$ coordinates does not guarantee (with probability 1) that all the variables of f are projected to different variables. Therefore, it may happen that f is k-linear and the projection gives a $(k-1)$-linear* function.

For the lower bound for one-sided testers see Subsect. 4.1.

3   The Testers for k-Linear* and k-Linear

In this section we give the non-adaptive distribution-free one-sided tester for k-Linear* and the non-adaptive distribution-free two-sided tester for k-Linear.

3.1   Notations

In this subsection, we give some notations that we use throughout the paper.

Denote $[n] = \{1, 2, \ldots, n\}$. For $X \subset [n]$ we denote by $\{0,1\}^X$ the set of all binary strings of length $|X|$ with coordinates indexed by $i \in X$. For $x \in \{0,1\}^n$ and $X \subseteq [n]$ we write $x_X \in \{0,1\}^X$ to denote the projection of x over coordinates in X. We denote by 1_X and 0_X the all-one and all-zero strings in $\{0,1\}^X$, respectively. For a variable x_i and a set X, we denote by $(x_i)_X$ the string x' over coordinates in X where for every $j \in X$, $x'_j = x_i$. For $X_1, X_2 \subseteq [n]$ where $X_1 \cap X_2 = \varnothing$ and $x \in \{0,1\}^{X_1}, y \in \{0,1\}^{X_2}$ we write $x \circ y$ to denote their concatenation, i.e., the string in $\{0,1\}^{X_1 \cup X_2}$ that agrees with x over coordinates in X_1 and agrees with y over coordinates in X_2. For $X \subseteq [n]$ we denote $\overline{X} = [n] \backslash X = \{x \in [n] | x \notin X\}$.

For example, if $n = 7$, $X_1 = \{1, 3, 5\}$, $X_2 = \{2, 7\}$, y_2 is a variable and $z = (z_1, z_2, z_3, z_4, z_5, z_6, z_7) \in \{0,1\}^7$ then $(y_2)_{X_1} \circ z_{X_2} \circ 0_{\overline{X_1 \cup X_2}} = (y_2, z_2, y_2, 0, y_2, 0, z_7)$.

3.2   The Tester

Consider the tester **Test-Linear**$_k^*$ for k-Linear* in Fig. 1. The tester uses three procedures. The first is **Self-corrector** that for an input $x \in \{0,1\}^n$ chooses a uniform random $z \in \{0,1\}^n$ and returns $f(x + z) + f(z)$. The procedure **BLR** is a non-adaptive uniform-distribution one-sided ϵ-tester for Linear. BLR makes c_1/ϵ queries for some constant c_1, [5]. The third procedure is **Hofmeister's Algorithm** (N, K), a deterministic non-adaptive algorithm that exactly learns K-Linear* over N coordinates from black box queries. Hofmeister's Algorithm makes $c_2 K \log N$ queries for some constant c_2, [15].

To test k-Linear we use the same tester but change step 11 to:
(11) If the output is not in k-Linear then reject
We call this tester **Test-Linear**$_k$.

3.3   Correctness of the Tester

In this section we prove

Theorem 5. Test-Linear$_k$ *is a non-adaptive distribution-free two-sided ϵ-tester for k-Linear that makes $O(k \log k + 1/\epsilon)$ queries.*

Theorem 6. Test-Linear$_k^*$ *is a non-adaptive distribution-free one-sided ϵ-tester for k-Linear* that makes $O(k \log k + 1/\epsilon)$ queries.*

Proof. Since there is no stage in the tester that uses the answers of the queries asked in previous ones, the tester is non-adaptive.

In Stage 1 the tester makes $O(1/\epsilon') = O(k \log k)$ queries. In stage 2.1, $O(k)$ queries. In stage 2.2, $O(k \log r) = O(k \log k)$ queries and in stage 3, $O(1/\epsilon)$ queries. Therefore, the query complexity of the tester is $O(k \log k + 1/\epsilon)$.

We will assume that $k \geq 12$. For $k < 12$, the non-adaptive tester of k-Junta with the BLR tester and the self-corrector gives a non-adaptive testers that makes $O(1/\epsilon) = O(k \log k + 1/\epsilon)$ queries.

Test-Linear$_k^*$
Input: Oracle that accesses a Boolean function f
Output: Either "Accept" or "Reject"

Procedures
    Self-corrector $g(x) := f(x + z) + f(z)$ for uniform random $z \in \{0,1\}^n$.
    BLR A procedure that ϵ-tests Linear using c_1/ϵ queries.
    Hofmeister's Algorithm(N, K) for learning K-Linear* over
               N coordinates using $c_2 K \log N$ queries.

Stage 1. BLR
1.   Run BLR on f with $\epsilon' = 1/(12(16k + c_2 k \log(256k^2))))$
2.   If BLR rejects then reject.
Stage 2.1. Testing if g is in Linear\\$8k$-Linear*
3.   Choose a uniform random partition $X_1, \ldots, X_{16k}$
4. $Count \leftarrow 0$;
5.   Choose a uniform random $z \in \{0,1\}^n$.
6.   For $i = 1$ to $16k$
7.   if $g(z_{X_i} \circ 0_{\overline{X_i}}) = 1$ then $Count \leftarrow Count + 1$
8.   If $Count > k$ then reject.
Stage 2.2. Testing if g is in k-Linear assuming it is in $8k$-Linear*
9.   Choose a uniform random partition $X_1, \ldots, X_r$ for $r = 256k^2$
10. Run Hofmeister's algorithm $(N = r, K = 8k)$ in order
        to learn $F = g((y_1)_{X_1} \circ (y_2)_{X_2} \circ \cdots \circ (y_r)_{X_r})$ as a function in $(y_1, \ldots, y_r)$.
11. If the output is not in k-Linear* then reject
        /* In **Test-Linear$_k$** (for testing k-Linear) we replace (11) with:
        /* 11. If the output is not in k-Linear then reject
Stage 3. Consistency test
12. Choose a sample $x^{(1)}, \ldots, x^{(t)}$ according to $\mathcal{D}$ of size $t = 4/\epsilon$.
13. For $i = 1$ to t.
14.   If $f(x^{(i)}) \neq g(x^{(i)})$ then reject.
15. Accept.

Fig. 1. An optimal two-sided tester for k-Linear.

Completeness: We first show the completeness for **Test-Linear$_k$** that tests k-Linear. Suppose $f \in k$-Linear. Then for every x we have $g(x) = f(x+z) + f(z) = f(x) + f(z) + f(z) = f(x)$. Therefore, $g = f$. In stage 1, BLR is one-sided and therefore it does not reject. In stage 2.1, since $X_1, \ldots, X_{16k}$ are pairwise disjoint, the number of functions $g(x_{X_i} \circ 0_{\overline{X_i}})$, $i = 1, 2, \ldots, 16k$, that are not identically zero is at most k and therefore stage 2.1 does not reject. In stage 2.2, with probability at least $1 - \binom{k}{2}/(256k^2) \geqslant 2/3$, the relevant coordinates of f fall into different X_i and then $F = g((y_1)_{X_1} \circ (y_2)_{X_2} \circ \cdots \circ (y_r)_{X_r}) = f((y_1)_{X_1} \circ (y_2)_{X_2} \circ \cdots \circ (y_r)_{X_r})$ is k-linear. Then, Hofmeister's algorithm returns a k-linear function. Therefore, with probability at least $2/3$ the tester does not reject. Stage 3 does not reject since $f = g$.

Now for the tester **Test-Linear**$_k^*$, in stage 2.2, with probability 1 the function F is in k-Linear*. In fact, if t relevant coordinates falls into the set X_i then the coordinate i (that correspond to the variable y_i) will be relevant in F if and only if t is odd. Therefore, the tester does not reject.

Notice that **Test-Linear**$_k^*$ is one-sided and **Test-Linear**$_k$ is two-sided. **Soundness:** We prove the soundness for **Test-Linear**$_k$. The same proof also works for **Test-Linear**$_k^*$. Suppose f is ϵ-far from k-Linear w.r.t. the distribution $\mathcal{D}$. We have four cases

Case 1: f is ϵ'-far from Linear w.r.t. the uniform distribution.
Case 2: f is ϵ'-close to $g \in$ Linear and g is in Linear\$8k$-Linear*.
Case 3: f is ϵ'-close to $g \in$ Linear and g is in $8k$-Linear*\k-Linear.
Case 4: f is ϵ'-close to $g \in$ Linear, g is in k-Linear and f is ϵ-far from k-Linear w.r.t. $\mathcal{D}$.

For Case 1, if f is ϵ'-far from Linear then, in stage 1, BLR rejects with probability $2/3$.

For Cases 2 and 3, since f is ϵ'-close to g, for any fixed $x \in \{0,1\}^n$ with probability at least $1 - 2\epsilon'$ (over a uniform random z), $f(x+z) + f(z) = g(x+z) + g(z) = g(x)$. Since stages 2.1 and 2.2 makes $(16k + c_2 k \log r)$ queries (to g), with probability at least $1 - (16k + c_2 k \log r)2\epsilon' \geq 5/6$, $g(x)$ is computed correctly for all the queries in stages 2.1 and 2.2.

For Case 2, consider stage 2.1 of the tester. If g is in Linear\$8k$-Linear* then g has more than $8k$ relevant coordinates. The probability that less than or equal to $4k$ of $X_1, \ldots, X_{16k}$ contains relevant coordinates of g is at most (since $\binom{n}{k} \leq \left(\frac{en}{k}\right)^k$)

$$\binom{16k}{4k}\frac{1}{4^{8k}} \leq \left(\frac{e16k}{4k}\right)^{4k}\frac{1}{4^{8k}} \leq \frac{1}{12}.$$

If X_i contains the relevant coordinates $i_1, \ldots, i_\ell$ then $g(x_{X_i} \circ 0_{\overline{X_i}}) = x_{i_1} + \cdots + x_{i_\ell}$ and therefore, for a uniform random $z \in \{0,1\}^n$, with probability at least $1/2$, $g(z_{X_i} \circ 0_{\overline{X_i}}) = 1$. Therefore, if at least $4k$ of $X_1, \ldots, X_{16k}$ contains relevant coordinates then, by Chernoff bound, with probability at least $1 - e^{-k/4} \geq 11/12$, the counter "*Count*" is greater than k. Therefore, for Case 2, if g is in Linear\$8k$-Linear* then, with probability at least $1 - (1/6 + 1/12 + 1/12) = 2/3$, the tester rejects.

For Case 3, consider stage 2.2. If g is in $8k$-Linear*\$k$-Linear then g has at most $8k$ relevant coordinates. Then with probability at least $1 - \binom{8k}{2}/(256k^2) \geq 5/6$, the relevant coordinates of g fall into different X_i and then Hofmeister's algorithm returns a linear function with the same number of relevant coordinates as g. Therefore stage 2.2 rejects with probability at least $2/3$.

For Case 4, if g is in k-Linear and f is ϵ-far from k-Linear w.r.t. $\mathcal{D}$, then f is ϵ-far from g w.r.t. $\mathcal{D}$. Then for uniform random z and $x \sim \mathcal{D}$,

$$
\begin{aligned}
\mathbf{Pr}_{\mathcal{D},z}[f(x) \neq g(x)] &\geqslant \mathbf{Pr}_{\mathcal{D},z}[f(x) \neq g(x) | g(x) = f(x+z) + f(z)] \\
&\quad \cdot \mathbf{Pr}_{\mathcal{D},z}[g(x) = f(x+z) + f(z)] \\
&= \mathbf{Pr}_{\mathcal{D}}[f(x) \neq g(x)] \mathbf{Pr}_z[g(x) = f(x+z) + f(z)] \\
&\geqslant \epsilon(1 - \epsilon') \geqslant \epsilon/2.
\end{aligned}
$$

Therefore, with probability at most $(1 - \epsilon/2)^t = (1 - \epsilon/2)^{4/\epsilon} \leqslant 1/3$, stage 3 does not reject.

4 Lower Bound

In this section we prove

Theorem 7. *Any non-adaptive uniform-distribution one-sided $1/8$-tester for k-Linear must make at least $\tilde{\Omega}(k \log n)$ queries.*

Theorem 8. *Any adaptive uniform-distribution one-sided $1/8$-tester for k-Linear must make at least $\tilde{\Omega}(\sqrt{k} \log n)$ queries.*

4.1 Lower Bound for Non-Adaptive Testers

We first show the result for non-adaptive testers.

Suppose there is a non-adaptive uniform-distribution one-sided $1/8$-tester $A(s, f)$ for k-Linear that makes q queries, where s is the random seed of the tester and f is the function that is tested. The algorithm has access to f through a black box queries.

Consider the set of linear functions $C = \{g^{(0)}\} \cup \{g^{(\ell)} = x_n + \cdots + x_{n-\ell+1} | \ell = 1, \ldots, k-1\} \subseteq (k-1)$-Linear* where $g^{(0)} = 0$. Any k-linear function is $1/2$-far from every function in C w.r.t. the uniform distribution. Therefore, using the tester A, with probability at least $2/3$, A can distinguish between any k-linear function and functions in C. We boost the success probability to $1 - 1/(2k)$ by running A, $\log(2k)/\log 3$ times, and accept if and only if all accept. We get a tester A' that asks $O(q \log k)$ queries and satisfies

1. If $f \in k$-Linear then with probability 1, $A'(s, f)$ accepts.
2. If $f \in C$ then, with probability at least $1 - 1/(2k)$, $A'(s, f)$ rejects.

Therefore, the probability that for a uniform random s, $A'(s, f)$ accepts for some $f \in C$ is at most $1/2$. Thus, there is a seed s_0 such that $A'(s_0, f)$ rejects for all $f \in C$ (and accept for all $f \in k$-Linear). This implies that there exists a deterministic non-adaptive algorithm $B(= A'(s_0, *))$ that makes $q' = O(q \log k)$ queries such that

1. If $f \in k$-Linear then $B(f)$ accepts.
2. If $f \in C$ then $B(f)$ rejects.

Let $a^{(i)}$, $i = 1, \ldots, q'$ be the queries that B makes. Let M be a $q' \times n$ binary matrix that its i-th row is $a^{(i)}$. Let $x^f \in \{0,1\}^n$ where $x_i^f = 1$ iff i is relevant coordinate in f. Then the vector of answers to the queries of $B(f)$ is Mx^f. If $Mx^f = Mx^g$ for some $g \in C$, that is, the answers of the queries to f are the same as the answers of the queries to g, then $B(f)$ rejects. Therefore, for every $f \in k$-Linear and every $g \in C$ we have $Mx^f \neq Mx^g$. Now since $\{x^f | f \in k\text{--Linear}\}$ is the set of all strings of weight k, the sum (over the field F_2) of every k columns of M is not equal to 0 (zero string) and not equal to the sum of the last ℓ columns of M, for all $\ell = 1, \ldots, k-1$. In particular, if M_i is the ith column of M, for every $i_1, \ldots, i_{k-\ell} \leqslant n-k+1$, $M_{i_1} + \cdots + M_{i_{k-\ell}} + M_{n-\ell+1} + \cdots + M_n \neq M_{n-\ell+1} + \cdots + M_n$ and therefore $M_{i_1} + \cdots + M_{i_{k-\ell}} \neq 0$. That is, the sum of every less or equal k columns of the first $n-k+1$ columns of M is not equal to zero. We then show in Lemma 2 that such matrix has at least $q' = \Omega(k \log n)$ rows. This implies that $q = \Omega((k/\log k) \log n)$.

Let $\pi(n,k)$ be the minimum integer q such that there exists a $q \times n$ matrix over F_2 that the sum of any of its less than or equal k columns is not 0. We have proved

Lemma 1. *Any non-adaptive uniform-distribution one-sided $1/8$-tester for k-Linear must make at least $\Omega(\pi(n - k + 1, k)/\log k)$ queries.*

Now to show that $\Omega(\pi(n-k+1, k)/\log k) = \Omega(k \log n)$ we prove the following result. This lemma follows from Hamming's bound in coding theory. We give the proof for completeness

Lemma 2. *(Hamming's Bound) We have*

$$\pi(n, k) \geqslant \log \sum_{i=0}^{\lfloor \frac{k}{2} \rfloor} \binom{n}{i} = \Omega(k \log(n/k)).$$

Proof. Let M be a $\pi(n,k) \times n$ matrix over F_2 that the sum of any of its less than or equal k columns is not 0. Let $m = \lfloor k/2 \rfloor$ and $S = \{M_{i_1} + \cdots + M_{i_t} \mid t \leqslant m \text{ and } 1 \leqslant i_1 < \cdots < i_t \leqslant n\} \subseteq \{0,1\}^{\pi(n,k)}$ be a multiset. The strings in S are distinct because, if for the contrary, we have two strings in S that satisfies $M_{i_1} + \cdots + M_{i_t} = M_{j_1} + \cdots + M_{j_{t'}}$, then $M_{i_1} + \cdots + M_{i_t} + M_{j_1} + \cdots + M_{j_{t'}} = 0$ (equal columns are cancelled) and $t + t' \leqslant k$, which is a contradiction. Therefore, $2^{\pi(n,k)} \geqslant |S| = \sum_{i=0}^{m} \binom{n}{i}$ and $\pi(n,k) \geqslant \log |S|$.

4.2   Lower Bound for Adaptive Testers

For the lower bound for adaptive testers we take $C = \{g^{(\ell)}\}$ for some $\ell \in \{0, 1, \ldots, k - 1\}$ and get an adaptive algorithm A that makes q queries and satisfies

1. If $f \in k$-Linear then with probability 1, $A(s, f)$ accepts.
2. If $f = g^{(\ell)}$ then, with probability at least $2/3$, $A(s, f)$ rejects.

This implies that there exists a deterministic adaptive algorithm $B = A(s_0, *)$ that makes q queries such that

1. If $f \in k$-Linear then $B(f)$ accepts.
2. If $f = g^{(\ell)}$ then $B(f)$ rejects.

Then, by the same argument as in the case of non-adaptive tester, we get a $q \times n$ matrix M that the sum of every $k - \ell$ columns of the first $n - \ell$ columns of M is not zero. Let $\Pi(n, k)$ be the minimum integer q such that there exists a $q \times n$ matrix over F_2 that the sum of any of its k columns is not 0. Then, we have proved that

Lemma 3. *Any adaptive uniform-distribution one-sided $1/8$-tester for k-Linear must make at least $\Omega(\max_{1 \leqslant \ell \leqslant k} \Pi(n - k, \ell))$ queries.*

In the full paper, [7], we prove

Lemma 4. *We have $\max_{1 \leqslant \ell \leqslant k} \Pi(n, \ell) = \tilde{\Omega}(\sqrt{k} \log n)$.*

References

1. Blais, E.: Improved bounds for testing Juntas. In: Goel, A., Jansen, K., Rolim, J.D.P., Rubinfeld, R. (eds.) APPROX/RANDOM -2008. LNCS, vol. 5171, pp. 317–330. Springer, Heidelberg (2008). https://doi.org/10.1007/978-3-540-85363-3_26
2. Blais, E.: Testing juntas nearly optimally. In: Proceedings of the 41st Annual ACM Symposium on Theory of Computing, STOC 2009, Bethesda, MD, USA, 31 May–2 June 2009, pp. 151–158 (2009). https://doi.org/10.1145/1536414.1536437
3. Blais, E., Brody, J., Matulef, K.: Property testing lower bounds via communication complexity. In: Proceedings of the 26th Annual IEEE Conference on Computational Complexity, CCC 2011, San Jose, California, USA, 8–10 June 2011, pp. 210–220 (2011). https://doi.org/10.1109/CCC.2011.31
4. Blais, E., Kane, D.M.: Tight bounds for testing k-linearity. In: Approximation, Randomization, and Combinatorial Optimization. Algorithms and Techniques - 15th International Workshop, APPROX 2012, and 16th International Workshop, RANDOM 2012, Cambridge, MA, USA, 15–17 August 2012, Proceedings, pp. 435–446 (2012). https://doi.org/10.1007/978-3-642-32512-0_37
5. Blum, M., Luby, M., Rubinfeld, R.: Self-testing/correcting with applications to numerical problems. J. Comput. Syst. Sci. **47**(3), 549–595 (1993). https://doi.org/10.1016/0022-0000(93)90044-W
6. Bshouty, N.J.: Almost optimal distribution-free junta testing. In: 34th Computational Complexity Conference, CCC 2019, 18–20 July 2019, New Brunswick, NJ, USA, pp. 2:1–2:13 (2019). https://doi.org/10.4230/LIPIcs.CCC.2019.2
7. Bshouty, N.H.: An optimal tester for k-linear. Electron. Colloquium Comput. Complex. **27**, 123 (2020). https://eccc.weizmann.ac.il/report/2020/123
8. Buhrman, H., García-Soriano, D., Matsliah, A., de Wolf, R.: The non-adaptive query complexity of testing k-parities. Chicago J. Theor. Comput. Sci. **2013**, 1–11 (2013). http://cjtcs.cs.uchicago.edu/articles/2013/6/contents.html

9. Fischer, E., Kindler, G., Ron, D., Safra, S., Samorodnitsky, A.: Testing juntas. In: 43rd Symposium on Foundations of Computer Science (FOCS 2002), 16–19 November 2002, Vancouver, BC, Canada, Proceedings, pp. 103–112 (2002). https://doi.org/10.1109/SFCS.2002.1181887

10. Goldreich, O.: On testing computability by small width OBDDs. In: Approximation, Randomization, and Combinatorial Optimization. Algorithms and Techniques, 13th International Workshop, APPROX 2010, and 14th International Workshop, RANDOM 2010, Barcelona, Spain, 1–3 September 2010. Proceedings, pp. 574–587 (2010). https://doi.org/10.1007/978-3-642-15369-3_43

11. Goldreich, O. (ed.): Property Testing. LNCS, vol. 6390. Springer, Heidelberg (2010). https://doi.org/10.1007/978-3-642-16367-8

12. Goldreich, O.: Introduction to Property Testing. Cambridge University Press, Cambridge (2017). http://www.cambridge.org/us/catalogue/catalogue.asp?isbn=9781107194052, https://doi.org/10.1017/9781108135252

13. Goldreich, O., Goldwasser, S., Ron, D.: Property testing and its connection to learning and approximation. J. ACM 45(4), 653–750 (1998). https://doi.org/10.1145/285055.285060

14. Halevy, S., Kushilevitz, E.: Distribution-free property-testing. SIAM J. Comput. 37(4), 1107–1138 (2007). https://doi.org/10.1137/050645804

15. Hofmeister, T.: An application of codes to attribute-efficient learning. In: Fischer, P., Simon, H.U. (eds.) EuroCOLT 1999. LNCS (LNAI), vol. 1572, pp. 101–110. Springer, Heidelberg (1999). https://doi.org/10.1007/3-540-49097-3_9

16. Ron, D.: Property testing: a learning theory perspective. Found. Trends Mach. Learn. 1(3), 307–402 (2008). https://doi.org/10.1561/2200000004

17. Ron, D.: Algorithmic and analysis techniques in property testing. Found. Trends Theor. Comput. Sci. 5(2), 73–205 (2009). https://doi.org/10.1561/0400000029

18. Rubinfeld, R., Shapira, A.: Sublinear time algorithms. SIAM J. Discrete Math. 25(4), 1562–1588 (2011). https://doi.org/10.1137/100791075

19. Rubinfeld, R., Sudan, M.: Robust characterizations of polynomials with applications to program testing. SIAM J. Comput. 25(2), 252–271 (1996). https://doi.org/10.1137/S0097539793255151

20. Saglam, M.: Near log-convexity of measured heat in (discrete) time and consequences. In: 59th IEEE Annual Symposium on Foundations of Computer Science, FOCS 2018, Paris, France, 7–9 October 2018, pp. 967–978 (2018). https://doi.org/10.1109/FOCS.2018.00095

Machine Learning Advised Ski Rental Problem with a Discount

Arghya Bhattacharya[1]([✉]) and Rathish Das[2]

[1] Stony Brook University, Stony Brook, NY, USA
argbhattacha@cs.stonybrook.edu
[2] University of Waterloo, Waterloo, ON, Canada
rathish.das@uwaterloo.ca

Abstract. Traditional online algorithms are designed to make decisions online in the face of uncertainty to perform well in comparison with the optimal offline algorithm for the worst-case inputs. On the other hand, machine learning algorithms try to extrapolate the pattern from the past inputs to predict the future and take decisions online on basis of the predictions to perform well for the average-case inputs. There have been recent studies to augment traditional online algorithms with machine learning oracles to get better performance for all the possible inputs. The machine learning augmented online algorithms perform provably better than the traditional online algorithms when the error of the machine learning oracle is low for the worst-case inputs and all other average-case inputs.

In this paper, we integrate the advantages of the traditional online algorithms and the machine learning algorithms in the context of a novel variant of the ski rental problem. Firstly, we propose the ski rental problem with a discount: in this problem, the rent of the ski, instead of being fixed over time, varies as a function of time. Secondly, we discuss the design and performance evaluation of the online algorithms with machine learning advice to solve the ski rental problem with a discount. Finally, we extend this study to the situation where multiple independent machine learning advice is available. This algorithm design framework motivates to redesign of several online algorithms by augmenting them with one or more machine learning oracles to improve the performance.

Keywords: Ski rental problem · Online algorithm · ML advice

1 Introduction

The conventional online algorithms try to find a decision-making strategy that is good for all possible situations. Hence, the goal of such an algorithm is to provide a worst-case guarantee for every possible input. The usual way to understand the effectiveness of an online algorithm is to compare its performance with the optimal offline algorithm that is visionary and hence can see the future. Traditionally we define *competitive ratio* of an online algorithm as the ratio of the

© Springer Nature Switzerland AG 2022
P. Mutzel et al. (Eds.): WALCOM 2022, LNCS 13174, pp. 213–224, 2022.
https://doi.org/10.1007/978-3-030-96731-4_18

performance of the online algorithm to the same of the optimal offline algorithm for the worst possible input [3,8].

Machine learning (ML) systems, on the contrary, are fundamentally different and serve distinct purposes altogether [17,18]. ML systems learn from past situations and extrapolate patterns encountered in the past to predict the future. They come with guarantees of *expected generalization error*: expected error in the prediction of the ML model for input data that the model has not seen yet. Hence, ML systems are "trained" to be good on the average-case inputs minimizing the expected loss. The interest lies in optimizing the prediction for most of the inputs at the expense of a few outliers. However, there is no theoretical guarantee on how "bad" the predictions are for those few outliers. The machine learning approach has another shortcoming: the expected generalization error guarantees are valid only if the training and testing data belong to the same probability distribution.

While machine learning approaches try to perform well for the average-case inputs, the conventional online algorithms compete with the optimal offline algorithm for the worst-case inputs. Online algorithms do not assume anything about the input sequence and stay oblivious of the future inputs. While designing the online algorithms, the objective is to retain the competitive ratio as small as possible. Hence, the online algorithms become overly cautious about the "bad" inputs where the algorithms would perform much worse than the optimal offline algorithm. It leads us to a higher competitive ratio for supposedly "easy" inputs—hence in a simulation of a satisfactorily reasonable real-world input sequence, unsurprisingly, the conventional algorithms are likely to perform worse on average than the ML algorithms. Average-case competitive analysis has been used to evaluate and capture this performance drop of traditional online algorithms [4,19].

Another recent line of work tends to improve the competitive ratio by augmenting traditional online algorithms with supportive information from machine-learned oracles. These algorithms found applications in several problems such as competitive paging [13], job scheduling problem [10,11,16], etc. In 2017, Medina and Vassilvitskii [14] initiated the study on machine-learned online algorithms and used them to find the expected price of bids for revenue optimization. Lykouris and Vassilvitskii [13] extended the framework of designing online algorithms with ML advice models and formulated two important functional aspects of such systems: *consistency* and *robustness*. A consistent system performs well when the ML predictions are fairly correct, and a robust system refrains from performing exceedingly poorly as the predictions degrade. They designed a competitive caching algorithm based on the prediction of the ML model for the subsequent arrival time of an element as the elements come online from the input sequence.

Ski Rental Problem. The classical *ski rental problem* [8], a canonical example of the large class of online *rent-or-buy problem*, has been of interest in this new avenue of research [1,5,10]. The rent-or-buy problem represents a scenario where an online decision maker faces a situation where it has to optimize between two choices: one is to *rent*, with a small recurrent cost, another is to *buy*, by paying up a comparatively large cost upfront with no expenditure thereafter. The uncertainty stems from the fact that the span of usage is unknown in advance. For instance, in case of the ski rental problem, the skier is unaware of the length of the ski session until it ends, and her objective is to optimize between two choices: she either rents a ski at a unit price every day or buys the ski for a higher price to ski for free from then on. Intuitively, for a long span of usage, purchasing makes sense, but for the short term renting is more optimized.

Kumar et al. [10] introduced ML augmented online algorithms for the classical ski rental problem instance, where the ML oracle predicts the number of days to ski. Gollapudi and Panigrahi [5] extended this notion to multiple experts, where multiple ML oracles augment the traditional online algorithm and predict the future. Anand et al. [1] worked to answer another complimentary question in this context: whether we can redesign the ML algorithms to serve the online algorithms well in the context of an online rent-or-buy problem.

The ski rental problem is well-studied in economics and computer science. Researchers studied several generalizations to the problem to imbibe real-life scenarios such as deciding whether to buy or rent a house, taking corporate decisions on whether to invest in a new office, data-center, or rent utilities. Irani and Ramanathan [6] studied a variant of this problem where the rental price remains fixed, but the purchasing price fluctuates. Furthermore, Meyerson [15] proposed the **parking permit problem** that introduces multiple purchasing options with different time intervals; Khanafer et al. [9] extended this notion to rent optimization for cloud servers. Lotker et al. [12] proposed **multi-slope ski rental problem**: the skier has to optimize among several choices for skis of separate rental and purchasing price. Zhang et al. [20] explored a generalization of the parking permit problem with multiple discount options available for the skis. However, in all these extensions, the price of any individual rental options is considered to be fixed.

Ski Rental Problem with a Discount. In this paper, we design and evaluate algorithms for the *ski rental problem with a discount*: the purchasing price is constant, but as the skier rents for a longer period she gets a discount.

El-Yaniv et al. took the first step in this direction by introducing a fixed nominal interest rate on the rental price, where the purchasing price remains fixed. The interest effectively introduces a discount on the rental price where the discount is a fraction of the rent of the first day. The discount remains fixed with respect to time, i.e. the rental price decreases linearly. They performed the traditional competitive analysis of the problem. We further generalize the problem by considering the discount rate to be arbitrary than fixed.

Ski rental problem with a discount has an arbitrary non-increasing function for the rental price of the ski: we do not restrict the rental price, except that the price does not increase from what it was on the previous day; see Sect. 3 for the detailed problem formulation. The above scenario turns up on several occasions in real life. For example, we need a computing resource for an unknown period, and we have two choices, either to buy with a fixed purchasing cost upfront or to rent where the rental cost reduces as a function of time. The approach to improve the performance is by aiding the online algorithm with additional information from one or multiple oracles.

One interesting aspect of the ski rental problem with a discount is that it approaches to simulate the *net present value* of an asset. That is, it takes into account the market interest rate, which is an essential part of any financial model. For example, let the rent of a ski be 1 dollar per day, the price of the ski be 10 dollars, and the number of days of usage of the ski is 8. Additionally, let the market interest rate be 2% for a span of less than 5 days and 1% for 5 days or more. In case the skier purchases the ski on the first day, she spends 10 dollars. In case she rents it for all 8 days, she spends $8 - (1 * 0.02)(1 + 2 + 3 + 4) - (1 * 0.01)(5 + 6 + 7) = 7.62$ dollars. Effectively, the rental price can be modeled as a variable that initially is 1 dollar, and reduces by 2% of the initial value till the 5th day, then onward by 1% of the initial value each day until the rent becomes zero. Furthermore, contiguous and non-contiguous usage patterns do not matter in the case of the traditional ski rental problem. However, for the ski rental problem with a discount, a gap period of nonusage by the skier matters as the rent changes during this period.

Results. Our goal is to improve the conventional online algorithms by designing algorithms with ML advice to solve the ski rental problem with a discount. If the algorithm blindly follows the ML oracle, the competitive ratio of the algorithm is unbounded (Lemma 1). However, instead of following the oracle blindly if the algorithm has a trust parameter λ, it has improved robustness of $1 + 1/\lambda$ without losing consistency. This algorithm has an overall competitive ratio $\min\{1 + \frac{1}{\lambda}, 1 + \frac{r\eta}{(1-\lambda)OPT}\}$ (Lemma 2). Finally, we extend the problem to the situation where two separate independent ML advisers are present. If one of the advisers is correct, the algorithm is ϕ-competitive; where ϕ is the golden ratio (see Sect. 3.2 and Lemma 3). In case both the ML advisers are erroneous, the competitive ratio of the algorithm is $\max\{\phi + 1, 1/(\phi - 1 - r\eta/b)\}$; where η is the prediction error of the relatively better predictor, r is the rent for the first day and b is the purchasing price (Lemma 4).

Paper Overview. The rest of the paper is organized as follows. Section 2 explains the traditional ski rental problem, classical competitive analysis, and the framework to analyze the classical online algorithms with machine learning augmentation. Section 3 introduces the ski rental problem with a discount, explains the classical algorithms and design framework of the online algorithms with single and multiple machine learning advisers. Finally, Sect. 4 concludes by stating some future directions of this work.

2   Preliminaries

In order to discuss the online problems and the existing and potential solutions, it is imperative to study the basics of machine learning and classical competitive analysis, and define a framework to analyze algorithms augmented by a machine learning oracle.

Classical Competitive Analysis. The classical competitive analysis is a tool to analyze the performance of an online algorithm. There is an adversary that decides the input sequence for an online problem. A conventional online algorithm makes decisions without any futuristic knowledge as each element from the input sequence arrives. We compare the performance of the online algorithm with the same of the optimal offline algorithm, which has the advantage of hindsight for all the sequences possible and finds a worst-case guarantee for the online algorithm [2]. For an optimization problem Π and a set of instances Σ_Π, τ is an input sequence where $\tau \in \Sigma_\Pi$, $OPT(\tau)$ is the cost incurred by the optimal offline algorithm, $ALG(\tau)$ is the cost incurred by an online algorithm, the online algorithm has a competitive ratio of α if $ALG(\tau) \leq \alpha \times OPT(\tau) \; \forall \; \tau$.

Ski Rental Problem. In the ski rental problem [8], a skier's goal is to optimize by either renting a ski at a unit price every day or buy the ski for a higher price b, and ski for free from then on. The skier does not know the length of the ski season in advance and only knows it when the season ends. Intuitively, if the ski session lasts for x days and $x > b$, the skier should buy the ski; otherwise, she should keep renting. The optimal offline algorithm being a visionary will do the same, thus incur a cost of $\min\{x, b\}$.

The ski rental problem and the potential strategies for the skier are widely studied in the last two decades [8,9,12]. The well-known best deterministic strategy for the skier, the **break-even algorithm**, is to rent the ski for $(b-1)$ days and to buy the ski on the next day. The break-even algorithm is 2-competitive, and no other deterministic algorithm can do better than that. Furthermore, it is worthy to note that there is a randomized algorithm [7,8] that is $(e/e-1)$-competitive (roughly 1.58-competitive).

Machine Learning Basics. This section explains the basics of a machine learning approach to solve a problem. In the ML problems, we have a **feature space** Σ, which says the salient attributes associated with each item and a set of corresponding **labels**, Δ. An example is a tuple (σ, δ), where $\sigma \in \Sigma$ describes the specific attributes of the sample item, and $\delta \in \Delta$ provides the corresponding label [13,16]. A **hypothesis** is a mapping (either deterministic or probabilistic) $h : \Sigma \rightarrow \Delta$, where $h \in H$, i.e. h belongs to a class of predictors or mappings, H. Depending on that, the predictor predicts δ ($\delta \in \Delta$) as the label of σ ($\sigma \in \Sigma$) with a deterministic or probabilistic strategy. The **loss function** is defined to quantify the error in the prediction of the machine learning oracle by $l : \Delta \times \Delta \rightarrow \mathcal{R}^{\geq 0}$.

Augmentation by Machine Learning Oracle. This section revisits the framework to analyze an algorithm with ML augmentation developed by Lykouris and Vassilvitskii [13]. As mentioned earlier, we assume a feature space Σ and a label space Δ and a predictor $h : \Sigma \rightarrow \Delta$ where $h \in H$ where H is a class of predictors. For an optimization problem Π, a given input sequence is $\tau = \{\tau_1, \tau_2, \ldots \tau_{|\tau|}\}$, where the length of the sequence is $|\tau|$, τ_i has feature $\sigma_i \in \Sigma$ and a label $\delta_i \in \Delta$, whereas the predicted label is $h(\sigma_i)$. The error in prediction or the loss function for the input sequence τ, $\eta_l(h, \tau)$, is defined as the difference between the predicted label and the true label: $\eta_l(h, \tau) = \sum_i l(\delta_i, h(\sigma_i))$.

We define $H_l(\varepsilon)$ to be the class of ε-accurate predictors for loss function l if and only if h is ε-accurate $\forall h \in H_l(\varepsilon)$, i.e. $\eta_l(h, \tau) \leq \varepsilon \times OPT_\Pi(\tau)$ $\forall$ input τ where the cost of the optimal offline algorithm is $OPT_\Pi(\tau)$. Algorithm ALG is ε-assisted if it has access to an ε-accurate predictor h. The competitive ratio of an ε-assisted algorithm ALG, $\alpha_{ALG,l}(\varepsilon)$, is expressed as $\alpha_{ALG,l}(\varepsilon) = \max_{\tau, h \in H_l(\varepsilon)} \alpha_{ALG,h}(\tau)$.

The goal of the ML oracle is twofold: the consistency goal where it improves the performance of the online algorithm when the predictions by the ML oracles are fairly accurate and the robustness goal where the performance of the online algorithm does not degrade significantly as the predictions degrade [5, 10]. We define an algorithm ALG to be β-consistent if the algorithm works no worse than β times of the optimal offline algorithm when the best prediction has no error, i.e. $\alpha_{ALG,l}(0) = \beta$. On the other hand where the best case prediction has an error ε, ALG is γ-robust if $\alpha_{ALG,l}(\varepsilon) = O(\gamma(\varepsilon))$. These goals have to be met by the online algorithm without the knowledge of the quality of predictions. For the worst-case prediction, we define an algorithm ALG to be α-competitive if $\alpha_{ALG,l}(\varepsilon) \leq \alpha$ $\forall \varepsilon$.

3   Ski Rental Problem with a Discount

This section introduces the ski rental problem with a discount in a mathematical setting. This problem is an extension of the traditional ski rental problem. Next, we explain algorithms augmented with ML oracle and analyze their performance and juxtapose them with the same of the conventional online algorithm.

In our notion of the ski rental problem with a discount, the initial daily rent of the ski is r, i.e. it takes r unit price to rent on the first day. The cost to buy the ski is b, and the actual number of days to snow in the season is x. However, the skier does not know x until it does not snow anymore. There is a discount on the rent of the ski; the rent on the tth day, $rent(t)$, is defined such that $rent(t) \geq 0$, and $rent(t + 1) \leq rent(t) \forall t$—it embeds that the rental price can be any arbitrary non-increasing function. The way to calculate the total rent up to the tth day is to add the rents from the first day till the tth day; we refer to this as $rent(1 \ldots t)$ where $rent(1 \ldots t) = \sum_{i=0}^{t} rent(i)$.

We define t_b such that $rent(1 \ldots t_b - 1) < b$ and $rent(1 \ldots t_b) \geq b$. Note that the skier knows both the $rent$ function and the purchasing price in advance. Hence, she can calculate t_b any time but the source of uncertainty comes from the actual number of days it snows.

Optimal Offline Algorithm. The optimal offline algorithm being a visionary, for $x < t_b$ it keeps renting the ski, and for $x \geq t_b$ it buys the ski on the first day itself. Hence, the cost of the optimal offline algorithm can be written as $OPT = \min\{rent(1 \ldots x), b\}$.

Break-Even Algorithm. The best deterministic strategy, the break-even algorithm, works the same way as it does in the case of the conventional ski rental problem. It rents the ski till day $t_b - 1$, and buys the ski on day t_b. Note that the worst-case appears for $x = t_b$. In this case, $OPT = b$ and $ALG = rent(1 \ldots t_b - 1) + b \le 2b$. This shows that the break-even algorithm has a competitive ratio of 2 for the ski rental problem with a discount.

3.1   Algorithm Augmented with a Single ML Adviser

This section introduces ML augmented online algorithms to solve the ski rental problem with a discount. First we propose a naive ML advised algorithm. Next, we redesign the algorithm with a "trust" parameter and evaluate the improved robustness.

Naive ML Advised Algorithm. First, we present Algorithm 1 based on an ML advice [10]. The ML based predictor estimates the number of days it is going to snow as y. Such a prediction can be made reasonably well, for instance, by building models based on weather forecasts and the past behavior of other skiers. The error in the prediction can be quantified as $\eta = |y - x|$. Algorithm 1 is 1-consistent when $\eta = 0$, i.e. the algorithm performs optimally when the prediction is correct. However, the algorithm is not robust; hence the competitive ratio is unbounded and can be arbitrarily large for poor predictions. The performance of Algorithm 1 is analyzed by the Lemma 1.

Algorithm 1. Naive ML Augmented Algorithm

Require: a sequence of past input $X = \{x_i\}$ and current input x
  Train the ML Oracle using X
  ML Oracle predicts y
  if $y \ge t_b$ **then**
    buy the ski on the first day
  else if $y < t_b$ **then**
    keep renting the ski
  end if
  Train the loss

Lemma 1. *Let ALG be the cost of the naive ML advised algorithm, OPT be the cost of the optimal offline algorithm, η be the error in the prediction of the ML model, and r be the rent of the ski on the first day. For all the possible inputs, the following bound is valid.*

$$ALG < \max\{OPT(\eta + 1), OPT + r\eta\}.$$

Proof. The proof is given in the full version of the paper.

Corollary 1. *Let ALG and OPT be the cost of the naive ML advised algorithm and the optimal offline algorithm respectively, the error in the ML prediction model be η where $\eta = k.OPT$. The cost of the naive algorithm, ALG, is unbounded.*

Proof. $ALG < k \cdot OPT^2 + OPT < k \cdot b^2 + b$, where $k \geq 1$ and k is unbounded. On the other hand, $ALG < OPT(1 + r \cdot k) < b(1 + r \cdot k)$ where $0 < k < 1$.

ML Advised Algorithm with Trust Parameter. The shortcoming of the naive ML advised algorithm is that it puts too much trust in the ML oracle. Hence, as the ML oracle fails and incurs a large prediction error, the performance degrades harshly leading to a high competitive ratio in the worst-case input. As per our definition of robustness in Sect. 2, the naive ML advised algorithm is not robust; hence we aim to design a more robust algorithm.

Algorithm 2. ML advised algorithm with trust parameter

Require: a sequence of past input $X = \{x_i\}$ and current input x
    Trust factor: $\lambda \in [0, 1]$
    Train the ML oracle using X
    if $y \geq t_b$ **then**
        buy on $\lceil \lambda t_b \rceil$-th day
    else if $y < t_b$ **then**
        buy on $\lceil t_b / \lambda \rceil$-th day
    end if
    Train the loss

Algorithm 2 improves by acquiring more robustness over Algorithm 1 while not losing much in terms of consistency. It introduces a hyperparameter, *trust factor* $\lambda \in [0, 1]$; this parameter introduces a tradeoff between the traditional algorithm and the ML advice. This helps the algorithm to acquire a tradeoff between consistency and robustness. Note that $\lambda \to 0$ refers to greater trust in the ML predictor. It leads to better consistency, i.e. the algorithm performs better than the naive ML augmented algorithm when the prediction is good. On the other hand, $\lambda \to 1$ refers to the case when the algorithm has lesser trust in the ML predictor. Algorithm 2 retains a good consistency compared to the naive algorithm while achieving more robustness. That is, the algorithm performs better when the quality of prediction is not so good. The performance of Algorithm 2 is analyzed by the Lemma 2.

Lemma 2. *Let ALG be the cost of the ML advised algorithm with trust parameter λ, OPT be the cost of the optimal offline algorithm, η be the error in the prediction of the ML adviser, and r be the rent of the ski on the first day. Then,*

$$ALG < \min\{OPT(1 + \frac{1}{\lambda}), OPT + \frac{r\eta}{(1 - \lambda)}\}.$$

Proof. The complete proof is given in the full version. In the case where $\lceil \lambda t_b \rceil \leq x < t_b$ and $y \geq t_b$, the worst-case appears at $x = \lceil \lambda t_b \rceil$.

$$ALG - OPT = b - rent(\lceil \lambda t_b \rceil) \leq b \leq \frac{1}{\lambda}(\lambda b) \leq \frac{1}{\lambda} rent(1 \ldots \lceil \lambda t_b \rceil) \leq \frac{OPT}{\lambda}.$$

If the algorithm has high trust on the ML oracle, it buys the ski on the first day instead of on $\lceil \lambda t_b \rceil$. In the case where $t_b \leq x \leq \lceil t_b/\lambda \rceil$ and $y < t_b$, the worst-case appears at $x = \lceil t_b/\lambda \rceil$.

$$ALG - OPT \leq b + r(\lceil t_b/\lambda \rceil - t_b) \leq b + b(1/\lambda - 1) \implies ALG = OPT(1 + \frac{1}{\lambda}).$$

For $x > \lceil t_b/\lambda \rceil$ and $y < t_b$, $ALG = b + rent(1 \ldots \lceil t_b/\lambda \rceil - 1)$, and $OPT = b$.

$$\eta > \lceil t_b/\lambda \rceil - t_b = t_b(1/\lambda - 1) \implies ALG \leq OPT + r(\frac{t_b}{\lambda}) \leq OPT + \frac{r\eta}{1 - \lambda}.$$

From this analysis, we see $ALG \leq \min\{OPT(1 + \frac{1}{\lambda}), OPT + \frac{r\eta}{(1-\lambda)}\}$.

3.2   Algorithm Augmented with Two ML Advisers

This section explains the variant of the problem where multiple independent ML advisers are present. Gollapudi and Panigrahi [5] showed that the presence of more advisers helps to improve the performance as the chance of good predictions increases. They also observed that as the number of ML advisers increases, the competitive ratio gets better, but the analysis remains the same. In this paper, we will restrict ourselves to the analysis where two advisers are present simultaneously, leading to two different predictions. In this case, ALG is γ-competitive if and only if $ALG \leq \gamma.OPT \; \forall \, x, \eta$, where x is the actual number of days it snows and η is the absolute error in prediction y by the best predictor, $\eta = \min\{|y - x|\}$.

One of the Predictions is Correct. The first scenario we study is where one of the two predictions is correct. The challenge here is to choose the correct one from one of these two predictions without the knowledge of its identity. Let us consider, the two predictions are y_1 and y_2, where $y_1 \geq y_2$, and the naive strategy is to blindly follow one of them. In case of $y_1 \geq y_2 \geq t_b$, the algorithm buys the ski on the first day, i.e., $ALG = OPT$. On the other hand, when $y_2 \leq y_1 < t_b$, the algorithm keeps renting the ski forever, resulting in $ALG = OPT$ again. In case of the other two possibilities, $y_1 \geq t_b$ and $y_2 < t_b$. If y_1 is correct and the algorithm blindly follows y_2, it keeps renting and the competitive ratio in this case becomes $\gamma = \frac{b}{rent(1...y_1)}$. Otherwise, if y_2 is correct and the algorithm trusts y_1, it buys the ski on the first day, leading to the competitive ratio, $\gamma = \frac{b}{rent(1...y_2)}$. In both the latter cases, γ is unbounded. We approach to modify the strategy so that when y_1 is correct and we trust y_2, instead of keeping renting forever, we rent till y_2 and then buy. Hence, the competitive ratio $= \gamma = \frac{rent(1...y_2)+b}{b} \leq 2$. Hence, although we get rid of the unbounded competitive ratio, we make no improvement from the traditional online algorithm. Necessarily, the constraint here is that the algorithm should make the decision to buy before y_2; in fact, it is the best to buy on the first day itself.

The solution lies in striking a balance between the two predictions and finding a tradeoff between them. Gollapudi and Panigrahi [5] developed such a strategy based on the golden ratio. We solve the equation $\frac{b}{z} = \frac{b+z}{b}$ and find $z = (\phi - 1)b$ where ϕ is the golden ratio (roughly equals 1.618). We define t_ϕ such that $rent(1 \ldots t_\phi - 1) < (\phi - 1)b$ and $rent(1 \ldots t_\phi) \geq (\phi - 1)b$. The algorithm decides to buy based on the value of t_ϕ as shown in Algorithm 3.

Algorithm 3. ML advised algorithm with two advisers

Require: a sequence of past input $X = \{x_i\}$ and current input x
  Train the ML oracle using X
  Predictions are y_1 and y_2 where $y_1 > y_2$
  if $y_1, y_2 \geq t_b$ **then**
    buy on the first day
  else if $y_1, y_2 < t_b$ **then**
    keep renting
  else if $t_\phi \leq y_2 \leq t_b$ and $y_1 > t_b$ **then**
    buy on the first day
  else if $y_2 < t_\phi$ and $y_1 > t_b$ **then**
    rent till y_2, buy on the next day
  end if
  Train the loss

Lemma 3. *Let the cost of the augmented algorithm with two ML advisers be ALG and one of the advisers is correct in the prediction, OPT be the cost of the optimal offline algorithm, and ϕ be the golden ratio. Then,*

$$ALG < \phi \cdot OPT.$$

Proof. The worst-case may have two possibilities. There are two possible cases: $y_2 = t_\phi$ and y_2 is correct, and in the second case $y_2 < t_\phi$ and y_1 is correct respectively.

$$\gamma = \frac{b}{rent(1 \ldots t_\phi)} \leq \frac{b}{(\phi - 1)b} = \phi \text{ and } \gamma = \frac{rent(1 \ldots y_2) + b}{b} \leq \frac{(\phi - 1)b + b}{b}.$$

From the above analysis, it shows that Algorithm 3 holds ϕ-competitiveness.

Both Predictions Have Non-zero Errors. This section generalizes the problem with two ML advisers to the case where both the predictors have non-zero prediction errors. This represents a more realistic situation where we assume that the best prediction has an error η. We modify Algorithm 3 and design Algorithm 4 to avoid unbounded error for the case $y_1, y_2 < t_b$. Next, we evaluate the performance of Algorithm 4 and observe the performance bounds. The predictions of the two experts are y_1 and y_2 where $y_2 \leq y_1$.

Lemma 4. *Let ALG be the cost of the augmented algorithm with two ML advisers where both the advisers have prediction errors. The best predictor has*

Algorithm 4. ML advised algorithm with two erroneous advisers

---

Require: a sequence of past input $X = \{x_i\}$ and current input x
  Train the ML oracle using X
  Predictions are y_1 and y_2 where $y_1 > y_2$
  if $y_2 \geq t_\phi$ **then**
    buy on the first day
  else if $y_2 < t_\phi$ **then**
    rent till y_2, buy on the next day
  end if
  Train the loss

---

an error η, the rent of the ski for the first day is r, and the price of the ski b. OPT is the cost of the optimal offline algorithm and ϕ is the golden ratio. Then,

$$ALG \leq \max\{\frac{OPT}{\phi - 1 - r\eta/b}, (\phi + 1)OPT\}.$$

Proof. The worst-case inputs for Algorithm 4 appear in two scenarios. Firstly, when $y_1 > t_b$, $y_2 \geq t_\phi$ and the actual number of days $x = t_\phi - \varepsilon$, $ALG \leq OPT(\frac{1}{\phi-1-r\eta/b})$. On the other hand, when $y_1 > t_b$, $y_2 < t_\phi$, and $x = t_\phi + \varepsilon < t_b$, $ALG \leq OPT(\phi + 1)$. The detailed proof is given in the full version of the paper.

4 Conclusion

We proposed the ski rental problem with a discount in a mathematical setting and designed online algorithms with single and multiple ML advice. The future direction of this study would be to explore the effect of several distributions of the prediction error of the ML advisers on the competitive ratio. We would also like to explore designing randomized algorithms useful for this setup. We also want to test empirically the interplay between a randomly chosen trust parameter and the robustness of Algorithm 2 over realistic data-sets.

Acknowledgments. This research was supported in part by NSF grant CCF-1617618, the Canada Research Chairs Program, and NSERC Discovery Grants.

References

1. Anand, K., Ge, R., Panigrahi, D.: Customizing ml predictions for online algorithms. In: Proceedings of the 37th International Conference on Machine Learning. PMLR (2020)
2. Awerbuch, B., Azar, Y.: Buy-at-bulk network design. In: Symposium on Foundations of Computer Science (1997)
3. Borodin, A., El-Yaniv, R.: Online Computation and Competitive Analysis. Cambridge University Press, Cambridge (1998)
4. Fujiwara, H., Iwama, K.: Average-case competitive analyses for ski-rental problems. Algorithmica **42**, 95–107 (2005)

5. Gollapudi, S., Panigrahi, D.: Online algorithms for rent-or-buy with expert advice. In: Proceedings of the 36th International Conference on Machine Learning (ICML). ACM (2019)
6. Irani, S., Ramanathan, D.: The problem of renting versus buying. Private communication, August 1998
7. Karlin, A.R., Kenyon, C., Randall, D.: Dynamic TCP acknowledgement and other stories about $e/(e-1)$. Algorithmica **36**(3), 209–224 (2003)
8. Karlin, A.R., Manasse, M.S., McGeoch, L.A., Owicki, S.: Competitive randomized algorithms for nonuniform problems. Algorithmica **11**(6), 542–571 (1994)
9. Khanafer, A., Kodialam, M., Puttaswamy, K.P.N.: The constrained ski-rental problem and its application to online cloud cost optimization. In: Proceedings of IEEE INFOCOM, Turin, Italy, pp. 1492–1500. INFOCOM, IEEE (2013)
10. Kumar, R., Purohit, M., Svitkina, Z.: Improving online algorithms via ML predictions. In: Proceedings of 32nd Conference on Neural Information Processing Systems (NeurIPS) (2018)
11. Lattanzi, S., Lavastida, T., Moseley, B., Vassilvitskii, S.: Online scheduling via learned weights. In: Proceedings of the 2020 ACM-SIAM Symposium on Discrete Algorithms (SODA), pp. 1859–1877 (2020)
12. Lotker, Z., Patt-Shamir, B., Rawitz, D.: Rent, lease or buy: randomized algorithms for multislope ski rental. In: Proceedings of 25th Annual Symposium on the Theoretical Aspects of Computer Science, Bordeaux, France, pp. 503–514. STACS (2008)
13. Lykouris, T., Vassilvitskii, S.: Competitive caching with machine learned advice. In: Proceedings of the 35th International Conference on Machine Learning (2018)
14. Medina, A.M., Vassilvitskii, S.: Revenue optimization with approximate bid predictions. In: 31st Conference on Neural Information Processing Systems (NeurIPS) (2017)
15. Meyerson, A.: The parking permit problem. In: 46th Annual IEEE Symposium on Foundations of Computer Science (FOCS), pp. 274–284 (2005)
16. Mitzenmacher, M.: Scheduling with predictions and the price of misprediction. In: Proceedings of Innovations in Theoretical Computer Science (ITCS) (2020)
17. Rakhlin, A., Sridharan, K.: Online learning with predictable sequences. In: Proceedings of the 26th Annual Conference on Learning Theory (COLT) (2013)
18. Sculley, D., et al.: Hidden technical debt in machine learning systems. In: Proceedings of the 28th International Conference on Neural Information Processing Systems (NIPS), pp. 2503–2511. MIT Press (2015)
19. Xu, Y., Xu, W., Li, H.: On the on-line rent-or-buy problem in probabilistic environments. J. Glob. Optim. **38**, 1–20 (2007)
20. Zhang, G., Poon, C.K., Xu, Y.: The ski-rental problem with multiple discount options. Inf. Process. Lett. **111**(18), 903–906 (2011)

Parameterized Complexity

On the Harmless Set Problem Parameterized by Treewidth

Ajinkya Gaikwad and Soumen Maity$^{(\boxtimes)}$

Indian Institute of Science Education and Research, Pune, India
ajinkya.gaikwad@students.iiserpune.ac.in, soumen@iiserpune.ac.in

Abstract. Given a graph $G = (V, E)$, a threshold function $t : V \rightarrow \mathbb{N}$ and an integer k, we study the HARMLESS SET problem, where the goal is to find a subset of vertices $S \subseteq V$ of size at least k such that every vertex v in V has less than $t(v)$ neighbors in S. We enhance our understanding of the problem from the viewpoint of parameterized complexity. Our focus lies on parameters that measure the structural properties of the input instance. We show that the HARMLESS SET problem with majority thresholds is W[1]-hard when parameterized by the treewidth of the input graph. On the positive side, we obtain a fixed-parameter tractable algorithm for the problem with respect to neighbourhood diversity.

Keywords: Parameterized Complexity · FPT · W[1]-hard · treewidth

1 Introduction

Social networks are used not only to stay in touch with friends and family, but also to spread and receive information on specific products and services. The spread of information through social networks is a well-documented and well-studied topic. Kempe, Kleinberg, and Tardos [15] initiated a model to study the spread of influence through a social network. One of the most well known problems that appear in this context is TARGET SET SELECTION introduced by Chen [6] and defined as follows. We are given a graph, modeling a social network, where each node v has a (fixed) threshold $t(v)$, the node will adopt a new product if $t(v)$ of its neighbors adopt it. Our goal is to find a small set S of nodes such that targeting the product to S would lead to adoption of the product by a large number of nodes in the graph. This problem may occur for example in the context of disease propagation, viral marketing or even faults in distributed computing [12,19]. This problem received considerable attention in a series of

A. Gaikwad—The first author gratefully acknowledges support from the Ministry of Human Resource Development, Government of India, under Prime Minister's Research Fellowship Scheme (No. MRF-192002-211).

S. Maity—The second author's research was supported in part by the Science and Engineering Research Board (SERB), Govt. of India, under Sanction Order No. MTR/2018/001025.

© Springer Nature Switzerland AG 2022
P. Mutzel et al. (Eds.): WALCOM 2022, LNCS 13174, pp. 227–238, 2022.
https://doi.org/10.1007/978-3-030-96731-4_19

papers from classical complexity [5,7,12,20], polynomial time approximability [1,6], parameterized approximability [3], and parameterized complexity [4,8,18]. A natural research direction considering this fact is to look for the complexity of variants or constrained version of this problem. Bazgan and Chopin [2] followed this line of research and introduced the notion of harmless set. Throughout this article, $G = (V, E)$ denotes a finite, simple and undirected graph. We denote by $V(G)$ and $E(G)$ its vertex and edge set respectively. For a vertex $v \in V$, we use $N(v) = \{u : (u, v) \in E(G)\}$ to denote the (open) neighbourhood of v in G. The degree $d(v)$ of a vertex $v \in V(G)$ is $|N(v)|$. For a subset $S \subseteq V(G)$, we use $N_S(v) = \{u \in S : (u, v) \in E(G)\}$ to denote the (open) neighbourhood of vertex v in S. The degree $d_S(v)$ of a vertex $v \in V(G)$ in S is $|N_S(v)|$. A harmless set consists of a set S of vertices with the property that no propagation occurs if any subset of S gets activated. In other words, a harmless set is defined as a converse notion of a target set. More formally,

Definition 1. [2] A set $S \subseteq V$ is a *harmless set* of $G = (V, E)$, if every vertex $v \in V$ has less than $t(v)$ neighbours in S.

Note that in the definition of harmless set, the threshold condition is imposed on every vertex, including those in the solution S. As mentioned in [2], another perhaps more natural definition could have been a set S such that every vertex $v \notin S$ has less than $t(v)$ neighbours in S. This definition creates two problems. First, it makes HARMLESS SET problem meaningless as the whole set of vertices of the input graph would be a trivial solution. Second, there might be some propagation steps inside S if some vertices are activated in S. In this paper, we consider the HARMLESS SET problem under structural parameters. We define the problem as follows:

HARMLESS SET
Input: A graph $G = (V, E)$, a threshold function $t : V \to \mathbb{N}$ where $1 \leq t(v) \leq d(v)$ for every $v \in V$, and an integer k.
Question: Is there a harmless set $S \subseteq V$ of size at least k?

The majority threshold is $t(v) = \lceil \frac{d(v)}{2} \rceil$ for all $v \in V$. We now review the concept of a tree decomposition, introduced by Robertson and Seymour in [21]. Treewidth is a measure of how "tree-like" the graph is.

Definition 2. [10] A *tree decomposition* of a graph $G = (V, E)$ is a tree T together with a collection of subsets X_t (called *bags*) of V labeled by the nodes t of T such that $\bigcup_{t \in T} X_t = V$ and (1) and (2) below hold:

1. For every edge $uv \in E(G)$, there is some t such that $\{u, v\} \subseteq X_t$.
2. (Interpolation Property) If t is a node on the unique path in T from t_1 to t_2, then $X_{t_1} \cap X_{t_2} \subseteq X_t$.

Definition 3. [10] The *width* of a tree decomposition is the maximum value of $|X_t| - 1$ taken over all the nodes t of the tree T of the decomposition. The *treewidth* $tw(G)$ of a graph G is the minimum width among all possible tree decompositions of G.

For the standard concepts in parameterized complexity, see the recent textbook by Cygan et al. [9].

1.1   Our Results

Our results are as follows:

- the HARMLESS SET problem with general thresholds is FPT when parameterized by the neighbourhood diversity.
- the HARMLESS SET problem with majority thresholds is W[1]-hard when parameterized by the treewidth of the graph.

1.2   Known Results

Bazgan and Chopin [2] studied the parameterized complexity of HARMLESS SET and the approximation of the associated maximization problem. When the parameter is k, they proved that the HARMLESS SET problem is W[2]-complete in general and W[1]-complete if all thresholds are bounded by a constant. When each threshold is equal to the degree of the vertex, they showed that HARMLESS SET is fixed-parameter tractable for parameter k and the maximization version is APX-complete. They gave a polynomial-time algorithm for graphs of bounded treewidth and a polynomial-time approximation scheme for planar graphs. The parametric dual problem $(n - k)$-HARMLESS SET asks for the existence of a harmless set of size at least $n - k$. The parameter is k and n denotes the number of vertices in the input graph. They showed that the parametric dual problem $(n - k)$-HARMLESS SET is fixed-parameter tractable for a large family of threshold functions.

2   FPT Algorithm Parameterized by Neighbourhood Diversity

In this section, we present an FPT algorithm for the HARMLESS SET problem parameterized by neighbourhood diversity. We say that two (distinct) vertices u and v have the same *neighborhood type* if they share their respective neighborhoods, that is, when $N(u) \setminus \{v\} = N(v) \setminus \{u\}$. If this is so we say that u and v are *twins*. It is possible to distinguish true-twins (those joined by an edge) and false-twins (in which case $N(u) = N(v)$).

Definition 4. [16] A graph $G = (V, E)$ has *neighbourhood diversity* at most d, if there exists a partition of V into at most d sets (we call these sets *type classes*) such that all the vertices in each set have the same neighbourhood type.

If neighbourhood diversity of a graph is bounded by an integer d, then there exists a partition $\{C_1, C_2, \ldots, C_d\}$ of $V(G)$ into d type classes. We would like to point out that it is possible to compute the neighborhood diversity of a graph in linear time using fast modular decomposition algorithms [23]. Notice that each type class could either be a clique or an independent set by definition and two type classes are either joined by a complete bipartite graph or no edge between vertices of the two types is present in G. For algorithmic purpose it is often useful to consider a *type graph* H of graph G, where each vertex of H is a type class in G, and two vertices C_i and C_j are adjacent iff there is a complete bipartite clique between these type classes in G. The key property of graphs of bounded neighbourhood diversity is that their type graphs have bounded size. The following result explains why the vertices with low thresholds are inside the solution.

Lemma 1. Let $C_i = \{v_1, \ldots, v_{|C_i|}\}$ be a type class in G such that $t(v_1) \leq t(v_2) \leq \ldots \leq t(v_{|C_i|})$. Let S be a maximum size harmless set in G and $x_i = |S_i| = |C_i \cap S|$. Then $S' = (S \setminus S_i) \cup \{v_1, v_2, \ldots, v_{x_i}\}$ is also a maximum size harmless set in G.

Proof. Clearly, $|S| = |S'|$. To show S' is a harmless set, it is enough to show that each vertex v in C_i has less than $t(v)$ neighbours in S'. For each $v \in \{v_1, \ldots, v_{x_i}\}$, we have

$$d_{S'}(v) = \begin{cases} d_S(v) & \text{if } v \in S \\ d_S(v) - 1 & \text{if } v \notin S \end{cases}$$

Therefore, every $v \in \{v_1, \ldots, v_{x_i}\}$ satisfies the threshold condition $d_{S'}(v) \leq d_S(v) < t(v)$. Let u be an arbitrary vertex in $\{v_{x_i+1}, v_{x_i+2} \ldots, v_{|C_i|}\}$. If $u \notin S$ then $d_{S'}(u) = d_S(u) < t(u)$. If $u \in S$ then, by definition of S', some vertex $v \in \{v_1, v_2, \ldots, v_{x_i}\} \setminus S$ must have replaced u as $t(v) \leq t(u)$. We have $d_S(v) = d_S(u) + 1$ and also $d_{S'}(u) = d_S(u) + 1$. It implies that $d_{S'}(u) = d_S(u) + 1 = d_S(v) < t(v) \leq t(u)$. Therefore, S' is a harmless set. $\square$

In this section, we prove the following theorem:

Theorem 1. The HARMLESS SET problem with general thresholds is FPT when parameterized by the neighbourhood diversity.

Given a graph $G = (V, E)$ with neighbourhood diversity $nd(G) \leq d$, we first find a partition of the vertices into at most d type classes $C_1, \ldots, C_d$. Let $\mathcal{C}$ be the set of all clique type classes and $\mathcal{I}$ be the set of all independent type classes. The case where some C_i are singletons can be considered as cliques or independent sets. For simplicity, we consider singleton type classes as independent sets.

ILP Formulation: Our goal here is to find a largest harmless set S of G. For each C_i, we associate a variable x_i that indicates $|S \cap C_i| = x_i$. As the vertices in C_i have the same neighbourhood, the variables x_i determine S uniquely, up to isomorphism. The threshold $t(C_i)$ of a type class C_i is defined to be

$$t(C_i) = \min\{t(v) \mid v \in C_i\}.$$

Let $\alpha(C_i)$ be the number of vertices in C_i with threshold value $t(C_i)$. We define $\mathcal{C}_1 = \{C_i \in \mathcal{C} \mid x_i < \alpha(C_i)\}$ and $\mathcal{C}_2 = \{C_i \in \mathcal{C} \mid x_i \geq \alpha(C_i)\}$. We next guess if a clique type class C_i belongs to $\mathcal{C}_1$ or $\mathcal{C}_2$. There are at most 2^d guesses as each clique type class C_i has two options: either it is in $\mathcal{C}_1$ or in $\mathcal{C}_2$. We reduce the problem of finding a maximum harmless set to at most 2^d integer linear programming problems with d variables. Since integer linear programming is fixed-parameter tractable when parameterized by the number of variables [17], we conclude that our problem is FPT when parameterized by the neighbourhood diversity d. We consider the following cases based on whether C_i is in $\mathcal{I}, \mathcal{C}_1$ or $\mathcal{C}_2$:

Case 1: Assume C_i is in $\mathcal{I}$.

Lemma 2. *Let C_i be an independent type class and $x_i \in \{0, 1, \ldots, |C_i|\}$. Let u_0 be a vertex in C_i with threshold $t(C_i)$. Then every vertex u in C_i has less than $t(u)$ neighbours in S if and only if u_0 has less than $t(C_i)$ neighbours in S.*

Proof. Suppose each $u \in C_i$ has less than $t(u)$ neighbours in S. Then obviously $u_0 \in C_i$ has less than $t(u_0) = t(C_i)$ neighbours in S. Conversely, suppose u_0 has less than $t(C_i)$ neighbours in S. Let u be an arbitrary vertex of C_i. As u and u_0 are two vertices in the same type class C_i, we have $d_S(u) = d_S(u_0)$. Moreover, for each $u \in C_i$, we have $t(C_i) \leq t(u)$ by definition of $t(C_i)$. Therefore, $d_S(u) = d_S(u_0) < t(C_i) \leq t(u)$. $\qquad\square$

Here $d_S(u_0) = \sum\limits_{C_j \in N_H(C_i)} x_j$. By Lemma 2, every vertex u in C_i has less than $t(u)$ neighbours in S if and only if

$$\sum_{C_j \in N_H(C_i)} x_j < t(C_i).$$

Case 2: Assume C_i is in $\mathcal{C}_1$. That is, C_i is a clique type class and $x_i < \alpha(C_i)$. Assuming $x_i < \alpha(C_i)$ ensures that there exists at least one vertex in $S^c \cap C_i$ with threshold $t(C_i)$.

Lemma 3. *Let $C_i \in \mathcal{C}_1$ and u_0 be a vertex in $S^c \cap C_i$ with threshold $t(C_i)$. Then every vertex u in C_i has less than $t(u)$ neighbours in S if and only if u_0 has less than $t(C_i)$ neighbours in S.*

Proof. Suppose every vertex u in C_i has less than $t(u)$ neighbours in S. Then obviously u_0 has less than $t(u_0) = t(C_i)$ neighbours in S. Conversely, suppose u_0 has less than $t(C_i)$ neighbours in S. Let y be an arbitrary vertex of $S \cap C_i$. Lemma 1 and the condition $x_i < \alpha(C_i)$ ensure that every vertex in $S \cap C_i$ has threshold value $t(C_i)$; hence y has threshold $t(C_i)$. Note that $d_S(y) = d_S(u_0) - 1 < t(C_i) - 1 < t(C_i) = t(y)$. Now, let z be an arbitrary element of $S^c \cap C_i$. It is easy to note that $d_S(z) = d_S(u_0) < t(C_i) \leq t(z)$. $\qquad\square$

Here $d_S(u_0) = x_i + \sum\limits_{C_j \in N_H(C_i)} x_j$. By Lemma 3, every vertex u in C_i has less than $t(u)$ neighbours in S if and only if

$$x_i + \sum_{C_j \in N_H(C_i)} x_j < t(C_i).$$

Case 3: Assume that C_i is in C_2. That is, C_i is a clique type class and $x_i \geq \alpha(C_i)$. By Lemma 1, all the vertices with threshold $t(C_i)$ are inside the solution.

Lemma 4. Let $C_i \in C_2$ and u_0 be a vertex in $S \cap C_i$ with threshold $t(C_i)$. Then every vertex u in C_i has less than $t(u)$ neighbours in S if and only if u_0 has less than $t(C_i)$ neighbours in S.

Proof. Suppose every vertex u in C_i has less than $t(u)$ neighbours in S. Then obviously u_0 has less than $t(u_0) = t(C_i)$ neighbours in S. Conversely, suppose u_0 has less than $t(C_i)$ neighbours in S. Let y be an arbitrary vertex of $S \cap C_i$. Note that $d_S(y) = d_S(u_0) < t(C_i) \leq t(y)$. Thus we showed that y has less than $t(y)$ neighbours in S. Let z be an arbitrary element of $S^c \cap C_i$. Note that such an element may not always exist, it is possible that all vertices in C_i are included in S (that is, $x_i = |C_i|$). Let us assume that such z exists. Since z is outside the solution and all the vertices with threshold $t(C_i)$ are inside the solution, it implies that $t(z) \geq t(C_i)+1$. It is easy to note that $d_S(z) = d_S(u_0)+1 < t(C_i)+1 \leq t(z)$. $\square$

Here $d_S(u_0) = (x_i - 1) + \sum\limits_{C_j \in N_H(C_i)} x_j$. By Lemma 4, every vertex u in C_i has less than $t(u)$ neighbours in S if and only if

$$(x_i - 1) + \sum_{C_j \in N_H(C_i)} x_j < t(C_i).$$

The next lemma follows readily from the three lemmas above and the definition of the sequence $(x_1, x_2, \ldots, x_d)$ and the harmless set.

Lemma 5. Let $G = (V, E)$ be a graph such that V can be partitioned into at most d type classes $C_1, \ldots, C_d$. The sequence $(x_1, x_2, \ldots, x_d)$ represents a harmless set S of G if and only if $(x_1, x_2, \ldots, x_d)$ satisfies

1. $x_i \in \{0, 1, \ldots, |C_i|\}$ for $i = 1, 2, \ldots, d$
2. $\sum\limits_{C_j \in N_H(C_i)} x_j < t(C_i)$ for all $C_i \in \mathcal{I}$.
3. $x_i + \sum\limits_{C_j \in N_H(C_i)} x_j < t(C_i)$ and $x_i < \alpha(C_i)$ for all $C_i \in C_1$
4. $(x_i - 1) + \sum\limits_{C_j \in N_H(C_i)} x_j < t(C_i)$ and $\alpha(C_i) \leq x_i \leq |C_i|$ for all $C_i \in C_2$.

In the following, we present an ILP formulation for the HARMLESS SET problem parameterized by neighbourhood diversity for a guess:

$$\text{Maximize} \sum_{C_i} x_i$$

Subject to

$$x_i \in \{0, 1, \ldots, |C_i|\} \quad \text{for } i = 1, 2, \ldots, d$$

$$\sum_{C_j \in N_H(C_i)} x_j < t(C_i), \quad \text{for all } C_i \in \mathcal{I},$$

$$x_i + \sum_{C_j \in N_H(C_i)} x_j < t(C_i) \quad \text{and } x_i < \alpha(C_i) \text{ for all } C_i \in \mathcal{C}_1$$

$$(x_i - 1) + \sum_{C_j \in N_H(C_i)} x_j < t(C_i) \quad \text{and } \alpha(C_i) \le x_i \le |C_i| \quad \text{for all } C_i \in \mathcal{C}_2$$

Solving the ILP. Lenstra [17] showed that the feasibility version of p-ILP is FPT with running time doubly exponential in p, where p is the number of variables. Later, Kannan [14] proved an algorithm for p-ILP running in time $p^{O(p)}$. In our algorithm, we need the optimization version of p-ILP rather than the feasibility version. We state the minimization version of p-ILP as presented by Fellows et al. [13].

p-VARIABLE INTEGER LINEAR PROGRAMMING OPTIMIZATION (p-OPT-ILP): Let matrices $A \in Z^{m \times p}$, $b \in Z^{p \times 1}$ and $c \in Z^{1 \times p}$ be given. We want to find a vector $x \in Z^{p \times 1}$ that minimizes the objective function $c \cdot x$ and satisfies the m inequalities, that is, $A \cdot x \ge b$. The number of variables p is the parameter. Then they showed the following:

Proposition 1. [13] p-OPT-ILP can be solved using $O(p^{2.5p+o(p)} \cdot L \cdot \log(MN))$ arithmetic operations and space polynomial in L. Here L is the number of bits in the input, N is the maximum absolute value any variable can take, and M is an upper bound on the absolute value of the minimum taken by the objective function.

In the formulation for HARMLESS SET problem, we have at most d variables. The value of the objective function is bounded by n and the value of any variable in the integer linear programming is also bounded by n. The constraints can be represented using $O(d^2 \log n)$ bits. Proposition 1 implies that we can solve the problem with the guess $\mathcal{P}$ in FPT time. There are at most 2^d guesses, and the ILP formula for a guess can be solved in FPT time. Thus Theorem 1 holds.

3   W[1]-Hardness Parameterized by Treewidth

In this section we show that the HARMLESS SET problem with majority thresholds is W[1]-hard when parameterized by the treewidth. To show W[1]-hardness of HARMLESS SET with majority thresholds, we reduce from the following problem, which is known to be W[1]-hard parameterized by the treewidth of the graph [22]:

MINIMUM MAXIMUM OUTDEGREE

Input: An undirected graph G whose edge weights are given in unary, and a positive integer r.

Question: Is there an orientation of the edges of G such that, for each $v \in V(G)$, the sum of the weights of outgoing edges from v is at most r?

In MINIMUM MAXIMUM OUTDEGREE problem, every edge weight $\omega(u,v)$ of G is given in unary, that is, every edge weight $\omega(u,v)$ is polynomially bounded in $|V(G)|$. In a weighted undirected graph G, the weighted degree of a vertex v, is defined as the sum of the weights of the edges incident to v in G. In this section, we prove the following theorem:

Theorem 2. The HARMLESS SET problem with majority thresholds is W[1]-hard when parameterized by the treewidth of the graph.

Proof. Let $G = (V, E, \omega)$ and a positive integer $r \geq 3$ be an instance I of MINIMUM MAXIMUM OUTDEGREE. We construct an instance $I' = (G', t, k)$ of HARMLESS SET the following way. See Fig. 1 for an illustration.

1. For each weighted edge $(u,v) \in E(G)$, we introduce two sets of new vertices $V_{uv} = \{u_1^v, \ldots, u_{\omega(u,v)}^v\}$ and $V_{vu} = \{v_1^u, \ldots, v_{\omega(u,v)}^u\}$ into G'. Make u and v adjacent to every vertex of V_{uv} and V_{vu}, respectively. The vertices of $\bigcup_{(u,v) \in E(G)} V_{uv} \cup V_{vu}$ are called *type 1* vertices.

2. For each $1 \leq i \leq \omega(u,v) - 1$, we introduce $x_{(u_i^v, v_i^u)}$ into G' and make it adjacent to u_i^v and v_i^u; introduce $x_{(u_i^v, v_{i+1}^u)}$ and make it adjacent to u_i^v and v_{i+1}^u; introduce $x_{(u_{i+1}^v, v_i^u)}$ and make it adjacent to u_{i+1}^v and v_i^u. We also add $x_{(u_{\omega(u,v)}^v, v_{\omega(u,v)}^u)}$ into G' and make it adjacent to $u_{\omega(u,v)}^v$ and $v_{\omega(u,v)}^u$. We call such vertices, the vertices of *type 2*.

3. For every vertex x of *type 2*, we add a triangle (cycle of length 3) and make x adjacent to exactly one vertex of this triangle. For every vertex x of *type 1*, let $n(x)$ be the number of neighbours of x in $V(G)$ and in the set of *type 2* vertices. Note that $2 \leq n(x) \leq 4$. We add $n(x)+1$ many triangles corresponds to vertex x and make x adjacent to exactly one vertex of each triagle.

4. The weighted degree of a vertex $x \in V$ in G is denoted by $d_\omega(x; G)$. We partition the vertices of $V(G)$ based on whether $\lceil \frac{d_\omega(x;G)}{2} \rceil \leq r + 1$ or $\lceil \frac{d_\omega(x;G)}{2} \rceil > r + 1$. A vertex x in G with $\lceil \frac{d_\omega(x;G)}{2} \rceil \leq r + 1$ is called a vertex of *low-degree-type*. For each $x \in V(G)$ of *low-degree-type*, we add $2\left[(r + 1) - \lceil \frac{d_\omega(x;G)}{2} \rceil\right]$ triangles and make x adjacent to exactly one vertex of each triangle. A vertex $x \in V(G)$ with $\lceil \frac{d_\omega(x;G)}{2} \rceil > r + 1$ is called a vertex of *high-degree-type*. For each $x \in V(G)$ of *high-degree-type*, we add a set $V_x^\triangle = \{v_x^{1\triangle}, \ldots, v_x^{\alpha\triangle}\}$ of $\alpha = d_\omega(x; G) - r$ many vertices and make them

adjacent to x. For each $v \in V_x^\triangle$, we add two triangles and make v adjacent to exactly one vertex of each triangle. For each *high-degree-type* vertex x, we also add a set of $(r+2)$ many triangles and make x adjacent to exactly one vertex of each triangle.

This completes the construction of graph G'.

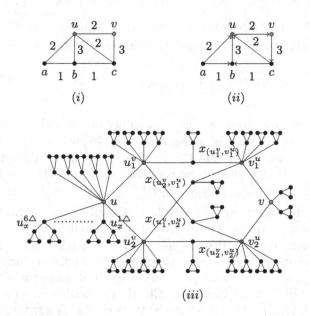

(i) (ii)

(iii)

Fig. 1. (i) An instance (G, r) of Minimum Maximum Outdegree with $r = 3$. (ii) A valid orientation of G when $r = 3$. (iii) An illustration of the reduction algorithm in Theorem 2 using an edge (u, v) with $\omega(u, v) = 2$. Note that u is a *high-degree-type* vertex and v is a *low-degree-type vertex*.

We set $k = n + W + \sum\limits_{(u,v) \in E(G)} (3\omega(u,v) - 2) + \sum\limits_{x \in high\text{-}degree\text{-}type} (d_\omega(x; G) - r)$

where $W = \sum\limits_{(u,v) \in E(G)} \omega(u,v)$. As every edge weight $\omega(u,v)$ of G and the integer r are polynomially bounded in $|V(G)|$, the number of vertices in G' is also polynomially bounded in $|V(G)|$. Therefore, it is clear that G' can be constructed in polynomial time. It can be proved that the treewidth of G' is at most the treewidth of G plus eight.

Now we show that our reduction is correct. That is, we prove that (G, w, r) is a yes instance of Minimum Maximum Outdegree if and only if I' is a yes instance of Harmless Set. Let D be the directed graph obtained by an orientation of the edges of G such that for each vertex the sum of the weights of outgoing edges is at most r. We claim that the set

$$H = V(G) \bigcup_{(u,v)\in E(D)} V_{uv} \bigcup_{x\in high\text{-}degree\text{-}type} V_x^{\triangle}$$

$$\bigcup_{(u,v)\in E(G)} \left\{ x_{(u_i^v,v_i^u)}, x_{(u_i^v,v_{i+1}^u)}, x_{(u_{i+1}^v,v_i^u)}, x_{(u_{\omega(u,v)}^v,v_{\omega(u,v)}^u)} \mid 1 \le i \le \omega(u,v)-1 \right\}$$

is harmless set of size at least k. Next, we show that all the vertices in H satisfy the threshold condition. It is easy to verify that each $u \in \bigcup_{(u,v)\in E(G)} (V_{uv} \cup V_{vu})$ satisfies the threshold condition as u has $n(u)$ neighbours in H and $n(u) + 1$ neighbours outside H, that is, u has less than $\lceil \frac{d_{G'}(u)}{2} \rceil$ neighbours in H. Each

$$x \in \bigcup_{(u,v)\in E(G)} \left\{ x_{(u_i^v,v_i^u)}, x_{(u_i^v,v_{i+1}^u)}, x_{(u_{i+1}^v,v_i^u)}, x_{(u_{\omega(u,v)}^v,v_{\omega(u,v)}^u)} \mid 1 \le i \le \omega(u,v)-1 \right\}$$

satisfies the threshold condition as x has only one neighbour in H and two neighbours outside H, that is, x has less than $\lceil \frac{d_{G'}(x)}{2} \rceil = \lceil \frac{3}{2} \rceil = 2$ neighbours in H. It is also easy to see that the vertices of triangles satisfy the threshold condition. Let u be an arbitrary vertex of *low-degree-type*. If the weighted degree of u in G is $d_\omega(u; G)$ then its degree in G' is $d_\omega(u; G) + 2\left[(r+1) - \lceil \frac{d_\omega(u;G)}{2} \rceil\right]$. Observe that the neighbours of u inside H are all of *type 1* which is equal to outdegree of u and we know outdegree of u is bounded by r. Thus each *low-degree-type* vertex has less than $\lceil \frac{d_{G'}(u)}{2} \rceil = r + 1$ neighbours in H. Therefore each *low-degree-type* vertex satisfies the threshold condition. Next, let x be an arbitrary vertex of *high-degree-type*. If the weighted degree of x in G is $d_\omega(x; G)$ then its degree in G' is $2d_\omega(x; G) + 2$. Clearly the neighbours of x inside H are at most $r + (d_\omega(x; G) - r)$. Therefore the vertices of *high-degree-type* satisfy the threshold condition. This implies that I' is a yes instance.

Conversely, assume that G' admits a harmless set H of size at least k. We make the following observations: (i) let C be the set of all triangles introduced in the reduction algorithm, then C does not intersect with H. This is true because any vertex with degree 2 has threshold equal to 1. This implies that both the neighbours of that vertex have to be outside the solution as otherwise the vertex will fail to satisfy the threshold condition, (ii) for each $(u,v) \in E(G)$ the set $V_{uv} \cup V_{vu}$ contributes at most half vertices in H as otherwise $x_{(u_i^v,v_i^u)}$ for some $1 \le i \le w(u,v)$ will fail to satisfy the threshold condition. Note that the total number of vertices in $\bigcup_{(u,v)\in E(G)} V_{uv} \cup V_{vu}$ is $2W$. The above observations imply that the size of harmless set H is at most $|V(G')| - W - 3|C| = n + W + \sum_{(u,v)\in E(G)} (3\omega(u,v) - 2) + \sum_{x\in high\text{-}degree\text{-}type} (d_\omega(x; G) - r)$, which is equal to k. It implies that either $V_{uv} \subseteq H$ or $V_{vu} \subseteq H$ for all $(u,v) \in E(G)$ as otherwise some vertex $x_{(u_i^v,v_{i+1}^u)}$ will fail to satisfy the threshold condition. Hence the harmless set is of the form

$$H - V(G) \bigcup_{(u,v) \in E(G)} (V_{uv} \text{ or } V_{vu}) \bigcup_{x \in high\text{-}degree\text{-}type} V_x^\triangle$$

$$\bigcup_{(u,v) \in E(G)} \left\{ x_{(u_i^v, v_i^u)}, x_{(u_i^v, v_{i+1}^u)}, x_{(u_{i+1}^v, v_i^u)}, x_{(u_{\omega(u,v)}^v, v_{\omega(u,v)}^u)} \mid 1 \le i \le \omega(u,v) - 1 \right\}.$$

Next, we define a directed graph D by $V(D) = V(G)$ and

$$E(D) = \left\{ (u,v) \mid u, v \in V(D) \text{ and } V_{uv} \subseteq H \right\} \bigcup \left\{ (v,u) \mid u, v \in V(D) \text{ and } V_{vu} \subseteq H \right\}.$$

Let us assume that there exists a vertex $x \in V(G)$ of *low-degree-type* such that the outdegree is more than r. We can easily see that $d_H(x) \ge \lceil \frac{d_{G'}(x)}{2} \rceil$ which is a contradiction. Let us assume that there exists a vertex $x \in V(G)$ of *high-degree-type* such that the outdegree is more than r. We can easily see that $d_H(x) \ge d_G^w(x) \ge \lceil \frac{d_{G'}(x)}{2} \rceil$ which is a contradiction. This implies that I is a yes-instance.

$\square$

4   Conclusion

The main contributions in this paper are that the HARMLESS SET problem with general thresholds is FPT when parameterized by the neighbourhood diversity and the HARMLESS SET problem with majority thresholds is W[1]-hard when parameterized by the treewidth of the graph. Drange, Muzi and Reidl [11] have independently achieved similar results, although in a completely different manner. It would be interesting to consider the parameterized complexity with respect to twin cover. The modular width parameter also appears to be a natural parameter to consider here. The parameterized complexity of the HARMLESS SET problem remains unsettled when parameterized by other structural parameters like feedback vertex set number, pathwidth, treedepth and clique-width.

Acknowledgement. We thank the anonymous reviewers for their constructive comments and suggestions, which helped us to improve the manuscript.

References

1. Aazami, A., Stilp, K.: Approximation algorithms and hardness for domination with propagation. SIAM J. Discrete Math. **23**(3), 1382–1399 (2009)
2. Bazgan, C., Chopin, M.: The complexity of finding harmless individuals in social networks. Discrete Optim. **14**(C), 170–182 (2014)
3. Bazgan, C., Chopin, M., Nichterlein, A., Sikora, F.: Parameterized approximability of maximizing the spread of influence in networks. In: Du, D.-Z., Zhang, G. (eds.) COCOON 2013. LNCS, vol. 7936, pp. 543–554. Springer, Heidelberg (2013). https://doi.org/10.1007/978-3-642-38768-5_48
4. Ben-Zwi, O., Hermelin, D., Lokshtanov, D., Newman, I.: Treewidth governs the complexity of target set selection. Discrete Optim. **8**(1), 87–96. Parameterized Complexity of Discrete Optimization (2011)

5. Centeno, C.C., Dourado, M.C., Penso, L.D., Rautenbach, D., Szwarcfiter, J.L.: Irreversible conversion of graphs. Theor. Comput. Sci. **412**(29), 3693–3700 (2011)
6. Chen, N.: On the approximability of influence in social networks. SIAM J. Discrete Math. **23**(3), 1400–1415 (2009)
7. Chiang, C.-Y., Huang, L.-H., Li, B.-J., Wu, J., Yeh, H.-G.: Some results on the target set selection problem. J. Comb. Optim. **25**(4), 702–715 (2013)
8. Chopin, M., Nichterlein, A., Niedermeier, R., Weller, M.: Constant thresholds can make target set selection tractable. Theory Comput. Syst. **55**(1), 61–83 (2014)
9. Cygan, M., et al.: Parameterized Algorithms. Springer, Cham (2015). https://doi.org/10.1007/978-3-319-21275-3
10. Downey, R.G., Fellows, M.R.: Parameterized Complexity. Springer, Heidelberg (2012)
11. Drange, P.G., Muzi, I., Reidl, F.: Kernelization and hardness of harmless sets in sparse classes. CoRR, abs/2111.11834 (2021)
12. Dreyer, P.A., Roberts, F.S.: Irreversible k-threshold processes: graph-theoretical threshold models of the spread of disease and of opinion. Discrete Appl. Math. **157**(7), 1615–1627 (2009)
13. Fellows, M.R., Lokshtanov, D., Misra, N., Rosamond, F.A., Saurabh, S.: Graph layout problems parameterized by vertex cover. In: Hong, S.-H., Nagamochi, H., Fukunaga, T. (eds.) ISAAC 2008. LNCS, vol. 5369, pp. 294–305. Springer, Heidelberg (2008). https://doi.org/10.1007/978-3-540-92182-0_28
14. Kannan, R.: Minkowski's convex body theorem and integer programming. Math. Oper. Res. **12**(3), 415–440 (1987)
15. Kempe, D., Kleinberg, J., Tardos, E.: Maximizing the spread of influence through a social network. In: Proceedings of the Ninth ACM SIGKDD International Conference on Knowledge Discovery and Data Mining, KDD 2003, New York, NY, USA, pp. 137–146 (2003). Association for Computing Machinery
16. Lampis, M.: Algorithmic meta-theorems for restrictions of treewidth. Algorithmica **64**, 19–37 (2012)
17. Lenstra, H.W.: Integer programming with a fixed number of variables. Math. Oper. Res. **8**(4), 538–548 (1983)
18. Nichterlein, A., Niedermeier, R., Uhlmann, J., Weller, M.: On tractable cases of target set selection. Soc. Netw. Anal. Min. **3**(2), 233–256 (2013)
19. Peleg, D.: Local majorities, coalitions and monopolies in graphs: a review. Theor. Comput. Sci. **282**(2), 231–257 (2002)
20. Reddy, T., Rangan, C.: Variants of spreading messages. J. Graph Algorithms Appl. **15**(5), 683–699 (2011)
21. Robertson, N., Seymour, P.: Graph minors. III. Planar tree-width. J. Comb. Theory Ser. B **36**(1), 49–64 (1984)
22. Szeider, S.: Not so easy problems for tree decomposable graphs. CoRR, abs/1107.1177 (2011)
23. Tedder, M., Corneil, D., Habib, M., Paul, C.: Simpler linear-time modular decomposition via recursive factorizing permutations. In: Aceto, L., Damgård, I., Goldberg, L.A., Halldórsson, M.M., Ingólfsdóttir, A., Walukiewicz, I. (eds.) ICALP 2008. LNCS, vol. 5125, pp. 634–645. Springer, Heidelberg (2008). https://doi.org/10.1007/978-3-540-70575-8_52

Isomorphism Testing for T-graphs in FPT

Deniz Ağaoğlu Çağırıcı[(✉)] and Petr Hliněný

Masaryk University, Brno, Czech Republic
agaoglu@mail.muni.cz, hlineny@fi.muni.cz

Abstract. A T-graph (a special case of a chordal graph) is the intersection graph of connected subtrees of a suitable subdivision of a fixed tree T. We deal with the isomorphism problem for T-graphs which is *GI-complete* in general – when T is a part of the input and even a star. We prove that the T-graph isomorphism problem is in FPT when T is the fixed parameter of the problem. This can equivalently be stated that isomorphism is in FPT for chordal graphs of (so-called) bounded leafage. While the recognition problem for T-graphs is not known to be in FPT wrt. T, we do *not* need a T-representation to be given (a promise is enough). To obtain the result, we combine a suitable isomorphism-invariant decomposition of T-graphs with the classical tower-of-groups algorithm of Babai, and reuse some of the ideas of our isomorphism algorithm for S_d-graphs [MFCS 2020].

Keywords: chordal graph · H-graph · leafage · graph isomorphism · parameterized complexity

1 Introduction

Two graphs G and H are called *isomorphic*, denoted by $G \simeq H$, if there is a bijection $f : V(G) \to V(H)$ such that for every pair $u, v \in V(G)$, $\{u, v\} \in E(G)$ if and only if $\{f(u), f(v)\} \in E(H)$. The well-known *graph isomorphism problem* asks whether two input graphs are isomorphic, and it can be solved efficiently for various special graph classes [1,9,12,14,18,23]. On the other hand, it is still unknown whether this problem is polynomial-time solvable or not (though, it is not expected to be NP-hard) in the general case, and a problem is said to be *GI-complete* if it is polynomial-time equivalent to the graph isomorphism.

We now briefly introduce two complexity classes of parameterized problems. Let k be the parameter, n be the input size, f and g be two computable functions, and c be some constant. A decision problem is in the *class FPT* (or *FPT-time*) if there exists an algorithm solving that problem correctly in time $\mathcal{O}(f(k) \cdot n^c)$. Similarly, a decision problem is in the *class XP* if there exists an algorithm solving that problem correctly in time $\mathcal{O}(f(k) \cdot n^{g(k)})$. Some parameters which yield to *FPT-* or *XP*-time algorithms for the graph isomorphism problem can be listed as tree-depth [10], tree-width [21], maximum degree [7] and genus [23].

This work was supported by the Czech Science Foundation, project no. 20-04567S.

P. Mutzel et al. (Eds.): WALCOM 2022, LNCS 13174, pp. 239–250, 2022.
https://doi.org/10.1007/978-3-030-96731-4_20

In this paper, we consider the parameterized complexity of the graph isomorphism problem for special instances of intersection graphs which we introduce next.

The *intersection graph* for a finite family of sets is an undirected graph G where each set is associated with a vertex of G, and each pair of vertices in G are joined by an edge if and only if the corresponding sets have a non-empty intersection. Chordal and interval graphs are two of the most well-known intersection graph classes related to our research.

A graph is *chordal* if every cycle of length more than three has a chord. They are also defined as the intersection graphs of subtrees of some (non-fixed) tree T [16]. Chordal graphs can be recognized in linear time, and they have linearly many maximal cliques which can be listed in polynomial time [24]. Deciding the isomorphism of chordal graphs is a *GI-complete* problem [25]. A graph G is an *interval graph* if it is the intersection graph of a set of intervals on the real line. Interval graphs form a subclass of chordal graphs. They can also be recognized in linear time, and interval graph isomorphism can be solved in linear time [9].

A subdivision of a graph G is the operation of replacing selected edge(s) of G by new induced paths (informally, putting new vertices to the middle of an edge). For a fixed graph H, an *H-graph* is the intersection graph of connected subgraphs of a suitable subdivision of the graph H [8], and they generalize many types of intersection graphs. For instance, interval graphs are K_2-graphs, their generalization called circular-arc graphs are K_3-graphs, and chordal graphs are the union of T-graphs where T ranges over all trees. We, however, consider *T-graphs* where T is a fixed tree. Even though chordal graphs can be recognized in linear time [24], deciding whether a given chordal graph is a T-graph is NP-complete when T is on the input [19]. In [11], Chaplick et al. gave an *XP*-time algorithm to recognize T-graphs parameterized by the size of T.

S_d-graphs form a subclass of T-graphs where S_d is the star with d rays. The isomorphism problem for S_d-graphs, and therefore for T-graphs, was shown to be GI-complete [25] with d on the input. In [3], we have proved by algebraic means that S_d-graph isomorphism can be solved in *FPT*-time parameterized by d, and then in [5] we have extended this approach to an *XP*-time algorithm for the isomorphism problem of T-graphs parameterized by the size of T. We have also considered in [5] the special case of isomorphism of proper T-graphs with a purely combinatorial *FPT*-time algorithm.

New Contribution. In this paper, we show that the graph isomorphism problem for T-graphs can be solved in *FPT*-time parameterized by the size of T. Our algorithm does not assume or rely on T-representations of the input graphs to be given, and in fact it uses only some special properties of T-graphs.

Moreover, our result can be equivalently reformulated as an *FPT*-time algorithm for testing isomorphism of chordal graphs of bounded leafage, where the *leafage* of a chordal graph G can be defined as the least number of leaves of a tree T such that G is a T-graph. Since there is only a bounded number of trees

T of a given number of leaves, modulo subdivisions, the correspondence of the two formulations is obvious.

Highly informally explaining our approach (which is different from [5]), we use chordality and properties of assumed T-representations of input graphs G and G' to efficiently compute their special hierarchical canonical decompositions into so-called fragments (Sect. 2). Each fragment will be an interval graph, and the isomorphism problem of interval graphs is well understood. Then we use some classical group-computing tools (Sect. 3, Babai's tower-of-groups approach) to compute possible "isomorphisms" between the decompositions of G and of G' (Sect. 4); each such isomorphism mapping between the fragments of the two decompositions, and simultaneously between the neighborhood sets of fragments in other fragments "higher up" in the decomposition.

We remark that the same problem has been independently and concurrently solved by Arvind, Nedela, Ponomarenko and Zeman [2],[1] using different means (by reducing the problem to automorphisms of colored order-3 hypergraphs with bounded sizes of color classes).

Due to restricted space, statements marked with an asterisk (*) have proofs only in the full paper [4] (an arXiv preprint).

2   Structure and Decomposition of T-graphs

In this section, we give a procedure to "extract" a bounded number of special interval subgraphs (called *fragments*) of a T-graph G in a way which is invariant under automorphisms and does not require a T-representation on input. Informally, the fragments can be seen as suitable "pieces" of G which are placed on the leaves of T in some representation, and their most important aspects are their simplicity and limited number. We use this extraction procedure repeatedly (and recursively) to obtain the full decomposition of a T-graph.

Structure of Chordal Graphs. We now give several useful terms and facts related to chordal graphs. A vertex v of a graph G is called *simplicial* if its neighborhood corresponds to a clique of G. It is known that every chordal graph contains a simplicial vertex and, by removing the simplicial vertices of a chordal graph repeatedly, one obtains an empty graph.

A *weighted clique graph* $\mathcal{C}_G$ of a graph G is the graph whose vertices are the maximal cliques of G and there is an edge between two vertices in $\mathcal{C}_G$ whenever the corresponding maximal cliques have a non-empty intersection. The edges in $\mathcal{C}_G$ are weighted by the cardinality of the intersection of the corresponding cliques.

A *clique tree* of G is any maximum-weight spanning tree of $\mathcal{C}_G$ which may not be unique. An edge of $\mathcal{C}_G$ is called *indispensable* (resp. *unnecessary*) if it appears

---

[1] To be completely accurate, our paper was first time submitted to a conference at the beginning of July 2021, and [2] appeared on arXiv just two weeks later, without mutual influence regarding the algorithms.

in every (resp. none) maximum-weight spanning tree of $\mathcal{C}_G$. If G is chordal, every maximum-weight spanning tree T of $\mathcal{C}_G$ is also a T-representation of G, e.g. [22].

For a graph G and two vertices $u \neq v \in V(G)$, a subset $S \subseteq V(G)$ is called a u-v *separator* (or u-v *cut*) of G if u and v belong to different components of $G - S$. When $|S| = 1$, then S is called a *cutvertex*. S is called *minimal* if no proper subset of S is a u-v separator. Minimal separators of a graph are the separators which are minimal for some pair of vertices. Chordal graphs, thus T-graphs, have linearly many minimal vertex separators [17].

A *leaf clique* of a T-graph G is a maximal clique of G which can be a leaf of some clique tree of G (informally, it can be placed on a leaf of T in some T-representation of G). We use the following lemma in our algorithm:

Lemma 2.1 (Matsui et al. [22]). *A maximal clique C of a chordal graph G can be a leaf of a clique tree if and only if C satisfies (1) C is incident to at most one indispensable edge of $\mathcal{C}_G$, and (2) C is not a cutvertex in $\mathcal{C}'_G$ which is the subgraph of $\mathcal{C}_G$ which includes all edges except the unnecessary ones. The conditions can be checked in polynomial time.*

Decomposing T-graphs. The overall goal now is to recursively find a unique decomposition of a given T-graph G into levels such that each level consists of a bounded number of interval fragments.

For an illustration, a similar decomposition can be obtained directly from a T-representation of G: pick the interval subgraphs of G which are represented exclusively on the leaf edges of T, forming the outermost level, and recursively in the same way obtain the next levels. Unfortunately, this is not a suitable solution for us, not only that we do not have a T-decomposition at hand, but mainly because we need our decomposition to be *canonical*, meaning invariant under automorphisms of the graph, while this depends on a particular representation.

The contribution of this section is to compute such a decomposition the right canonical way. As sketched above, the core task is to canonically determine in the given graph G one bounded-size collection of fragments which will form the outermost level of the decomposition, and then the rest of the decomposition is obtained in the same way from recursively computed collections of fragments in the rest of the graph, which is also a T-graph[2].

For a chordal graph G and a (fixed) collection $Z_1, Z_2, \ldots, Z_s \subseteq G$ of distinct cliques, we write $Z_i \preceq Z_j$ if there exists $k \in \{1, \ldots, s\} \setminus \{i, j\}$ such that Z_j *separates* Z_i from Z_k in G (meaning that there is no path from $Z_i \setminus Z_j$ to $Z_k \setminus Z_j$ in $G - Z_j$), and say that $Z_i \preceq Z_j$ is *witnessed* by Z_k. Note that $\preceq$ is transitive, and hence a preorder. Let $Z_i \precneqq Z_j$ mean that $Z_i \preceq Z_j$ but $Z_j \not\preceq Z_i$. We also write $Z_i \approx Z_j$ if there exists $k \in \{1, \ldots, s\} \setminus \{i, j\}$ such that both $Z_i \preceq Z_j$ and

---

[2] Since the requirement of canonicity of our collection does not allow us to relate this collection to a particular T-representation of G, we cannot say whether the rest of G (after removing our collection of fragments) would be a T_1-graph for some strict subtree $T_1 \subsetneq T$, or only a T-graph again. That is why we speak about T-graphs for the same T (or, we could say graphs of bounded leafage here) throughout the whole recursion. In particular, we cannot directly use this procedure to recognize T-graphs.

$Z_j \preceq Z_i$ hold and are witnessed by Z_k. Note that $Z_j \approx Z_i$ is stronger than just saying '$Z_i \preceq Z_j$ and $Z_j \preceq Z_i$,' and that $Z_i \cap Z_j$ then separates $Z_i \Delta Z_j$ from Z_k.

Lemma 2.2. (*) *Let T be a tree with d leaves, and G be a T-graph. Assume that $Z_1, \ldots, Z_s \subseteq G$ are distinct cliques of G such that one of the following holds:*
a) for each $1 \leq i \leq s$, the set Z_i is a maximal clique in G, and for any $1 \leq i \neq j \leq s$, neither $Z_i \nsucceq Z_j$ nor $Z_i \approx Z_j$ is true, or
b) for each $1 \leq i \leq s$, the set Z_i is a minimal separator in G cutting off a component F of $G - Z_i$ such that F contains a simplicial vertex of a leaf clique of G, and that F is disjoint from all Z_j, $j \neq i$.
Then $s \leq d$.

Let $Z \subseteq G$ be a minimal separator in G and $F \subseteq G$ a connected component of $G - Z$. Then Z is a clique since G is chordal, and whole Z is in the neighborhood of F by minimality. We call a *completion of F* (in implicit G) the graph F^+ obtained by contracting all vertices of G not in $V(F) \cup Z$ into one vertex l (the neighborhood of l is thus Z) and joining l with a new leaf vertex l', called the *tail* of F^+. Since F determines Z in a chordal graph G, the term F^+ is well defined.

We call a collection of disjoint nonempty induced subgraphs (not necessarily connected) $X_1, X_2, \ldots, X_s \subseteq G$, such that there are no edges between distinct X_i and X_j, a *fragment collection* of G of size s. We first give our procedure for computing a fragment collection, and subsequently formulate (and prove) the crucial properties of the computed collection and the whole decomposition.

Procedure 2.3 Let T be a tree with d leaves and no degree-2 vertex. Assume a T-graph G on the input. We compute an induced (and canonical) fragment collection $X_1, X_2, \ldots, X_s \subseteq G$ of G of size $0 < s \leq 2d$ as follows.

1. List all maximal cliques in G (using a simplicial decomposition) and compute the weighted clique graph $\mathcal{C}_G$ of G. Compute the list $\mathcal{L}$ of all possible leaf cliques of G by Lemma 2.1; in more detail, using [22, Algorithm 2] for computation of the indispensable edges in $\mathcal{C}_G$.
2. For every pair $L_1, L_2 \in \mathcal{L}$ such that $L_1 \nsucceq L_2$, remove L_2 from the list. Let $\mathcal{L}_0 \subseteq \mathcal{L}$ be the resulting list of cliques, which is nonempty since $\nsucceq$ is acyclic.
3. Let $\mathcal{L}_1 := \{L \in \mathcal{L}_0 : \forall L' \in \mathcal{L}_0 \setminus \{L\}. \ L \not\approx L'\}$ be the subcollection of cliques incomparable with others in $\approx$. By Lemma 2.2(a) we have $|\mathcal{L}_1| \leq d$. If $\mathcal{L}_1 \neq \emptyset$, then **output** the following fragment collection of G: for each $L \in \mathcal{L}_1$, include in it the set $F \subseteq L$ of all simplicial vertices of L in the graph G.
4. Now, for each $L \in \mathcal{L}_0$ we have $L' \in \mathcal{L}_0 \setminus \{L\}$ such that $L \approx L'$ (and so $L \cap L'$ is a separator in G). For distinct $L_1, L_2 \in \mathcal{L}_0$ such that $L_1 \approx L_2$, we call a set $Z \subseteq L_1 \cap L_2$ a *joint separator for L_1, L_2* if Z separates $L_1 \Delta L_2$ from $L \setminus Z$ for some (any) $L \in \mathcal{L}_0 \setminus \{L_1, L_2\}$. We compute the family $\mathcal{Z}$ of all inclusion-minimal sets Z which are joint separators for some pair $L_1 \approx L_2 \in \mathcal{L}_0$ as above, over all such pairs L_1, L_2. This is efficient since all minimal separators in chordal graphs can be listed in linear time. Note that no set $Z \in \mathcal{Z}$ contains any simplicial vertex of G, and so $V(G) \nsubseteq \bigcup \mathcal{Z}$.

5. Let $\mathcal{C}$ be the family of the connected components of $G - \bigcup \mathcal{Z}$, and $\mathcal{C}_0 \subseteq \mathcal{C}$ consist of such $F \in \mathcal{C}$ that F is incident to just one set $Z_F \in \mathcal{Z}$. Note that $\mathcal{C}_0 \neq \emptyset$, since otherwise the incidence graph between $\mathcal{C}$ and $\mathcal{Z}$ would have a cycle and this would in turn contradict chordality of G. Let $\mathcal{Z}_0 := \{Z_F \in \mathcal{Z} : F \in \mathcal{C}_0\}$. Moreover, by Lemma 2.2(b), $|\mathcal{Z}_0| \leq d$.
6. We make a collection $\mathcal{C}_0'$ from $\mathcal{C}_0$ by the following operation: for each $Z \in \mathcal{Z}_0$, take all $F \in \mathcal{C}_0$ such that $Z_F = Z$ and every vertex of F is adjacent to whole Z, and join them into one graph in $\mathcal{C}_0'$ (note that there can be arbitrarily many such F for one Z). Remaining graphs of $\mathcal{C}_0$ stay in $\mathcal{C}_0'$ without change. Then, we denote by $\mathcal{C}_1 \subseteq \mathcal{C}_0'$ the subcollection of those $F \in \mathcal{C}_0'$ such that the completion F^+ of F (in G) is an interval graph.[3]
7. If $\mathcal{C}_1 \neq \emptyset$, then **output** $\mathcal{C}_1$ as the fragment collection. (As we can show from Lemma 2.2(b), $|\mathcal{C}_1| \leq d + |\mathcal{Z}_0| \leq 2d$.)
8. Otherwise, for each graph $F \in \mathcal{C}_0'$, we call this procedure recursively on the completion F^+ of F (these calls are independent since the graphs in $\mathcal{C}_0'$ are pairwise disjoint). Among the fragments returned by this call, we keep only those which are subgraphs of F.[4] We **output** the fragment collection formed by the union of kept fragments from all recursive calls.

One call to Procedure 2.3 clearly takes only polynomial time (in some steps this depends on G being chordal – e.g., listing all cliques or separators). Since the possible recursive calls in the procedure are applied to pairwise disjoint parts of the graph (except the negligible completion of F to F^+), the overall computation of Procedure 2.3 takes polynomial time regardless of d. Regarding correctness, we are proving that $s \leq 2d$, which is in the respective steps 3 and 7 indicated as a corollary of Lemma 2.2, except in the last (recursive) step 8 where it can be derived in a similar way from Lemma 2.2 applied to the final collection. We leave the remaining technical details for the full paper [4].

The last part is to prove a crucial fact that the collection $X_1, X_2, \ldots, X_s \subseteq G$ is indeed canonical, which is precisely stated as follows:

Lemma 2.4. (*) *Let G and G' be isomorphic T-graphs. If Procedure 2.3 computes the canonical collection $X_1, \ldots, X_s$ for G and the canonical collection $X_1', \ldots, X_{s'}'$ for G', then $s = s'$ and there is an isomorphism between G and G' matching in some order $X_1, \ldots, X_s$ to $X_1', \ldots, X_s'$.*

Levels, Attachments and Terminal Sets. Following Procedure 2.3, we now show how the full decomposition of a T-graph G is completed.

For every fragment X of the canonical collection computed by Procedure 2.3, we define the list of *attachment sets* of X in $G - X$ as follows. If $X = F$ is obtained in step 3, then it has one attachment set $L \setminus F$. Otherwise (steps 6 and 7), the attachment sets of $X = F$ are all subsets A of the corresponding separator Z

---

[3] Informally, $F^+ \in \mathcal{C}_1$ iff F has an interval representation (on a horizontal line) to which its separator Z_F can be "attached from the left" on the same line.

[4] Note that, e.g., the separator and tail of F^+ may also be involved in a recursively computed fragment.

(of F) such that some vertex of X has the neighborhood in Z equal to A. Observe that the attachment sets of X are always cliques contained in the completion X^+, as defined above. Moreover, it is important that the attachment sets of X form a chain by the set inclusion, since G is chordal, and hence they are *uniquely determined* independently of automorphisms of X^+.

Procedure 2.5. Given a T-graph G, we determine a **canonical decomposition** of G recursively as follows. Start with $i = 1$ and $G_0 := G$.

1. Run Procedure 2.3 for G_{i-1}, obtaining the collection $X_1, \ldots, X_s$.
2. We call the special interval subgraphs $X_1, \ldots, X_s$ *fragments* and their family $\mathcal{X}_i := \{X_1, \ldots, X_s\}$ a *level* (of number i) of the constructed decomposition.
3. Let $G_i := G - \big(V(X_1) \cup \ldots \cup V(X_s)\big)$. Mark every attachment set of each X_j in G_i as a *terminal set*. These terminal sets will be further refined when recursively decomposing G_i; namely, further constructed fragments of G_i will inherit induced subsets of marked terminal sets as their terminal sets.
4. As long as G_i is not an interval graph, repeat this from step 1 with $i \leftarrow i + 1$.

Regarding this procedure, we stress that the obtained levels are numbered "from outside", meaning that the first (outermost) level is of the least index. The rule is that fragments from lower levels have their attachment sets as terminal sets in higher levels. As it will be made precise in the next section, an isomorphism between two T-graphs can be captured by a mapping between their canonical decompositions, which relates pairwise isomorphic fragments and preserves the incidence (i.e., identity) between the attachment sets of mapped fragments and the terminal sets of fragments in higher levels. See also Fig. 1.

3 Group-Computing Tools

We first recall the notion of the *automorphism group* which is closely related to the graph isomorphism problem. An *automorphism* is an isomorphism of a graph G to itself, and the *automorphism group* of G is the group $Aut(G)$ of all automorphisms of G. There exists an isomorphism from G_1 to G_2 if and only if the automorphism group of the disjoint union $H := G_1 \uplus G_2$ contains a permutation exchanging the vertex sets of G_1 and G_2. We work with automorphism groups by means of their generators; a subset A of elements of a group Γ is called a *set of generators* if the members of A together with the operation of Γ can generate each element of Γ.

There are two related classical algebraic tools which we shall use in the next section. The first one is an algorithm performing computation of a subgroup of an arbitrary group, provided that we can efficiently test the membership in the subgroup and the subgroup is not "much smaller" than the original group:

Theorem 3.1. (Furst, Hopcroft and Luks [15, Cor. 1]**)** *Let Π be a permutation group given by its generators, and Π_1 be any subgroup of Π such that one can test in polynomial time whether $\pi \in \Pi_1$ for any $\pi \in \Pi$ (membership test). If the ratio $|\Pi|/|\Pi_1|$ is bounded by a function of a parameter d, then a set of generators of Π_1 can be computed in FPT-time (with respect to d).*

The second tool, known as Babai's "tower-of-groups" procedure (cf. [6]), will not be used as a standalone statement, but as a mean of approaching the task of computation of the automorphism group of our object H (e.g., graph). This procedure can be briefly outlined as follows; imagine an inclusion-ordered chain of groups $\Gamma_0 \supseteq \Gamma_1 \supseteq \ldots \supseteq \Gamma_{k-1} \supseteq \Gamma_k$ such that

- Γ_0 is a group of some unrestricted permutations on the ground set of our H,
- for each $i \in \{1, \ldots, k\}$, we "add" some further restriction (based on the structure of H) which has to be satisfied by all permutations of Γ_i,
- the restriction in the previous point is chosen such that the ratio $|\Gamma_{i-1}|/|\Gamma_i|$ is guaranteed to be "small", and
- in Γ_k, we get the automorphism group of our object H.

Then Theorem 3.1 can be used to compute Γ_1 from Γ_0, then Γ_2 from Γ_1, and so on until we get the automorphism group Γ_k.

Automorphism Group of a Decomposition. Here we are going to apply the above procedure in order to compute the automorphism group of a special object which combines the decompositions (cf. Procedure 2.5) of given T-graphs G_1 and G_2, but abstracts from precise structure of the fragments as interval graphs.

Consider canonical decompositions of the graphs G_1 and G_2, as produced by Procedure 2.5 in the form of level families $\mathcal{X}_1^1, \ldots, \mathcal{X}_\ell^1$ and $\mathcal{X}_1^2, \ldots, \mathcal{X}_{\ell'}^2$, respectively. We may assume that $\ell = \ell'$ since otherwise we immediately answer 'not isomorphic'. A combined decomposition of $H = G_1 \uplus G_2$ hence consists of the levels $\mathcal{X}_i := \mathcal{X}_i^1 \cup \mathcal{X}_i^2$ for $i = 1, \ldots, \ell$ and their respective terminal sets. More precisely, let $\mathcal{X} := \mathcal{X}_1 \cup \ldots \cup \mathcal{X}_\ell$. Let $\mathcal{A}[X]$ for $X \in \mathcal{X}_k$ be the family of all terminal sets in X (as marked by Procedure 2.5 and then restricted to $V(X)$), and specially $\mathcal{A}^i[X] \subseteq \mathcal{A}[X]$ be those terminal sets in X which come from attachment sets of fragments on level $i < k$. Let $\mathcal{A}_k := \bigcup_{X \in \mathcal{X}_k} \mathcal{A}[X]$ and $\mathcal{A}_k^i := \bigcup_{X \in \mathcal{X}_k} \mathcal{A}^i[X]$ for $k = 1, \ldots, \ell$, and let $\mathcal{A} := \mathcal{A}_1 \cup \ldots \cup \mathcal{A}_\ell$.

Recall, from Sect. 2, the definition of the completion X^+ of any $X \in \mathcal{X}_i$ which, in the current context, is defined with respect to the subgraph of H induced on the union U of vertex sets of $\mathcal{X}_{i+1} \cup \ldots \cup \mathcal{X}_\ell$ (of the higher levels from X). This is, exactly, the completion of X defined by the call to Procedure 2.3 on the level i which defined X as a fragment. Recall also the attachment sets of X which are subsets of U (in X^+) and invariant on automorphisms of X^+.

The *automorphism group of such a decomposition of H* (Fig. 1) acts on the ground set $\mathcal{X} \cup \mathcal{A}$, and consists of permutations ϱ of $\mathcal{X} \cup \mathcal{A}$ which, in particular, map $\mathcal{X}_i$ onto $\mathcal{X}_i$ and $\mathcal{A}_i$ onto $\mathcal{A}_i$ for all $i = 1, \ldots, \ell$. Overall, we would like the permutation ϱ correspond to an actual automorphism of the graph H, for which purpose we introduce the following definition. A permutation ϱ of $\mathcal{X} \cup \mathcal{A}$ is an *automorphism of the decomposition of H* if the following hold true;

(A1) for each $X \in \mathcal{X}_i$ ($i \in \{1, \ldots, \ell\}$), we have $\varrho(X) \in \mathcal{X}_i$, and there is a graph isomorphism from the completion X^+ to the completion $\varrho(X)^+$ mapping the tail of X^+ to the tail of $\varrho(X)^+$ and the terminal sets in $\mathcal{A}^j[X]$ to the terminal sets in $\mathcal{A}^j[\varrho(X)]$ for each $1 \leq j < i$, and

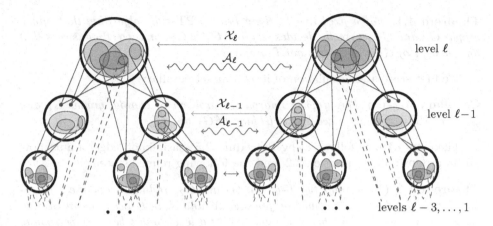

Fig. 1. An illustration of a (combined) canonical decomposition of the graph $H = G_1 \uplus G_2$ into ℓ levels, with the collections of fragments $\mathcal{X}$ (thick black circles) and of terminal sets $\mathcal{A}$ (colored ellipses inside them). The arrows illustrate an automorphism of this decomposition: straight arrows show the possible mapping between isomorphic fragments on the same level, as in (A1), and wavy arrows indicate preservation of the incidence between attachment sets and the corresponding terminal sets, as stated by condition (A2). (Color figure online)

(A2) for every $X \in \mathcal{X}_i$ and $A \in \mathcal{A}_k^i$ where $i \in \{1, \ldots, \ell\}$ and $k \in \{i+1, \ldots, \ell\}$, we have that if A is an attachment set of the fragment X (so, $A \subseteq X^+$), then $\varrho(A) \subseteq \varrho(X)^+$ is the corresponding attachment set of the fragment $\varrho(X)$.

Notice the role of the last two conditions. While (A1) speaks about consistency of ϱ with the actual graph H on the same level, (A2) on the other hand ensures consistency "between the levels". Right from this definition we get:

Proposition 3.2. (*) *Let $H = G_1 \uplus G_2$ and its canonical decomposition (Procedure 2.5) formed by families $\mathcal{X}$ and $\mathcal{A}$ be as above. A permutation ϱ of $\mathcal{X} \cup \mathcal{A}$ is an automorphism of this decomposition, if and only if there exists a graph automorphism of H which acts on $\mathcal{X}$ and on $\mathcal{A}$ identically to ϱ.*

4   Main Algorithm

We are now ready to present our main result which gives an *FPT*-time algorithm for isomorphism of T-graphs (without need for a given decomposition). The algorithm is based on Proposition 3.2, and so on efficient checking of the conditions (A1) and (A2) in the combined decomposition of two graphs. Stated precisely:

Theorem 4.1. *For a fixed tree T, there is an FPT-time algorithm that, given graphs G_1 and G_2, correctly decides whether $G_1 \simeq G_2$, or correctly answers that one or both of G_1 and G_2 are not T-graphs.*[5]

We first state a reformulation of it as a direct corollary.

Corollary 4.2. (*) *The graph isomorphism problem of chordal graphs G_1 and G_2 is in FPT parameterized by the leafage of G_1 and G_2.*

Theorem 4.1 now follows using Procedure 2.5, basic knowledge of automorphism groups and Proposition 3.2, and the following refined statement.

Theorem 4.3. (*) *Assume two T-graphs G_1 and G_2, and their combined decomposition (Procedure 2.5) formed by families $\mathcal{X}$ and $\mathcal{A}$ in ℓ levels, as in Sect. 3. Let $s = \max_{1 \le i \le \ell} |\mathcal{X}_i|$ be the maximum size of a level, and t be an upper bound on the maximum antichain size among the terminal set families $\mathcal{A}^i[X]$ over each $X \in \mathcal{X}$. Then the automorphism group of the decomposition, defined by (A1) and (A2) above, can be computed in FPT-time with the parameter $s + t$.*

Notice that, in our situation, the parameter $s + t$ indeed is bounded in terms of $|T|$; we have $s \le 2d$ and $t \le d$ directly from the arguments in Procedure 2.3. Due to space limits, we give only a sketch of proof in this short paper.

Proof (sketch). First, we outline that the condition (A1) can be dealt with (in step 1 below) efficiently w.r.t. the parameter t: the arguments combine the known and nice description of interval graphs via so-called PQ-trees [9,13], with an *FPT*-time algorithm [3] for the automorphism group of set families with bounded-size antichain (where the latter assumption is crucial for this to work).

Using the previous, we prove the rest as a commented algorithm outline:

1. For every level $k \in \{1, \ldots, \ell\}$ of the decomposition of $H = G_1 \uplus G_2$ we compute the following permutation group Λ_k acting on $\mathcal{X}_k \cup \mathcal{A}_k$.
   a) We partition $\mathcal{X}_k$ into classes according to the isomorphism condition (A1); i.e., $X_1, X_2 \in \mathcal{X}_k$ fall into the same class iff there is a graph isomorphism from X_1^+ to X_2^+ preserving the tail and bijectively mapping $\mathcal{A}^i[X_1]$ to $\mathcal{A}^i[X_2]$ for all $1 \le i < k$. We add the bounded-order symmetric subgroup on each such class of $\mathcal{X}_k$ to Λ_k.
   b) Now, for every permutation $\varrho \in \Lambda_k$ of $\mathcal{X}_k$ and all $X \in \mathcal{X}_k$, and for any chosen isomorphism $\iota_X : X^+ \to \varrho(X)^+$ conforming to (A1), we add to Λ_k the permutation of $\mathcal{A}_k$ naturally composed of partial mappings of the terminal sets induced by the isomorphisms ι_X over $X \in \mathcal{X}_k$.
   c) For every $X \in \mathcal{X}_k$, we compute generators of the automorphism subgroup of X^+ which maps $\mathcal{A}^i[X]$ to $\mathcal{A}^i[X]$ for every $1 \le i < k$, and we add to Λ_k the action of each such generator on $\mathcal{A}[X] \subseteq \mathcal{A}_k$ (as a new generator of Λ_k). This is a nontrivial algorithmic task and we provide the details in the full paper [4].

---

[5] The latter outcome ('not a T-graph') happens when some of the assertions assuming a T-graph in Procedure 2.3 fails.

2. We let $\Gamma_0 = \Lambda_1 \times \ldots \times \Lambda_\ell$ be the direct product of the previous subgroups. Notice that Γ_0 is formed by the permutations conforming to condition (A1).

3. Finally, we apply Babai's tower-of-groups procedure [6] to Γ_0 in order to compute the desired automorphism group of the decomposition. We loop over all pairs $1 \le i < j \le \ell$ of levels and over all cardinalities r of terminal sets in $\mathcal{A}_j$, which is $\mathcal{O}(n^3)$ iterations, and in iteration $k = 1, 2 \ldots$ compute:

   * $\Gamma_k \subseteq \Gamma_{k-1}$ consisting of exactly those automorphisms which conform to the condition (A2) for every component $X \in \mathcal{X}_i$ and every terminal set $A \in \mathcal{A}_j^i$ such that $|A| = r$. Then Γ_k forms a subgroup of Γ_{k-1} (i.e., closed on a composition) thanks to the condition (A1) being true in Γ_{k-1}, and so we can compute Γ_k using Theorem 3.1.

4. We output the group Γ_m of the last iteration $k = m$ of step 3 as the result.

Correctness of the outcome of this algorithm is self-explanatory from the outline; Γ_m satisfies (A1) and (A2) for all possible choice of X and A.

We finish with a brief argument of why the computation in step 3 via Theorem 3.1 is indeed efficient. Observe that for all i, j, $|\mathcal{X}_i| \le s$ and the number of $A \in \mathcal{A}_j^i$ such that $|A| = r$ is at most st. By standard algebraic means (counting cosets of Γ_k in Γ_{k-1}), we get that $|\Gamma_{k-1}|/|\Gamma_k|$ is bounded from above by the order of the subgroup "induced" on $\mathcal{X}_i$ times the order of the subgroup on considered sets A of cardinality r. The latter number is at most $s! \cdot (st)!$ regardless of Γ_{k-1}, and hence bounded in the parameter. $\qquad\square$

5   Conclusions

We have provided an FPT-time algorithm to solve the isomorphism problem for T graphs with a fixed parameter $|T|$ and for chordal graphs of bounded leafage. There seems to be little hope to further extend this result for more general classes of chordal graphs since already for split graphs of unbounded leafage the isomorphism problem is GI-complete. Though, we may combine our result with that of Krawczyk [20] for circular-arc graphs isomorphism to possibly tackle the case of H-graphs for which H contains exactly one cycle.

On the other hand, an open question remains whether a similar decomposition technique as that in Sect. 2 can be used to solve the recognition problem of T-graphs in FPT-time, since the currently best algorithm [11] works in XP-time.

References

1. Aho, A.V., Hopcroft, J.E., Ullman, J.D.: The Design and Analysis of Computer Algorithms. Addison-Wesley, Reading (1974)
2. Arvind, V., Nedela, R., Ponomarenko, I., Zeman, P.: Testing isomorphism of chordal graphs of bounded leafage is fixed-parameter tractable. CoRR abs/2107.10689 (2021)
3. Ağaoğlu, D., Hliněný, P.: Isomorphism problem for S_d-graphs. In: MFCS 2020. LIPIcs, vol. 170, pp. 4:1–4:14. Schloss Dagstuhl - Leibniz-Zentrum für Informatik (2020)
4. Ağaoğlu, D., Hliněný, P.: Isomorphism testing for T-graphs in FPT. CoRR abs/2111.10910 (2021)

5. Ağaoğlu, D., Hliněný, P.: Efficient isomorphism for S_d-graphs and T-graphs. CoRR abs/1907.01495 (2021)
6. Babai, L.: Monte Carlo algorithms in graph isomorphism testing. Technical report, 79–10, Université de Montréal, 42 p. (1979)
7. Babai, L., Luks, E.M.: Canonical labeling of graphs. In: Proceedings of the Fifteenth Annual ACM Symposium on Theory of Computing, STOC 1983, pp. 171–183 (1983)
8. Biró, M., Hujter, M., Tuza, Z.: Precoloring extension. I. interval graphs. Discrete Mathe. **100**, 267–279 (1992)
9. Booth, K.S., Lueker, G.S.: Testing for the consecutive ones property, interval graphs, and graph planarity using PQ-tree algorithms. J. Comput. Syst. Sci. **13**, 335–379 (1976)
10. Bouland, A., Dawar, A., Kopczyński, E.: On tractable parameterizations of graph isomorphism. In: Thilikos, D.M., Woeginger, G.J. (eds.) IPEC 2012. LNCS, vol. 7535, pp. 218–230. Springer, Heidelberg (2012). https://doi.org/10.1007/978-3-642-33293-7_21
11. Chaplick, S., Töpfer, M., Voborník, J., Zeman, P.: On H-topological intersection graphs. CoRR abs/1608.02389 (2016). https://link.springer.com/article/10.1007/s00453-021-00846-3
12. Colbourn, C.J.: On testing isomorphism of permutation graphs. Networks **11**(1), 13–21 (1981)
13. Colbourn, C.J., Booth, K.S.: Linear time automorphism algorithms for trees, interval graphs, and planar graphs. SIAM J. Comput. **10**(1), 203–225 (1981)
14. Curtis, A.R., et al.: Isomorphism of graph classes related to the circular-ones property. Discret. Math. Theor. Comput. Sci. **15**, 157–182 (2013)
15. Furst, M.L., Hopcroft, J.E., Luks, E.M.: Polynomial-time algorithms for permutation groups. In: 21st Annual Symposium on Foundations of Computer Science (FOCS), pp. 36–41 (1980)
16. Gavril, F.: The intersection graphs of subtrees in trees are exactly the chordal graphs. J. Comb. Theor. Ser. B **16**(1), 47–56 (1974)
17. Golumbic, M.C.: Algorithmic graph theory and perfect graphs. Ann. Discrete Math. **57** (2004)
18. Hopcroft, J.E., Wong, J.K.: Linear time algorithm for isomorphism of planar graphs. In: STOC, pp. 172–184 (1974)
19. Klavík, P., Kratochvíl, J., Otachi, Y., Saitoh, T.: Extending partial representations of subclasses of chordal graphs. Theoret. Comput. Sci. **576**, 85–101 (2015)
20. Krawczyk, T.: Testing isomorphism of circular-arc graphs - Hsu's approach revisited. CoRR abs/1904.04501 (2019)
21. Lokshtanov, D., Pilipczuk, M., Pilipczuk, M., Saurabh, S.: Fixed-parameter tractable canonization and isomorphism test for graphs of bounded treewidth. SIAM J. Comput., 161–189 (2017)
22. Matsui, Y., Uehara, R., Uno, T.: Enumeration of the perfect sequences of a chordal graph. Theoret. Comput. Sci. **411**(40), 3635–3641 (2010)
23. Miller, G.: Isomorphism testing for graphs of bounded genus. In: Proceedings of the Twelfth Annual ACM Symposium on Theory of Computing, STOC 1980, pp. 225–235 (1980)
24. Rose, D., Lueker, G., Tarjan, R.E.: Algorithmic aspects of vertex elimination on graphs. SIAM J. Comput. **5**(2), 266–283 (1976)
25. Zemlyachenko, V.N., Korneenko, N.M., Tyshkevich, R.I.: Graph isomorphism problem. J. Sov. Math. **29**, 1426–1481 (1985)

Parameterized Algorithms for Steiner Tree and Dominating Set: Bounding the Leafage by the Vertex Leafage

Celina M. H. de Figueiredo[1], Raul Lopes[2(✉)], Alexsander A. de Melo[1], and Ana Silva[2]

[1] Universidade Federal do Rio de Janeiro, Rio de Janeiro, Brazil
{celina,aamelo}@cos.ufrj.br
[2] Universidade Federal do Ceará, Fortaleza, Brazil
raul@alu.ufc.br, anasilva@mat.ufc.br

Abstract. Chordal graphs are intersection graphs of subtrees of a tree, while interval graphs are intersection graphs of subpaths of a path. Undirected path graphs are an intermediate class of graphs, defined as the intersection graphs of paths of a tree. It is known that DOMINATING SET, CONNECTED DOMINATING SET, and STEINER TREE are W[2]-hard on chordal graphs, when parameterized by the size of the solution, and are polynomial-time solvable on interval graphs. As for the undirected path graphs, all these problems are known to be NP-complete, and when parameterized by the size of the solution, no classification in the parameterized complexity theory is known apart from the trivial XP classification. We prove that DOMINATING SET, CONNECTED DOMINATING SET, and STEINER TREE are FPT for undirected path graphs when parameterized by the size of the solution, and that they continue to be FPT for general chordal graphs when parameterized by the size of the solution plus the vertex leafage of the graph, provided a tree model with optimal vertex leafage is given. We show a relation between the parameterization of MIN-LC-VSP problems by the leafage of the graph versus the vertex leafage plus the size of a solution.

Keywords: Chordal graphs · Undirected Path graphs · Dominating Set · Steiner Tree · FPT algorithms

1 Introduction

Given a graph G and a family of subsets $\mathcal{S} = \{S_u\}_{u \in V(G)}$ of a set U, we say that G is the *intersection graph* of $\mathcal{S}$ if $uv \in E(G)$ if and only if $S_u \cap S_v \neq \emptyset$, and that $(U, \mathcal{S})$ is a *model* of G. *Chordal* graphs are defined as graphs having

This study was financed in part by the Coordenação de Aperfeiçoamento de Pessoal de Nível Superior - Brasil (CAPES) - Finance Code 001, CNPq grants 140399/2017-8, 407635/2018-1, and 303803/2020-7, FAPERJ grant E-26/202.793/2017, STIC-AMSUD 88881.197438/2018-01, and FUNCAP/CNPq PNE-0112-00061.01.00/16.

P. Mutzel et al. (Eds.): WALCOM 2022, LNCS 13174, pp. 251–262, 2022.
https://doi.org/10.1007/978-3-030-96731-4_21

no induced cycle of size bigger than three, but it is known that they are also the intersection graphs of subtrees of a characteristic tree [13]. Nested subclasses of chordal graphs are defined by putting constraints in either the characteristic tree, or the subtrees. *Interval* graphs are the intersection graphs of subpaths of a path [6]; *rooted directed path* graphs are the intersection graphs of directed paths of an out-branching [14] (an oriented rooted tree with all vertices being reachable from the root); *directed path* graphs are the intersection graphs of directed paths of an oriented tree [20]; and *undirected path* graphs are the intersection graphs of paths of a tree [12]. The cited papers give polynomial-time recognition algorithms that also provide models for these classes, called *tree models*.

A set $D \subseteq V(G)$ is *dominating* if, for every vertex $v \in V(G) \setminus D$, we have that v has a neighbor in D. Given a graph G and a positive integer κ, the DOMINATING SET problem consists of deciding whether G has a dominating set of size at most κ, while the CONNECTED DOMINATING SET asks the same but requires additionally that $G[D]$ is connected. Given also a subset $X \subseteq V(G)$, called set of *terminals*, the STEINER TREE problem consists of deciding whether there exists a subset $S \subseteq V(G) \setminus X$, called *Steiner set*, such that $|S| \leq \kappa$ and $G[S \cup X]$ is connected—and hence $G[S \cup X]$ has a spanning tree T, called a *Steiner tree* of G for X. It is known that CONNECTED DOMINATING SET and STEINER TREE have the same complexity for chordal graphs and subclasses [23]. The natural parameter of all these problems is κ.

DOMINATING SET is considered the canonical problem in the class W[2]-hard when parameterized by κ, which explains the great interest in it (see e.g. [15]). When restricted to chordal graphs (and even to split graphs), DOMINATING SET, as well as CONNECTED DOMINATING SET, are still W[2]-hard when parameterized by κ [21]. However, they become polynomial-time solvable on interval graphs, and more generally on rooted directed path graphs [5,23], which brings the natural question about whether they are also polynomial-time solvable on undirected path graphs. This unfortunately is not the case, as both are NP-complete on these graphs [5,10]. Up to our knowledge, it is not known whether (CONNECTED) DOMINATING SET is solvable in polynomial time on directed path graphs. Nevertheless, it could still happen that they are FPT when parameterized by κ on undirected path graphs, and indeed this is one of our results. This classification closes all the parameterized complexity open entries for undirected path graphs presented in [10].

Undirected path graphs can also be seen as intersection graphs of subtrees of a tree where each subtree has at most 2 leaves. A natural generalization therefore is to investigate intersection of subtrees with at most ℓ leaves, which leads to the definition of vertex leafage of a chordal graph. Given a tree model $\mathcal{T} = (T, \{T_u\}_{u \in V(G)})$ of a chordal G, the *vertex leafage of* $\mathcal{T}$ is the maximum number $v\ell(\mathcal{T})$ of leaves in a subtree T_u, while the *vertex leafage of* G is the minimum vertex leafage over all of its tree models [8]; we denote the parameter by $v\ell(G)$. Undirected path graphs are exactly the chordal graphs with vertex leafage 2. Recall that DOMINATING SET and CONNECTED DOMINATING SET are NP-complete on undirected path graphs [5,10], which gives us that they are NP-complete on chordal graphs with vertex leafage k for every fixed $k \geq 2$.

This fact prevents the existence of FPT algorithms parameterized by the vertex leafage of chordal graphs unless P = NP.

In this work we prove that CONNECTED DOMINATING SET and DOMINATING SET are FPT on chordal graphs when parameterized by $\kappa + v\ell(G)$, as long as a tree model with optimal vertex leafage is provided. Since a tree model with optimal vertex leafage can be computed in polynomial time for undirected path graphs [12], we get that these problems are FPT when parameterized by κ on these graphs, which is best possible by the mentioned results.

Theorem 1. *Let G be a chordal graph. If a tree model $\mathcal{T}$ such that $v\ell(\mathcal{T}) = v\ell(G)$ is provided, then* DOMINATING SET, CONNECTED DOMINATING SET *and* STEINER TREE *are* FPT *when parameterized by $\kappa + v\ell(G)$. In particular, when restricted to undirected path graphs, then* DOMINATING SET *can always be solved in time $O^*(2^{2\kappa(1+\log \kappa)})$, while* CONNECTED DOMINATING SET *and* STEINER TREE *can be solved in time $O^*(4^\kappa)$.*

A closely related parameter is the *leafage of G*, denoted by $\ell(G)$, which is the minimum number of leaves $\ell(\mathcal{T})$ in the tree of a tree model $\mathcal{T}$ of G [19]. Surprisingly enough, a tree model with $\ell(G)$ leaves can be computed in polynomial time [16]. This unfortunately is not the case for the vertex leafage parameter, as it is known [8] that it is NP-complete to decide whether a chordal graph G has vertex leafage at most 3; they also give an algorithm to compute $v\ell(G)$ in time $n^{\ell(G)}$, which is XP when parameterized by $\ell(G)$. In [11] they provide an FPT algorithm for DOMINATING SET when parameterized by $\ell(G)$. Since $v\ell(G)$ is a *weaker* parameter than $\ell(G)$, the algorithm provided in [11] is not readily applicable to DOMINATING SET parameterized by κ and $v\ell(G)$. Nevertheless, we show that positive instances of DOMINATING SET and CONNECTED DOMINATING SET must have bounded leafage, which brought us to the question about whether the same holds for generalizations of DOMINATING SET. Indeed, we have found that the broader class of problems, called MIN-LC-VSP problems [7,11], have the same property. Given a graph G and subsets $\sigma, \rho \subseteq \{0, \cdots, n-1\}$, a subset $S \subseteq V(G)$ is a (σ, ρ)-*set* if: $|N(v) \cap S| \in \sigma$ for every $v \in S$, and $|N(v) \cap S| \in \rho$ for every $v \in V(G) \setminus S$. Fixing σ, ρ, and given a graph G and an integer κ, the MIN-LC-VSP$_{\sigma,\rho}$ problem consists in deciding whether there exists a (σ, ρ)-set S of size at most κ. Observe that if $0 \in \rho$, then the answer is always yes since taking the empty set satisfies the constraints; this is why we suppose $0 \notin \rho$ in what follows. MIN-LC-VSP problems generalize a number of optimization problems, as e.g. DOMINATING SET, d-DOMINATING SET, TOTAL DOMINATING SET, INDUCED d-REGULAR SUBGRAPH, etc. [7]. We state our result and its corollary obtained from $v\ell(G) \le \ell(G)$.

Theorem 2. *Let $\sigma, \rho \subseteq \{0, \cdots, n-1\}$ be such that $0 \notin \rho$, G a chordal graph and κ a positive integer. If (G, κ) is a* YES *instance of* MIN-LC-VSP$_{\sigma,\rho}$, *then $\ell(G) \le \kappa \cdot v\ell(G)$.*

Corollary 1. *Let $\sigma, \rho \subseteq \{0, \cdots, n-1\}$, G be a chordal graph and κ be a positive integer. If* MIN-LC-VSP$_{\sigma,\rho}$ *is* FPT *when parameterized by $v\ell(G)$, then* MIN-LC-VSP$_{\sigma,\rho}$ *is also* FPT *when parameterized by $\ell(G)$. And if* MIN-LC-VSP$_{\sigma,\rho}$

is FPT *when parameterized by* $\ell(G)$ *and a tree model* T *with* $v\ell(T) = v\ell(G)$ *is provided, then* MIN-LC-VSP$_{\sigma,\rho}$ *is also* FPT *when parameterized by* $\kappa + v\ell(G)$.

We mention that MAX-LC-VSP$_{\sigma,\rho}$ can also be defined (in this case, the problem consists in deciding whether there exists a (σ,ρ)-set S such that $|S| \geq \kappa$), but our proof cannot be applied to these problems. Nevertheless, many of the MAX-LC-VSP$_{\sigma,\rho}$ problems cited in [7] are known to be polynomial-time solvable in chordal graphs, e.g. INDEPENDENT SET, MAXIMUM INDUCED MATCHING, MAXIMUM EFFICIENT EDGE DOMINATING SET and MAXIMUM DOMINATING INDUCED MATCHING, STRONG STABLE SET, etc. (see for instance [18]).

Another parameter of interest is the mim-width of G [22], since many problems can be solved in XP time when parameterized by mim-width [2,7,17], and rooted directed path graphs have mim-width 1 [17]. One could therefore ask whether undirected path graphs also have bounded mim-width. Up to our knowledge, no explicit construction of undirected path graphs with unbounded mim-width is known, but the fact that LC-VSP problems can be solved in polynomial time on graphs with bounded mim-width [7], combined with the NP-hardness of DOMINATING SET on undirected path graphs, give evidence that undirected path graphs do not have bounded mim-width, unless P = NP.

2   Preliminaries

A *parameterized problem* is a language $\Pi \subseteq \Sigma \times \mathbb{N}$, where Σ is a fixed finite alphabet. A pair $(I,\kappa) \in \Sigma \times \mathbb{N}$ is called an *instance* of Π with *parameter* κ, and we say that it is a YES *instance* if $(I,\kappa) \in \Pi$. Given instances $(I,\kappa),(I',\kappa')$ of the same parameterized problem Π, it is said that they are *equivalent* if (I,κ) is a YES instance of Π if and only if so does (I',κ'). A *reduction rule* for Π is a polynomial-time computable function that maps an instance (I,κ) to another instance (I',κ'). It is *safe* if (I,κ) and (I',κ') are equivalent and $\kappa' \leq g(\kappa)$, where g is a computable function. We refer the reader to [9] for further background on parameterized complexity.

We denote by $T = (T, \{T_u\}_{u \in V(G)})$ a tree model of G. Given a node $t \in V(T)$, we denote by V_t the set $\{u \in V(G): t \in V(T_u)\}$. We say that $u \in V(G)$ is a *leafy vertex* of G (with respect to T) if $V(T_u) = \{\ell_u\}$ and ℓ_u is a leaf in T; denote by $\mathcal{L}(G,T)$ the set of leafy vertices of G with respect to T, and for each $u \in \mathcal{L}(G,T)$, denote by ℓ_u the unique node in T_u. We omit (G,T) when it is clear from the context.

A tree model $(T, \{T_u\}_{u \in V(G)})$ of G is said to be *minimal* if there are no two adjacent nodes $t,t' \in V(T)$ such that $V_t \subseteq V_{t'}$. It is known that such a tree model can be computed in polynomial time [12]. Even though obtaining a minimal tree model, given a tree model of G, is a standard operation, we prove it explicitly in the appendix in order to show that also the vertex leafage does not increase.

Proposition 1 ([12]). *Let* G *be a chordal graph, and* $T = (T, \{T_u\}_{u \in V(G)})$ *be a tree model of* G. *Then, a minimal tree model* $T' = (T', \{T'_u\}_{u \in V(G)})$ *of* G *with* $v\ell(T') \leq v\ell(T)$ *and* $\ell(T') \leq \ell(T)$ *can be computed in polynomial time.*

The following lemma directly implies Theorem 2 and will also be useful in the following sections.

Lemma 1. *Let G be a chordal graph, $\mathcal{T} = (T, \{T_u\}_{u \in V(G)})$ a minimal tree model of G such that $v\ell(\mathcal{T}) = v\ell(G)$, κ a positive integer and $S \subseteq V(G)$ such that $N[u] \cap S \neq \emptyset$ for every leafy vertex $u \in \mathcal{L}$. If $|S| \leq \kappa$, then $\ell(G) \leq \kappa \cdot v\ell(G)$.*

Proof. By contradiction, let $\ell_1, \ldots, \ell_k$ be the leaves of T, with $k \geq \kappa \cdot v\ell(G) + 1$. Since $\mathcal{T}$ is minimal, for each $i \in \{1, \ldots, k\}$, there exists $v_i \in V_{\ell_i}$ such that $V(T_{v_i}) = \{\ell_i\}$, as otherwise we would have $V_{\ell_i} \subseteq V_{t_i}$ where t_i is the neighbor of ℓ_i in T. For each $u \in S$, let $D_u = \{v_i \mid u \in N[v_i]\}$. Observe that if $v_i \in S$, then $D_{v_i} = \{v_i\}$ since $\ell_1, \ldots, \ell_k$ are all distinct leaves of T (i.e., $\{v_1, \ldots, v_k\}$ is an independent set). Note also that if $u \in S \setminus \{v_1, \ldots, v_k\}$, then $|D_u| \leq v\ell(G)$. By assumption, we know that $N[v_i] \cap S \neq \emptyset$ for every $v_i \in \{v_1, \ldots, v_k\} \setminus S$, which means that $\bigcup_{u \in S} D_u = \{v_1, \ldots, v_k\}$. However, we know that $|\bigcup_{u \in S} D_u| \leq \sum_{u \in S} |D_u| \leq |S| \cdot v\ell(G) \leq \kappa \cdot v\ell(G)$, a contradiction since $k > \kappa \cdot v\ell(G)$. $\square$

Since in Theorem 2 we have $0 \notin \rho$, we get directly that a solution S to MIN-LC-VSP$_{\sigma,\rho}$ applied to (G, κ) must be such that $N[u] \cap S \neq \emptyset$ for every $u \in V(G)$, and in particular for every leafy vertex. Hence, Theorem 2 follows from the above lemma. Additionally, it is known that DOMINATING SET can be solved in time $2^{O(\ell^2)} \cdot n^{O(1)}$ on a chordal graph G, where $\ell = \ell(G)$ and $n = |V(G)|$ [11]. Since DOMINATING SET is equivalent to MIN-LC-VSP$_{\sigma,\rho}$ with $\sigma = \{0, \ldots, n-1\}$ and $\rho = \{1, \ldots, n-1\}$, we get that Corollary 1 implies that DOMINATING SET can be solved in FPT time on a chordal graph G when parameterized by $\kappa + v\ell(G)$, provided the appropriate model is given. To finish the proof of Theorem 1, we need to investigate the complexity of STEINER TREE and CONNECTED DOMINATING SET, and to present the claimed algorithm for DOMINATING SET when restricted to undirected path graphs. This is done in Sects. 3 and 4, respectively.

3   Connected Dominating Set and Steiner Tree

In this section, we present FPT algorithms for CONNECTED DOMINATING SET and STEINER TREE parameterized by $\kappa + v\ell(G)$. For simplicity, in what follows we denote an instance of STEINER TREE and CONNECTED DOMINATING SET parameterized by $\kappa + v\ell(G)$ simply by (G, X, κ) and (G, κ), respectively, since $v\ell(G)$ depends on G and hence appears implicitly in the notation. We start by solving STEINER TREE, and at the end of the section we prove that CONNECTED DOMINATING SET is equivalent to STEINER TREE applied to $(G, \mathcal{L})$, where $\mathcal{L}$ is the set of leafy vertices in a given model of G. And to solve STEINER TREE, we apply two reduction rules that allows us to consider only instances $(G, \mathcal{L}, \kappa)$. We start by getting rid of the leafy vertices that are not in X.

Reduction Rule 1. *Let (G, X, κ) be an instance of STEINER TREE where G is chordal, and $\mathcal{T} = (T, \{T_u\}_{u \in V(G)})$ be a tree model of G. If there exists $v \in \mathcal{L} \setminus X$, then* **delete** *$v$, obtaining the instance $(G - v, X, \kappa)$.*

Proof of safeness. Removing vertices clearly cannot increase the vertex leafage; hence we just need to prove that (G, X, κ) is a YES instance if and only if $(G - v, X, \kappa)$ also is. Clearly a solution for $(G - v, X, \kappa)$ is also a solution for (G, X, κ) since $G - v \subseteq G$. Conversely, let $S \subseteq V(G)$ be a Steiner set for (G, X) such that $|S| \leq \kappa$. By definition $H = G[S \cup X]$ is connected. If $v \notin S$, then $H \subseteq G - v$, so suppose otherwise. Observe that, by definition of tree model and since $V(T_v) = \{\ell\}$ for some leaf ℓ of T, we get that $N(v)$ is a clique of G. This clearly implies that v cannot be a cut-vertex in H, i.e., that $H - v$ is still connected, which means that $S - v$ is a solution for $(G - v, X, \kappa)$. $\square$

We now show that it is enough to consider terminal vertices that are leafy vertices. We cannot, however, simply delete the set $X \setminus \mathcal{L}$ of non-leafy terminal vertices since they might be useful to connect terminal leafy vertices, without making an impact on the size of the Steiner set. Thus we use the bypass operation to eliminate vertices in $X \setminus \mathcal{L}$ while maintaining the connectivity that is gained by including these vertices in the induced subgraph $G[S \cup X]$. The *bypass* operation of a vertex $v \in V(G)$ consists of removing v from $V(G)$, and adding uw for every pair u, w of neighbors of v (such that $uw \notin E(G)$ to avoid multiple edges). Before we apply the bypass reduction, we prove the following lemma.

Lemma 2. *Let G be a chordal graph, $\mathcal{T} = (T, \{T_u\}_{u \in V(G)})$ be a tree model of G such that $\mathcal{L} \cap V_\ell \neq \emptyset$ for every leaf $\ell \in V(T)$, and $\emptyset \neq X \subseteq V(G)$ be such that $\mathcal{L} \subseteq X$. If S is a Steiner set for (G, X), then V_t contains some vertex of $S \cup X$ for every $t \in V(T)$.*

Proof. If $V(T) = \{t\}$ it follows trivially because $X \neq \emptyset$; so suppose $|V(T)| > 1$. The lemma also holds trivially for the leaves of T since $\mathcal{L} \subseteq X$ and $\mathcal{L} \cap V_\ell \neq \emptyset$ for every leaf $\ell \in V(T)$. So consider a non-leaf node t of T. Note that t must be within a path between two leaves ℓ_1 and ℓ_2 of T; let $v_1, v_2 \in V(G)$ be such that $V(T_{v_i}) = \{\ell_i\}$ for each $i \in \{1, 2\}$ (they exist by assumption). Since $G[S \cup X]$ is connected and $\{v_1, v_2\} \subseteq \mathcal{L} \subseteq X$, there is a path P in $G[S \cup X]$ between v_1, v_2. Because G is chordal, we get that V_q separates v_1 from v_2 in G for every internal node q in the ℓ_1, ℓ_2-path Q in T. Therefore, we get that P must contain a vertex of V_q for internal node q of Q, in particular it must contain a vertex of V_t. $\square$

Reduction Rule 2. *Let (G, X, κ) be an instance of* STEINER TREE *where G is chordal, $\mathcal{T} = (T, \{T_u\}_{u \in V(G)})$ be a tree model of G such that $\mathcal{L} \cap V_\ell \neq \emptyset$ for every leaf $\ell \in V(T)$, and suppose that Reduction 1 cannot be applied. If there exists $v \in X \setminus \mathcal{L}$, then* bypass *$v$, obtaining the instance $(G', X - v, \kappa)$.*

Proof of safeness. First we show that the vertex leafage cannot increase by constructing a tree model of G' from $\mathcal{T}$. Consider $\mathcal{T}' = (T', \{T'_u\}_{u \in V(G')})$ obtained as follows.

1. T' is the tree obtained from T by contracting T_v to a single vertex, t_v; and
2. For each $u \in V(G')$, if $V(T_v) \cap V(T_u) = \emptyset$, then T_u remains the same; otherwise, T'_u is the subtree of T' containing exactly the vertices in $(V(T_u) \setminus V(T_v)) \cup \{t_v\}$.

To see that the vertex leafage does not increase, just observe that edge contractions of trees cannot increase the number of leaves. It remains to argue that T' is indeed a tree model of G'. For this, we must have $uw \in E(G')$ if and only if $V(T'_u) \cap V(T'_w) \neq \emptyset$. To see that it holds it suffices to observe that $t_v \in V(T'_u)$ if and only if $u \in N(v)$. Now, because the value of κ remains the same, and since the bypass operation can be clearly applied in polynomial time, it remains to show that (G, X, κ) and $(G', X - v, \kappa)$ are equivalent. Denote $X - v$ by X', and by $\mathcal{L}'$ the set $\mathcal{L}(G', T')$. Note that the existence of v implies that $|\mathcal{L}| \geq 2$. Since Reduction 1 cannot be applied, we get that $\mathcal{L} \subseteq X$.

First, consider a solution S for (G, X, κ). We argue that S is also a Steiner set for (G', X'), i.e., that $S \cup X'$ induces a connected subgraph of G'. Indeed, if $u, w \in S \cup X' \subset S \cup X$, then there exists a u, w-path P in $G[S \cup X]$. If $v \notin V(P)$, then P still exists in G'; and otherwise, v is an intermediate vertex in P that is replaced by an edge in G', i.e., u, w are still connected in $G'[S \cup X']$.

Now, let S be a solution for (G', X', κ), which means that $H' = G'[S \cup X']$ is connected. We want to prove that $H = G[S \cup X]$ is also connected. For this, first observe that H' is obtained from $H - v$ by turning $N(v) \cap V(H')$ into a clique. Therefore, the only way H could be disconnected is if v is an isolated vertex in H; we show that this cannot occur. Indeed, note that contracting T_v into a single vertex t_v maintains the property that each leaf of T' must contain a leafy vertex, i.e., that $\mathcal{L}' \cap V_\ell \neq \emptyset$ for every leaf $\ell \in V(T')$. Hence, by Lemma 2 we must have $(S \cup X') \cap V_{\ell_v} \neq \emptyset$, i.e., v has some neighbor in H. $\square$

We are finally ready to prove the main result of this section.

Theorem 3. *Let G be a chordal graph on n vertices and m edges, $X \subseteq V(G)$, and κ be a positive integer. STEINER TREE can be solved on (G, X, κ) in time $O^*(2^{\kappa \cdot v\ell(G)})$, provided a tree model with optimal vertex leafage is given. In particular, if G is an undirected path graph, then STEINER TREE can always be solved in time $O(4^\kappa n^2 + nm)$.*

Proof. Let (G, X, κ) be an instance of STEINER TREE where G is a chordal graph. If $\ell(G) = 1$ or $|X| = 1$, then $G[X]$ is a complete graph and thus $S = \emptyset$ is a solution. Thus we now assume that $\ell(G) \geq 2$ and $|X| \geq 2$. First, compute a minimal tree-model of G; this can be done in polynomial time [12]. Observe that a minimal tree model satisfies the condition of Reduction Rule 2. By iteratively applying Reduction Rules 1 and 2, and Proposition 1 to maintain a minimal tree model, we obtain in polynomial time an equivalent instance (G', X', κ) such that $v\ell(G') \leq v\ell(G)$, and X' is the set of leafy vertices of G' (related to a tree model $T' = (T', \{T'_u\}_{u \in V(G)})$). Now, let $S \subseteq V(G')$ be a Steiner set for (G', X') such that $|S| \leq \kappa$. The connected components of $G[X']$ are exactly the cliques $V_\ell \cap X'$, ℓ a leaf of T'. So, we get that either $N[u] = N[v]$ or $N[u] \cap N[v] = \emptyset$ for every pair of leafy vertices $u, v \in X'$. Hence, we get $N(u) \cap S \neq \emptyset$ for every $u \in X'$, and by Lemma 1 we get $\ell(G') \leq \kappa \cdot v\ell(G') \leq \kappa \cdot v\ell(G)$. We can solve $(G', \mathcal{L}', \kappa)$ in the claimed time using the algorithm given in [3] for STEINER TREE which runs in this instance in time $O(2^{\kappa \cdot v\ell(G)} n^2 + nm)$ time, and in particular if G is an undirected path graph, the starting tree model with optimal vertex leafage can be found in polynomial time [12]. □

Finally, our result for CONNECTED DOMINATING SET is obtained by proving equivalence to STEINER TREE on $(G, \mathcal{L}, \kappa)$. Our proof is necessary since the complexity equivalence proved in [23] concerns only classical complexity.

Theorem 4. *Let G be a chordal graph on n vertices and m edges, and κ be a positive integer.* CONNECTED DOMINATING SET *can be solved on (G, κ) in time $O^*(2^{\kappa \cdot v\ell(G)})$, provided a tree model with optimal vertex leafage is given. In particular, if G is an undirected path graph, then* CONNECTED DOMINATING SET *can always be solved in time $O(4^{\kappa} n^2 + nm)$.*

Proof. We prove that S is a connected dominating set of G if and only if S is a Steiner set for $(G, \mathcal{L})$, where $\mathcal{L}$ denotes the set of leafy vertices in a tree model $(T, \{T_u\}_{u \in V(G)})$ of G. The theorem follows by Theorem 3.

Let S be a connected dominating set of G. So $G[S]$ is a connected subgraph of G, and since it is also dominating, we get that $N(u) \cap S \neq \emptyset$ for every $u \in \mathcal{L}$. It follows that $G[S \cup \mathcal{L}]$ is also connected, and hence S is a Steiner set for $(G, \mathcal{L})$.

On the other hand, if S is a Steiner set for $(G, \mathcal{L})$, then by Lemma 2 we know that $V_t \cap (S \cup \mathcal{L}) \neq \emptyset$ for every $t \in V(T)$, which in turn implies that every $u \in V(G)$ has a neighbor in $S \cup \mathcal{L}$. To finish the proof, just recall that if v is a leafy vertex, then $N(v)$ is a clique. Hence, if $u \in V(G)$ is adjacent to $v \in \mathcal{L}$, then u is also adjacent to $w \in S \cap N(v)$ (which exists since S is a Steiner set for $(G, \mathcal{L})$ and $\mathcal{L}$ is a collection of disjoint cliques). □

4   Dominating Set

In this section, we present an FPT algorithm for DOMINATING SET parameterized by κ restricted to undirected path graphs. Although we believe that our method can be extended to any chordal graph, when parameterized by $\kappa + v\ell(G)$, we remark that the expected running time of such approach is worse than simply applying the algorithm given in [11] after bounding the leafage of the input graph. Thus we refrain from discussing this extension and focus only on the particular case of undirected path graphs since, in this case, our proof is self-contained, simpler, and the $O^*(2^{2\kappa(1+\log \kappa)})$ running time beats the $2^{O(\kappa^2)} n^{O(1)}$ running time provided by applying the algorithm in [11].

In the B-DOMINATING SET, we are given a graph G, a positive integer κ, and a subset $B \subseteq V(G)$ (called set of black vertices), and the goal is to decide if there is a set $D \subseteq V(G)$ with $|D| \leq \kappa$ such that $N[b] \cap D \neq \emptyset$ for every $b \in B$. In other words, the goal is to find a set of at most κ vertices that dominates every black vertex of the instance. We say that such a set D is a *B-dominating set* (in G). Clearly, solving DOMINATING SET on (G, κ) is equivalent to solving B-DOMINATING SET on $(G, V(G), \kappa)$.

From this point on, we assume that G is an undirected path graph, and that $\mathcal{T} = (T, \{P_u\}_{u \in V(G)})$ is a tree model of G where each P_u is a subpath of T (this can be computed in polynomial time [12]). We also denote by $\mathcal{L}$ the set $\mathcal{L}(G, \mathcal{T})$. As in the previous section, we solve this problem by first applying a series of reduction rules, the first of which is analogous to Reduction Rule 1.

Reduction Rule 3. *Let* (G, B, κ) *be an instance of* B-DOMINATING SET. *If there exists* $v \in \mathcal{L} \setminus B$, *then* **delete** v, *obtaining the instance* $(G - v, B, \kappa)$. *And if there exists* $v \in \mathcal{L} \cap B$ *such that* v *is an isolated vertex, then* **delete** v, *obtaining the instance* $(G - v, B - v, \kappa - 1)$.

Proof of safeness. Deleting a vertex clearly does not increase the vertex leafage, so we just need to prove the equivalence between instances. For the first case, clearly a B-dominating set in $G - v$ is also a B-dominating set in G. So let S be a B-dominating set in G. If $v \notin S$, then there is nothing to prove. Otherwise, since v is a leafy vertex, we get that $N[v]$ is a clique, which means that any $b \in B$ dominated by v can be dominated by any $u \in N(v)$ instead. The second part is analogous. □

Now we can assume that every leafy vertex v of G is black and is not isolated. The following rule allows us to bound the number of leaves in T.

Reduction Rule 4. *If* $B = \emptyset$, *then output* YES. *And if* $B \neq \emptyset$ *and either* $\kappa \leq 0$ *or* T *has more than* 2κ *leaves, then output* No.

Safeness. Follows from the assumption that every leafy vertex v is black and from Lemma 1. □

Thus, we assume that T has at most 2κ leaves. Furthermore, if $|V(T)| = 1$, then G is the complete graph and any vertex dominates B; so from now on we assume that T has at least 2 leaves. Our next operation is not a reduction rule, but a branching rule instead. More specifically, we create a number of smaller instances in order to solve the problem. The amount of instances created is bounded by a function of κ, thanks to the fact that T has at most 2κ leaves.

Given nodes t, t' of T, denote by $P(t, t')$ the t, t'-path in T. Also, given a subpath P of T, denote by V_P the set $\{u \in V(G) \mid P_u \subseteq P\}$. Say that $u \in V_P$ is P-*maximal* if there is no $v \in V_P$ such that P_u is a proper subpath of P_v.

Branching Rule. *Let* $\mathcal{I} = (G, B, \kappa)$ *be an instance of* B-DOMINATING SET. *Let* $\ell \in V(T)$ *be a leaf of* T, *and* $u \in V(G)$ *be such that* $V(P_u) = \{\ell\}$. *For each leaf* $t \in V(T)$, $t \neq \ell$, *do the following:*

1. *Choose* $v \in V_{P(\ell, t)}$ *to be a* $P(\ell, t)$-*maximal vertex such that* $\ell \in V(P_v)$;
2. *Define* $G' = G - V_{P_v}$ *and* $B' = B \setminus N_G[v]$;
3. *Create the instance* $\mathcal{I}(u, t) = (G', B', \kappa - 1)$.

We remark that $\{u, v\} \subseteq V_{P_v}$ and thus those two vertices are not in G'.

Correctness of the Branching Rule. First, observe that a minimal tree model of G' can again be obtained by applying Proposition 1 to the tree model $\mathcal{T}$ restricted to G'. Therefore, it remains to show that $\mathcal{I}$ is a YES instance of B-DOMINATING

SET if and only if there exists a leaf t of T distinct from ℓ such that the instance $\mathcal{I}(u,t)$ is also a YES instance.

For the necessity, let S be a B-dominating set of G. By our assumption that Reduction Rule 3 is not applicable, we get that $u \in B$, and $N(u) \neq \emptyset$. Note that, since $V(P_u) = \{\ell\}$ we get that $N(u)$ is a clique. This means that if $u \in S$, then $(S \setminus \{u\}) \cup \{v\}$ is also B-dominating, for any $v \in N(u)$. Therefore, we can assume that $u \notin S$. Now, let v be the neighbor of u in S. Also, let t' be the endpoint of P_v distinct from ℓ, and let t be any leaf separated from ℓ by the edge of P_v incident to t' (it might happen that $t = t'$). Then, either v is $P(\ell,t)$-maximal, or there exists $x \in V_\ell$ which is $P(\ell,t)$-maximal. If the latter occurs, we get that $P_v \subseteq P_x$, which in turn gives us that $N[v] \subseteq N[x]$ and that $(S \setminus \{v\}) \cup \{x\}$ is a B-dominating set of G. We can therefore, without loss of generality, suppose that v is $P(\ell,t)$-maximal. Now, let $\mathcal{I}(u,t)$ be the instance of B-DOMINATING SET constructed as in the statement of the Branching Rule. Observe that if v' is also $P(\ell,t)$-maximal such that $\ell \in V(P_{v'})$, then $P_{v'} = P_v$ and the constructed instance is the same, so we can suppose that indeed v is the iterated $P(\ell,t)$-maximal vertex. It remains to prove that $S' = S \setminus \{v\}$ is a B'-dominating set of $\mathcal{I}(u,t)$. For this, let $b \in B'$. By construction $b \in B \setminus N_G[v]$. Therefore, b has a neighbor in $S \setminus \{v\}$, as we wanted to show.

For the sufficiency, let $\mathcal{I}(u,t) = (G', B', \kappa - 1)$ be the instance given by the Branching Rule, and let S' be a B'-dominating set of G'. Because every $b \in B'$ is dominated by S', and $B \setminus B' = N_G[v]$, we get that $S = S' \cup \{v\}$ is a B-dominating set in G, as we wanted. $\square$

The last part of Theorem 1 follows by bounding the number of instances, since each instance is solved in polynomial time.

Theorem 5. *Let G be an undirected path graph. Then* DOMINATING SET *can be solved in time* $O^*(2^{\mathcal{O}(\kappa \log \kappa)})$.

Proof. We start by obtaining a tree model with optimal vertex leafage for G by applying the polynomial algorithm in [12]. Then, we iteratively apply Reduction Rules 3 and 4 (also applying Proposition 1 to maintain a minimal tree model), until we reach the need to apply the Branching Rule. The latter is then applied for every leaf of the current tree model, which generates at most $(2\kappa)^2 = 4\kappa^2$ new instances. The process then starts over on each of the generated instances. Finally, since the budget for the size of the solution decreases by 1 after applying the Branching Rule, we get that a new application of the rule would generate at most $(2\kappa - 2)^2$ new instances, and so on. Observe that this cascade can be done at most κ times, since at each application we keep one vertex in the dominating set that is being constructed. Therefore, in the worst case scenario, we get that the total number of generated instances is: $(2\kappa)^2 \cdot (2\kappa - 2)^2 \cdot \cdots \cdot (2\kappa - (2\kappa - 2))^2$ $= [(2\kappa) \cdot (2\kappa - 2) \cdot \cdots \cdot (2)]^2 = O([(2\kappa)^\kappa]^2) = O(2^{2\kappa(\log \kappa + 1)})$. Observe that if an instance eventually ends up with a non-empty set of black vertices and a budget of 0 (base case of the branching procedure), then Reduction Rule 4 will output NO. Because the applications of Reduction Rules 3 and 4 and of Proposition 1 are done in polynomial time, we get the claimed running time. $\square$

5 Conclusion

We have investigated the complexity of DOMINATING SET, CONNECTED DOMI-NATING SET and STEINER TREE when parameterized by the size of the solution plus the vertex leafage $(\kappa + v\ell(G))$ of a given chordal graph G. We have found that they are all FPT, provided that a tree model with optimal vertex leafage of G is given. Since such a tree model can be found in polynomial time if G is an undirected path graph (which are graphs with vertex leafage 2), we get that they are all FPT on these graphs when parameterized by the size of the solution. A question is whether the condition about the provided tree model can be lifted. Because positive instances have leafage bounded by a function of κ and $v\ell(G)$, we know that if computing $v\ell(G)$ is FPT when parameterized by $\ell(G)$, then we would have a complete fixed-parameter algorithm. Another option could be to provide a tree model which is not very far from an optimal one, i.e., that has vertex leafage at most $c \cdot v\ell(G)$ for some constant c. This would increase only the constants in our complexities, and we would again have complete algorithms. We ask whether this is achievable. We recall the reader that deciding $v\ell(G) \leq 3$ is NP-complete, but that the vertex leafage can be computed in time $n^{O(\ell(G))}$ [8].

The inequality $v\ell(G) \leq \ell(G)$ says that the vertex leafage of G is a weaker parameter, i.e., that if a problem is FPT when parameterized by $v\ell(G)$, then it is also FPT when parameterized by $\ell(G)$. However, we have also seen that if some MIN-LC-VSP problem is FPT when parameterized by $\ell(G)$, then we get also parameterization by $\kappa + v\ell(G)$. In [11] they provide a fixed-parameter algorithm for DOMINATING SET when parameterized by $\ell(G)$. A question is whether their result can be generalized to all MIN-LC-VSP problems. Given the complexity of the algorithm given in [11], this seems to be a very challenging problem.

Recall the definitions of undirected path, rooted directed path and directed path graphs given in the introduction. It is known that undirected path graphs and rooted directed path graphs are separated by DOMINATING SET, STEINER TREE, CONNECTED DOMINATING SET and GRAPH ISOMORPHISM [1,4,5,10,23], while directed path and rooted directed path graphs are separated by GRAPH ISOMORPHISM [1]. Therefore we ask whether any of the investigated problems also separates these classes. More generally, is there a problem that separates directed path graphs from undirected path graphs?

Finally, we also leave as open the question of whether STEINER TREE and DOMINATING SET admit polynomial kernels with relation to the parameter κ when restricted to undirected path graphs.

References

1. Babel, L., Ponomarenko, I.N., Tinhofer, G.: The isomorphism problem for directed path graphs and for rooted directed path graphs. J. Algorithms **21**(3), 542–564 (1996)
2. Belmonte, R., Vatshelle, M.: Graph classes with structured neighborhoods and algorithmic applications. Theoret. Comput. Sci. **511**, 54–65 (2013)

3. Björklund, A., Husfeldt, T., Kaski, P., Koivisto, M.: Fourier meets Möbius: fast subset convolution. In: Proceedings of the Thirty-Ninth Annual ACM Symposium on Theory of Computing, pp. 67–74 (2007)
4. Booth, K.S., Colbourn, C.J.: Problems polynomially equivalent to graph isomorphism. Technical report. CS-77-04, Computer Science Department, University of Waterloo, Waterloo, Ontario (1979)
5. Booth, K.S., Johnson, J.H.: Dominating sets in chordal graphs. SIAM J. Comput. **11**(1), 191–199 (1982)
6. Booth, K.S., Lueker, G.S.: Testing for the consecutive ones property, interval graphs, and graph planarity using PQ-tree algorithms. J. Comput. Syst. Sci. **13**(3), 335–379 (1976)
7. Bui-Xuan, B.M., Telle, J.A., Vatshelle, M.: Fast dynamic programming for locally checkable vertex subset and vertex partitioning problems. Theoret. Comput. Sci. **511**, 66–76 (2013)
8. Chaplick, S., Stacho, J.: The vertex leafage of chordal graphs. Discret. Appl. Math. **168**, 14–25 (2014)
9. Cygan, M., et al.: Parameterized Algorithms. Springer, Cham (2015). https://doi.org/10.1007/978-3-319-21275-3
10. de Figueiredo, C.M.H., de Melo, A.A., Sasaki, D., Silva, A.: Revising Johnson's table for the 21st century. Discrete Appl. Math. (2021). https://doi.org/10.1016/j.dam.2021.05.021
11. Fomin, F.V., Golovach, P.A., Raymond, J.F.: On the tractability of optimization problems on H-graphs. Algorithmica **89**(2), 1–42 (2020)
12. Gavril, F.: A recognition algorithm for the intersection graphs of paths in trees. Discret. Math. **23**(3), 211–227 (1978)
13. Gavril, F.: The intersection graphs of subtrees in trees are exactly the chordal graphs. J. Comb. Theory Ser. B **16**(1), 47–56 (1974)
14. Gavril, F.: A recognition algorithm for the intersection graphs of directed paths in directed trees. Discret. Math. **13**(3), 237–249 (1975)
15. Goddard, W., Henning, M.A.: Independent domination in graphs: a survey and recent results. Discret. Math. **313**(7), 839–854 (2013)
16. Habib, M., Stacho, J.: Polynomial-time algorithm for the leafage of chordal graphs. In: Fiat, A., Sanders, P. (eds.) ESA 2009. LNCS, vol. 5757, pp. 290–300. Springer, Heidelberg (2009). https://doi.org/10.1007/978-3-642-04128-0_27
17. Jaffke, L., Kwon, O.J., Strømme, T.J.F., Telle, J.A.: Mim-width III. Graph powers and generalized distance domination problems. Theor. Comput. Sci. **796**, 216–236 (2019)
18. Johnson, D.S.: The NP-completeness column: an ongoing guide. J. Algorithms **6**(3), 434–451 (1985)
19. Lin, I.J., McKee, T.A., West, D.B.: The leafage of a chordal graph. Discussiones Mathematicae Graph Theory **18**, 23–48 (1998)
20. Monma, C.L., Wei, V.K.: Intersection graphs of paths in a tree. J. Comb. Theory Ser. B **41**(2), 141–181 (1986)
21. Raman, V., Saurabh, S.: Short cycles make W-hard problems hard: FPT algorithms for W-hard problems in graphs with no short cycles. Algorithmica **52**(2), 203–225 (2008)
22. Vatshelle, M.: New width parameters of graphs. Ph.D. thesis, The University of Bergen (2012)
23. White, K., Farber, M., Pulleyblank, W.: Steiner trees, connected domination and strongly chordal graphs. Networks **15**, 109–124 (1985)

Parameterized Complexity
of Reconfiguration of Atoms

Alexandre Cooper[1], Stephanie Maaz[2], Amer E. Mouawad[3,4(✉)],
and Naomi Nishimura[2]

[1] Institute of Quantum Computing, University of Waterloo,
Waterloo, Ontario, Canada
alexandre.cooper@uwaterloo.ca
[2] David R. Cheriton School of Computer Science,
University of Waterloo, Waterloo, Ontario, Canada
{smaaz,nishi}@uwaterloo.ca
[3] Department of Computer Science, American University of Beirut, Beirut, Lebanon
aa368@aub.edu.lb
[4] University of Bremen, Bremen, Germany

Abstract. Our work is motivated by the challenges presented in preparing arrays of atoms for use in quantum simulation [10]. The recently-developed process of loading atoms into traps results in approximately half of the traps being filled. To consolidate the atoms so that they form a dense and regular arrangement, such as all locations in a grid, atoms are rearranged using moving optical tweezers. Time is of the essence, as the longer that the process takes and the more that atoms are moved, the higher the chance that atoms will be lost in the process. Viewed as a problem on graphs, we wish to solve the problem of reconfiguring one arrangement of tokens (representing atoms) to another using as few moves as possible. Because the problem is NP-complete on general graphs as well as on grids [4], we focus on the parameterized complexity for various parameters, considering both undirected and directed graphs, and tokens with and without labels. For unlabelled tokens, the problem is fixed-parameter tractable when parameterized by the number of tokens, the number of moves, or the number of moves plus the number of vertices without tokens in either the source or target configuration, but intractable when parameterized by the difference between the number of moves and the number of differences in the placement of tokens in the source and target configurations. When labels are added to tokens, however, most of the tractability results are replaced by hardness results.

A. Cooper—This research was undertaken thanks in part to funding from the Canada First Research Excellence Fund.

A. E. Mouawad—Research supported by the Alexander von Humboldt Foundation and partially supported by URB project "A theory of change through the lens of reconfiguration".

N. Nishimura and S. Maaz—Research supported by the Natural Sciences and Engineering Research Council of Canada.

P. Mutzel et al. (Eds.): WALCOM 2022, LNCS 13174, pp. 263–274, 2022.
https://doi.org/10.1007/978-3-030-96731-4_22

1 Introduction

To maximize the probability of success in arranging atoms, approaches need to minimize the probability of atoms being lost during the time between the array being loaded and the atoms being arranged. The lifetime of trapped atoms is short and limited, and the process of moving an atom may result in the loss of the atom. Previous work [10,20] has focused on minimizing the total time required, including both the generation and the execution of the sequence of steps, and consequently has aimed to minimize the number of moves.

The rearrangement of atoms can be framed as a reconfiguration problem; the *reconfiguration framework* [2,3,5,12–15,18] characterizes the transformation between *configurations* by means of a sequence of *reconfiguration steps*. By representing atoms as tokens, we can define each configuration of unlabelled tokens as a subset of vertices of a graph, indicating that there is a token placed on each vertex in the subset; for tokens with labels, a *labelled configuration* can be represented as a mapping of distinct labels to a subset of the vertices. One configuration can be transformed into another by a sequence of moves, where in each move a token is moved from one vertex to another along a token-free path.

Since finding a shortest sequence of moves between configurations is NP-hard, even when restricted to grids [4], we turn to the field of parameterized complexity [6,8,11,17], which studies the impact of one or more *parameters* on the running time of algorithms. A problem is in *FPT* if there exists an algorithm with worst-case running time bounded by $f(k) \cdot n^{O(1)}$ for n the size of the instance, k the size of the parameters, and f a computable function; analogous to NP-hardness in the realm of classical complexity are the classes of intractable problems in the *W-hierarchy* such as W[1] and W[2].

We explore the fixed-parameter tractability of the TOKING MOVING problem for unlabelled and labelled tokens on undirected and directed graphs, with respect to various parameters, namely, the number of tokens (k), the number of moves (ℓ), the number of token-less vertices outside the source and target configurations (f), and the number of moves exceeding the minimum possible for any instance (namely, the number of differences between the source and target configurations). Our results are summarized in Table 1.

Table 1. Summary of results for **U**nlabelled/**L**abelled and **U**ndirected/**D**irected **T**oken **M**oving problem variants

	k	ℓ	$\ell + f$	$\ell - \lvert S \setminus T \rvert$
UUTM	FPT (Corollary 1)	FPT (Theorem 1)	FPT (Theorem 1)	W[2]-hard (Theorem 4)
UDTM	FPT (Corollary 1)	FPT (Theorem 3)	FPT (Theorem 3)	W[2]-hard (Theorem 4)
LUTM	Open	W[1]-hard (Theorem 6)	W[1]-hard (Theorem 6)	W[2]-hard (Theorem 4)
LDTM	Open	W[1]-hard (Theorem 5)	W[1]-hard (Theorem 5)	W[2]-hard (Theorem 4)

Our work constitutes a first step in a larger interdisciplinary project to develop a toolkit for use by physicists, wherein the synergy between theory and practice will be used to determine the direction of theoretical underpinnings as well as implementations. Algorithms currently in use rely on the assumption that one atom can be moved at a time, where the goal is to arrange a small number of atoms in a grid. Our goal is to incorporate into our algorithms both the physical constraints of how optical tweezers can be used to move atoms between optical traps (which can be encoded by adding directions to the underlying graphs), as well as the loss probabilities of atoms due to movement and the passage of time (which can be encoded by adding labels to tokens), and to consider arbitrary graphs. In future work, we will investigate simultaneous movement of multiple atoms; labels on tokens, representing loss probabilities, will change based on motion.

The results in this paper, as well as the choice of parameters, reflect aspects of both current practice and future plans for atom rearrangement. Algorithms based on parameters k and ℓ have direct application to situations in which the total number of atoms or number of moves is small. The number of free vertices f captures the density of the placement of atoms with respect to the target arrangement, and the results for $\ell - |S \setminus T|$ indicate that bounding the difference between the number of moves and those that are required (differing in the source and target configurations) does not provide a good strategy for finding algorithms. By considering both directed and undirected graphs and both labelled and unlabelled tokens, our work can be compared to other results as well as serve as the basis for future research.

2   Terminology

We formulate our problems in terms of the moving of tokens in a graph, using the notation $G = (V(G), E(G))$ for an undirected graph and $D = (V(D), E(D))$ for a directed graph. The reader is directed to a standard textbook on graph theory [7] for definitions of graph classes and other terminology.

We define a *move* as a pair (s, t), where s is the *source vertex* of the move and t is the *target vertex* of the move. Note that s and t are not necessarily adjacent. The *execution* of the move (s, t) results in the change from a configuration containing s to a configuration containing t, where the same label is mapped to s and t, with the rest of the configuration remaining unchanged. In order to ensure that atoms do not collide, a move cannot pass through a vertex that contains a token. A vertex is *free* if there is no token on it, and a path (directed or undirected) is *free* if all intermediate vertices in the path are free. A move (s, t) in a sequence is *valid* if, after the execution of the previous moves in the sequence, there is a token on s, t is free, and there is a free path from s to t, which we designate as the *path for the move*. For a sequence of valid moves α in a graph G (respectively, D), we use G_α (D_α) to denote the graph induced on the union of edges in the paths for the moves in α.

The execution of a sequence of valid moves *transforms* a configuration S into another configuration T if executing the moves starting from S results in tokens being placed as in configuration T. We will call such a sequence of valid moves a *transforming sequence for S and T* or, when S and T are clear from context, simply a *transforming sequence*. In defining a sequence of indices, we use $[n]$ to denote $\{1, \ldots, n\}$.

We define the four problems UNLABELLED UNDIRECTED TOKEN MOVING (UUTM), UNLABELLED DIRECTED TOKEN MOVING (UDTM), LABELLED UNDIRECTED TOKEN MOVING (LUTM), and LABELLED DIRECTED TOKEN MOVING (LDTM) as follows:

Input (UUTM). An undirected graph G and configurations $S \subseteq V(G)$ and $T \subseteq V(G)$ such that $|S| = |T| = k$ and an integer ℓ

Input (UDTM). A digraph D and configurations $S \subseteq V(D)$ and $T \subseteq V(D)$ such that $|S| = |T| = k$ and an integer ℓ

Input (LUTM). An undirected graph G and labelled configurations $S \subseteq V(G)$ and $T \subseteq V(G)$ such that $|S| = |T| = k$ and an integer ℓ

Input (LDTM). A digraph D and labelled configurations $S \subseteq V(D)$ and $T \subseteq V(D)$ such that $|S| = |T| = k$ and an integer ℓ

Output (all problems). A transforming sequence of length at most ℓ

Unless specified otherwise, all definitions apply to instances (G, S, T, ℓ) of UUTM and LUTM as well as instances (D, S, T, ℓ) of UDTM and LDTM. We refer to S as the *source configuration*, T as the *target configuration*, $O = S \cap T$ as the set of *obstacles*, and $S \Delta T$ as the *symmetric difference* of S and T. In a *clearing move*, the source vertex is an obstacle and in a *filling move*, the target vertex is an obstacle. When discussing parameters, we use k to denote $|S| = |T|$ and f to denote $|V(G)| - |S \cup T|$ or $|V(D)| - |S \cup T|$ (the number of vertices without tokens in either S or T). We will refer to an instance (G, S, T) (respectively, (D, S, T)) when discussing the length of a shortest transforming sequence from S to T in G (D). Two instances are *equivalent* if the lengths of the shortest transforming sequences of the instances are equal. For a (directed) path between a pair of vertices s and t, any token on a vertex other than s or t is said to *block* that path. If all paths between a pair of vertices s and t are blocked, then we say that the move (s, t) is *blocked*. For shorthand, when the presence of a token on a vertex v results in a move m being blocked, we'll say that m *is blocked by* v. The observations below follow from the definitions:

Observation 1. *If a vertex v blocks move (s, t) in G (respectively, D), then there exists a path (resp., directed path) from v to t.*

Observation 2. *Suppose that $m = (s, t)$ is the last move in a transforming sequence α from S to T. The sequence α' formed from α by removing m is a transforming sequence from S to a configuration T', where T' differs from T by having a token on s instead of on t.*

Observation 3. *For α, α', $m = (s, t)$, S, and T as in Observation 2, suppose there exists a transforming sequence γ from S to a configuration U, where U differs from T' by a single token, where U has a token on a vertex u and T' has a token on a vertex $v \neq s$. Then if m is blocked after the execution of γ on S, there must be a free (directed, if in D) path from u to t.*

3   Fixed-Parameter Tractability Results

3.1   Preliminaries

We first establish properties of shortest transforming sequences that allow for clean proofs of our results. Călinescu et al. [4] have shown that in any unlabelled undirected graph, it is possible to transform any configuration into another by a single move of each token; in Lemma 1, we show that even for shortest transforming sequences (i.e., ignoring ℓ which is an upper bound on the length of a sequence), we can assume no token moves twice.

Lemma 1. *For any instance (G, S, T, ℓ) of* UUTM *or instance (D, S, T, ℓ) of* UDTM, *there exists a shortest transforming sequence in which no token moves more than once.*

We use Lemma 1 to show that we can find an equivalent instance in which every vertex is in $S \cup T$; we refer to such an instance as a *contracted instance*. In other words, in a contracted instance (G, S, T, ℓ), we have $V(G) = S \cup T$ and hence $|V(G)| \le 2k$. We use contracted instances to form algorithms parameterized by $k = |S| = |T|$, the number of tokens.

Lemma 2. *For any instance (G, S, T, ℓ) of* UUTM *or any instance (D, S, T, ℓ) of* UDTM, *we can form an equivalent contracted instance (G', S', T', ℓ) or (D', S', T', ℓ).*

Corollary 1. UUTM *and* UDTM *admit an $O(k)$ vertex-kernel when parameterized by k. Moreover, the problems can be solved in $k^{O(\ell)} \cdot n^{O(1)}$ time.*

3.2   Unlabelled Undirected Token Moving

Our algorithm for UNLABELLED UNDIRECTED TOKEN MOVING relies on the characterization of the graph G_α of a transforming sequence α of the minimum length of a contracted instance. In Lemmas 3 and 4, we show that there exists α such that G_α is a forest of minimum Steiner trees. By considering all possible ways of partitioning vertices in $S \triangle T$ into trees, and counting the number of moves required by each choice, in Theorem 1 we are able to obtain an FPT algorithm for the UUTM problem parameterized by ℓ (the upper bound on the number of moves) on contracted instances, and hence by Lemma 2, for all instances.

Lemma 3. *For any contracted instance (G, S, T) of* UUTM, *there exists a transforming sequence α of minimum length such that G_α is a forest.*

Lemma 4. *For a contracted instance (G, S, T) of* UUTM *and a transforming sequence α of minimum length such that G_α is a forest, each tree in the forest is a minimum Steiner tree with terminals and leaves in $S \triangle T$ and internal vertices in $S \cup T$, and such that each internal vertex in O is the source vertex of a move.*

Theorem 1. UUTM *is fixed-parameter tractable when parameterized by ℓ.*

Proof. We first form an equivalent contracted instance (Lemma 2), and then attempt all possible partitions of vertices in $S\Delta T$ into at most ℓ Steiner trees, starting first with a single tree, then two, and so on.

When considering the use of d trees, we first consider all possible ways of partitioning the vertices of $S\Delta T$ into d groups, where each group has an equal number of vertices in $S \setminus T$ and $T \setminus S$, and then run the FPT Steiner tree algorithm [9] on each such set of vertices. Because in each Steiner tree each token must move, the number of moves associated with each tree $\mathcal{T}$ will be $|(S \setminus T) \cap V(\mathcal{T})| + |O \cap V(\mathcal{T})|$. If the total number of moves is at most ℓ, we have a yes-instance. If we have not succeeded using d trees to verify that the instance is a yes-instance, we run the algorithm again with the next value of d. If the procedure fails on all values of d, we conclude that (G, S, T, ℓ) is a no-instance.

The correctness of the algorithm follows from Lemmas 3 and 4.   □

3.3   Unlabelled Directed Token Moving

Like in the case of undirected graphs, our algorithm for the directed case relies on the characterization of the graph D_α of a transforming sequence α of the minimum length of a contracted instance. We show, in Lemma 8, that for any yes-instance there exists an α such that D_α is a directed forest. As a replacement for the Steiner tree approach of Sect. 3.2, the fact that we can bound the size of D_α suggests the use of the machinery of color coding, introduced by Alon et al. [1] (similar to a result obtained by Plehn and Voigt [19]), to determine whether D contains a labelled subgraph of the correct form to be D_α for a contracted yes-instance. Unfortunately, Theorem 2 cannot be used directly; in Theorem 3 we adapt the technique for our purposes.

Theorem 2 ([1]). *Let H be a directed forest on q vertices. Let $D = (V, E)$ be a directed n-vertex graph. A subgraph of D isomorphic to H, if one exists, can be found in $2^{O(q)} \cdot n^2 \cdot \log n$ worst-case time. Moreover, if a real-weight function $\beta : E \to \mathcal{R}$ is defined on the edges of D, then the algorithm can be adapted to find the copy of H in D with the maximal total weight.*

After removing extraneous vertices (Lemma 5), we demonstrate that D_α forms a forest (Lemma 8); to show that we can ignore cycles, we focus on a minimal graph containing a cycle as a counterexample. More formally, we call a directed graph D a *circle graph* if D is connected and the vertices in $V(D)$ can be partitioned into *cycle vertices*, forming a simple cycle C in the underlying undirected graph, and *forest vertices*, forming a forest of trees attached to the cycle vertices, where in each tree either all arcs are directed towards the root or all arcs are directed away from the root. An instance (D, S, T, ℓ) of UDTM is said to be a *contracted circle instance* whenever D is a circle graph and $S \cup T = V(D)$.

Lemma 5. *For (D, S, T, ℓ) a contracted instance of UDTM and $v \in S \cap T$, (D, S, T, ℓ) is a yes-instance if and only if $(D - v, S \setminus \{v\}, T \setminus \{v\}, \ell)$ is a yes-instance when any of the following conditions hold:*

1. There is no directed path from any vertex in $S \setminus T$ to v.
2. There is no directed path from v to any vertex in $T \setminus S$.
3. Every directed path from any vertex in $S \setminus T$ to v contains at least $\ell + 1$ obstacles and every directed path from v to any vertex in $T \setminus S$ contains at least $\ell + 1$ obstacles.

Lemma 6. *If there exist instances (D, S, T) of UDTM such that for every transforming sequence α of minimum length, D_α is not a forest, then at least one of those instances must be a contracted circle instance.*

We use Lemma 6 in the proof of Lemma 8, where we use the structure of a circle graph to form a transforming sequence. We number the cycle vertices as v_1, v_2, ..., v_q in clockwise order, observe that for each i, either (v_i, v_{i+1}) or (v_{i+1}, v_i) is an arc in D. We say that a subsequence of vertices $v_a, v_{a+1}, \ldots, v_b$ is a *forward cycle segment* if (v_i, v_{i+1}) is an arc for each $a \leq i < b$ and a *backward cycle segment* if (v_{i+1}, v_i) is an arc for each $a \leq i < b$. Since the paths of moves are directed paths, no move can use edges in more than one cycle segment.

Of particular interest are the *junction vertices* shared by two consecutive cycle segments, where each is either a source or a sink in both cycle segments. We refer to the *in-pool* of a source junction vertex v as the set containing v and any tree vertex that can reach v by a directed path, and the *out-pool* of a sink junction vertex v as the set containing v and any tree vertex that can be reached by a directed path from v. Since each cycle segment starts and ends with a junction vertex, we can refer without ambiguity to the in-pool and out-pool of a cycle segment. By partitioning the moves by cycle segments, we form a sequence of directed trees, thereby allowing us to apply Lemma 7.

Lemma 7. *Given a directed tree D, two configurations S and T of D such that every leaf of D is in $S \Delta T$, and a one-to-one mapping μ from S to T such that there is a directed path from each $s \in S$ to $\mu(s) \in T$ (and $s \neq \mu(s)$ for all s), then there exists a transformation from S to T in D.*

Lemma 8. *For any contracted yes-instance (D, S, T, ℓ) of UDTM, there exists a transforming sequence α of minimum length such that D_α is a directed forest.*

To check if (D, S, T, ℓ) is a contracted yes-instance, it suffices to determine whether or not a labelled version of D contains a subgraph of the correct form to be D_α. We assign labels to vertices of D such that the vertices of $S \setminus T$ are labelled from s_1 to s_Δ, the vertices of $T \setminus S$ are labelled from t_1 to t_Δ, and all other vertices are assigned label Δ. Thus, we have $|S \Delta T| + 1$ distinct labels. We say that D is *($\Delta + 1$)-labelled* and use $\mathrm{lab}(v)$ to denote the label of vertex v. In D_α, all vertices not in $S \Delta T$ receive label Δ.

When (D, S, T, ℓ) is a contracted yes-instance with α a transforming sequence of minimum length, D_α has at most $|S \Delta T| + \ell - |S \setminus T| = \ell + |S \setminus T| \leq 2\ell$ vertices, as otherwise $\ell + 1$ moves are required. Thus, we can enumerate all possible $(\Delta + 1)$-labelled directed graphs of size at most 2ℓ and check whether any one of them implies a yes-instance and can be found as a subgraph of D. For H a $(\Delta + 1)$-labelled directed forest, we let $S' = \{v \in V(H) \mid \mathrm{lab}(v) \in \{s_1, \ldots, s_\Delta, \Delta\}\}$ and

$T' = \{v \in V(H) \mid \text{lab}(v) \in \{t_1, \ldots, t_\Delta, \Delta\}\}$. Then, H is said to be a *witness* for (D, S, T, ℓ) if (H, S', T', ℓ) is a yes-instance and a subgraph isomorphic to H can be found in D such that the labelling of the vertices in $S \Delta T$ is respected.

Theorem 3. UDTM *is in FPT when parameterized by* ℓ.

4   Hardness Results

4.1   Preliminaries

To strengthen some of our hardness results, we prove that for any instance of UDTM we can find an equivalent instance that is of degree at most three or 2-degenerate.

Lemma 9. *For any instance* (D, S, T) *of* UDTM, *we can form an equivalent instance* (D', S', T') *such that* D' *has maximum degree three.*

Lemma 10. *For any instance* (G, S, T) *of* UUTM *or any instance* (D, S, T) *of* UDTM, *we can form an equivalent instance* (G', S', T') *or* (D', S', T') *such that* G' *or* D' *is 2-degenerate.*

4.2   Parameter $\ell - |S \setminus T|$

Theorem 4. *The problems* UUTM, UDTM, LUTM, *and* LDTM *are W[2]-hard when parameterized by* $\ell - |S \setminus T|$.

We use Lemmas 9 and 10 (combined with Theorem 4) to obtain the following results.

Corollary 2. UUTM *is W[2]-hard when parameterized by* $\ell - |S \setminus T|$, *even when restricted to 2-degenerate graphs.*

Corollary 3. UDTM *is W[2]-hard when parameterized by* $\ell - |S \setminus T|$, *even when restricted to 2-degenerate graphs of maximum degree three.*

4.3   Parameter $\ell + f$

We now show that LUTM and LDTM are W[1]-hard parameterized by ℓ or $\ell + f$ on general graphs. Recall that f denotes the number of token-less vertices outside the source and target configurations. We give reductions from the MULTICOLORED SUBGRAPH ISOMORPHISM problem, which determines whether there is a subgraph of a vertex-colored graph G_M that is isomorphic to a vertex-colored graph H. The problem is W[1]-hard when parameterized by solution size, even when H is a 3-regular connected bipartite graph [16]. We define H to be a connected 3-regular bipartite graph such that $V(H) = [c]$, and use $\text{col}(u) \in [c]$ to denote the color of a vertex $u \in V(G_M)$.

Both of our reductions create a *node-vertex* $v(w)$ for each $w \in V(G_M)$ and then use the structure of H to group node-vertices by color to form *super-nodes*. By judicious assignment of labels to tokens moving between supernodes, we ensure that reconfiguration can occur only if we can clear tokens on a set of node-vertices that form a subgraph of G_M isomorphic to H. We start with the directed case and then explain modifications for the undirected case.

Labelled Directed Token Moving. We explain the reduction in three steps: (1) forming a DAG H' from H to provide the structure of D, (2) creating and connecting supernodes, and (3) adding gadgets to constrain movement of tokens.

- **Step 1.** To form a DAG H', we create a vertex $h_i \in V(H')$ for each vertex $i \in V(H)$, making use of breadth-first search to assign each vertex to a level. Choosing an arbitrary vertex r of H for the sole vertex h_r at level 0 in H', we assign each remaining vertex h_i to the minimum level p such that there is a path of p vertices from r to i in H. By adding the edges forming the breadth-first search tree and directing them from vertices at smaller levels to larger levels, we form a directed acyclic graph. By our construction, each arc connects vertices in adjacent levels.
- **Step 2.** We can safely ignore any edge in G_M between vertices of the same color or between colors not connected by an edge in H, as no such edge can form part of a subgraph of G_M isomorphic to H. For each color i, supernode D_i consists of all node-vertices $v(w)$ such that $\mathrm{col}(w) = i$ in G_M; we consider D_i to be at the same level as h_i in H'. To form D, we add an arc between any vertex $v(x) \in D_i$ and $v(y) \in D_j$ such that (h_i, h_j) is an arc in H' and (x, y) is an edge in G_M. When there exists an arc between vertices in D_i and D_j, we say that D_i is a *super-in-neighbor* of D_j and that D_j is a *super-out-neighbor* of D_i.
- **Step 3.** In order to ensure that each edge is traversed, we associate a token with each arc in H', where for arc (h_i, h_j), in S the token is assigned to a *source gadget* in D_i, and in T, the token is assigned to a *target gadget* in D_j. Accordingly, we choose token labels so that for each supernode, there is one token in its source gadget for each super-out-neighbor, and there is one vertex in its target gadget for each super-in-neighbor. For each token in $S \setminus T$, its source vertex and target vertex are in consecutive levels.

For LDTM, the source gadget attached to a supernode consists of a set of vertices with tokens connected by arcs into each node-vertex in the supernode and the target gadget consists of a set of vertices without tokens connected by arcs from each node-vertex in the supernode. In addition, associated with each supernode is a *storage gadget* consisting of a single vertex that is the in-neighbor and out-neighbor of every node-vertex in the supernode. As their names suggest, the union of the vertices in the source gadgets equals $S \setminus T$ and the union of the vertices in the target gadgets equals $T \setminus S$.

To complete the construction for LDTM, we set $S \setminus T$ to all vertices in source gadgets, $T \setminus S$ to all vertices in target gadgets, and $S \cap T$ to all node-vertices. By construction, $|S \setminus T| = |E(H)|$ and $f = |V(H)|$; we set $\ell = |E(H)| + 2k$.

Lemma 11. *If (G_M, H) is a yes-instance of* MULTICOLORED SUBGRAPH ISO-MORPHISM, *then (D, S, T, ℓ) is a yes-instance of* LDTM.

Lemma 12. *If (D, S, T, ℓ) is a yes-instance of* LDTM, *then (G_M, H) is a yes-instance of* MULTICOLORED SUBGRAPH ISOMORPHISM.

Combining Lemmas 11 and 12 with the fact that $\ell + f = O(|V(H)| + |E(H)|) = O(k)$, we obtain the following theorem.

Theorem 5. LDTM *is W[1]-hard when parameterized by $\ell + f$.*

Labelled Undirected Token Moving. We use the same basic structure as in the previous reduction, but need extra machinery to ensure that a token moving from D_i to D_j is unable to find a route that avoids all edges corresponding to edges in G_M with endpoints of colours i and j. To this end, we introduce superedges and a clock gadget, defined below, and for each arc $(h_i, h_j) \in E(H')$, specify the numbers of tokens to move from D_i to D_j and D_j to D_i based on the level. Source gadgets, target gadgets, and storage gadgets all consist of vertices connected by a single undirected edge to all node-vertices in a supernode.

To construct G for LUTM, we construct supernodes in the same way as in D, and although G is undirected, use the terms level, super-in-neighbor, and super-out-neighbor based on the structure of H'. In addition, we form an *edge-path* $p(x, y)$ of length K (to be defined later) for each edge (x, y) in G_M, and for each arc (h_i, h_j) in H', we form a *superedge* $D_{i,j}$ consisting of all edge-paths $p(x, y)$ such that $\mathrm{col}(x) = i$ and $\mathrm{col}(y) = j$. The level of $D_{i,j}$ is considered to be the same as the level of D_i. To connect the node-vertices and edge-paths in G, for each superedge $D_{i,j}$ and each edge-path $p(x, y)$ in $D_{i,j}$, we add an edge from $v(x)$ to one end of $p(x, y)$ and an edge from the other end of $p(x, y)$ to $v(y)$.

For convenience, we refer to the tokens in the source gadget attached to D_i destined for the target gadget attached to D_j as the $D_i - D_j$ *tokens*. Our proofs hinge on showing that the $D_i - D_j$ moves pass through superedge $D_{i,j}$. To limit possible paths, for each supernode $D_{i,j}$ at level r, we create ℓ_r $D_i - D_j$ tokens and ℓ_r $D_j - D_i$ tokens, where for the last level z, $\ell_z = Q = 3k^2/2$ and for any level y, $\ell_y = Q \cdot \ell_{y+1}$.

The *clock gadget* is designed to allow the freeing of edge-paths in superedges one at a time in increasing order of level, which we call the *clock numbering* (or just *numbering*) of the superedges. For large values K and L to be defined later, the clock gadget consists of $|E(H)|$ *storage paths* $K_1, \ldots, K_{|E(H)|}$, each of length K, and $|E(H)| - 1$ *linking paths* $L_1, \ldots, L_{|E(H)|-1}$, each of length L. Referring to the two endpoints of each storage path as the *top end* and the *bottom end* and the two endpoints of each linking path as the *left end* and the *right end*, for each i, we add edges connecting the bottom end of K_i to the right end of L_{i-1} (if it exists) and to the left end of L_i (if it exists).

We view tokens for the clock gadget in S as being grouped into $|E(H)| - 1$ *clock segments*, each containing $K + L$ tokens, such that K_1 is free before the movement of the tokens in the first clock segment from source to target vertices, K_2 is free between the movements of the first and second clock segments, and

$K_{|E(H)|}$ is free only after the movement of all of the clock segments. All vertices in K_1 are in $T \setminus S$, all vertices in $K_{|E(H)|}$ are in $S \setminus T$, and for clock segment i, the tokens are on L_i and K_{i+1} in S, in order from left to right and bottom to top, and on K_i and L_i in T, in order from top to bottom and left to right. We say that the *clock is at position* p whenever the top end of K_p does not have a token. Finally, we connect the top end of each K_i to the middle vertex in each edge-path in the superedge numbered i.

To complete the construction, we set S to consist of vertices in the source gadgets, source vertices in the clock gadget, and all node-vertices and edge-paths, and T to consist of vertices the target gadgets, target vertices in the clock gadget, and all node-vertices and edge-paths. The number of free vertices in G is thus the total size of the storage gadgets, or $f = k$.

We now choose large enough values of K and L to control token movement. We set $K = 2Q^*+k+1$, where Q^* is the total number of vertices in source gadgets in all supernodes. To determine the number of moves ℓ, we sum $(|E(H)|-1)(K+L)$ moves for the $|E(H)| - 1$ clock segments, Q^* moves of tokens from source to target gadgets, $2k$ moves to clear and fill the k node-vertices, and $2K|E(H)|$ to clear and fill one edge-path in each superedge, or $\ell = (|E(H)| - 1)(K + L) + Q^* + 2k + 2K|E(H)|$. We let $L = (|E(H)| - 1)K + Q^* + 2k + 2K|E(H)| + 1$, so that written in a more convenient form, we have $\ell = |E(H)|L - 1$.

Lemma 13. *If (G_M, H) is a yes-instance of* MULTICOLORED SUBGRAPH ISO-MORPHISM, *then (G, S, T, ℓ) is a yes-instance of* LUTM.

Before we prove the reverse direction, we first show that for a yes-instance of LUTM, the clock gadget behaves as required: the clock can be in only one position at a time, "time" cannot go backwards, and at any point in the transformation, we can have at most one superedge with an edge-path free of tokens.

Lemma 14. *If (G, S, T, ℓ) is a yes-instance of* LUTM *then we cannot have K_p and $K_{p'}$ such that the top ends of both are free of tokens.*

Lemma 15. *If (G, S, T, ℓ) is a yes-instance of* LUTM, *then after the clock reaches position p it can never go back to position $p-1$ (or any earlier position).*

Lemma 16. *If (G, S, T, ℓ) is a yes-instance of* LUTM *then at any point in the transformation we can have at most one superedge with an edge-path free of tokens. Moreover, whenever superedge numbered p has a free edge-path, then the clock must be at position p.*

Lemma 17. *If (G, S, T, ℓ) is a yes-instance of* LUTM, *then (G_M, H) is a yes-instance of* MULTICOLORED SUBGRAPH ISOMORPHISM.

Combining Lemmas 13 and 17 with the fact that $f = O(k)$ and $\ell = k^{O(k)}$, we obtain the following theorem.

Theorem 6. LUTM *is W[1]-hard when parameterized by $\ell + f$.*

References

1. Alon, N., Yuster, R., Zwick, U.: Color coding. In: Kao, M. (ed.) Encyclopedia of Algorithms - 2008 Edition. Springer, New York (2008). https://doi.org/10.1007/978-0-387-30162-4_76
2. Bartier, V., Bousquet, N., Heinrich, M.: Recoloring graphs of treewidth 2. Discret. Math. **344**(12), 112553 (2021). https://doi.org/10.1016/j.disc.2021.112553
3. Bonamy, M., Dorbec, P., Ouvrard, P.: Dominating sets reconfiguration under token sliding. Discret. Appl. Math. **301**, 6–18 (2021). https://doi.org/10.1016/j.dam.2021.05.014
4. Călinescu, G., Dumitrescu, A., Pach, J.: Reconfigurations in graphs and grids. SIAM J. Discret. Math. **22**(1), 124–138 (2008). https://doi.org/10.1137/060652063
5. Cereceda, L., van den Heuvel, J., Johnson, M.: Connectedness of the graph of vertex-colourings. Discret. Math. **308**(56), 913–919 (2008)
6. Cygan, M., et al.: Parameterized Algorithms. Springer, Cham (2015). https://doi.org/10.1007/978-3-319-21275-3
7. Diestel, R.: Graph Theory, 4th Edn. Graduate Texts in Mathematics, vol. 173. Springer, Heidelberg (2012)
8. Downey, R.G., Fellows, M.R.: Parameterized Complexity. Springer-Verlag, New York (1997)
9. Dreyfus, S.E., Wagner, R.A.: The Steiner Problem in Graphs, vol. 1, pp. 195–207 (1972)
10. Ebadi, S., et al.: Quantum phases of matter on a 256-atom programmable quantum simulator. Nature **595**(7866), 227–232 (2021). https://doi.org/10.1038/s41586-021-03582-4
11. Flum, J., Grohe, M.: Parameterized Complexity Theory. Texts in Theoretical Computer Science. An EATCS Series, Springer, Heidelberg (2006)
12. van den Heuvel, J.: The complexity of change. Surv. Comb. **2013**(409), 127–160 (2013)
13. Ito, T., et al.: On the complexity of reconfiguration problems. Theoret. Comput. Sci. **412**(12–14), 1054–1065 (2011). https://doi.org/10.1016/j.tcs.2010.12.005
14. Ito, T., Kamiński, M., Demaine, E.D.: Reconfiguration of list edge-colorings in a graph. Discret. Appl. Math. **160**(15), 2199–2207 (2012)
15. Lokshtanov, D., Mouawad, A.E.: The complexity of independent set reconfiguration on bipartite graphs. ACM Trans. Algorithms **15**(1), 7:1–7:19 (2019). https://doi.org/10.1145/3280825
16. Marx, D.: Can you beat treewidth? Theory Comput. **6**(1), 85–112 (2010). https://doi.org/10.4086/toc.2010.v006a005
17. Niedermeier, R.: Invitation to Fixed-Parameter Algorithms. Oxford Lecture Series in Mathematics and Its Applications. Oxford University Press, Oxford (2006)
18. Nishimura, N.: Introduction to reconfiguration. Algorithms **11**(4), 52 (2018). https://doi.org/10.3390/a11040052
19. Plehn, J., Voigt, B.: Finding minimally weighted subgraphs. In: Möhring, R.H. (ed.) WG 1990. LNCS, vol. 484, pp. 18–29. Springer, Heidelberg (1991). https://doi.org/10.1007/3-540-53832-1_28
20. Schymik, K.N., et al.: Enhanced atom-by-atom assembly of arbitrary tweezer arrays. Phys. Rev. A **102**, 063107 (2020)

Parameterized Complexity
of Immunization in the Threshold Model

Gennaro Cordasco[1]($\boxtimes$), Luisa Gargano[2], and Adele A. Rescigno[2]

[1] Department of Psychology, University of Campania "L. Vanvitelli", Caserta, Italy
gennaro.cordasco@unicampania.it
[2] Department of Computer Science, University of Salerno, Fisciano, Italy
{lgargano,arescigno}@unisa.it

Abstract. We consider the problem of controlling the spread of harmful items in networks, such as the contagion proliferation of diseases or the diffusion of fake news. We assume the linear threshold model of diffusion where each node has a threshold that measures the node's resistance to the contagion. We study the parameterized complexity of the problem: Given a network, a set of initially contaminated nodes, and two integers k and ℓ, is it possible to limit the diffusion to at most k other nodes of the network by immunizing at most ℓ nodes? We consider several parameters associated with the input, including the bounds k and ℓ, the maximum node degree Δ, the treewidth, and the neighborhood diversity of the network. We first give $W[1]$ or $W[2]$-hardness results for each of the considered parameters. Then we give fixed-parameter algorithms for some parameter combinations.

Keywords: Parameterized complexity · Contamination minimization · Threshold model

1 Introduction

The problem of controlling the spread of harmful items in networks, such as the contagion proliferation of diseases or the diffusion of fake news, has recently attracted much interest from the research community. The goal is to try to limit as much as possible the spreading process by adopting immunization measures. One such a measure consists in intervening on the network topology by either blocking some links so that they cannot contribute to the diffusion process [24] or by immunizing some nodes [11]. In this paper we focus on the second strategy: Limit the spread to a small region of the network by immunizing a bounded number of nodes in the network. We study the problem in the linear threshold model where each node has a threshold, measuring the node resistance to the diffusion [22]. A node gets influenced/contaminated if it receives the item from a number of neighbors at least equal to its threshold. The diffusion proceeds in rounds: Initially only a subset of nodes has the item and is contaminated. At each round the set of contaminated nodes is augmented with each node that has a number of already contaminated neighbors at least equal to its threshold.

P. Mutzel et al. (Eds.): WALCOM 2022, LNCS 13174, pp. 275–287, 2022.
https://doi.org/10.1007/978-3-030-96731-4_23

In the presence of an immunization campaign, the *immunization* operation on a node inhibits the contamination of the node itself.

Under this diffusion model, we perform a broad parameterized complexity study of the following problem: *Given a network, a spreader set, and two integers k and ℓ, is it possible to limit the diffusion to at most k other nodes of the network by immunizing at most ℓ nodes?*

Influence Diffusion: Related Work. During the past decade the study of spreading processes in complex networks has experienced a particular surge of interest across many research areas from viral marketing, to social media, to population epidemics. Several studies have focused on the problem of finding a small set of individuals who, given the item to be diffused, allow its diffusion to a vast portion of the network, by using the links among individuals in the network to transmit the item to their contacts [29]. Threshold models are widely adopted by sociologists to describe collective behaviours [19] and their use to study the propagation of innovations through a network was first considered in [22]. The linear threshold model has then been widely used in the literature to study the problem of influence maximization, which aims at identifying a small subset of nodes that can maximize the influence diffusion [2,4–6,8,22]. The related Target set selection problem, which aims at selecting a smallest possible set of nodes, whose activation eventually leads to influence all the nodes in the network, has also been widely studied; see for example [2,4,6,7,22].

Recently, some attention has been devoted to the important issue of developing strategies for reducing the spread of negative things through a network. In particular several studies considered the problem of which structural changes can be made to the network topology in order to block negative diffusion processes. Contamination minimization in the linear threshold model by blocking some links was studied in [23,24]. Strategies for reducing the spread size by immunizing/removing nodes were considered in several papers. As an example [1,30] consider a greedy heuristic that immunizes nodes in decreasing order of out-degree. When all the node thresholds are 1, the immunization can be obtained by a cut of the network. Some papers dealing with this problem are [3,20,21] in case of edge cuts and [15] in case of node cuts. Another conceptually related problem is the Firefighter problem [13]; this is a diffusion process with thresholds equal to 1, but it assumes that at each round of the burning process one can defend (immunize) up to d new nodes, instead of ℓ nodes overall, in order to contain the fire.

Parameterized Complexity. Parameterized complexity is a refinement to classical complexity theory in which one takes into account not only the input size, but also other aspects of the problem given by a parameter p. We recall that a problem with input size n and parameter p is called *fixed parameter tractable (FPT)* if it can be solved in time $f(p) \cdot n^c$, where f is a computable function only depending on p and c is a constant.

We study the parameterized complexity of the studied problem, formally defined in Sect. 2. We consider several parameters associated to the input: the bounds k and ℓ, the number ζ related to initially contaminated nodes, and some parameters of the underlying network: The maximum degree Δ, the treewidth $\mathtt{tw}$ [32], and the neighborhood diversity $\mathtt{nd}$ [28]. The two last parameters, formally defined in Sects. 3.4 and 5 respectively, are two incomparable parameters of a graph that can be viewed as representing sparse and dense graphs, respectively [28]; they received much attention in the literature [2,6,8,16,18,26]. Due to space constraints, some proofs are omitted but appear in the full version of the paper available at https://arxiv.org/pdf/2102.03537.pdf.

2 Problem Statement

Let $G = (V, E, t)$ be an undirected graph where V is the set of nodes, E is the set of edges, and $t : V \to \mathbb{N}$ is a node threshold function. We use n and m to denote the number of nodes and edges in the graph, respectively. The degree of a node v is denoted by $d_G(v)$. The neighborhood of v is denoted by $\Gamma_G(v) = \{u \in V \mid (u,v) \in E\}$. In general, the neighborhood of a set $V' \subseteq V$ is denoted by $\Gamma_G(V') = \{u \in V \mid (u,v) \in E, \ v \in V', \ u \notin V'\}$. The graph induced by a node set V' in G is denoted $G[V'] = (V', E', t')$ where $E' = \{(u,v) \mid u,v \in V', \ (u,v) \in E\}$ and $t'(v) = t(v)$ for each $v \in V'$.

Given the network and a spreader set S, after one diffusion round, the influenced nodes are all those which are influenced by the nodes in S, that is, have a number of neighbors in S at least equal to their threshold. Noticing that nodes in S are already contaminated and cannot be immunized, we can then model the diffusion process by a graph which represents the network except the spreader set. Namely, we consider the graph $G = (V, E, t)$ where: V is the set of nodes of the network excluding those in the spreader set, $E \subseteq V \times V$ is the edge set, and t is the threshold function $t : V \to \mathbb{N}$ where $t(v)$ is equal to the original threshold of the node v in the network decreased by the number of its neighbors in S (if the difference is negative then $t(v)$ is set to 0). Hence in G, the diffusion process can be seen as starting at the nodes of threshold 0. Each node in V, including those of threshold 0, may be immunized.

Definition 1. *The diffusion process in $G = (V, E, t)$ in the presence of a set $Y \subseteq V$ of immunized nodes is a sequence of node subsets $\mathsf{D}_{G,Y}[1] \subseteq \cdots \subseteq \mathsf{D}_{G,Y}[\tau] \subseteq \cdots \subseteq V$ with*

- $\mathsf{D}_{G,Y}[1] = \{u \mid u \in V - Y, \ t(u) = 0\}$, *and*
- $\mathsf{D}_{G,Y}[\tau] = \mathsf{D}_{G,Y}[\tau - 1] \cup \left\{ u \mid u \in V - Y, \ |\Gamma_G(u) \cap \mathsf{D}_{G,Y}[\tau - 1]| \geq t(u) \right\}$.

The process ends at τ^* s.t. $\mathsf{D}_{G,Y}[\tau^*] = \mathsf{D}_{G,Y}[\tau^* + 1]$. We set $\mathsf{D}_{G,Y} = \mathsf{D}_{G,Y}[\tau^*]$.

We omit the subscript Y when no node is immunized, that is, $\mathsf{D}_G = \mathsf{D}_{G,\emptyset}$. Moreover, we assume that for the input graph it holds $\mathsf{D}_G = V$; indeed, we could otherwise remove all the nodes that cannot be influenced, since they are irrelevant to the immunization problem. In particular, each remaining node $v \in V$ has $t(v) \leq d_G(v)$, otherwise it could not be influenced. An example is given in Fig. 1 (a).

We are now ready to formally define our problem.

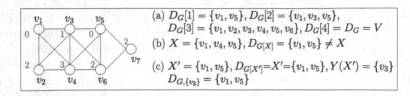

Fig. 1. A graph G (node thresholds appear in red). (a) The diffusion process in G. (b) An example of X whose $G[X]$ includes nodes not influenced. (c) An example of immunizing set $Y(X') = \{v_3\}$, which enables to confine the diffusion to $X' = \{v_1, v_5\}$. (Color figure online)

Influence-Immunization Bounding (IIB): Given a graph $G = (V, E, t)$ and bounds k and ℓ, is there a set Y such that $|Y| \leq \ell$ and $|D_{G,Y}| \leq k$?

For a given set Y we are partitioning the nodes into three subsets: The set $D_{G,Y}$ which contains the nodes that get influenced, the *immunizing* set Y, which has the property that, if all its nodes are immunized then the diffusion process is contained to $D_{G,Y}$, and the set $V - Y - D_{G,Y}$ of the nodes that, by immunizing Y, are not influenced. We will refer to the nodes in the above subsets as *influenced, immunized* and *safe*, respectively.

In some cases it will be easier to deal with a different formulation of IIB based on the set of nodes to which one wants to confine the diffusion. Given $X \subseteq V$, we define the *immunizing set $Y(X)$* of X as the set that contains each node in $V - X$ that can be influenced in one round by those in $D_{G[X]}$, that is, only by its influenced neighbors in X, namely

$$Y(X) = \{u \mid u \in V - X, |\Gamma_G(u) \cap D_{G[X]}| \geq t(u)\}. \tag{1}$$

By the above definitions, we have

$$D_{G[X]} = D_{G,Y(X)} = D_{G[V-Y(X)]} \subseteq X; \tag{2}$$

the influenced, immunized, and safe sets are $D_{G[X]}, Y(X), V - Y(X) - D_{G[X]}$.

For some X, some nodes in $G[X]$ may be not influenced, even though they would in the whole graph G (see Fig. 1 (b)). However, it is easy to see that for each X the set $X' = D_{G[X]} \subseteq X$ is such that $D_{G[X']} = X'$ and $Y(X') = \{u \mid u \in V - X', |\Gamma_G(u) \cap D_{G[X']}| \geq t(u)\} = Y(X)$. In the following, we will refer as *minimal* to a set X such that $D_{G[X]} = X$ (see Fig. 1 (c)).

Fact 1 (IIB equivalent) $\langle G, k, \ell \rangle$ *is a* YES *instance iff there is a minimal X s.t.*

$$|X| = |D_{G[X]}| \leq k \text{ and } |Y(X)| \leq \ell. \tag{3}$$

Summary of Results. We prove INFLUENCE-IMMUNIZATION BOUNDING is:
- W[1]-hard with respect to any of the parameters k, tw or nd
- W[2]-hard with respect to the pairs (ℓ, Δ), or $(\ell, \zeta = |\{v|v \in V, t(v) = 0\}|)$;
- FPT with respect to any of the pairs $(k, \ell), (k, \zeta), (k, \text{tw}), (\Delta, \text{tw}), (k, \text{nd})$, (ℓ, nd).

In Sects. 3 and 4, we will focus on parameters k, ℓ, Δ, ζ, and tw; the results for the neighbourhood diversity parameter will be briefly described in Sect. 5.

3 Hardness

3.1 Parameter k

Theorem 1. IIB *is* $W[1]$-*hard with respect to* k, *the size of the influenced set.*

Proof. We give a reduction from the CUTTING AT MOST k VERTICES WITH TERMINAL (CVT-k) problem studied in [15]: *Given a graph* $H = (V(H), E(H))$, $s \in V(H)$, *and two integers* k *and* ℓ, *is there a set* $X_H \subseteq V(H)$ *such that* $s \in X_H$, $|X_H| \leq k$, *and* $|\Gamma_H(X_H)| \leq \ell$?

To this aim, construct the instance $\langle G, k-1, \ell \rangle$ of IIB where $G = H[V(H) - \{s\}]$ and $t(v) = 0$ for each node $v \in \Gamma_H(s)$ and $t(v) = 1$ for each node $v \in V(H) - \{s\} - \Gamma_H(s)$.

Lemma 1. $\langle H, s, k, \ell \rangle$ *is a* YES *instance of CVT-k iff* $\langle G, k-1, \ell \rangle$ *is a* YES *instance of* IIB.

The theorem follows, since Theorem 3 in [15] proves that CVT-k is $W[1]$-hard with respect to k. The same reduction, recalling that Theorem 5 in [15] proves that CVT-k is $W[1]$-hard with respect to ℓ, also gives that IIB is $W[1]$-hard with respect to ℓ; however, a stronger result is given in the next section.

3.2 Parameters ζ and ℓ

Theorem 2. IIB *is* $W[2]$-*hard with respect to the pair of parameters* ζ, *the number of nodes with threshold* 0, *and* ℓ, *the size of the immunized set.*

Proof. We give a reduction from HITTING SET (HS), which is $W[2]$-complete in the size of the hitting set: *Given a collection* $\{S_1, \ldots, S_m\}$ *of subsets of a set* $A = \{a_1, \ldots, a_n\}$ *and an integer* $h > 0$, *is there a set* $H \subseteq A$ *such that* $H \cap S_i \neq \emptyset$, *for each*[1] $i \in [m]$ *and* $|H| \leq h$?

Given an instance $\langle \{S_1, \ldots, S_m\}, A = \{a_1, \ldots, a_n\}, h \rangle$ of HS, we construct an instance $\langle G, n+1, h \rangle$ of IIB. The graph $G = (V, E, t)$ has node set $V = I \cup A \cup S$, where $I = \{v_0, \ldots, v_h\}$ is a set of $h+1$ independent nodes, $A = \{a_1, \ldots, a_n\}$ is the ground set, and $S = \{s_1, \ldots, s_m\}$ (each s_j represents the set S_j), edge set
$$E = \{(v_i, a_j) \mid v_i \in I, \ a_j \in A\} \cup \{(a_j, s_t) \mid a_j \in A, \ s_t \in S, \ a_j \in S_t\},$$
and threshold function defined by

$$t(v) = \begin{cases} 0 & \text{if } v \in I \\ 1 & \text{if } v \in A \\ |S_t| = d_G(s_t) & \text{if } v = s_t \in S. \end{cases}$$

Trivially, $\mathsf{D}_G[1] = I$, $\mathsf{D}_G[2] = I \cup A$, and $\mathsf{D}_G[3] = I \cup A \cup S = V$.

Lemma 2. $\langle \{S_1, \ldots, S_m\}, A, h \rangle$ *is a* YES *instance of HS iff* $\langle G, n+1, h \rangle$ *is a* YES *instance of* IIB.

---

[1] For a positive integer a, we use $[a]$ to denote the set of integers $[a] = \{1, 2, \ldots, a\}$.

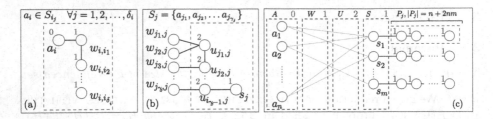

Fig. 2. (a) The *expansion* gadget. (b) The *reduction* gadget. (c) The graph G.

3.3   Parameters Δ and ℓ

Theorem 3. IIB *is $W[2]$-hard with respect to the parameter ℓ, the size of the immunized set, even when the maximum degree of the graph is at most 3.*

Given an instance $\langle\{S_1,\ldots,S_m\}, A = \{a_1,\ldots,a_n\}, h\rangle$ of HS, we construct an instance $\langle G, k, \ell\rangle$ of IIB, where the maximum node degree is 3. We start the construction of G by inserting the nodes in $A \cup W \cup U \cup S$ where $A = \{a_1,\ldots,a_n\}$ is the ground set and $S = \{s_1,\ldots,s_m\}$ (each s_j represents the set S_j), while W and U are two auxiliary sets, of at most nm nodes each, that will be used to keep the degree bounded and, at the same time, simulating a complete bipartite connection between A and S (depicted using gray connection in Fig. 2 (c)). We then add the following *expansion, reduction* and *path* gadgets.

Expansion Gadgets. For each $i \in [n]$, if the sets containing a_i are exactly $S_{i_1}, S_{i_2}, \ldots, S_{i_{\delta_i}}$ then we encode these relationships with a gadget. Namely, we add δ_i nodes $\{w_{i,i_1}, w_{i,i_2}, \ldots w_{i,i_{\delta_i}}\}$ and the edges (a_i, w_{i,i_1}) and $(w_{i,i_j}, w_{i,i_{j+1}})$ for $j \in [\delta_i - 1]$. See Fig. 2 (a).

Reduction Gadgets. For each $j \in [m]$, if $S_j = \{a_{j_1}, a_{j_2}, \ldots, a_{j_{\gamma_j}}\}$ then we encode this relationships with a gadget. Namely, we add $\gamma_j - 1$ nodes $\{u_{j_1,j}, u_{j_2,j}, \ldots, u_{j_{\gamma_j}-1,j}\}$ and the edges $(w_{j_{r+1},j}, u_{j_r,j}), (u_{j_r,j}, u_{j_{r+1},j})$, for $r \in [\gamma_j - 2]$ and $(w_{j_1,j}, u_{j_1,j}), (w_{j_{\gamma_j},j}, u_{j_{\gamma_j}-1,j})$ and $(u_{j_{\gamma_j}-1,j}, s_j)$. The *reduction* gadget is presented in Fig. 2 (b).

Path Gadgets. We complete the construction by adding m paths each on of $p = n + 2nm$ nodes, which depart from each $s_j \in S$. See Fig. 2 (c).

Notice that, by construction the degree of nodes is upper bounded by 3. We set now the thresholds of the nodes in G as: $t(v) = 0$ for each node $v \in A$, $t(v) = 2$ for each node $v \in U$ and $t(v) = 1$ for all the remaining nodes.

Lemma 3. $\langle\{S_1,\ldots,S_m\}, A, h\rangle$ *is a* YES *instance of HS iff* $\langle G, p, h\rangle$ *is a* YES *instance of IIB.*

3.4   Graphs of Bounded Treewidth

Definition 2. *A tree decomposition of a graph $G = (V, E)$ is a pair $(T, \{W_u\}_{u\in V(T)})$, where T is a tree where each node u is assigned a node subset*

$W_u \subseteq V$ such that:

1. $\bigcup_{u \in V(T)} W_u = V$.
2. For each $e = (v, w) \in E$, there exists u in T s.t. W_u contains both v and w.
3. For each $v \in V$, the set $T_v = \{u \in V(T) : v \in W_u\}$, induces a connected subtree of T.

The width of a tree decomposition $(T, \{W_u\}_{u \in V(T)})$ of a graph G, is $\max_{u \in V(T)} |W_u| - 1$. The treewidth of G, denoted by $\mathtt{tw}(G)$, is the minimum width over all tree decompositions of G.

Theorem 4. IIB *is* $W[1]$-*hard with respect to the treewidth of the input graph.*

In order to prove Theorem 4, we present a reduction from MULTI-COLORED CLIQUE (MQ): *Given a graph* $G = (V, E)$ *and a proper vertex-coloring* $\mathbf{c} : V \to [q]$ *for* G, *does* G *contain a clique of size* q?
Given an instance $\langle G, q \rangle$ of MQ, we construct an instance $\langle G' = (V', E'), k, \ell \rangle$ of IIB. We denote by $n' = |V'|$ the number of nodes in G'. For a color $c \in [q]$, we denote by V_c the class of nodes in G of color c and for a pair of distinct $c, d \in [q]$, we let E_{cd} be the subset of edges in G between a node in V_c and one in V_d.

Our goal is to guarantee that any optimal solution of IIB in G' encodes a clique in G and vice-versa. Following some ideas in [2], we construct G' using the following gadgets:

Parallel-Paths Gadget: A parallel-paths gadget of size h, between nodes x and y, consists of h disjoint paths each built by a connection node which is adjacent to both x and y. In order to avoid cluttering, we draw such a gadget as an edge with label h (cf. Fig. 3 (a)).

Selection Gadgets: The selection gadgets encode the selection of nodes (node-selection gadgets) and edges (edge-selection gadgets):

  Node-Selection Gadget: For each $c \in [q]$, we construct a c-node-selection gadget which consists of a node x_v for each $v \in V_c$; these nodes are referred to as node-selection nodes. We then add a guard node g_c that is connected to all the other nodes in the gadget; thus the gadget is a star centered at g_c.
  Edge-Selection Gadget: For each $c, d \in [q]$ with $c \neq d$, we construct a $\{c, d\}$-edge-selection gadget which consists of a node $x_{u,v}$ for every edge $(u, v) \in E_{cd}$; these nodes are referred to as edge-selection nodes. We then add a guard node g_{cd} that is connected to all the other nodes in the gadget; thus the gadget is a star centered at g_{cd}.

Overall there are n node-selection nodes with q guard nodes and m edge-selection nodes with $\binom{q}{2}$ guard nodes (cf. Fig. 3 (b)).

Validation Gadgets: We assign to every node $v \in V(G)$ two unique identifier numbers, $low(v)$ and $high(v)$, with $low(v) \in [n]$ and $high(v) = 2n - low(v)$. For every pair of distinct $c, d \in [q]$, we construct two validation gadgets. One between the c-node-selection gadget and the $\{c, d\}$-edge-selection gadget and one between the d-node-selection gadget and the $\{c, d\}$-edge-selection gadget. We describe the

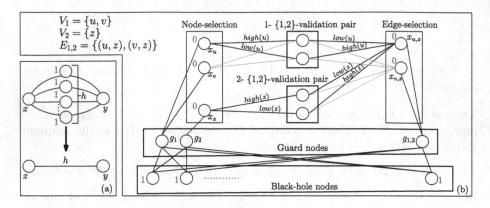

Fig. 3. (a) Parallel-paths gadget. (b) Representation of the graph G' for a trivial instance of the MQ problem $\langle G = (V_1 \cup V_2, E_{1,2}), 2 \rangle$.

validation gadget between the c-node-selection and $\{c,d\}$-edge-selection gadgets. It consists of two nodes. The first one is connected to each node x_v, for $v \in V_c$, by parallel-paths gadgets of size $high(v)$, and to each edge-selection node $x_{u,v}$, for $(u,v) \in E_{cd}$ and $v \in V_c$, by parallel-paths gadgets of size $low(v)$. The other node is connected to each node x_v, for $v \in V_c$, by parallel-paths gadgets of size $low(v)$, and to each edge-selection node $x_{u,v}$, for $(u,v) \in E_{cd}$ and $v \in V_c$, by parallel-paths gadgets of size $high(v)$. Overall, there are $q(q-1)$ validation gadgets, each composed of two nodes.

Black-Hole Gadget: We add a set B of $|B| = (n-q)(2nq - 2n + 1) + \left(m - \binom{q}{2}\right)(4n+1)$ independent nodes and a complete bipartite graph between nodes in B and the guard nodes.

To complete the construction, we specify the thresholds of the nodes in G'

$$t(x) = \begin{cases} 0 & \text{if } x \text{ is a selection node} \\ 1 & \text{if } x \text{ is a connection node or } x \in B \\ d_{G'}(x) - 2n + 1 & \text{if } x \text{ is a validation node} \\ |V_c| & \text{if } x = g_c \text{ is a guard node for some } c \in [q] \\ |E_{cd}| & \text{if } x = g_{cd} \text{ is a guard node for some } c, d \in [q] \end{cases}$$

The complete construction of G' for an instance of the MQ problem appears in Fig. 3 (b).

Lemma 4. $\langle G, q \rangle$ *is a* YES *instance of MQ if and only if* $\langle G', k, \ell \rangle$, *where* $k = (n-q)(2nq - 2n + 1) + \left(m - \binom{q}{2}\right)(4n+1)$ *and* $\ell = q + \binom{q}{2}$ *is a* YES *instance of* IIB. *G' has treewidth* $O(q^2)$.

4   FPT Algorithms

In this section, we present FPT algorithms for several pairs of parameters.

Parameters k and ℓ

Theorem 5. IIB *can be solved in time* $2^{k+\ell}(k+\ell)^{O(\log(k+\ell))} \cdot n^{O(1)}$.

The fixed parameter tractability of IIB with respect to $k + \ell$ can be proved by the arguments used in Theorem 1 in [15] for the problem CUTTING AT MOST k VERTICES WITH TERMINAL.

Parameters k and ζ

Theorem 6. IIB *can be solved in time* $O(\zeta^{3k}n^5)$, *where* $\zeta=|\{v \in V \mid t(v) = 0\}|$.

Proof. Let $\langle G, k, \ell\rangle$ be the input instance of IIB. Suppose $v_1, \ldots v_\zeta$ are the nodes in G having threshold 0 and let Δ denote the maximum degree of a node in G. Consider the graph $G' = (V', E')$ obtained from G by adding the internal nodes and the edges of a Δ-ry tree whose leaves are $v_1, \ldots v_\zeta$. Assume $\langle G, k, \ell, \rangle$ is a YES instance of IIB. We notice that in G, the solution set X (cfr. (3)) can be disconnected but any of its connected components must include at least one node of threshold 0. Hence, in G' the nodes in X are now connected through a path in the Δ-ry tree. This implies that there exists $X' \subseteq V'$ such that: $X \subseteq X'$, $(X' - X) \subseteq V' - V$, and $G'[X']$ is connected. In particular, if s is the root of the tree, we can assume that $s \in X'$. In the worst case, all the paths within the Δ-ry tree go through the root s, hence $|X'| \le |X|\log_\Delta \zeta + 1$.

Let $k' = k\log_\Delta \zeta + 1$. We use the following result [27], Lemma 2: *There are at most $4^{k'}\Delta^{k'}$ connected subgraphs that contain s and have order at most k'. Furthermore, these subgraphs can be enumerated in $O(4^{k'}\Delta^{k'}(|V'| + |E'|))$ time.* This can be done in time $O(\zeta^{3k}n^3)$ noticing that $(4\Delta)^{k'}(|V'| + |E'|) \le 4\Delta\zeta^{k+\frac{k}{\log_\Delta \Delta}}(2n^2) \le 8\zeta^{3k}n^3$. We can then apply the result in [27] to enumerate all the connected subgraphs of G' of size up to k'. For each candidate set X' (the node set of the current connected subgraph) one has to determine whether $X' \cap V$ is a solution according to (3), which can be done in $O(n^2)$ time. □

Parameters k (or Δ) and Treewidth

We present a dynamic programming algorithm, which exploiting a tree decomposition, enables to solve a minimization version of IIB, namely the

Influence Diffusion Minimization (IDM): Given a graph $G = (V, E, t)$ and a budget ℓ, find a set Y such that $|Y| \le \ell$ and $|D_{G,Y}|$ is minimized.

We use the nice tree decomposition [25] and the fact that if a graph G admits a tree decomposition of width at most $\mathtt{tw}$, then it admits a nice tree decomposition of width at most $\mathtt{tw}$. Moreover, given a tree decomposition of size $\mathtt{tw}$, one can compute in polynomial time a nice tree decomposition of width at most $\mathtt{tw}$ that has $O(\mathtt{tw}|V(G)|)$ nodes [25].

Consider a graph $G = (V, E)$ with treewidth $\mathtt{tw}$ and nice tree decomposition $(T, \{W_u\}_{u \in V(T)})$. Let T be rooted at node r and denoted by $T(u)$ the subtree of T rooted at u, for any node u of T. Moreover, we denote by $W(u)$ the union of all the bags in $T(u)$, i.e., $W(u) = \bigcup_{v \in T(u)} W_v$. We will denote by $s_u = |W_u|$ the size of W_u.

We recursively compute the solution of IDM. The algorithm exploits a dynamic programming strategy and traverses the input tree T in a breadth-first fashion. Fix a node u in T, in order to be able to recursively reconstruct the solution, we calculate optimal solutions under different hypotheses based on the following considerations:

– For each node $v \in W_u$ we have three cases: v gets influenced, v is immunized, or v is safe. We are going to consider all the 3^{s_u} combinations of such states with respect to some solution of the problem. We denote each combination with a vector $\mathcal{C}$ of size s_u indexed by the elements of W_u, where the element indexed by $v \in W_u$ denotes the state influenced (0), immunized (1), safe (2) of node v. The configuration $\mathcal{C} = \emptyset$ denotes the vector of length 0 corresponding to an empty bag. We denote by $\mathbb{C}_u$ the family of all the 3^{s_u} possible state vectors of the s_u nodes in W_u.

– Let U be a subset of $V(G)$. Let us first notice that by 3) of Definition 2, all the edges between nodes in $V - W(u)$ and $W(u)$ connect a node in $V - W(u)$ with a node in W_u (the bag corresponding to the root of $T(u)$). We are going to consider all the possible contribution to the diffusion process, of nodes in $V - W(u)$; that is, for each $v \in W_u$, we consider all the possible thresholds among $t(v), t(v) - 1, \ldots, t(v) - \min\{t(v), k\}$ (recall that at most k nodes belong to X and can therefore reduce the threshold of v). We notice that, for each node v, it is possible to bound the number of thresholds to be considered by the value $\min\{t(v), k\}$. Moreover, since no node with $t(v) > d_G(v)$ can be influenced and we can purge such nodes from G in a preprocessing step, we can assume that in G it holds $(\max_{v \in V} t(v)) \leq \Delta$. Hence, we will have up to μ^{s_u} threshold combinations, where $\mu = \min\{k, \Delta\}$. We will denote each possible threshold combination with a vector $\mathcal{T}$, indexed by the s_u elements in W_u, where the element indexed by v belongs to $\{\max\{0, t(v) - k\}, \ldots, t(v)\}$ and denotes the threshold of $v \in W_u$. The configuration $\mathcal{T} = \emptyset$ denotes the vector of length 0 corresponding to an empty bag. We denote by $\mathbb{T}_u$ the family of all the possible threshold combinations of nodes in W_u.

The following definition introduces the values that will be computed by the algorithm in order to keep track of all the above cases:

Definition 3. *For each node $u \in T$, each $j = 0, \ldots, \ell$, $\mathcal{C} \in \mathbb{C}_u$ and $\mathcal{T} \in \mathbb{T}_u$ we denote by $X_u(j, \mathcal{C}, \mathcal{T})$ the minimum number of influenced nodes one can attain in $G[W(u)]$ by immunizing at most j nodes in $W(u)$, where the states and the thresholds of nodes in W_u are given by $\mathcal{C}$ and $\mathcal{T}$.*

Considering that the root r of a nice tree decomposition has $W_r = \emptyset$,

Lemma 5. *For each $u \in T$, the computation of $X_u(j, \mathcal{C}, \mathcal{T})$, for each $j \in \{0, \ldots, \ell\}$, state configuration $\mathcal{C} \in \mathbb{C}_u$, and threshold configuration $\mathcal{T} \in \mathbb{T}_u$ comprises $O(\ell 3^{\mathtt{tw}} \mu^{\mathtt{tw}})$ values, where $\mu = \min\{k, \Delta\}$, each of which can be computed recursively in time $O(2^{\mathtt{tw}} + \ell)$.*

Hence, using [25], Lemma 18, we have that $X_r(\ell, \emptyset, \emptyset)$, which corresponds to the solution of the IDM instance $\langle G, \ell \rangle$, can be computed in time $O(n\mathtt{tw}(2^{\mathtt{tw}} +$

$\ell)\ell 3^{\mathtt{tw}}\mu^{\mathtt{tw}}$). The optimal set X can be computed in the same time by standard backtracking techniques. As a consequence,

Theorem 7. IDM *is solvable in time* $O(n\mathtt{tw}(2^{\mathtt{tw}} + \ell)\ell 3^{\mathtt{tw}}\mu^{\mathtt{tw}})$, *where* $\mu = \min\{k, \Delta\}$.

5 Neighbourhood Diversity

Given a graph $G = (V, E)$, two nodes $u, v \in V$ are said to have the same *type* if $\Gamma_G(v) \setminus \{u\} = \Gamma_G(u) \setminus \{v\}$. The *neighborhood diversity* of a graph G, introduced by Lampis in [28] and denoted by $\mathtt{nd}(G)$, is the minimum number $\mathtt{nd}$ of sets in a partition $V_1, V_2, \ldots, V_{\mathtt{nd}}$, of the node set V, such that all the nodes in V_i have the same type, for $i \in [\mathtt{nd}]$.

We are able to prove both positive and W[1]-hardness results for IIB on graphs of bounded neighborhood diversity. Namely, we have the following results: (i) IIB is W[1]-hard with respect to $\mathtt{nd}$; (ii) IIB can be solved in time $O(n^2\,2^{k+nd-1})$; (iii) IIB can be solved in time $O(n^2\,2^{\ell+nd-1})$, where $\mathtt{nd}$ be the neighborhood diversity of the input graph G.

6 Conclusion

We introduced the influence immunization problem on networks under the threshold model and analyzed its parameterized complexity. We considered several parameters and showed that the problem remains intractable with respect to each one. We have also shown that for some pairs (e.g., (ζ, ℓ) and (Δ, ℓ)) the problem remains intractable. On the positive side, the problem was shown to be FPT for some other pairs: (k, ℓ), (k, ζ), $(k, \mathtt{tw})$, $(\Delta, \mathtt{tw})$, $(k, \mathtt{nd})$ and $(\ell, \mathtt{nd})$. It would be interesting to assess the parameterized complexity of IIB with respect to the remaining pairs of parameters; in particular with respect to k and Δ.

References

1. Albert, R., Jeong, H., Barabási, A.-L.: Error and attack tolerance of complex networks. Nature **404**, 378–382 (2000)
2. Ben-Zwi, O., Hermelin, D., Lokshtanov, D., Newman, I.: Treewidth governs the complexity of target set selection. Discrete Optim. **8**(1), 87–96 (2011). ISSN 1572–5286. https://doi.org/10.1016/j.disopt.2010.09.007
3. Chen, P., David, M., Kempe, D.: Better vaccination strategies for better people. In: Proceedings 11th ACM Conference on Electronic Commerce (EC-2010), Cambridge, Massachusetts, USA, 7–11 June (2010)
4. Cordasco, G., Gargano, L., Mecchia, M., Rescigno, A.A., Vaccaro, U.: A fast and effective heuristic for discovering small target sets in social networks. In: Lu, Z., Kim, D., Wu, W., Li, W., Du, D.-Z. (eds.) COCOA 2015. LNCS, vol. 9486, pp. 193–208. Springer, Cham (2015). https://doi.org/10.1007/978-3-319-26626-8_15
5. Cordasco, G., Gargano, L., Rescigno, A.A.: Influence propagation over large scale social networks. In: Proceedings of ASONAM 2015, pp. 1531–1538 (2015)

6. Cordasco, G., Gargano, L., Rescigno, A.A.: On finding small sets that influence large networks. Soc. Netw. Anal. Min. **6**(1), 1–20 (2016). https://doi.org/10.1007/s13278-016-0408-z

7. Cordasco, G., Gargano, L., Rescigno, A.A.: Active influence spreading in social networks. Theoret. Comput. Sci. **764**, 15–29 (2019)

8. Cordasco, G., Gargano, L., Rescigno, A.A., Vaccaro, U.: Evangelism in social networks: algorithms and complexity. Networks **71**(4), 346–357 (2018)

9. Cygan, M., et al.: Parameterized Algorithms. Springer, Cham (2015). https://doi.org/10.1007/978-3-319-21275-3

10. Downey, R.G., Fellows, M.R.: Fundamentals of Parameterized Complexity. TCS, Springer, London (2013). https://doi.org/10.1007/978-1-4471-5559-1

11. Ehard, S., Rautenbach, D.: Vaccinate your trees! Theoret. Comput. Sci. **772**, 46–57 (2019). ISSN 0304–3975. https://doi.org/10.1016/j.tcs.2018.11.018

12. Feige, U., Krauthgamer, R., Nissim, K.: On cutting a few vertices from a graph. Discret. Appl. Math. **127**, 643–649 (2003)

13. Finbow, S., MacGillivray, G.: The firefighter problem: a survey of results, directions and questions. Australas. J. Comb. **43**, 57–77 (2009)

14. Feige, U., Kogan, S.: Target Set Selection for Conservative Population CoRR abs/1909.03422 (2019)

15. Fomin, F.V., Golovach, P.A., Korhonen, J.H.: On the parameterized complexity of cutting a few vertices from a graph. In: Chatterjee, K., Sgall, J. (eds.) MFCS 2013. LNCS, vol. 8087, pp. 421–432. Springer, Heidelberg (2013). https://doi.org/10.1007/978-3-642-40313-2_38

16. Gargano, L., Rescigno, A.A.: Complexity of conflict-free colorings of graphs. Theoret. Comput. Sci. **566**, 39–49 (2015)

17. Garey, M., Johnson, D.: Computers and Intractability: A Guide to the Theory of NP-Completeness. Freeman, San Francisco (1979)

18. Gavenciak, T., Knop, D., Koutecký, M.: Integer Programming in Parameterized Complexity: Three Miniatures. In Proceedings of IPEC 2018 (2018). https://doi.org/10.4230/LIPIcs.IPEC.2018.21

19. Granovetter, M.: Threshold models of collective behaviors. Am. J. Sociol. **83**(6), 1420–1443 (1978)

20. Hayrapetyan, A., Kempe, D., Pál, M., Svitkina, Z.: Unbalanced graph cuts. In: Brodal, G.S., Leonardi, S. (eds.) ESA 2005. LNCS, vol. 3669, pp. 191–202. Springer, Heidelberg (2005). https://doi.org/10.1007/11561071_19

21. Hanaka, T., Bodlaender, H.L., van der Zanden, T.C., Ono, H.: On the maximum weight minimal separator. Theor. Comput. Sci. **796**, 294–308 (2019)

22. Kempe, D., Kleinberg, J., Tardos, E.: Maximizing the spread of influence through a social network. In: Proceedings of the 9th ACM SIGKDD International Conference on Knowledge Discovery and Data Mining, Washington, USA, pp. 137–146 (2003)

23. Khalil, E.B., Dilkina, B., Song, L.: CuttingEdge: influence minimization in networks. In: Methods, Models, and Applications at NIPS, Workshop on Frontiers of Network Analysis (2013)

24. Kimura, M., Saito, K., Motoda, H.: Blocking links to minimize contamination spread in a social network. ACM Trans. Knowl. Discovery Data **3**(2) (2009)

25. Kloks, T. (ed.): Treewidth. LNCS, vol. 842. Springer, Heidelberg (1994). https://doi.org/10.1007/BFb0045375ISSN: 0302-9743

26. Knop, D., Koutecký, M., Masařík, T., Toufar, T.: Simplified algorithmic metatheorems beyond MSO: treewidth and neighborhood diversity. Logical Methods Comput. Sci. **15**(4) (2019)

27. Komusiewicz, C., Sorge, M.: Finding dense subgraphs of sparse graphs. In: Thilikos, D.M., Woeginger, G.J. (eds.) IPEC 2012. LNCS, vol. 7535, pp. 242–251. Springer, Heidelberg (2012). https://doi.org/10.1007/978-3-642-33293-7_23
28. Lampis, M.: Algorithmic meta-theorems for restrictions of treewidth. Algorithmica **64**, 19–37 (2012)
29. Menczer, F., Fortunato, S., Davis, C.A.: A First Course in Network Science, 1st edn. Cambridge University Press (2020)
30. Newman, M.E.J., Forrest, S., Balthrop, J.: Email networks and the spread of computer viruses. Phys. Rev. E **66** (2002)
31. Niedermeier, R.: Invitation to Fixed-Parameter Algorithms. Oxford University Press (2006)
32. Robertson, N., Seymour, P.D.: Graph minors. II. Algorithmic aspects of tree-width. J. Algorithms **7**(3), 309–322 (1986)

Parameterized Complexity of Minimum Membership Dominating Set

Akanksha Agrawal[1], Pratibha Choudhary[2], N. S. Narayanaswamy[1],
K. K. Nisha[1($\boxtimes$)], and Vijayaragunathan Ramamoorthi[1]

[1] Department of Computer Science and Engineering, IIT Madras, Chennai, India
{akanksha,swamy,kknisha,vijayr}@cse.iitm.ac.in
[2] Faculty of Information Technology, Czech Technical University in Prague,
Prague, Czech Republic
pratibha.choudhary@fit.cvut.cz

Abstract. Given a graph $G = (V, E)$ and an integer k, the MINIMUM MEMBERSHIP DOMINATING SET (MMDS) problem seeks to find a dominating set $S \subseteq V$ of G such that for each $v \in V$, $|N[v] \cap S|$ is at most k. We investigate the parameterized complexity of the problem and obtain the following results about MMDS:

1. W[1]-hardness of the problem parameterized by the pathwidth (and thus, treewidth) of the input graph.
2. W[1]-hardness parameterized by k on split graphs.
3. An algorithm running in time $2^{\mathcal{O}(\mathbf{vc})} |V|^{\mathcal{O}(1)}$, where **vc** is the size of a minimum-sized vertex cover of the input graph.
4. An ETH-based lower bound showing that the algorithm mentioned in the previous item is optimal.

1 Introduction

For a graph $G = (V, E)$, a set $S \subseteq V$ is a *dominating set* for G, if for each $v \in V$, either $v \in S$, or a neighbor of v in G is in S. The DOMINATING SET problem takes as input a graph $G = (V, E)$ and an integer k, and the objective is to test if there is a dominating set of size at most k in G. The DOMINATING SET problem is a classical NP-hard problem [9], which together with its variants, is a well-studied problem in Computer Science. It is also known under standard complexity theoretic assumption that, DOMINATING SET cannot admit any algorithm running in time $f(k) \cdot |V|^{\mathcal{O}(1)}$ time, where k is the size of dominating set.[1] A variant of DOMINATING SET that is of particular interest to us in this paper, is the one where we have an additional constraint that the number of closed neighbors that a vertex has in a dominating set is bounded by a given integer as input.[2] As DOMINATING SET is a notoriously hard problem in itself, so naturally, the

---

[1] More formally, in the framework of parameterized complexity, the problem is W[2]-hard, and thus we do not expect any FPT algorithm for the problem, when parameterized by the solution size.

[2] For a vertex v in a graph $G = (V, E)$, the closed neighborhood of v in G, $N_G[v]$, is the set $\{u \in V \mid \{a, b\} \in E\} \cup \{v\}$.

© Springer Nature Switzerland AG 2022
P. Mutzel et al. (Eds.): WALCOM 2022, LNCS 13174, pp. 288–299, 2022.
https://doi.org/10.1007/978-3-030-96731-4_24

above condition does not make the problem any easier. The above variant has been studied in the literature, and several hardness results are known for it [11]. Inspired by such negative results, in this paper, we remove the size requirement of the dominating set that we are seeking, and attempt to study the complexity variation for such a simplification. We call this version of the DOMINATING SET problem as MINIMUM MEMBERSHIP DOMINATING SET (MMDS, for short). For a graph $G = (V, E)$, a vertex $u \in V$ and a set $S \subseteq V$, the *membership* of u in S is $M(u, S) = |N[u] \cap S|$. Next we formally define the MMDS problem.

MINIMUM MEMBERSHIP DOMINATING SET (MMDS)
Input: A graph $G = (V, E)$ and a positive integer k.
Parameter: k.
Question: Does there exist a dominating set S of G such that $\max_{u \in V} M(u, S) \leq k$?

We refer to a solution of MMDS as a k-membership dominating set (k-mds). Unless, otherwise specified, for MMDS, by k we mean the membership. The term "membership" is borrowed from a similar version of the SET COVER problem by Kuhn *et al.* [10], that was introduced to model reduction in interference among transmitting base stations in cellular networks.

Our Results. We prove that the MMDS problem is NP-Complete and study the problem in the realm of parameterized complexity.

Theorem 1. *The MMDS problem is* NP-*complete on planar bipartite graphs for* $k = 1$.

This shows that the MMDS problem for the parameter k is Para-NP-hard, even for planar bipartite graphs. In other words, for every polynomial time computable function f, there is no $O(n^{f(k)})$-time algorithm for the MMDS problem. Further, our reduction also shows that the MMDS restricted to planar bipartite graphs does not have a $(2 - \epsilon)$ approximation for any $\epsilon > 0$.
Having proved the NP-Completeness property of MMDS, we study the problem parameterized by the pathwidth and treewidth of the input graph. (Please see [3] for formal definitions of treewidth and pathwidth). We note that DOMINATING SET parameterized by the treewidth admits an algorithm running in time $3^{tw}|V|^{\mathcal{O}(1)}$ [3]. In contrast to the above, we show that such an algorithm cannot exist for MMDS.

Theorem 2. MMDS *is* W[1]-hard when parameterized by the pathwidth of the input graph.

We note that the pathwidth of a graph is at least as large as its treewidth, and thus the above theorem implies that MMDS parameterized by the treewidth does not admit any FPT algorithm. We prove Theorem 2 by demonstrating an appropriate parameterized reduction from a well-known W[1]-hard problem called MULTI-COLORED CLIQUE (see [8] for its W[1]-hardness). We note that, an algorithm with running time $k^{\mathbf{tw}}|V|^{\mathcal{O}(1)}$, where $\mathbf{tw}$ is the treewidth of the input graph, follows from Theorem 2 of Chapelle [2].

Next we study MMDS for split graphs, and prove the following theorem.

Theorem 3. MMDS *is* W[1]-*hard on split graphs when parameterized by* k.

We prove the above theorem by giving a parameterized reduction from MULTI-COLORED INDEPENDENT SET, which is known to be W[1]-hard [8]. Our reduction is inspired by the known parameterized reduction from MULTI-COLORED INDEPENDENT SET to DOMINATING SET, where we carefully incorporate the membership constraint and remove the size constraint on the dominating set. We would like to note that DOMINATING SET is known to be W[2]-complete for split graphs [13].

Next we study MMDS parameterized by the vertex cover number of the input graph and show that it admits an FPT algorithm.

Theorem 4. MMDS *admits an algorithm running in time* $2^{\mathcal{O}(vc)}|V|^{\mathcal{O}(1)}$, *where* vc *is the size of a minimum-sized vertex cover of the input graph.*

We prove the above theorem by exhibiting an algorithm which is obtained by "guessing" the portion of the vertex cover that belongs to the solution, and for the remainder of the portion, solving an appropriately created instance of INTEGER LINEAR PROGRAMMING. To complement our Theorem 4, we obtain a matching algorithmic lower bound as follows.

Theorem 5. *Assuming ETH,* MMDS *does not admit an algorithm running in time* $2^{o(vc)}|V|^{\mathcal{O}(1)}$, *where* vc *is the size of a minimum-sized vertex cover of the input graph.*

For the graph theoretic terminology used in this paper, we refer to Diestel [4]. Further, for parameterized complexity related terminology, we refer to books of Cygan *et al.* [3] and Downey and Fellows [6]. We refer the reader to the Arxiv version of the paper [1] for related works, and proofs of some claims and lemmas.

2 The MMDS Problem on Planar Bipartite Graphs Is NP-complete

We show that the MMDS problem is NP-hard for $k = 1$ even when restricted to planar bipartite graphs. The NP-hardness is proved by a reduction from PLANAR POSITIVE 1-IN-3 SAT as follows. Let ϕ be a boolean formula with no negative literals on n variables $X = \{x_1, x_2, \ldots, x_n\}$ having m clauses $C = \{C_1, C_2, \ldots, C_m\}$. Further we consider the restricted case when the graph encoding the variable-clause incidence is planar. Such a boolean formula is naturally associated with a planar bipartite graph $G_\phi = (C \cup X, E)$ where $X = \{x_1, x_2, \ldots, x_n\}$, $C = \{C_1, C_2, \ldots, C_m\}$ and $E = \{(x_i, C_j) \mid \text{variable } x_i \text{ appears in the clause } C_j\}$.

> **PP1in3SAT (Planar Positive 1-in-3 SAT)**
> **Input :** A boolean formula $\phi(X)$ without negative literals and that G_ϕ is planar
> **Decide:** Does there exist an assignment of values $a_1, a_2, \ldots, a_n$ to the variables $x_1, x_2, \ldots, x_n$ such that exactly one literal in each clause is set to true?

It is known that PP1in3SAT is NP-complete [12]. A reduction from PP1in3SAT to the MMDS problem is shown to prove that the MMDS problem is NP-hard. The hardness reduction is given in the full version [1].

Remark: The reduction also shows that the MMDS problem does not have a polynomial time $(2 - \epsilon)$ approximation algorithm unless P = NP. This is because such an algorithm can solve the MMDS problem for $k = 1$.

3   W[1]-Hardness with Respect to Pathwidth

We prove Theorem 2 by a reduction from the MULTI-COLORED CLIQUE problem to the MMDS problem. It is well-known that the MULTI-COLORED CLIQUE problem is W[1]-hard for the parameter solution size [5].

> MULTI-COLORED CLIQUE
> Input: A positive integer k and a k-colored graph G.
> Parameter: k
> Question: Does there exists a clique of size k with one vertex from each color class?

Let $(G = (V, E), k)$ be an instance of the MULTI-COLORED CLIQUE problem. Let $V = (V_1, \ldots, V_k)$ denote the partition of the vertex set V. By a partition, we mean the set of all vertices of same color. We assume, without loss of generality, $|V_i| = n$ for each $i \in [k]$. We usually use n to denote number of vertices in the input graph. However, we use n here to denote the number of vertices in each color class. For each $1 \leq i \leq k$, let $V_i = \{u_{i,\ell} \mid 1 \leq \ell \leq n\}$.

3.1   Gadget Based Reduction from Multi-colored Clique

For an input instance (G, k) of the MULTI-COLORED CLIQUE problem, the reduction outputs an instance (H, k') of the MMDS problem where $k' = n + 1$. The graph H is constructed using two types of gadgets, $\mathcal{D}$ and I (illustrated in Fig. 1). The gadget I is the primary gadget and the gadget $\mathcal{D}$ is secondary gadget that is used to construct the gadget I.

Gadget of type $\mathcal{D}$. For two vertices u and v, the gadget $\mathcal{D}_{u,v}$ is an interval graph consisting of vertices u, v and $n + 2$ additional vertices that form an independent set. The vertices u and v are adjacent, and both u and v are adjacent to every other vertex. We refer to the vertices u and v as *heads* of the gadget

$\mathcal{D}_{u,v}$. Intuitively, for any feasible solution S, and for any gadget $\mathcal{D}_{u,v}$, either u or v should be in S. Otherwise, remaining $n + 2$ vertices must be in S which contradicts the optimality of S because membership for both u and v is at least $n + 2$.

Observation 6. *The pathwidth of the gadget $\mathcal{D}$ is two. Indeed, it is an interval graph with maximum clique of size three and thus, by definition, has pathwidth 2.*

Gadget of Type I. Let $n \geq 1$ be an integer. The gadget has two vertices h_1 and h_2, and two disjoint sets: $A = \{a_1, \ldots, a_n\}$ and $D = \{d_1, \ldots, d_n\}$. For each $i \in [n]$, vertices a_i and d_i are connected by the gadget $\mathcal{D}_{a_i,d_i}$. Let h_2 and h_1 be two additional vertices which are adjacent. The vertices in the sets A and D are adjacent to h_2 and h_1, respectively. For each $1 \leq i \leq n$, a_i and h_1 are connected by the gadget $\mathcal{D}_{a_i,h_1}$, and d_i and h_2 are connected by the gadget $\mathcal{D}_{d_i,h_2}$. In the reduction a gadget of type I is denoted by the symbol I and an appropriate subscript.

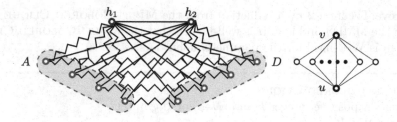

Fig. 1. To the left is the type-I gadget for $n = 4$ and to the right is the type-D gadget. The zigzag edges between vertices u and v represent the gadget $\mathcal{D}_{u,v}$.

Claim 7. $\Diamond$[3] *The pathwidth of a gadget type I is at most four.*

Whenever we refer to a gadget we mean the primary gadget I unless the gadget $\mathcal{D}$ is specified. For each vertex and edge in the given graph, our reduction has a corresponding gadget in the instance output by the reduction.

Description of the Reduction. For $1 \leq i < j \leq k$, let $E_{i,j}$ denote the set of edges with one end point in V_i and the other in V_j, that is $E_{i,j} = \{xy \mid x \in V_i, y \in V_j\}$.

For each vertex and edge in G, the reduction uses a gadget of type I. For each $1 \leq i < j \leq k$, the graph H has an induced subgraph H_i corresponding to V_i, and has an induced subgraph $H_{i,j}$ for the edge set $E_{i,j}$. We refer to H_i as a vertex-partition block and $H_{i,j}$ as an edge-partition block. Inside block[4] H_i, there is a gadget of type I for each vertex in V_i, and in the block $H_{i,j}$ is a gadget for each edge in $E_{i,j}$. For a vertex $u_{i,x}$, I_x denotes the gadget corresponding to $u_{i,x}$ in the partition V_i, and for an edge e, I_e denotes the gadget corresponding to e. Finally, the blocks are connected by the connector vertices which we describe below. We

---

[3] Proofs of results marked with $\Diamond$ can be found in the full version of the paper [1].

[4] Not to be confused with "block" of a 1-vertex-connected graph.

next define the structure of a block which we denote by B. The definition of the block applies to both the vertex-partition block and the edge-partition block. A block B consists of the following gadgets, additional vertices, and edges.

- The block B corresponding to the vertex-partition block H_i for any $i \in [k]$ is as follows: for each $\ell \in [n]$, add a gadget I_ℓ to the vertex-partition block H_i, to represent the vertex $u_{i,\ell} \in V_i$.
- The block B corresponding to the edge-partition gadget $H_{i,j}$ for any $1 \leq i < j \leq k$ is as follows: for each $e \in E_{i,j}$, add a gadget I_e in the edge-partition block $H_{i,j}$, to represent the edge e.
- In addition to the gadgets, we add $(n+1)(n+3) + 2$ vertices to the block B as follows (A figure is illustrated in full version): Let $C(B)$ denote the set $\{f, f', c_1, c_2, \ldots, c_{n+1}, b_1, b_2, \ldots, b_{(n+1)(n+2)}\}$, which is the set of additional vertices that are added to the block B. Let $C'(B)$ denote the subset $\{c_1, c_2, \ldots, c_{n+1}\}$. For each gadget I in B, and for each $t \in [n]$, a_t in I is adjacent to f, and the vertex f is adjacent to f'. Further, the vertex f' is adjacent to each vertex c_p for $p \in [n+1]$. Finally, for each $p \in [n+1]$ and $(p-1)(n+2) < q \leq p(n+2)$, c_p is adjacent to b_q.

Next, we introduce the connector vertices to connect the edge-partition blocks and vertex-partition blocks. Let $R = \{r_{i,j}^i, s_{i,j}^i, r_{i,j}^j, s_{i,j}^j \mid 1 \leq i < j \leq k\}$ be the connector vertices. The blocks are connected based on the following exclusive and exhaustive cases, and is illustrated in Fig. 2:
For each $i \in [k]$, each $i < j \leq k$ and each $\ell \in [n]$, the edges are described below.

- for each $1 \leq t \leq \ell$, the vertex a_t in the gadget I_ℓ of H_i is adjacent to the vertex $s_{i,j}^i$
- for each $\ell \leq t \leq n$, the vertex a_t in the gadget I_ℓ of H_i is adjacent to the vertex $r_{i,j}^i$

For each $i \in [k]$, each $1 \leq j < i$ and each $\ell \in [n]$,

- for each $1 \leq t \leq \ell$, the vertex a_t in the gadget I_ℓ of H_i is adjacent to the vertex $s_{j,i}^i$
- for each $\ell \leq t \leq n$, the vertex a_t in the gadget I_ℓ of H_i is adjacent to the vertex $r_{j,i}^i$

The gadgets for the edges are also connected to the connector vertices similarly (explained in full version). This completes construction of the graph H with $\mathcal{O}(mn^2)$ vertices and $\mathcal{O}(mn^3)$ edges. We next bound the pathwidth of the graph H as a polynomial function of k.

Claim 8. $\Diamond$ *The pathwidth of a block B is at most five.*

Lemma 1. $\Diamond$ *The pathwidth of the graph H is at most $4\binom{k}{2} + 5$.*

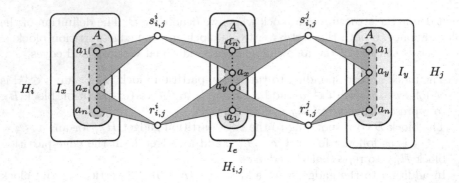

Fig. 2. An illustration of the connector vertices $s_{i,j}^i$, $r_{i,j}^i$, $s_{i,j}^j$ and $r_{i,j}^j$ connect the blocks H_i and $H_{i,j}$, and, H_j and $H_{i,j}$ for some $1 \le i < j \le k$. The edge e represented in the gadget I_e is $u_{i,x}u_{j,y} \in E_{i,j}$.

Properties of a Feasible Solution for the MMDS Instance (H, k'). Let S be a feasible solution for the MMDS instance (H, k'). We state the following properties of the set S. In all the arguments below, we crucially use the property that for each $u \in V(H)$, $M(u, S) \le n + 1$.

Claim 9 $\diamondsuit$ *For each block B in the graph H, $C'(B) \subseteq S$.*

Claim 10 $\diamondsuit$ *For each block B in H, the vertices f and f' in B are not in the set S.*

Claim 11 $\diamondsuit$ *For each gadget of type I in each block B in the graph H, either $A \cap S = A$ or $A \cap S = \emptyset$.*

Claim 12 $\diamondsuit$ *For each block B in the graph H, there exists a unique gadget of type I in the block B such that the set A in the gadget is in S.*

Using these properties in the following two lemmas, we prove the correctness of the reduction.

Lemma 2. $\diamondsuit$ *If (G, k) is a YES-instance of the MULTI-COLORED CLIQUE problem, then (H, k') is a YES-instance of the MMDS problem.*

Lemma 3. $\diamondsuit$ *If (H, k') is a YES-instance of the MMDS problem, then (G, k) is a YES-instance of the MULTI-COLORED CLIQUE problem.*

Thus, the proofs of Lemmas 2 and 3 complete the proof of Theorem 2. A detailed proof is given in full version.

4   W[1]-Hardness in Split Graphs

In this section we prove that MMDS is W[1]-hard on split graphs when parameterized by the membership parameter k. We prove this result by demonstrating a parameterized reduction from MULTI-COLORED INDEPENDENT SET (MIS) to MINIMUM MEMBERSHIP DOMINATING SET. MULTI-COLORED INDEPENDENT SET requires finding a colorful independent set of size k and is known to be W[1]-hard for the parameter solution size [8].

MULTI-COLORED INDEPENDENT SET
Input: A positive integer k, and a k-colored graph G.
Parameter: k
Question: Does there exist an independent set of size k with one vertex from each color class?

Let $(G = (V, E), k)$ be an instance of the MULTI-COLORED INDEPENDENT SET problem. Let $V = (V_1, \ldots, V_k)$ be the partition of the vertex set V, where vertices in set V_i belong to the ith color class, $i \in [k]$. We now show how to construct a split graph $H = (V' \cup V'', E')$ such that if (G, k) is a YES instance, then H has a dominating set with maximum membership k. V' refers to the clique partition of H and V'' consists of the partition containing a set of independent vertices.

Construction of graph $H = (V' \cup V'', E')$:

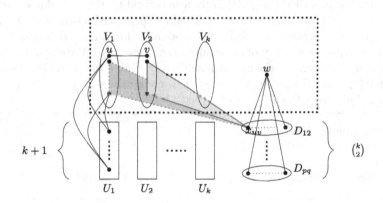

Fig. 3. Construction of graph H

For each vertex in V we introduce a vertex in the clique V' as in the input instance. Additionally we add a vertex w to V'. Edges are added among each pair of vertices in V'. The set V'' in H is an independent set, and it consists of a set of vertices denoted by U, a set of vertex sets denoted by $\mathcal{D} = \{D_{pq} \mid p, q \in [k], p < q\}$. The vertex set U comprises a partition of k vertex sets, $U = \{U_i \mid i \in [k]\}$, and $|U_i| = k + 1$. For each edge between a vertex $u \in V_p$ and $v \in V_q$ in G, we introduce a vertex x_{uv} in the set D_{pq}. Conceptually D_{pq} denotes the set of edges between V_p and V_q. Thus, the vertex set of H, $V(H) = V' \cup V''$, where V' induces a clique and V'' induces an independent set.

The remaining edges, other than those in clique V', are described as follows: $V_i \uplus U_i$ forms a complete bipartite graph, vertex w is made adjacent to all vertices in the set $\mathcal{D}$, each vertex $x_{uv} \in D_{pq}$ is made adjacent to every vertex in $V_p \setminus \{u\}$ and $V_q \setminus \{v\}$.

The above construction is depicted in Fig. 3. Next we show the correctness of the reduction from the instance (G, k) of MIS to the instance (H, k) of MMDS.

Lemma 4. $\diamond$ *If (G, k) is a YES instance of the* MULTI-COLORED INDEPENDENT SET *problem then (H, k) is a YES instance of the* MMDS *problem.*

Lemma 5. $\diamond$ *If (H, k) is a YES instance of the* MMDS *problem then (G, k) is a YES instance of the* MULTI-COLORED INDEPENDENT SET *problem.*

5   Parameterizing MMDS by Vertex Cover

First, we show that MMDS is FPT parameterized by vertex cover number, **vc**. We then show that conditioned on the truth of the ETH, MMDS does not have a subexponential algorithm in the size of vertex cover.

5.1   MMDS Is FPT Parameterized by Vertex Cover

In order to design an FPT algorithm parameterized by the size of a vertex cover of the input graph, we construct an FPT-time Turing reduction from MMDS to Integer Linear Programming (ILP, See Appendix for formal definition). In the reduced instance the number of constraints is at most twice the size of a minimum vertex cover. We then use the recent result by Dvořák *et al.* [7] which proves that ILP is FPT parameterized by the number of constraints. The following theorem directly follows from Corollary 9 of [7].

Integer Linear Programming
Input   : A matrix $A \in \mathbb{Z}^{m \times \ell}$ and a vector $b \in \mathbb{Z}^m$.
Parameter : m
Question   : Is there a vector $x \in \mathbb{Z}^\ell$ such that $A \cdot x \leq b$?

Theorem 13 (Corollary 9, [7]). *ILP is FPT in the number of constraints and the maximum number of bits for one entry.*

FPT Time Turing Reduction from MMDS to ILP: Let (G, k) be the input instance of MMDS. Compute a minimum vertex cover of G, denoted by C, in time FPT in $|C|$ [3]. Let I denote the maximum independent set $V \setminus C$. The following lemma is crucial to the correctness of the reduction.

Lemma 6. $\diamond$ *Let D be a k membership dominating set of G. Let $C_1 = D \cap C$, $I_1 = I \setminus (N(C_1) \cap I)$, and $R = N(C_1) \cap I \cap D$. Then, $I_1 \subseteq D$, and $C \setminus (N[C_1] \cup N(I_1))$ is dominated by R.*

As a consequence of this lemma, it is clear that the choice of C_1 immediately fixes I_1. Thus, to compute the set D, the task is to compute R. We pose this problem as the *constrained* MMDS problem. A CMMDS problem instance is a 4-tuple (G, k, C, C_1) where C is a vertex cover and C_1 is a subset of C. The decision question is whether there is a k membership dominating set D of G such that $D \cap C = C_1$.

From Lemma 6, we know that given an instance of (G, k, C, C_1), we know that C_1 immediately fixes $I_1 \subseteq I = V \setminus C$. Thus, to compute D, we need to compute R as defined in Lemma 6. We now describe the ILP formulation to compute R once C_1 (and thus I_1) is fixed. Since R is a subset of $I \setminus I_1$, it follows that the variables correspond to vertices in $I \setminus I_1$ which do not already have k neighbors in C_1; we use I_e to denote this set. It can be immediately checked if $C_1 \cup I_1$ can be part of a feasible solution- we check that for no vertex is the intersection of its closed neighborhood greater than k. We now assume that this is the case, and specify the linear constraints. The linear constraints in the ILP are associated with the vertices in C. For each vertex in C there are at most two constraints- if v is in $C \setminus (N[C_1] \cup N(I_1))$, then at least one neighbor and at most k neighbors from I_e must be chosen into R. On the other hand, for $v \in (N[C_1] \cap C) \cup N(I_1)$, we have the constraint that at most k neighbors must be in $C_1 \cup I_1 \cup R$. The choice of variables in I_e does not affect any other vertex in I, and thus there are no costraints among the vertices in I. To avoid notation, we assume that an instance of CMMDS(G, k, C, C_1), also denotes the ILP.

Lemma 7. $\Diamond$ *The CMMDS problem on an instance (G, k, C, C_1) can be solved in time which is FPT in the size of the vertex cover.*

5.2   Lower Bound Assuming ETH

We show that there is no sub-exponential-time parameterized algorithm for MMDS when the parameter is *the vertex cover number*, using a reduction from 3-SAT. By the ETH, 3-SAT does not have a sub-exponential-time algorithm, and thus the reduction proves the lower bound for MMDS.

Proof: [Proof of Theorem 5] Let ϕ be a boolean formula on n variables $X = \{x_1, x_2, \ldots, x_n\}$ having m clauses $C = \{C_1, C_2, \ldots, C_m\}$. We construct a graph $G = (V, E)$ from the input formula ϕ such that ϕ has a satisfying assignment if and only if G has a k membership dominating set.

Construction of Graph G

For each variable x_i, $1 \le i \le n$ in ϕ, create two vertices v_{x_i} and $v_{\overline{x_i}}$, denoting its literals, with an edge between them. Make both v_{x_i} and $v_{\overline{x_i}}$ adjacent to $k + 1$ degree-two vertices labelled $a_i^j : 1 \le j \le k + 1$ and another set of $k - 1$ vertices $b_i^j : 1 \le j \le k - 1$. Each b_i^j is in turn adjacent to $k + 1$ pendant vertices $d_{i,j}^t : 1 \le t \le k + 1$.

For each clause $C_l : 1 \le l \le m$, create a vertex v_{C_l}. For each clause C_l, make v_{C_l} adjacent to a vertex Y. Y is again connected to k more vertices $u_q : 1 \le q \le k$. Each u_q is in turn adjacent to $k + 1$ pendant vertices $r_q^p : 1 \le p \le k + 1$. Finally,

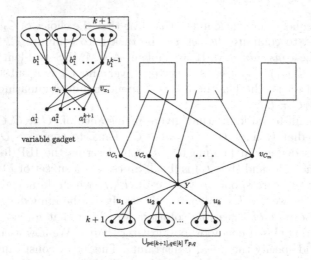

Fig. 4. Construction for reduction from 3-SAT to MMDS.

create edges between clause vertices and those literal vertices which are in the clause. The reduction is illustrated in Fig. 4.

Claim 14. ◇ *The vertex cover number of graph G is $(n+1)(k+1)$.*

Lemma 8. ◇ *If ϕ has a satisfying assignment then G has a dominating set with membership value k.*

Lemma 9. ◇ *If G has a dominating set with membership value k, then ϕ has a satisfying assignment.*

From Lemma 8 and Lemma 9, it follows that the 3-SAT can be reduced to MMDS parameterized by vertex cover number. Therefore a $2^{o(\mathbf{vc}(G))}n^{O(1)}$ algorithm for MMDS will give a $2^{o(n)}$ algorithm for 3-SAT which is a violation of ETH. Hence the proof.

6   Conclusion

In this paper we study the parameterized complexity of the MINIMUM MEMBERSHIP DOMINATING SET problem, which requires finding a dominating set such that each vertex in the graph is dominated minimum possible times. We start our analysis by showing that in spite of having no constraints on the size of the solution, unlike DOMINATING SET, MMDS turns out to be W[1]-hard when parameterized by pathwidth (and hence treewidth). We further show that the problem remains W[1]-hard for split graphs when the parameter is the size of the membership. For general graphs we prove that MMDS is FPT when parameterized by the size of vertex cover. Finally, we show that assuming ETH, the problem does not admit a sub-exponential algorithm when parameterized by the size of vertex cover, thus showing our FPT algorithm to be optimal. There are many related open problems that are yet to be explored. One such problem is analyzing the complexity of MMDS in chordal graphs. Other directions involve structural parameterization of MMDS with respect to other parameters such as maximum degree, distance to bounded degree graphs, bounded genus and maximum number of leaves in a spanning tree.

References

1. Agrawal, A., Choudhary, P., Narayanaswamy, N.S., Nisha, K.K., Ramamoorthi, V.: Parameterized complexity of minimum membership dominating set. CoRR abs/2110.06656 (2021)
2. Chapelle, M.: Parameterized complexity of generalized domination problems on bounded tree-width graphs. CoRR abs/1004.2642 (2010)
3. Cygan, M., et al.: Parameterized Algorithms. Springer, Cham (2015). https://doi.org/10.1007/978-3-319-21275-3
4. Diestel, R.: Graph Theory, 4th edn. Graduate Texts in Mathematics, vol. 173. Springer (2012)
5. Downey, R.G., Fellows, M.R.: Parameterized Complexity. Monographs in Computer Science. Springer, New York (1999). https://doi.org/10.1007/978-1-4612-0515-9
6. Downey, R.G., Fellows, M.R.: Fundamentals of Parameterized Complexity. TCS, Springer, London (2013). https://doi.org/10.1007/978-1-4471-5559-1
7. Dvořák, P., Eiben, E., Ganian, R., Knop, D., Ordyniak, S.: Solving integer linear programs with a small number of global variables and constraints. In: Proceedings of the Twenty-Sixth International Joint Conference on Artificial Intelligence, IJCAI-17, pp. 607–613 (2017). https://doi.org/10.24963/ijcai.2017/85
8. Fellows, M.R., Hermelin, D., Rosamond, F., Vialette, S.: On the parameterized complexity of multiple-interval graph problems. Theor. Comput. Sci. 410(1), 53–61 (2009). https://doi.org/10.1016/j.tcs.2008.09.065. https://www.sciencedirect.com/science/article/pii/S0304397508007329
9. Garey, M.R., Johnson, D.S.: Computers and Intractability: A Guide to the Theory of NP-Completeness. Freeman, W. H (1979)
10. Kuhn, F., von Rickenbach, P., Wattenhofer, R., Welzl, E., Zollinger, A.: Interference in cellular networks: the minimum membership set cover problem. In: Wang, L. (ed.) COCOON 2005. LNCS, vol. 3595, pp. 188–198. Springer, Heidelberg (2005). https://doi.org/10.1007/11533719_21
11. Meybodi, M.A., Fomin, F.V., Mouawad, A.E., Panolan, F.: On the parameterized complexity of $[1, j]$-domination problems. Theor. Comput. Sci. 804, 207–218 (2020). https://doi.org/10.1016/j.tcs.2019.11.032
12. Mulzer, W., Rote, G.: Minimum-weight triangulation is NP-hard. J. ACM 55(2), 11:1–11:29 (2008)
13. Raman, V., Saurabh, S.: Short cycles make W-hard problems hard: FPT algorithms for W-hard problems in graphs with no short cycles. Algorithmica 52(2), 203–225 (2008)

Graph Algorithms

Finding Popular Branchings in Vertex-Weighted Digraphs

Kei Natsui[1(✉)] and Kenjiro Takazawa[2]

[1] University of Tsukuba, Ibaraki 305-8577, Japan
s2120448@s.tsukuba.ac.jp
[2] Hosei University, Tokyo 184-8584, Japan
takazawa@hosei.ac.jp

Abstract. Popular matchings have been intensively studied recently as a relaxed concept of stable matchings. By applying the concept of popular matchings to branchings in directed graphs, Kavitha et al. introduced popular branchings. In a directed graph $G = (V_G, E_G)$, each vertex has preferences over its incoming edges. For branchings B_1 and B_2 in G, a vertex $v \in V_G$ prefers B_1 to B_2 if v prefers its incoming edge of B_1 to that of B_2, where having an arbitrary incoming edge is preferred to having none, and B_1 is more popular than B_2 if the number of vertices that prefer B_1 is greater than the number of vertices that prefer B_2. A branching B is called a popular branching if there is no branching more popular than B. Kavitha et al. proposed an algorithm for finding a popular branching when the preferences of each vertex are given by a strict partial order. The validity of this algorithm is proved by utilizing classical theorems on the duality of weighted arborescences. In this paper, we generalize popular branchings to weighted popular branchings in vertex-weighted directed graphs in the same manner as weighted popular matchings by Mestre. We give an algorithm for finding a weighted popular branching, which extends the algorithm of Kavitha et al., when the preferences of each vertex are given by a total preorder and the weights satisfy certain conditions. Our algorithm includes elaborated procedures resulting from the vertex-weights, and its validity is proved by extending the argument of the duality of weighted arborescences.

1 Introduction

Popular matchings provide a relaxed concept of stable matchings. Popular matchings were introduced by Gärdenfors [9], and have been attracting intensive attention recently since Abraham et al. [1] started studying their algorithmic aspects. In a bipartite graph, each vertex has preferences over its adjacent vertices, and a matching M is *more popular* than another matching N if the number of vertices that prefer the adjacent vertex in M to that in N is greater than the number of vertices that prefer the adjacent vertex in N to that in M. A matching M is called a *popular matching* if no matching is more popular than M.

The second author is supported by JSPS KAKENHI Grant Number JP20K11699, Japan.

P. Mutzel et al. (Eds.): WALCOM 2022, LNCS 13174, pp. 303–314, 2022.
https://doi.org/10.1007/978-3-030-96731-4_25

For popular matchings, several algorithms are known. Abraham et al. [1] were the first to give an efficient algorithm determining whether a popular matching exists and finding one if exists.

There have been various other studies on popular matching in recent years, including [2–4,8,10]. Among those, Mestre [12] provided an algorithm for weighted popular matching. In the weighted popular matching problem, weights are attached to the vertices and, instead of the number of vertices, the popularity of matchings is defined by the sum of the weights of the corresponding vertices. The algorithm [12] runs in polynomial time regardless of whether ties are allowed or not.

By applying the concept of popular matchings to branchings in directed graphs, Kavitha, Király, Matuschke, Schlotter, and Schmidt-Kraepelin [11] introduced *popular branchings*. In a directed graph $G = (V_G, E_G)$, each vertex has preferences over its incoming edges. Let B and B' be branchings in G. We say that a vertex $v \in V$ prefers B to B' if v prefers the incoming edge in B to that in B', where having an arbitrary incoming edge is preferred to having none. We say that B is more popular than B' if the number of vertices that prefer B is greater than the number of vertices that prefer B'. A branching B is called a popular branching if there is no other branching more popular than B.

Kavitha et al [11] proposed an algorithm for finding a popular branching when the preferences of the vertices are given by a strict partial order. This algorithm determines whether a popular branching exists, and if so, outputs one. Its validity is proved by a characterization of popular branchings which utilizes the duality of weighted arborescences. The algorithm constructs a directed graph D from G by adding a dummy vertex r as a root and an edge (r, v) for each $v \in V_G$. Each branching B in G is extended to an r-arborescence in D by adding an edge (r, v) for every vertex $v \in V_G$ with no incoming edge in B. They proved that an r-arborescence A in D is a popular arborescence if and only if it is a minimum cost arborescence with respect to edge weights defined in a certain manner.

For these edge weights, they further proved that, for an integral optimal solution $y \in \mathbb{R}^{2^{V_G}}$ of the dual problem, the laminar structure of the support $\mathcal{F}(y) = \{X \subseteq V_G : y(X) > 0\}$ has at most two layers. This structure leads to a one-to-one correspondence between the support $\mathcal{F}(y)$ and the vertex set V_G, and the concept of *safe edges*, which are candidates of the edges in a popular arborescence. The algorithm of Kavitha et al. [11] essentially relies on this structure.

In this paper, we generalize popular branchings to weighted popular branchings in the same manner as weighted popular matchings [12]. Each vertex $v \in V_G$ is assigned a positive integer weight $w(v)$. For an r-arborescence A in D and a vertex $v \in V_G$, let $A(v)$ denote the edge in A entering v. For two r-arborescences A and A' in D, we define an integer $\Delta_w(A, A')$ by

$$\Delta_w(A, A') = \sum_{v : A(v) \succ_v A'(v)} w(v) - \sum_{v : A'(v) \succ_v A(v)} w(v), \qquad (1)$$

where $e \succ_v f$ denotes that v prefers e to f. If $\Delta_w(A, A') > 0$, we say that A is more popular than A'. An r-arborescence A in D is a *popular arborescence* if no arborescence is more popular than A.

The main contribution of this paper is an algorithm for finding a popular arborescence in vertex-weighted directed graphs, which extends the algorithm of Kavitha et al. [11]. Its validity builds upon a characterization of weighted popular arborescences, which extends that of popular branchings [11], and our algorithm includes elaborated procedures resulting from the vertex-weights.

The following two points are specific to our algorithm. The first is that the preferences of each vertex are given by a total preorder, while they are given by a strict partial order in [11]. The second is that it requires an assumption on the vertex weights:

$$w(s) + w(t) > w(u) \quad (s, t, u \in V_G).$$

Under this assumption, we can derive that the laminar structure of the support of an integer dual optimal solution y has at most two layers. By virtue of this laminar structure, we can define the one-to-one correspondence between $\mathcal{F}(y)$ and V_G, and safe edges in the same manner as [11], which are essential in designing the algorithm.

Let us mention an application of popular branchings in the context of a voting system, and what is offered by the generalized model of weighted popular branchings. Kavitha et al. [11] suggested an application in a voting system called liquid democracy. This is a new voting system that lies between representative democracy and direct democracy. In liquid democracy, voters can choose to vote themselves or to delegate their votes to the judgment of others who they believe in, and their votes flow over a network, constructing a fluid voting system. In this system, a popular branching amounts to a reasonable delegation process. Here, if we take vertex weights into account, it represents a situation where there is a difference in voting power. That is, weighted popular branchings are of help when each voter has distinct voting power, and we want to make a decision based on the total voting power rather than the number of votes.

This paper is organized as follows. Section 2 formally defines weighted popular branchings. In Sect. 3, in preparation for algorithm design, we analyze some properties of weighted popular branchings and introduce safe edges. In Sect. 4, we present our algorithm for finding a weighted popular branching and prove its correctness.

2 Definition of Weighted Popular Branchings

Let $G = (V_G, E_G)$ be a directed graph, where every vertex has a positive integer weight $w(v)$ and preferences over its incoming edges. The preferences of each vertex v are given by a *total preorder* $\precsim_v$ on the set of edges that enter v.

Recall that a total preorder is defined in the following way. Let S be a finite set. A binary relation R on S is transitive if, for all $a, b, c \in S$, aRb and bRc imply aRc. Also, R is reflexive if aRa holds for all $a \in S$. A relation R is called a *preorder* if R is transitive and reflexive. In addition, R is a *total relation* if aRb or bRa holds for all $a, b \in S$. That is, a total preorder is a relation which is transitive, reflexive, and total. Note that a partial order is a preorder, whereas it is not necessarily a total preorder, and a total preorder is not necessarily a partial order.

Let e and f be two edges entering the same vertex v. Then, $e \precsim_v f$ means that f has more or the same priority than e. If both $e \precsim_v f$ and $f \precsim_v e$ holds, we denote it by $e \sim_v f$, indicating that v is indifferent between e and f. Note that $\sim_v$ is an equivalence relation. Furthermore, if $e \precsim_v f$ holds but $f \precsim_v e$ does not, we denote it by $e \prec_v f$. This indicates that vertex v strictly prefers f to e. If an edge f is strictly preferred to e, then we say that f *dominates* e.

Instead of discussing branchings in G, we mainly handle arborescences in an auxiliary directed graph D. The directed graph D is constructed from G by adding a dummy vertex r as a root and an edge (r, v) for each $v \in V_G$. That is, D is represented as $D = (V, E)$, where $V = V_G \cup \{r\}$ and $E = E_G \cup \{(r, v) : v \in V_G\}$. For each vertex $v \in V_G$, let $\delta^-(v) \subseteq E$ be the set of edges in D that enter v, and make (r, v) the least preferred incoming edge in $\delta^-(v)$. That is, every edge in $E_G \cap \delta^-(v)$ dominates (r, v) for each $v \in V_G$.

An r-arborescence in D is an out-tree with root r. For an r-arborescence A in D and $v \in V_G$, let $A(v)$ denote the edge in A entering v. For r-arborescences A and A' in D, we define $\Delta_w(A, A')$ by

$$\Delta_w(A, A') = \sum_{v : A(v) \succ_v A'(v)} w(v) - \sum_{v : A'(v) \succ_v A(v)} w(v). \tag{2}$$

An r-arborescence A is more popular than A' if $\Delta_w(A, A') > 0$. If no r-arborescence is more popular than A, then we say that A is a popular arborescence. Our primary goal is to find a popular arborescence in D.

3   Properties of Weighted Popular Branchings

3.1   Characterizing Weighted Popular Arborescences

In this subsection, by extending the argument in [11], we give a characterization of popular arborescences (Proposition 4) by utilizing the duality theory of weighted arborescences. We then investigate the structure of dual optimal solutions with a certain property (Proposition 8). Let A be an r-arborescence in D. For each edge $e = (u, v)$ in D, we define the cost $c_A(e)$ as follows:

$$c_A(e) = \begin{cases} 0 & (e \succ_v A(v)), \\ w(v) & (e \sim_v A(v)), \\ 2w(v) & (e \prec_v A(v)). \end{cases} \tag{3}$$

Since $c_A(e) = w(v)$ for every $e \in A$, we have $c_A(A) = w(V_G)$. For an arbitrary r-arborescence A' in D, the following holds:

$$c_A(A') = \sum_{v:A(v)\succ_v A'(v)} 2w(v) + \sum_{v:A(v)\sim_v A'(v)} w(v) + \sum_{v:A(v)\prec_v A'(v)} 0$$

$$= w(V_G) + \sum_{v:A(v)\succ_v A'(v)} w(v) - \sum_{v:A(v)\prec_v A'(v)} w(v)$$

$$= c_A(A) + \Delta_w(A, A').$$

We thus obtain the following proposition. An r-arborescence is called a *min-cost r-arborescence* if the sum of the costs of all edges is the smallest among the r-arborescences.

Proposition 1. *An r-arborescence A is popular if and only if it is a min-cost r-arborescence in D with respect to the edge costs c_A.*

Based on Proposition 1, consider the following linear program (LP1), which describes the min-cost r-arborescence problem, and its dual (LP2). For any non-empty set $X \subseteq V_G$, let $\delta^-(X) \subseteq E$ be the set of edges in D that enter X.

$$\text{(LP1)} \qquad \text{minimize} \qquad \sum_{e \in E} c_A(e) \cdot x(e) \tag{4}$$

$$\text{subject to} \qquad \sum_{e \in \delta^-(X)} x(e) \geq 1 \qquad \forall X \subseteq V_G, X \neq \emptyset, \tag{5}$$

$$x(e) \geq 0 \qquad \forall e \in E. \tag{6}$$

$$\text{(LP2)} \qquad \text{maximize} \qquad \sum_{X \subseteq V_G, X \neq \emptyset} y(X) \tag{7}$$

$$\text{subject to} \qquad \sum_{X : e \in \delta^-(X)} y(X) \leq c_A(e) \qquad \forall e \in E, \tag{8}$$

$$y(X) \geq 0 \qquad \forall X \subseteq V_G, X \neq \emptyset. \tag{9}$$

For any feasible solution y to (LP2), let $\mathcal{F}(y) = \{X \subseteq V_G : y(X) > 0\}$ be the support of y.

The following proposition is a direct consequence of the definition of the costs c_A. Its proof is described in the full version of this extended abstract.

Proposition 2. *For an r-arborescence A and a feasible solution y to (LP2), we have that*

$$\sum_{X : v \in X} y(X) \leq 2w(v) \qquad (v \in V_G).$$

A set family $\mathcal{F}$ is called *laminar* if for any two sets $X, Y \in \mathcal{F}$, at least one of the three sets $X \setminus Y, Y \setminus X, X \cap Y$ is empty. From now on, we deal with an optimal solution y with certain properties. The first property is described in the following lemma.

Lemma 3. ([5–7]). *If the costs c_A are integers, there exists an integral optimal solution y^* to (LP2) such that $\mathcal{F}(y^*)$ is laminar.*

Since the costs $c_A(e) \in \{0, w(v), 2w(v)\}$ are integers, it follows from Lemma 3 that there exists an integral optimal solution y_A^* to (LP2) such that $\mathcal{F}(y_A^*)$ is laminar. Furthermore, the following proposition can be derived from the duality of weighted arborescences, similarly as described in [11].

Proposition 4. *For an r-arborescence A, the following statements are equivalent.*

(i) *A is a popular arborescence.*
(ii) *$\sum_{X \subseteq V_G} y_A^*(X) = w(V_G)$.*
(iii) *$|A \cap \delta^-(X)| = 1$ for all $X \in \mathcal{F}(y_A^*)$ and $\sum_{X : e \in \delta^-(X)} y_A^*(X) = w(v)$ for all $e = (u, v) \in A$.*

Let $A \subseteq E$ be an r-arborescence and y_A^* be an optimal solution for (LP2). Here, we consider the properties of y_A^* and $\mathcal{F}(y_A^*)$. Let E° be the set of edges $e \in E$ satisfying $\sum_{X : e \in \delta^-(X)} y_A^*(X) = c_A(e)$ and let $D^\circ = (V, E^\circ)$. For a directed graph $D' = (V', E')$ and its vertex subset $X \subseteq V'$, the subgraph induced by X is denoted by $D'[X] = (X, E'[X])$. Similarly, for an edge subset $A' \subseteq E'$, the set of edges in A' induced by X is denoted by $A'[X]$. The following lemma applies to general weighted arborescences.

Lemma 5. *For an r-arborescence $A \subseteq E$, there exists an integral optimal solution y_A^* to (LP2) such that $\mathcal{F}(y_A^*)$ is laminar and $D^\circ[Y]$ is strongly connected for every $Y \in \mathcal{F}(y_A^*)$.*

In what follows, we denote by y_A^* the integer optimal solution to (LP2) described in Lemma 5. In addition to Lemma 5, if A is a popular arborescence, we can impose a stronger condition on y_A^*. See the full version for a proof.

Lemma 6. *For a popular arborescence A, there exists an integral optimal solution y_A^* to (LP2) such that $\mathcal{F}(y_A^*)$ is laminar, $D^\circ[Y]$ is strongly connected for every $Y \in \mathcal{F}(y_A^*)$, and the following is satisfied.*
For $Y \in \mathcal{F}(y_A^*)$, let $Y_1, \ldots, Y_k$ be the sets in $\mathcal{F}(y_A^*)$ that are maximal proper subsets of Y. Then,

$$|Y \setminus (Y_1 \cup \cdots \cup Y_k)| = 1. \tag{10}$$

For a popular arborescence, the following lemma also holds.

Lemma 7. *If A is a popular arborescence and a set $Y \in \mathcal{F}(y_A^*)$ is minimal in $\mathcal{F}(y_A^*)$, then $|Y| = 1$.*

Proof. Suppose to the contrary that $|Y| \geq 2$ for a minimal set Y in $\mathcal{F}(y_A^*)$. Since $Y \in \mathcal{F}(y_A^*)$, by Proposition 4(iii), we have $|A \cap \delta^-(Y)| = 1$. Also, since A is an r-arborescence, it holds that $|A \cap (\bigcup_{v \in Y} \delta^-(v))| = |Y|$, and hence $|A[Y]| = |A \cap (\bigcup_{v \in Y} \delta^-(v))| - |A \cap \delta^-(Y)| = |Y| - 1 \geq 1$. It then follows that $A[Y] \neq \emptyset$, and let e be an edge in $A[Y]$. Now Y is a minimal set in $\mathcal{F}(y_A^*)$, implying that $e \notin \delta^-(Y')$ for any $Y' \in \mathcal{F}(y_A^*)$. This contradicts Proposition 4(iii). $\qquad\square$

From Lemmas 6 and 7, the next proposition follows.

Proposition 8. *Let A be a popular arborescence. Then, there exists a one-to-one correspondence between the sets in $\mathcal{F}(y_A^*)$ satisfying (10) and the vertices in V_G. Moreover, for each $X \in \mathcal{F}(y_A^*)$ and the terminal vertex v of the edge $(u, v) \in A \cap \delta^-(X)$, we have $y_A^*(X) = w(v)$.*

Denote by Y_v the unique set in $\mathcal{F}(y_A^*)$ that is in correspondence with v in the sense of Proposition 8. Note that $\mathcal{F}(y_A^*) = \{Y_v : v \in V_G\}$ and the unique edge in A entering Y_v is $A(v)$. We thus refer to v as the *entry-point* of Y_v.

3.2 Weight Assumption and Safe Edges

In Kavitha et al.'s algorithm for finding popular branching [11], the laminar structure of $\mathcal{F}(y_A^*)$ has at most two layers:

$$|\{X \in \mathcal{F}(y_A^*) \mid v \in X\}| \leq 2 \quad (v \in V_G). \tag{11}$$

This structure plays an important role in the algorithm: it leads to a one-to-one correspondence between $\mathcal{F}(y_A^*)$ and the vertex set V_G, and the concept of safe edges.

   In the unweighted case, (11) follows from Proposition 2. However, when the weights $w(v)$ can be more than one, Proposition 2 alone does not rule out the case where $|\{X \in \mathcal{F}(y_A^*) \mid v \in X\}| \geq 3$. In order to maintain (11), as mentioned in Sect. 1, we impose an assumption on the vertex weights. Recall that the assumption is:

$$w(s) + w(t) > w(u) \quad (s, t, u \in V_G). \tag{12}$$

From this assumption, we can derive the following proposition.

Proposition 9. *Let A be a popular arborescence and let y_A^* satisfy (10). If the condition (12) holds for any three vertices $s, t, u \in V$, then $|\{X \in \mathcal{F}(y_A^*) : v \in X\}| \leq 2$ holds for every vertex $v \in V_G$.*

Proof. Assume to the contrary that $|\{X \in \mathcal{F}(y_A^*) \mid v \in X\}| \geq 3$ for some $v \in V_G$. In this case, there exist two vertices $a, b \in V_G \setminus \{v\}$ such that $v \in Y_a \cap Y_b \cap Y_v$. Since $y_A^*(Y_a) = w(a), y_A^*(Y_b) = w(b), y_A^*(Y_v) = w(v)$ from Proposition 8 and $w(v) < w(a) + w(b)$ from assumption (12), we have $c_A(r, v) \leq 2w(v) < y_A^*(Y_a) + y_A^*(Y_b) + y_A^*(Y_v)$, which contradicts the constraint (8) in (LP2). Therefore, under assumption (12), $|\{X \in \mathcal{F}(y_A^*) \mid v \in X\}| \leq 2$ holds for every $v \in V_G$. $\quad\square$

From Propositions 8 and 9, and the laminarity of $\mathcal{F}(y_A^*)$, the following corollary can be derived.

Corollary 10. *Under assumption (12), for y_A^* satisfying (10) and $v \in V_G$ with $|Y_v| \geq 2$, it holds that $Y_u = \{u\}$ for each $u \in Y_v \setminus \{v\}$.*

The safe edges used in our algorithm are defined in the same way as [11], described below. For $X \subseteq V_G$, an edge $(u, v) \in E[X]$ satisfying the following two conditions is called a *safe edge* in X, and the set of safe edges in X is denoted by $S(X)$.

1. (u, v) is not dominated by any edge in $E[X]$, i.e., $(u, v) \succsim_v (u', v)$ for all $(u', v) \in E[X]$.
2. (u, v) dominates each edge (t, v) with $t \notin X$, i.e., $(u, v) \succ_v (t, v)$ for all $(t, v) \in \delta^-(X)$.

Recall that the preference of each vertex is taken as the total preorder. Hence, if there exists $e \in \delta^-(v) \cap \delta^-(X)$ such that e is one of the most preferred edges in $\delta^-(v)$, it holds that $S(X) \cap \delta^-(v) = \emptyset$. Otherwise, $S(X) \cap \delta^-(v)$ is the set of the most preferred edges in $\delta^-(v)$.

The edges in a popular arborescence are basically chosen from safe edges, as shown in the next proposition. Its proof is described in the full version.

Proposition 11. *For any popular arborescence A and $X \in \mathcal{F}(y_A^*)$ satisfying (10), it holds that $A \cap E[X] \subseteq S(X)$.*

4   Weighted Popular Branching Algorithm

We are now ready to describe our algorithm for finding a weighted popular arborescence and prove its validity. The algorithm is described as follows.

1. For each $v \in V_G$ do:
   - let $X_v^0 = V_G, i = 0$;
   - while v does not reach all vertices in the graph $D_v^i = (X_v^i, S(X_v^i))$ do:
     $X_v^{i+1} =$ the set of vertices reachable from v in D_v^i; let $i = i + 1$.
   - let $X_v = X_v^i$ and $D_v = D_v^i$.
2. Let $\mathcal{X} = \{X_v : v \in V_G\}$, $\mathcal{X}' = \{X_v \in \mathcal{X} : X_v$ is maximal in $\mathcal{X}\}$, $E' = \emptyset$, and $D' = (\mathcal{X}' \cup \{r\}, E')$.
3. For each $X_v \in \mathcal{X}'$, let $\bar{X}_v$ be the strongly connected component of D_v such that no edge in $S(X_v)$ enters. For each $\bar{X}_v$ do:
   - if every $v' \in \bar{X}_v$ which has minimum weight in $\bar{X}_v$ satisfies the following condition, then go to STEP 5.
     There exist a vertex $s \in X_v \setminus \bar{X}_v$ and an edge $f \in (E[X_v] \setminus S(X_v)) \cap \delta^-(v')$ such that
     1. $w(s) < w(v')$,
     2. v' is reachable from s by the edges in $S(X_v) \cup \{f\}$,
     3. $f \succ_{v'} e$ holds for all $e \in \delta^-(v') \cap \delta^-(X_v)$.
   - otherwise, for every $v' \in \bar{X}_v$ which has minimum weight in $\bar{X}_v$, do:
     • for every $u \notin X_v$, if $e = (u, v')$ is not dominated by an edge in $\delta^-(X_v)$, then define an edge e' in D' by

$$e' = \begin{cases} (U, X_v) & (u \in U, U \in \mathcal{X}), \\ (r, X_v) & (u = r), \end{cases} \tag{13}$$

   and let $E' := E' \cup \{e'\}$.

4. If $D' = (\mathcal{X}' \cup \{r\}, E')$ does not contain an r-arborescence A', then go to STEP 5. Otherwise, do the following.
   - let $\tilde{A} = \{e : e' \in A'\}$;
   - let $R = \{v \in V_G : |X_v| \geq 2, \delta^-(v) \cap \tilde{A} \neq \emptyset\}$;
   - for each $v \in R$, let A_v be an v-arborescence in $(X_v, S(X_v))$;
   - return $A^* = \tilde{A} \cup \bigcup_{v \in R} A_v$.
5. Return "No popular arborescence in D."

A major difference from the algorithm without the vertex-weights [11] appears in STEP 3. If the condition shown in STEP 3 is satisfied, there exists no popular arborescence (see Lemma 18).

We now prove the validity of the algorithm described above by showing that

- if the algorithm returns an edge set A^*, then A^* is a popular arborescence in D (Theorem 13), and
- if D admits a popular arborescence, then the algorithm returns an edge set A^* (Theorem 19).

The following lemma in [11] is useful in our proof as well.

Lemma 12 ([11]). *For each $v \in V_G$, let X_v be the set defined in the algorithm STEP 1. Then, $\mathcal{X}$ is laminar, and $u \in X_v$ implies $X_u \subseteq X_v$.*

Theorem 13. *If the algorithm returns an edge set A^*, then A^* is a popular arborescence.*

Proof. First, we can show that A^* is an r-arborescence in D in the same manner as [11].

Next, we show that A^* is a popular arborescence. For $X \in \mathcal{X}'$ such that $|X| \geq 2$, let $v_X \in R$ be the terminal vertex of the edge in $A^* \cap \delta^-(X)$. Let $Y_{v_X} \subseteq X$ be a strongly connected component of the subgraph induced by $S(X) \cup \{e \in \delta^-(v_X): e \succ_{v_X} A^*(v_X)\}$ that contains v_X. For a vertex $t \in V_G$ such that $t \in X \setminus \{v_X\}$ for some $X \in \mathcal{X}$ with $|X| \geq 2$ or $\{t\} \in \mathcal{X}'$, let $Y_t = \{t\}$. Here, based on Proposition 8, we define

$$y(Y) = \begin{cases} w(v) & (Y = Y_v \text{ for some } v \in V_G), \\ 0 & \text{(otherwise)}. \end{cases} \tag{14}$$

It is clear that $\sum_{Y \subseteq V_G} y(Y) = w(V_G)$. By Proposition 4, the proof completes by showing that y is a feasible solution to (LP2) determined by A^*.

We show that y satisfies the constraint (8) of (LP2) for all edges. First, we consider the edges in $\delta^-(v_X)$ for each $X \in \mathcal{X}'$ with $|X| \geq 2$. The edges in $\delta^-(v_X) \cap E[Y_{v_X}]$ do not enter any set in $\mathcal{F}(y)$. For $e' \in \delta^-(v_X) \cap \delta^-(X)$, since $A^*(v_X)$ is not dominated by e' from Algorithm STEP 3, it follows that $c_A(e') \in \{w(v_X), 2w(v_X)\}$ holds. Since Y_{v_X} is the only set in $\mathcal{F}(y)$ that e' enters and $y(Y_{v_X}) = w(v_X)$, y satisfies the constraint (8) for e' in (LP2). Consider an edge $f' \in \delta^-(v_X) \cap \delta^-(Y_{v_X}) \cap E[X]$. By construction of Y_{v_X}, it must hold that $A^*(v_X) \succsim_{v_X} f'$, and hence $c_A(f') \in \{w(v_X), 2w(v_X)\}$. Since Y_{v_X} is the only set in $\mathcal{F}(y)$ that f' enters, it follows that y satisfies the constraint (8) for f' in (LP2).

Next, for $t \in Y_{v_X} \setminus \{v_X\}$, consider the edges in $\delta^-(t)$. By our algorithm, $A^*(t) \in S(X)$. By the definition of safe edges, for $g \in \delta^-(t) \cap E[Y_{v_X}]$, it holds that $g \precsim_t A^*(t)$, and hence $c_A(g) = \{w(t), 2w(t)\}$. Since Y_t is the only set in $\mathcal{F}(y)$ that g enters and $y(Y_t) = w(t)$, y satisfies the constraint (8) for g in (LP2).

Then, consider an edge $g' \in \delta^-(t) \cap \delta^-(Y_{v_X})$. Let $g' = (t_0, t)$. If $t_0 \notin X$, then $A^*(t) \succ_t g'$ holds by the definition of $S(X)$. If $t_0 \in X$, then $g' \notin S(X)$ holds by the definition of Y_{v_X}. Since $A^*(t) \in S(X)$ and the preferences are given by a total preorder, it follows that

$$A^*(t) \succ_t g' \tag{15}$$

for any $g' \in \delta^-(t) \cap \delta^-(Y_{v_X})$, and thus $c_{A^*}(g') = 2w(t)$.

If $t \in \bar{X}$, then $w(v_X) \leq w(t)$ follows from the fact that v_X is minimum weight in $\bar{X}$ by STEP 3 of the algorithm. Suppose that $t \in Y_{v_X} \setminus \bar{X}$. Since $t \notin \bar{X}$, v_X is unreachable by safe edges from t. Furthermore, by construction of Y_{v_X}, there exists a path P from t to v_X consisting of safe edges and the edges in $\delta^-(v_X)$ preferred to $A^*(v_X)$. In this path P, let (t', v_X) be the edge in $\delta^-(v_X)$. From our algorithm, v_X is a vertex which does not satisfy at least one condition 1, 2, or 3 in STEP 3, and v_X satisfies the condition 2 from $(t', s) \succ_{v_X} A^*(v_X)$. Also, since the path $P \subseteq S(X) \cap (t', s)$, the condition 3 holds for v_X and (t', v_X). Thus, it follows that v_X does not satisfy the condition 1, and hence $w(v_X) \leq w(t)$. From the above, $c_{A^*}(g') = 2w(t) \geq w(v_X) + w(t) = y(Y_{v_X}) + y(Y_t)$ holds. Thus, y satisfies the constraint (8) for g' in (LP2).

Lastly, for $u \in X \setminus Y_{v_X}$, consider the edges in $\delta^-(u)$. By our algorithm, $A^*(u) \in S(X)$. For an arbitrary edge $h \in \delta^-(u)$, $A^*(u)$ is not dominated by h by the definition of safe edges, and therefore $c_{A^*}(h) \in \{w(u), 2w(u)\}$. Since Y_u is the only set in $\mathcal{F}(y)$ that h enters and $y(Y_u) = w(u)$, it follows that y satisfies the constraint (8) for h in (LP2).

Therefore we have proved that y defined by (14) satisfies the constraints in (LP2) for all edges.   □

We remark that Theorem 13 does not hold if the preferences are given by a partial order. In the case of partial orders, (15) cannot be derived from $A^*(t) \in S(X)$ and $g' \notin S(X)$, because it allows for incomparable edges. This is the reason why we assume throughout the paper that the preferences are given by a total preorder.

Next, we prove that when a directed graph D has a popular arborescence, the algorithm always finds one of them. Lemmas needed for the proof are given below. Omitted proofs of the lemmas are presented in the full version.

Lemma 14 is shown similarly as Lemma 17 in [11], while our proof involves the vertex weights.

Lemma 14. *Let A be a popular arborescence, let $y_A^* = \{Y_v \mid v \in V\}$ be the dual optimal solution determined by A satisfying (10), and let $\mathcal{X}' = \{X_v : v \in V_G\}$ be the family of sets defined in the algorithm STEP 1. Then, $Y_v \subseteq X_v$ for each $v \in V_G$.*

Lemma 15. *Let A be a popular arborescence in D and let $X \in \mathcal{X}'$. Then, $A \cap \delta^-(X)$ contains only one edge, and $X = X_v$ holds for the terminal vertex v of that edge.*

Lemma 16. *Let A be an r-arborescence in D and let $X \in \mathcal{X}'$. If there exists an edge in $A[X] \setminus S(X)$, then there is an r-arborescence more popular than A.*

Proof. By Lemma 15, if $|A \cap \delta^-(X)| \neq 1$, then there is an r-arborescence more popular than A. We thus assume that $|A \cap \delta^-(X)| = 1$. Let $v \in X$ be the terminal vertex of the edge in $A \cap \delta^-(X)$. Again by Lemma 15, $X \neq X_v$ implies that there is an r-arborescence more popular than A, and hence we further assume that $X = X_v$.

Denote the edge in $A[X] \setminus S(X)$ by $f = (s, t)$. Then, it follows from $A(v) \in \delta^-(X)$ that $t \neq v$. Since $X = X_v$, we can construct a v-arborescence in a subgraph $(X, S(X))$. Define an r-arborescence A' in D by replacing the edges in A entering each vertex in $X \setminus \{v\}$ with this v-arborescence. Now, since $A'(t) \in S(X)$ and $f \notin S(X)$, we have $A'(t) \succ_t f$. Also, observe that $A'(t') \succsim_{t'} A(t')$ if $t' \in X \setminus \{t\}$ and $A'(t') = A(t')$ if $t \in V \setminus X$. Thus, $\Delta_w(A', A) \geq w(t) > 0$, implying that A' is more popular than A. $\square$

Lemma 17. *For a popular arborescence A and $X \in \mathcal{X}'$, let $\bar{X}$ be the strongly connected component in the subgraph $(X, S(X))$ that an edge in $S(X)$ does not enter. Then the terminal vertex v of the edge in $A \cap \delta^-(X)$ belongs to $\bar{X}$ and has minimum weight in $\bar{X}$.*

Lemma 18. *For some $v \in V_G$, if all the vertices of minimum weight in $\bar{X}_v$ satisfy the condition of Algorithm STEP 3, then there is no popular arborescence.*

Proof. Let $v \in V_G$ and $\bar{M}_v$ be the set of the vertices with the minimum weight in $\bar{X}_v$. Assume to the contrary that every vertex $v' \in \bar{M}_v$ satisfies the condition of STEP 3, i.e., there exist $s_{v'} \in X_v \setminus \bar{X}_v$ and $f_{v'} \in (E[X_v] \setminus S(X_v)) \cap \delta^-(v')$ satisfying the three conditions, and there exists a popular arborescence A.

First, consider the case when $|\bar{X}_v| = 1$, i.e., $v = v'$. By the definition of $\bar{X}_v$, there is no safe edge entering v' since all vertices in X_v are reachable by safe edges from v'. In this case, however, the condition 3 implies that the most preferred edge in $\delta^-(v) \cap E[X_v]$ satisfies the definition of safe edges, a contradiction.

Next, suppose that $|\bar{X}_v| \geq 2$. Note that $A[X_v] \subseteq S(X_v)$ by Lemma 16. Let u be the entry-point of X_v for A. By Lemma 17, we have $u \in \bar{M}_v$ and hence it follows that there exist $s_u \in X_v \setminus \bar{X}_v$ and $f_u \in (E[X_v] \setminus S(X_v)) \cap \delta^-(u)$ satisfying the three conditions in Algorithm STEP 3. Let B be the r-arborescence obtained from A by replacing the edges in $A[X_v] \cup A(u)$ with (r, s_u) and s_u-arborescence in $(X_v, S(X_v) \cup \{f_u\})$. It then follows from $\Delta_w(B, A) = w(u) - w(s_u) > 0$ that B is more popular than A, which contradicts that A is a popular arborescence. $\square$

Theorem 19. *If D admits a popular arborescence, then our algorithm finds one.*

Proof. Let A be a popular arborescence and let y_A^* be the dual optimal solution satisfying (10). By Lemma 15, for each $X \in \mathcal{X}'$, it holds that $|A \cap \delta^-(X)| = 1$.

Let $e_X = (u, v) \in A \cap \delta^-(X)$. Then $X = X_v$ holds by Lemma 15, and thus we have $v \in \bar{X}$.

First, we show that $e_X = (u, v)$ is not dominated by any edge $(u', v) \in \delta^-(X)$. By Lemma 14, we have $Y_v \subseteq X_v = X$. If $(u', v) \in \delta^-(X)$ dominates (u, v), then $c_A(u', v) = 0$. However, since $(u', v) \in \delta^-(Y_v)$ holds, this violates the constraint (8) in (LP2). Thus, $e_X = (u, v)$ is not dominated by an edge in $\delta^-(X) \cap \delta^-(v)$.

From Lemma 17, the vertex v has the minimum weight in $\bar{X}$. Furthermore, from Lemma 18, when D admits a popular arborescence, there exists a vertex that has minimum weight in $\bar{X}$ which does not satisfy the conditions of Algorithm STEP 3. For such a vertex v', our algorithm adds to E' an edge in $\delta^-(X) \cap \delta^-(v')$ that is not dominated by any edge in $\delta^-(X)$. Thus, we have $e_X \in E'$. Since there exists an r-arborescence in D and $e_X \in E'$ holds, for the graph D' constructed in STEP 4 of the algorithm, each $X \in \mathcal{X}'$ is reachable from r using the edges in E'. Hence E' contains an r-arborescence A' in D'. Therefore, the algorithm returns an edge set A^*, which is a popular arborescence in D by Theorem 13. □

References

1. Abraham, D.J., Irving, R.W., Kavitha, T., Mehlhorn, K.: Popular matchings. SIAM J. Comput. **37**, 1030–1045 (2007)
2. Biró, P., Irving, R.W., Manlove, D.F.: Popular matchings in the marriage and roommates problems. In: Calamoneri, T., Diaz, J. (eds.) CIAC 2010. LNCS, vol. 6078, pp. 97–108. Springer, Heidelberg (2010). https://doi.org/10.1007/978-3-642-13073-1_10
3. Cseh, Á., Huang, C.-C., Kavitha, T.: Popular matchings with two-sided preferences and one-sided ties. SIAM J. Discr. Math. **31**(4), 2348–2377 (2017)
4. Cseh, Á., Kavitha, T.: Popular edges and dominant matchings. Math. Programm. **172**(1), 209–229 (2017)
5. Edmonds, J., Giles, R.: A min-max relation for submodular functions on graphs. Stud. Integer Programm. Ann. Discr. Math. 185–204 (1977)
6. Frank, A.: Kernel systems of directed graphs. Acta Scientiarum Mathematicarum **41**, 63–76 (1979)
7. Fulkerson, D.R.: Packing rooted directed cuts in a weighted directed graph. Math. Programm. **6**(1), 1–13 (1974)
8. Fenza, Y., Kavitha, T.: Quasi-popular matchings, optimality, and extended formulations. In: Proceedings of the 31st Annual ACM-SIAM Symposium on Discrete Algorithms (SODA 2020), pp. 325–344 (2020)
9. Gärdenfors, P.: Match making: assignments based on bilateral preferences. Behav. Sci. **20**(3), 166–173 (1975)
10. Kavitha, T., Nasre, M.: Optimal popular matchings. Discr. Appl. Math. **157**(14), 3181–3186 (2009)
11. Kavitha, T., Király, T., Matuschke, J., Schlotter, I., Schmidt-Kraepelin, U.: Popular branchings and their dual certificates. In: Bienstock, D., Zambelli, G. (eds.) IPCO 2020. LNCS, vol. 12125, pp. 223–237. Springer, Cham (2020). https://doi.org/10.1007/978-3-030-45771-6_18
12. Mestre, J.: Weighted popular matchings. ACM Trans. Algor. **10**(1), 1–16 (2014)

Vertex-Weighted Graphs: Realizable and Unrealizable Domains

Amotz Bar-Noy[1], Toni Böhnlein[2(✉)], David Peleg[2], and Dror Rawitz[3]

[1] City University of New York (CUNY), New York, USA
amotz@sci.brooklyn.cuny.edu
[2] Weizmann Institute of Science, Rehovot, Israel
{toni.bohnlein,david.peleg}@weizmann.ac.il
[3] Bar Ilan University, Ramat-Gan, Israel
dror.rawitz@biu.ac.il

Abstract. Consider the following natural variation of the degree realization problem. Let $G = (V, E)$ be a simple undirected graph of order n. Let $f \in \mathbb{R}^n_{\geq 0}$ be a vector of vertex *requirements*, and let $w \in \mathbb{R}^n_{\geq 0}$ be a vector of *provided services* at the vertices. Then w *satisfies* f on G if the constraints $\sum_{j \in N(i)} w_j = f_i$ are satisfied for all $i \in V$, where $N(i)$ denotes the neighborhood of i. Given a requirements vector f, the WEIGHTED GRAPH REALIZATION problem asks for a suitable graph G and a vector w of provided services that satisfy f on G.

In [7] it is observed that any requirement vector where n is even can be realized. If n is odd, the problem becomes much harder. For the unsolved cases, the decision of whether f is realizable or not can be formulated as whether f_n (the largest requirement) lies within certain intervals. In [5] some intervals are identified where f can be realized, and their complements form $\frac{n-3}{2}$ connected intervals ("unknown domains") which we give odd indices $k = 1, 3, \dots, n - 4$. The unknown domain for $k = 1$ is shown to be unrealizable.

Our main result presents structural properties that a graph must have if it realizes a vector in one of these unknown domains for $k \geq 3$. The unknown domains are characterized by inequalities which we translate to graph properties. Our analysis identifies several realizable sub-intervals, and shows that each of the unknown domains has at least one sub-interval that cannot be realized.

Keywords: Graph realization · Degree sequence · Graph-algorithms

1 Introduction

Given an n-integer sequence d, the degree realization problem is to decide if there exists an n-vertex graph whose degree sequence is d, and if so, to construct such a realization (see [1,9,14,18,20,24,27–29]). The problem was well

Supported in part by a US-Israel BSF grant (2018043). Partly supported by ARL Cooperative Grant, ARL Network Science CTA, W911NF-09-2-0053.

© Springer Nature Switzerland AG 2022
P. Mutzel et al. (Eds.): WALCOM 2022, LNCS 13174, pp. 315–327, 2022.
https://doi.org/10.1007/978-3-030-96731-4_26

researched over the recent decades and plays an important role in the field of social networks and complex networks (cf. [8,11,13,21,26]). For additional graph realization problems see [2–4,10,12,15–17,19,22,25,33] and the surveys [6,23,30–32].

The following natural variation of the problem was introduced in [7]: Let $G = (V, E)$ be a simple undirected graph on $V = \{1, 2, \ldots, n\}$. Let $f \in \mathbb{R}_{\geq 0}^n$ be a vector of *vertex requirements*, and let $w \in \mathbb{R}_{\geq 0}^n$ be a vector of *vertex weights*. The weight vector w *satisfies* the requirement vector f on G if the constraints $\sum_{j \in N(i)} w_j = f_i$ are satisfied for all $i \in V$, where $N(i)$ denotes the (open) neighbourhood of vertex i. The vertex-weighted realization problem is now as follows: Given a requirement vector f, find a suitable graph G and a weight vector w that satisfy f on G (if exist). In the original degree realization problem, the requirements are integers in $[0, n - 1]$ and all vertex weights must equal one.

It is shown in [7] that any requirement vector f of even length can be realized by a matching graph (composed of $n/2$ independent edges). Each vertex u in a matching graph has a unique neighbour v, so the weights $w_u = f_v$ and $w_v = f_u$ realize f.

Theorem 1. *[7] If n is even, then any f can be realized using a perfect matching.*

The problem becomes significantly harder for odd n. In this case, f can be realized (denoting $[i, j] = \{i, \ldots, j\}$) if either $f_i = 0$ or $f_i = f_j$, for two distinct indices $i, j \in [1, n]$. Without loss of generality, we focus on the following domain:

$$\mathcal{F}_n \triangleq \{f \in \mathbb{R}_{\geq 0}^n : 0 < f_1 < f_2 < \cdots < f_n\}.$$

As an introductory example, consider the domain $\mathcal{F}_3$. Observe that any graph that potentially realizes some $f \in \mathcal{F}_3$ must be connected since $f_1 > 0$. The only two connected graphs on 3 vertices are the path P_3 (Fig. 1a) and the complete graph K_3 (Fig. 1c). The layout of a graph implies a system of equations that the requirements and weights must satisfy as displayed in Figs. 1b and 1d for P_3 and K_3, respectively.

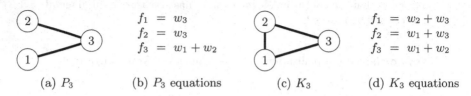

(a) P_3	(b) P_3 equations	(c) K_3	(d) K_3 equations

$$\begin{aligned} f_1 &= w_3 \\ f_2 &= w_3 \\ f_3 &= w_1 + w_2 \end{aligned} \qquad \begin{aligned} f_1 &= w_2 + w_3 \\ f_2 &= w_1 + w_3 \\ f_3 &= w_1 + w_2 \end{aligned}$$

Fig. 1. Graphs that realize vectors of $\mathcal{F}_3$ and their equation systems.

The system in Fig. 1b implies that $f_1 = f_2$ must hold. In general, P_3 implies that f must satisfy $f_i = f_j$ where i, j are the labels of the two vertices of degree one. As a consequence, P_3 cannot realize a vector $f \in \mathcal{F}_3$. For K_3, the labeling is immaterial due to the graph's symmetry. Solving this system for the weights yields:

$$w_1 = (f_2 + f_3 - f_1)/2, \qquad w_2 = (f_1 + f_3 - f_2)/2, \qquad w_3 = (f_1 + f_2 - f_3)/2.$$

By the problem definition, the weights need to be non-negative. Hence, each equation implies a constraint, yielding

$$0 \le f_2 + f_3 - f_1, \qquad 0 \le f_1 + f_3 - f_2, \qquad 0 \le f_1 + f_2 - f_3.$$

The first two equations are satisfied by any $f \in \mathcal{F}_3$; the third implies that $f \in \mathcal{F}_3$ can be realized if and only if $f_3 \le f_1 + f_2$.

The example demonstrates an approach (detailed in Sect. 3) that we use frequently in the course of this study: Given a graph G (or a family of graphs), we deduce constraints, and use them to define the domain realizable by G in a convenient way.

For general n, it is shown in [7] that a vector $f \in \mathcal{F}_n$ does not have a realization if it belongs to the *exponential growth* domain

$$\mathcal{D}_n^{\exp} = \left\{ f : \forall i \in [1, n], f_i > \sum\nolimits_{j<i} f_j \right\}.$$

However, f *can* be realized if it falls in the *sub-exponential growth* domain,

$$\mathcal{D}_n^{\mathrm{sub}} = \left\{ f : \exists i \in [1, n-1], f_i \le \sum\nolimits_{j<i} f_j \right\}.$$

Theorem 2. *[7] Let $n \ge 3$ be an odd integer. Then,*

1. *a requirements vector $f \in \mathcal{D}_n^{\exp}$ cannot be realized.*
2. *a requirements vector $f \in \mathcal{D}_n^{\mathrm{sub}}$ can be realized.*

Note that in the definition of $\mathcal{D}_n^{\mathrm{sub}}$, there is no inequality for bounding f_n. The "unknown domain" at this point, for which the realizability problem is unsettled, is the "almost exponential" domain

$$\mathcal{D}_n^{\exp-} = \left\{ f : \forall i \in [1, n-1], f_i > \sum\nolimits_{j<i} f_j \ \text{ and } \ f_n \le \sum\nolimits_{j<n} f_j \right\}.$$

Hence, subsequent analysis should concentrate on refining the domain $\mathcal{D}_n^{\exp-}$, resolve the status of its subdomains and thus narrow down the unknown regions.

Based on these results, the question whether a vector $f \in \mathcal{D}_n^{\exp-}$ can be realized or not, depends on the value of f_n in relation to the other requirements. Hence, subdomains of $\mathcal{D}_n^{\exp-}$ are typically defined in terms of *intervals* in the range of possible values for f_n. The situation at the two extremes of this range is clear. If f_n is larger than $\sum_{i<n} f_i$, then vector f cannot be realized due to Theorem 2. At the other end, if $f_n \le f_{n-1} + f_{n-2}$, then there exists a realization for f that uses K_3 and a matching graph as described in [7]. Consequently, our analysis concentrates on vectors $f \in \mathcal{D}_n^{\exp-}$ where f_n is in the intermediate range, $f_n \in [f_{n-1} + f_{n-2}, \sum_{i<n} f_i]$.

It is shown in [7] that parts of this interval can be realized by two types of domains which are called *windmill* and *kite* domains. In a subsequent study [5], these domains were extended, and it was shown that certain collections of these domains have pairwise overlapping intervals, i.e., they form a single, larger interval. The larger intervals are used to define *meta*-domains $\mathcal{D}_{n,\ell}^{\mathcal{M}}$, for every even integer $2 \leq \ell \leq n - 5$:

$$\mathcal{D}_{n,\ell}^{\mathcal{M}} = \left\{ f \in \mathcal{D}_n^{\exp-} : \sum_{j=\ell+1}^{n-1} (f_j - f_{\ell+1}) \leq f_n \leq \sum_{j=1}^{n-1} f_j - f_\ell \right\}, \text{ and}$$

$$\mathcal{D}_{n,0}^{\mathcal{M}} = \left\{ f \in \mathcal{D}_n^{\exp-} : \sum_{j=2}^{n-1} (f_j - f_1) \leq f_n \leq \sum_{j=1}^{n-1} f_j \right\}.$$

Theorem 3. [5] *Let $n \geq 5$ and let $\ell \leq n - 5$ be an even number. The vector f in the meta domain $\mathcal{D}_{n,\ell}^{\mathcal{M}}$ can be realized.*

We now describe the *unknown* domains located between the meta domains where it is an open question whether a vector can be realized or not. Let k be an odd index such that $k \leq n - 4$. The k-th unknown domain is:

$$\mathcal{D}_{n,k}^{\mathcal{U}} = \left\{ f \in \mathcal{D}_n^{\exp-} : \sum_{j=1}^{n-1} f_j - f_{k+1} < f_n < \sum_{j=k}^{n-1} (f_j - f_k) \right\}.$$

Our Results. Section 2 analyses the domains $\mathcal{D}_{n,k}^{\mathcal{U}}$, using a complementary approach to the earlier one (of generating a set of constraints from a graph). Assume that a vector $f \in \mathcal{D}_{n,k}^{\mathcal{U}}$ is realized by a graph G and weights w. Then f is subject to a set of constraints, namely, upper and lower bounds on f_n, and exponential growth constraints for $f_1, \ldots, f_{n-1}$ implied by the definition of $\mathcal{D}_n^{\exp-}$. From these, we deduce *structural* properties of G. For example, we show that vertex n plays a central role in G's layout: it must be adjacent to the $n - k + 1$ vertices with the largest weights, and moreover, these *large weight* vertices must have degree two.

Based on the exponential growth of $f_1, \ldots, f_{n-1}$, we show that each vertex has a neighbour with a *dedicated weight* ensuring that its requirement is met. This dependency is pairwise by deducing a one-to-one correspondence between weights and requirements, revealed by decomposing the graph G. The decomposition process can be viewed as removing pairs of vertices in $\frac{n-1}{2}$ many steps. Each pair of vertices is connected by an edge and all these edges form a matching in G, such that the matching partner of a vertex carries its dedicated weight.

The main outcome of Sect. 2 is a structural *pattern* (or *layout*), namely, some property that a graph must have in order to realize a requirements vector $f \in \mathcal{D}_{n,k}^{\mathcal{U}}$. All subsequent steps make heavy use of this structural result.

In Sect. 4, we put our pattern to use and find new graphs that realize subdomains of $\mathcal{D}_{n,k}^{\mathcal{U}}$, and define the domains that are based on these graphs in terms of bounds on f_n. Our approach is a bit more general, in the sense that we first define families of domains and then identify useful members of these families. This provides us with additional tools exploited in later sections. We obtain three new types of realizable subdomains of $\mathcal{D}_{n,k}^{\mathcal{U}}$, named the *double windmill*, *kite clique* and *extended kite clique* domains. However, these subdomains do not cover $\mathcal{D}_{n,k}^{\mathcal{U}}$ completely.

Some of these subdomains are then used (in Sect. 5) to show a negative result: For every k there is at least one interval $\hat{I}$ for f_n where $f \in \mathcal{D}_{n,k}^{\mathcal{U}}$ cannot be realized. This $\hat{I}$ has the same lower bound as the domain $\mathcal{D}_{n,k}^{\mathcal{U}}$. To find its upper bound, we had to find the realizable domain of $\mathcal{D}_{n,k}^{\mathcal{U}}$ with the smallest lower bound.

The existence of $\hat{I}$ is shown by contradiction. Assuming that a vector f is realized by graph G where f_n falls within $\hat{I}$, it follows that $f \in \mathcal{D}_{n,k}^{\mathcal{U}}$, which implies that G adheres to our pattern. This enables us to deduce an upper bound on the requirements, which eventually yields a contradiction with the lower bound of the realizable domains bordering $\hat{I}$.

Organization. Section 2 shows the structure a graph must have to realize a vector of $\mathcal{D}_{n,k}^{\mathcal{U}}$. In Sect. 3, we outline our general approach to define domains, which we apply then in Sect. 4. In Sect. 5, we show that each domain $\mathcal{D}_{n,k}^{\mathcal{U}}$ contains at least one unrealizable part.

2 Exploring the Domain $\mathcal{D}_{n,k}^{\mathcal{U}}$

We analyse the structure of a graph that realizes a requirement vector $f \in \mathcal{D}_{n,k}^{\mathcal{U}}$, where n and k are odd and $k \leq n - 4$. Let f be realized by a graph G and weights $w \in \mathbb{R}_{\geq 0}^n$. By definition of $\mathcal{D}_{n,k}^{\mathcal{U}}$, we have that $f \in \mathcal{D}_n^{\exp-}$ and

$$\sum_{j=1}^{n-1} f_j - f_{k+1} < f_n, \tag{1}$$

$$f_n < \sum_{j=k+1}^{n-1} (f_j - f_k). \tag{2}$$

Recall that such a domain exists only if $\sum_{j=1}^{k-1} f_j + (n-k)f_k < f_{k+1}$.

The Connected Component Containing n: Let H be the connected component of G that contains the vertex n, and define $h \triangleq |H|$. For the rest of this section, we focus on H. We first show that H is the only component with an odd number of vertices. Hence other components, if exist, play a minor role and can be thought of as a matching graph.

Lemma 1. *Let $f \in \mathcal{D}_{n,k}^{\mathcal{U}}$, where n and k are odd and $k \leq n - 4$. H is the only connected component of G that has an odd number of vertices.*

Proof. Suppose, towards a contradiction, that H' is an odd, connected component of G that does not contain n. The component H' induces a requirement vector $f|_{H'}$ on $V(H')$. Observe that $f|_{H'}$ is realized by H' and $w|_{H'}$. Since $f \in \mathcal{D}_n^{\exp-}$, it follows that $f|_{H'} \in \mathcal{D}_{|H'|}^{\exp}$. However, Theorem 2 implies that $f|_{H'}$ cannot be realized as H' has an odd number of vertices. We reach the desired contradiction. Since n is odd, G must have at least one odd component. We conclude that H is the only component of G which has an odd number of vertices. $\square$

Let $\bar{H}$ be the version of H where we renumber the vertices in $V(H)$ such that $V(\bar{H}) \triangleq [1, h]$. Furthermore, let $\bar{f}$ and $\bar{w}$ denote the induced requirement vector and weights on $V(\bar{H})$. It follows that $\bar{H}$ and $\bar{w}$ realize $\bar{f}$. Since $n \in H$, we have that $\bar{f}_h = f_n$. In addition, since $V(H) \subseteq V(G)$, it follows that $\sum_{j<i} \bar{f}_j < \bar{f}_i$, for every $i \in [1, h-1]$. Since $\bar{f}$ is realized by $\bar{H}$ and $\bar{w}$, it must be that $\bar{f}_h \leq \sum_{j=1}^{h-1} \bar{f}_j$. Hence, $\bar{f} \in \mathcal{D}_h^{\exp -}$.

Next, we show that H contains the largest $n - k$ requirements, i.e., the requirements of the set $[k + 1, n]$.

Lemma 2. *Let $f \in \mathcal{D}_{n,k}^{\mathcal{U}}$ where n and k are odd and $k \leq n - 4$. The vertices $[k + 1, n - 1]$ are contained in H.*

Proof. Suppose towards a contradiction that there exists $i \in [k + 1, n - 1]$ such that $i \notin V(H)$. Observe that

$$\sum_{j=1}^{h-1} \bar{f}_j \leq \sum_{j=1}^{n-1} f_j - f_i \leq \sum_{j=1}^{n-1} f_j - f_{k+1} \overset{(\star)}{<} f_n = \bar{f}_h \, ,$$

where $(\star)$ holds due to Eq. (1). It follows that $\bar{f} \in \mathcal{D}_n^{\exp}$. Hence, Theorem 2 implies that $\bar{f}$ is not realizable and we reach the desired contradiction. $\square$

Since H contains the vertices $[k + 1, n]$, we have that $h \geq n - k$. Moreover, since h is odd and $n - k$ is even, it must be that $h \geq n - k + 1$. This implies that, there is at least one vertex $i \in V(H)$, such that $f_i \leq f_k$. Let $\bar{k} \triangleq h - n + k$ be the number of such vertices in H. Observe that $\bar{k} \leq k$ and that $\bar{k}$ is odd. Next, we show that $\bar{f} \in \mathcal{D}_{h,\bar{k}}^{\mathcal{U}}$.

Lemma 3. *Let $f \in \mathcal{D}_{n,k}^{\mathcal{U}}$, where n and k are odd and $k \leq n - 4$. Then, $\bar{f} \in \mathcal{D}_{h,\bar{k}}^{\mathcal{U}}$.*

Proof. Lemma 2 implies that $\bar{f}_{\bar{k}+j} = f_{k+j}$ for $j \in [1, n - k]$. With Eq. (2) it follows that

$$\sum_{j=\bar{k}+1}^{h-1} (\bar{f}_j - \bar{f}_{\bar{k}}) \geq \sum_{j=k+1}^{n-1} (f_j - f_k) > f_n = \bar{f}_h \, . \tag{3}$$

In addition, due to Eq. (1)

$$\sum_{j=1}^{h-1} \bar{f}_j - \bar{f}_{\bar{k}+1} \leq \sum_{j=1}^{n-1} f_j - f_{k+1} < f_n = \bar{f}_h \, . \tag{4}$$

Finally, recall that $\bar{f} \in \mathcal{D}_h^{\exp -}$. The lemma follows. $\square$

We note that the lemma implies that the interval for $\bar{f}_h = f_n$ that corresponds to $\mathcal{D}_{h,\bar{k}}^{\mathcal{U}}$ is at least as large as the interval for f_n that corresponds to $\mathcal{D}_{n,k}^{\mathcal{U}}$.

The Partition of $\bar{H}$: Given a weight function $w : V(\bar{H}) \to \mathbb{R}_{\geq 0}^h$, let σ be a permutation of $V(\bar{H}) = [1, h]$ for which $\bar{w}_{\sigma(i)} \leq \bar{w}_{\sigma(i+1)}$ for every $i \in [1, h - 1]$. We call $\bar{w}_{\sigma(1)}, \ldots, \bar{w}_{\sigma(\bar{k}+1)}$ (resp., $\bar{w}_{\sigma(\bar{k}+2)}, \ldots, \bar{w}_{\sigma(h)}$) the *small* (resp., *large*) *weights*. Define

$$L \triangleq \left\{ \sigma^{-1}(i) \in \bar{H} : i > \bar{k} + 1 \right\} \qquad S \triangleq \left\{ \sigma^{-1}(i) \in \bar{H} : i \leq \bar{k} + 1 \right\}.$$

The set S contains small weight vertices, while the set L contains vertices with large weights. Note that $|L| = h - \bar{k} - 1$ is odd and $|S|$ is even.

We call $\bar{f}_1, \ldots, \bar{f}_{\bar{k}}$ *small requirements* and $\bar{f}_{\bar{k}+1}, \ldots, \bar{f}_{h-1}$ *large requirements*. Note that the requirement $\bar{f}_h$ is neither small nor large.

Define the sets X, Y, Z, and W as follows:

$$X \triangleq L \cap \{\bar{k}+1, \ldots, h-1\} \qquad Z \triangleq S \cap \{\bar{k}+1, \ldots, h-1\}$$
$$Y \triangleq L \setminus (X \cup \{h\}) \qquad W \triangleq S \setminus (Z \cup \{h\})$$

The set $X \cup Y$ contains large weight vertices. Vertices in X have large requirements while vertices in Y have small requirements. Similarly, Z and W partition the set of small weight vertices. Vertex h is not contained in any of X, Y, Z, or W.

The main result of this section is the next theorem which summarizes important structural properties of H. An illustration is given in Fig. 2.

Theorem 4. *Let $f \in \mathcal{D}_{n,k}^{\mathcal{U}}$, where n and $k \leq n - 4$ are odd. The vertices of $V(\bar{H})$ are partitioned into five subsets X, Y, Z, W, and $\{h\}$, which satisfy the following conditions:*

(1) $X \cup Z = [\bar{k}+1, h-1]$ and $Y \cup W = [1, \bar{k}]$.
(2) $X \cup Y \subseteq N(h)$.
(3) $\deg(i) = 2$, for every $i \in X \cup Y$.
(4) $|X|$ is even and $E \cap (X \times X)$ is a matching of size $|X|/2$.
(5) $|Y| = |Z|$ is odd and $E \cap (Y \times Z)$ is a matching of size $|Y|$.
(6) $|W|$ is even and $E \cap (W \times W)$ contains a matching of size $|W|/2$.

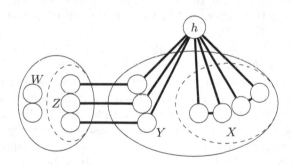

Fig. 2. The structure of $\bar{H}$. X and Y vertices cannot have more incident edges than depicted. $Z \cup W$ vertices may be adjacent to other vertices in $Z \cup W \cup \{h\}$.

Observe that item (1) of Theorem 4 is already implied by Lemma 2 and the definition of X and Z. The other conditions of Theorem 4 are shown in the full version of this paper.

3 General Approach to Defining Domains

Based on the results of [7], we defined the domain $\mathcal{D}_n^{\exp-}$ in the introduction. For a vector $f \in \mathcal{D}_n^{\exp-}$, we exhibit exponential growth for the indices $[1, n-1]$. Whether f is realizable or not depends therefore on the value of f_n. Hence, we ask, given a vector $f' \in \mathcal{D}_{n-1}^{\exp}$, which values for f_n *extend* f' to a realizable vector $f \in \mathcal{D}_n^{\exp-}$? We say that f extends f' if $f_i = f_i'$ for $i = 1, \ldots n-1$. Our approach to answering this question is by relating it to a 'realizing' graph. Let $f' \in \mathcal{D}_{n-1}^{\exp}$ and let G be a graph. Define their *realization domain* to be the collection of requirement vectors

$$\mathcal{D}_{real}(G, f') \triangleq \left\{ f \in \mathcal{D}_n^{\exp-} \; : \; f \text{ extends } f' \text{ and } \exists w \in \mathbb{R}_{\geq 0}^n \text{ s.t. } G \text{ and } w \text{ realize } f \right\}.$$

We now define a second collection of requirement vectors associated with each graph G and $f' \in \mathcal{D}_{n-1}^{\exp}$, namely, those vectors f whose last component lies in a specific range, i.e., $f_n \in [\mathcal{LB}, \mathcal{UB}]$. Formally, let

$$\mathcal{R}(G, f') \triangleq \{\hat{f} \; : \; \exists f \in \mathcal{D}_{real}(G, f') \text{ s.t. } f_n = \hat{f}\},$$

$$\mathcal{LB} \triangleq \min \mathcal{R}(G, f') \qquad \text{and} \qquad \mathcal{UB} \triangleq \max \mathcal{R}(G, f') .$$

Then the *range domain* of G and f' is defined to be

$$\mathcal{D}_{range}(G, f') \triangleq \{f \in \mathcal{D}_n^{\exp-} \; : \; f \text{ extends } f' \text{ and } \mathcal{LB} \leq f_n \leq \mathcal{UB}\} .$$

A-priori, it is unclear that every requirement vector $f \in \mathcal{D}_{range}(G, f')$ can be realized using G. It might be that the range $[\mathcal{LB}, \mathcal{UB}]$ contains "holes" in the form of requirement vectors f such that $f_n \in [\mathcal{LB}, \mathcal{UB}]$ and yet f does not belong to $\mathcal{D}_{real}(G, f')$. As it turns out, these mishaps do not happen for the graphs analysed in this paper, and in fact, their realization and range domains satisfy that $\mathcal{D}_{range}(G) = \mathcal{D}_{real}(G)$.

Our approach to show that $\mathcal{D}_{real}(G) = \mathcal{D}_{range}(G)$ for these graphs is as follows. Graph G yields a system of equations of the form $f_j = \sum_{i \in N(j)} w_i$, for each $i \in [1, n]$. For a given requirement vector f the unknowns in this system are the weights. The system can also be expressed using the adjacency matrix A_G of G, as $A_G \cdot w = f$. For the graphs that we consider in the following, this system has a unique solution, i.e., A_G has full rank. Consequently, we get an expression for each of the weights that is a linear combination of the requirements: $w_j = \sum_{i=1}^n a_{i,j} \cdot f_i$, for $j \in [1, n]$. It follows that f is realized by G and these weights. However, the weights are not necessarily non-negative.

Requiring that $0 \leq \sum_{i=1}^n a_{i,j} \cdot f_i$ yields a constraint for each $j \in [1, n]$. It follows that each requirement vector that satisfies all of these constraints is realized by G and non-negative weights.

For the graphs that we consider, it turns out that the constraints are either satisfied by any requirements vector (e.g. $f_1 < f_2$) or they can be formulated as a bound on f_n. A constraint can be rearranged as a bound on f_n if $a_{n,j} \neq 0$. In case $a_{n,j} < 0$, a constraint yields an upper bound while a constraint yields a lower bound if $a_{n,j} > 0$. We define $\mathcal{LB}$ to be the largest lower bound and $\mathcal{UB}$ to be the smallest upper bound[1] (it must hold that $\mathcal{LB} < \mathcal{UB}$). As a consequence, we showed that the weights which we found realize f based on G and that the weights are non-negative as long as $\mathcal{LB} \leq f_n \leq \mathcal{UB}$.

This approach allows us to restrict our attention to the definition of range domains, which is easier to manipulate and use than realization domains. Subsequently, our strategy is based on seeking specific graphs that turn out to be useful as realization graphs, and finding explicit expressions for $\mathcal{UB}$ and $\mathcal{LB}$ for those graphs.

4   Realizable Domains of $\mathcal{D}_{n,k}^{\mathcal{U}}$

In Sect. 2, we deduce a pattern that a graph must have if it realizes a vector $f \in \mathcal{D}_{n,k}^{\mathcal{U}}$. In this section, we describe realizable domains based on this pattern.

Let $\mathcal{G}_{n,k}^{U}$ be the set of all graphs that realize some vector $f \in \mathcal{D}_{n,k}^{\mathcal{U}}$. For a graph $G \in \mathcal{G}_{n,k}^{U}$, $V(G)$ contains the subsets X, Y, Z and W, but several aspects of G are not specified by Theorem 4, e.g., the connectivity of vertices in Z and W.

To describe domains, we first define domain families that are based on sub-families of $\mathcal{G}_{n,k}^{U}$. The sub-families are (mostly) characterized by constraints on the size of X, Y, Z and W. Second, we define domains that are useful for the following section. Here, we specify the vertex labelling as well as the connectivity of vertices in W and Z. Note that, the sets Y and W do not have to contain all of the small requirements: An even sub-sequence can be realized in a separate component, e.g., with a matching graph.

We use the notation $N_U(i)$ for the neighbourhood of vertex $i \in V(G)$ inter-sected with $U \subseteq V(G)$, and $\delta_U(i) \triangleq |N_U(i)|$. For vertices $i, j \in V(G)$ we use a binary indicator variable: $I_{i,j}(G) = 1$ if and only if i and j are adjacent in G.

In the extended abstract, we only present one domain family and one domain based on it.

The Windmill Family. The first family $\mathcal{G}_{n,k}^{\bowtie} \subseteq \mathcal{G}_{n,k}^{U}$ is called the *windmill family*. It contains all graphs such that (i) $|Y| = |Z| = 1$, (ii) for $i \in W$, $N_W(i) = \chi(i)$, and (iii) i and n are not adjacent. Let $Y = \{y\}$ and $Z = \{z\}$ where f_y and f_z can be any small and large requirement, respectively. Vertex z can be adjacent to any $i \in W$ and n. The remaining small and large requirements are contained in W and X, respectively. For a vertex $i \in X$, $N(i) = \{\varphi(i), n\}$, and $N(y) = \{z, n\}$. Figure 3 presents the layout of graphs in $\mathcal{G}_{n,k}^{\bowtie}$.

---

[1] The lower and upper bound depend on f, i.e., $\mathcal{LB}(f), \mathcal{UB}(f)$.

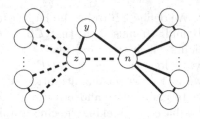

Fig. 3. Layout of a graph in the windmill family. The dashed edges are optional.

Let $G \in \mathcal{G}_{n,k}^{\bowtie}$. We show that $\mathcal{D}_{real}(G, f)$ can be defined in terms of bounds on f_n. To state the bound, let $\mu = \min \{i \in W : I_{i,z}(G) = 1\}$,

$$c^{\bowtie}(G) = \sum_{i \in W} I_{i,z}(G) I_{\chi(i),z}(G) + 2I_{z,n}(G) + |X|, \text{ and}$$

$$\Delta^{\bowtie}(G, f) \triangleq -\sum_{i \in W} I_{\chi(i),z}(G) \cdot f_i - (I_{z,n}(G) + |X|) \cdot f_y + \sum_{i \in X} f_i + f_z.$$

Theorem 5. *Let n, k be odd integer such that $n \geq 5$ and $k \leq n-4$. Requirement vector $f \in \mathcal{D}_{n,k}^{\mathcal{U}}$ can be realized based on graph $G \subset \mathcal{G}_{n,k}^{\bowtie}$ if*

$$\Delta^{\bowtie}(G, f) \leq f_n \leq \Delta^{\bowtie}(G, f) + c^{\bowtie}(G) \cdot f_\mu.$$

Note that $0 < c^{\bowtie}(G)$ and that we can assign any of the large requirements to z and get the same lower and upper bounds. The proof of Theorem 5 is presented in the full version. It is based on the outline in Sect. 3.

Useful Domains. We set up domains based on the families. Besides showing that parts of $\mathcal{D}_{n,k}^{\mathcal{U}}$ can be realized, this serves towards the goal of showing that there is at least one interval for f_n where $f \in \mathcal{D}_{n,k}^{\mathcal{U}}$ cannot be realized (in Sect. 5). This unrealizable interval has the same lower bound as the domain $\mathcal{D}_{n,k}^{\mathcal{U}}$. To find the upper bound of this interval, we need to find the realizable domain of $\mathcal{D}_{n,k}^{\mathcal{U}}$ which has the smallest lower bound.

One such domain is denoted $\mathcal{D}_{n,k}^{DW1}$, and it is based on the graph $G_{n,k}^{DW1} \in \mathcal{G}_{n,k}^{\bowtie}$ where $Y = \{k\}$ and $W = \{1, \ldots, k - 1\}$. Vertex z is adjacent to all vertices in W and to vertex n.

Theorem 6. *Let n, k be odd integer such that $n \geq 5$ and $k \leq n - 4$. $f \in \mathcal{D}_{n,k}^{DW1}$ can be realized.*

In the full version, we present two more domain families. Moreover, we define the domains $\mathcal{D}_{n,k}^{KC1}, \mathcal{D}_{n,k}^{KC2}$ and $\mathcal{D}_{n,k}^{KC^+}$ which are, in addition to $\mathcal{D}_{n,k}^{DW1}$, candidates for the smallest lower bound in $\mathcal{D}_{n,k}^{\mathcal{U}}$.

5   Unrealizable Domains of $\mathcal{D}_{n,k}^{\mathcal{U}}$

In this section, we show that parts of $\mathcal{D}_{n,k}^{\mathcal{U}}$ cannot be realized. In [5] it was shown that a vector $f \in \mathcal{D}_{n,1}^{\mathcal{U}}$ cannot be realized. For $k \geq 3$, there are $f \in \mathcal{D}_{n,k}^{\mathcal{U}}$ which can be realized, e.g., $f \in \mathcal{D}_{n,k}^{DW1}$. However, we can state that $f \in \mathcal{D}_{n,k}^{\mathcal{U}}$ cannot be realized based on graph $G_{n,k}^{DW1}$ if $f_n < \delta_{n,k}^{DW1}(f) + \sum_{i=k+1}^{n-1} f_i$, The next theorem shows that there is an interval for f_n that cannot be realized based on any graph. This unrealizable interval has the lower bound of $\mathcal{D}_{n,k}^{\mathcal{U}}$ as a left border. The right border, denoted $\eta(f)$, is the smallest, lower bound of the domains $\mathcal{D}_{n,k}^{DW1}, \mathcal{D}_{n,k}^{KC1}, \mathcal{D}_{n,k}^{KC2}$ or $\mathcal{D}_{n,k}^{KC^+}$, i.e.,

$$\eta(f) = \min\{\mathcal{LB}(\mathcal{D}_{n,k}^{DW1}, f), \mathcal{LB}(\mathcal{D}_{n,k}^{KC1}, f), \mathcal{LB}(\mathcal{D}_{n,k}^{KC2}, f), \mathcal{LB}(\mathcal{D}_{n,k}^{KC^+}, f)\}.$$

The relations between the requirements $f_1, \ldots, f_k$ determine which lower bound is the smallest. See Fig. 4 for an illustration.

The proof of Theorem 7 relies on the results of Sect. 2 which we use to derive an upper bound on the sum of the large requirements. Eventually, this upper bound will contradict Eq. (5). The proof is presented in the full version of the paper.

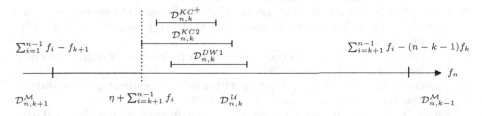

Fig. 4. The figure shows $\mathcal{D}_{n,k}^{\mathcal{U}}$ and a possible placement of the domains $\mathcal{D}_{n,k}^{KC2}, \mathcal{D}_{n,k}^{DW1}$ and $\mathcal{D}_{n,k}^{KC^+}$. Theorem 7 shows that the red interval cannot be realized.

Theorem 7. *Let* n, k *be odd integers such that* $n \geq 7$ *and* $3 \leq k \leq n - 4$. *A vector* $f \in \mathcal{D}_{n,k}^{\mathcal{U}}$ *cannot be realized if*

$$f_n < \eta(f) + \sum_{i=k+1}^{n-1} f_i. \tag{5}$$

6   Conclusion

Based on [5,7], we advanced the understanding of the WEIGHTED GRAPH REALIZATION problem. Our analysis concentrates on the domains $\mathcal{D}_{n,k}^{\mathcal{U}}$. We show that a graph which realizes a vector $f \in \mathcal{D}_{n,k}^{\mathcal{U}}$ must conform to a pattern, which lead us to discovering realizable subdomains and, qualitatively more important, enabled us to show that parts of the domains cannot be realized. Note that our positive results are based on constructive proofs. The full characterization for any odd n remains an open problem.

Additionally, [5] initiates studying vertex weighted realizations on specific graph classes, by giving full characterizations for paths and acyclic graphs.

References

1. Aigner, M., Triesch, E.: Realizability and uniqueness in graphs. Discr. Math. **136**, 3–20 (1994)
2. Althöfer, I.: On optimal realizations of finite metric spaces by graphs. Discr. Comput. Geometry **3**(2), 103–122 (1988). https://doi.org/10.1007/BF02187901
3. Bar-Noy, A., Böhnlein, T., Lotker, Z., Peleg, D., Rawitz, D.: The generalized microscopic image reconstruction problem. In: 30th ISAAC, pp. 1–15 (2019)
4. Bar-Noy, A., Böhnlein, T., Lotker, Z., Peleg, D., Rawitz, D.: Weighted microscopic image reconstruction. In: 47th SOFSEM (to appear) (2021)
5. Bar-Noy, A., Böhnlein, T., Peleg, D., Rawitz, D.: On vertex-weighted graph realizations. In: 12th Conference on Algorithms and Complexity (CIAC) (2021)
6. Bar-Noy, A., Choudhary, K., Peleg, D., Rawitz, D.: Realizability of graph specifications: characterizations and algorithms. In: 25th SIROCCO, pp. 3–13 (2018)
7. Bar-Noy, A., Peleg, D., Rawitz, D.: Vertex-weighted realizations of graphs. Theor. Comput. Sci. **807**, 56–72 (2020)
8. Blitzstein, J.K., Diaconis, P.: A sequential importance sampling algorithm for generating random graphs with prescribed degrees. Internet Math. **6**(4), 489–522 (2010)
9. Choudum, S.A.: A simple proof of the Erdös-Gallai theorem on graph sequences. Bull. Australian Math. Soc. **33**(1), 67–70 (1986)
10. Chung, F.R.K., Garrett, M.W., Graham, R.L., Shallcross, D.: Distance realization problems with applications to internet tomography. J. Comput. Syst. Sci. **63**(3), 432–448 (2001)
11. Cloteaux, B.: Fast sequential creation of random realizations of degree sequences. Internet Math. **12**(3), 205–219 (2016)
12. Culberson, J.C., Rudnicki, P.: A fast algorithm for constructing trees from distance matrices. Inf. Process. Lett. **30**(4), 215–220 (1989)
13. Erdös, D., Gemulla, R., Terzi, E.: Reconstructing graphs from neighborhood data. ACM Trans. Knowl. Discov. Data, **8**(4), 1–22 (2014)
14. Erdös, P., Gallai, T.: Graphs with prescribed degrees of vertices [hungarian]. Matematikai Lapok **11**, 264–274 (1960)
15. Feder, T., Meyerson, A., Motwani, R., O'Callaghan, L., Panigrahy, R.: Representing graph metrics with fewest edges. In: 20th STACS, pp. 355–366 (2003)
16. Frank, A.: Connectivity augmentation problems in network design. In: Mathematical Programming: State of the Art. University Michigan, pp. 34–63 (1994)
17. Frank, H., Chou, W.: Connectivity considerations in the design of survivable networks. IEEE Trans. Circuit Theory, CT-17, 486–490 (1970)
18. Hakimi, S.L.: On realizability of a set of integers as degrees of the vertices of a linear graph -I. SIAM J. Appl. Math. **10**(3), 496–506 (1962)
19. Hakimi, S.L., Yau, S.S.: Distance matrix of a graph and its realizability. Quart. Appl. Math. **22**, 305–317 (1965)
20. Havel, V.: A remark on the existence of finite graphs [in Czech]. Casopis Pest. Mat. **80**, 477–480 (1955)
21. Mihail, M., Vishnoi, N.: On generating graphs with prescribed degree sequences for complex network modeling applications. In: 3rd ARACNE (2002)
22. Nieminen, J.: Realizing the distance matrix of a graph. J. Inf. Process. Cybern. **12**(1/2), 29–31 (1976)

23. Rao, S.B.: A survey of the theory of potentially P-graphic and forcibly P-graphic degree sequences. In: Rao, Siddani Bhaskara (ed.) Combinatorics and Graph Theory. LNM, vol. 885, pp. 417–440. Springer, Heidelberg (1981). https://doi.org/10.1007/BFb0092288

24. Sierksma, G., Hoogeveen, H.: Seven criteria for integer sequences being graphic. J. Graph Theory **15**(2), 223–231 (1991)

25. Simões-Pereira, J.M.S.: An algorithm and its role in the study of optimal graph realizations of distance matrices. Discret. Math. **79**(3), 299–312 (1990)

26. Tatsuya, A., Nagamochi, H.: Comparison and enumeration of chemical graphs. Computat. Struct. Biotechnol. **5**, e201302004 (2013)

27. Tripathi, A., Tyagi, H.: A simple criterion on degree sequences of graphs. Discr. Appl. Math. **156**(18), 3513–3517 (2008)

28. Tripathi, A., Venugopalan, S., West, D.B.: A short constructive proof of the erdos-gallai characterization of graphic lists. Discr. Math. **310**(4), 843–844 (2010)

29. Tripathi, A., Vijay, S.: A note on a theorem of erdös & gallai. Discr. Math. **265**(1–3), 417–420 (2003)

30. Tyshkevich, R.I., Chernyak, A.A., Chernyak, Z.A.: Graphs and degree sequences: a survey. I. Cybernetics **23**, 734–745 (1987)

31. Tyshkevich, R.I., Chernyak, A.A., Chernyak, Z.A.: Graphs and degree sequences: a survey. II. Cybernetics **24**, 137–152 (1988)

32. Tyshkevich, R.I., Chernyak, A.A., Chernyak, Z.A.: Graphs and degree sequences: a survey. III. Cybernetics **24**, 539–548 (1988)

33. Varone, S.C.: A constructive algorithm for realizing a distance matrix. Eur. J. Oper. Res. **174**(1), 102–111 (2006)

Hypergraph Representation
via Axis-Aligned Point-Subspace Cover

Oksana Firman[(✉)][iD] and Joachim Spoerhase[iD]

Institut für Informatik, Universität Würzburg, Würzburg, Germany
{oksana.firman,joachim.spoerhase}@uni-wuerzburg.de

Abstract. We propose a new representation of k-partite, k-uniform hypergraphs (i.e. a hypergraph with a partition of vertices into k parts such that each hyperedge contains exactly one vertex of each type; we call them k-hypergraphs for short) by a finite set P of points in $\mathbb{R}^d$ and a parameter $\ell \leq d - 1$. Each point in P is covered by $k = \binom{d}{\ell}$ many axis-aligned affine ℓ-dimensional subspaces of $\mathbb{R}^d$, which we call ℓ-subspaces for brevity. We interpret each point in P as a hyperedge that contains each of the covering ℓ-subspaces as a vertex. The class of (d, ℓ)-*hypergraphs* is the class of k-hypergraphs that can be represented in this way, where $k = \binom{d}{\ell}$. The resulting classes of hypergraphs are fairly rich: Every k-hypergraph is a $(k, k-1)$-hypergraph. On the other hand, (d, ℓ)-hypergraphs form a proper subclass of the class of all $\binom{d}{\ell}$-hypergraphs for $\ell < d - 1$.

In this paper we give a natural structural characterization of (d, ℓ)-hypergraphs based on vertex cuts. This characterization leads to a polynomial-time recognition algorithm that decides for a given $\binom{d}{\ell}$-hypergraph whether or not it is a (d, ℓ)-hypergraph and that computes a representation if existing. We assume that the dimension d is constant and that the partitioning of the vertex set is prescribed.

1 Introduction

Motivation and Related Work. Geometric representations of graphs or hypergraphs is a wide and intensively studied field of research in particular in the geometric context. Well-known examples are geometric intersection or incidence graphs with a large body of literature [6,17,18]. The benefit of studying geometric graph representations is two-fold. On the one hand, knowing that a given graph can be represented geometrically may give new insights because the geometric perspective is often more intuitive. On the other hand, giving a graphical characterization for certain types of geometric objects may help pin down the essential combinatorial properties that can be exploited in the geometric setting.

One example of this interplay is the study of geometric set cover and hitting set problems [2,3,19]. In this important branch of geometric optimization, incidence relations of two types of geometric objects are studied where one object type is represented by vertices of a hypergraph whose hyperedges are, in turn, represented by the other object type. In this representation a vertex is contained

© Springer Nature Switzerland AG 2022
P. Mutzel et al. (Eds.): WALCOM 2022, LNCS 13174, pp. 328–339, 2022.
https://doi.org/10.1007/978-3-030-96731-4_27

in a hyperedge if and only if the corresponding geometric objects have a certain geometric relation such as containment or intersection. The objective is to find the minimum number of nodes hitting all hyperedges.[1] In this line of research, the goal is to exploit the geometry in order to improve upon the state of the art for general hypergraphs. This is known to be surprisingly challenging even in many seemingly elementary settings.

For example, in the well-studied point line cover problem [12,15] we are given a set of points in the plane and a set of lines. The goal is to identify a smallest subset of the lines to cover all the points. This problem can be cast as a hypergraph vertex cover problem. Points can be viewed as hyperedges containing the incident lines as vertices. The objective is to cover all the hyperedges by the smallest number of vertices.

It seems quite clear that point line cover instances form a heavily restricted subclass of general hypergraph vertex cover. For example, they have the natural intersection property that two lines can intersect in at most one point. However, somewhat surprisingly, in terms of approximation algorithms, no worst-case result improving the ratios for general hypergraph vertex cover [1,5,11] is known. In fact, it has been shown that merely exploiting the above intersection property in the hypergraph vertex cover is not sufficient to give improved approximations [16]. Giving a simple combinatorial characterization of the point line cover instances seems to be challenging.

In this paper, we study a representation of hypergraphs that arises from a natural variant of point line cover where we want to cover a given set P of points in $\mathbb{R}^d$ by *axis-parallel* lines, see Fig. 1 for an illustration. While the axis-parallel case of point line cover has been considered before [10] the known algorithms do not improve upon the general case of hypergraph vertex cover [1, 5]. More generally, we investigate the generalization where we are additionally given a parameter $\ell \leq d - 1$ and we would like to cover P by axis-aligned affine ℓ-dimensional subspaces of $\mathbb{R}^d$, which we call ℓ-subspaces. The resulting classes of hypergraphs is fairly rich as any k-partite k-uniform hypergraph (i.e. a hypergraph with a partition of vertices into k parts such that each hyperedge contains exactly one vertex of each type) can be represented by a set of points in $\mathbb{R}^k$ to be covered by $(k-1)$-subspaces. On the other hand for $\ell < d - 1$ we obtain proper subclasses of all k-hypergraphs.

We remark that our representation does not exploit the geometry of the Euclidean $\mathbb{R}^d$. Rather, the representation can also be considered on a hypercube X^d for some set X where subspaces are subsets of X^d fixing certain coordinates. We feel that the usage of the geometric language is more intuitive.

Related Work. The question of representing (hyper-)graphs geometrically is related to the area of *graph drawing*. For example, Evans et al. [8] study drawing hypergraphs in 3D by representing vertices as points and hyperedges as convex

---

[1] For the sake of presentation, we use here the representation as hitting set problem rather than the equivalent and maybe more common geometric set cover interpretation.

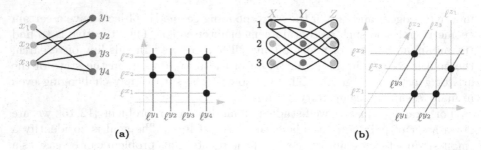

Fig. 1. A graph (a) and a hypergraph (b) and their representations in 2D and 3D, respectively.

polygons while preserving incidence relations. Note that in our work we study the "dual" where hyperedges are represented as points and vertices are represented as axis-parallel lines or affine subspaces. Another related problem in graph drawing has been introduced by Chaplick et al. [4]. They study drawings of graphs on the plane or in higher dimension so that the vertices (and the edges in some variants) of the graph can be covered by a minimum number of lines or hyperplanes.

Our Contribution and Outlook. To the best of our knowledge, we are the first to study the representation of k-hypergraphs via axis-aligned point subspace cover instances in this generality. Our main insight is that the axis-aligned case of point subspace cover allows for a natural, combinatorial characterization contrasting what is known for the non-aligned (see discussion above). The characterization is based on vertex cuts and can be leveraged to obtain a polynomial time recognition algorithm for such hypergraphs assuming the dimension d is a constant and that we are given the partition of the vertices (which is NP-hard to compute for $k \geq 3$ [14]).

We believe that it is an interesting research direction to exploit these combinatorial properties in order to obtain improved results for various optimization problems in hypergraphs such as hypergraph vertex cover or hypergraph matching. We also hope that our combinatorial characterization may help make progress on geometric problems. We conclude our paper by an outlook containing related open questions and some first small motivating results in this direction.

2 Point Line Cover and Hypergraph Representation

For the sake of an easier presentation, we first describe the result for the special case of point line covers, that is, for $(d, 1)$-hypergraphs. We later describe how the result generalizes to higher-dimensional axis-aligned affine subspaces. Generally, we use d, ℓ to denote the dimensions of the (sub-)space and $k = \binom{d}{\ell}$ to denote the number of parts in the corresponding hypergraph. For the special case of point line cover considered in this section we have $k = d$.

Let P be a finite set of points in $\mathbb{R}^k$. We define the k-hypergraph G_P as follows. The vertex set of G_P is the set of axis-parallel lines containing at least one point in P. The hyperedges in G_P correspond to the points in P where the hyperedge corresponding to some $p \in P$ contains the k axis-parallel lines incident on p as vertices. Note that G_P is k-partite and k-uniform (that is G_P is a k-hypergraph) where the k parts of the partition correspond to the k dimensions.

Our main task is to decide for a given hypergraph G whether there is a point line cover instance P such that G and G_P are isomorphic. We say that G is *represented* by P and, thus, *representable*. We assume that the partition of G into k parts is given.

More formally, we want to compute for a given k-hypergraph $G = (V = V_1 \cup V_2 \cup, \ldots, \cup V_k, E)$ a point line cover instance P such that each $e = (v_1, \ldots, v_k) \in E$ corresponds to some $p^e = (x_1^e, \ldots, x_k^e) \in P$ and where $v_i \in V_i$ corresponds to the line ℓ^{v_i} that is parallel to the i-th coordinate axis and contains p_e, that is, for all $j \neq i$, we fix the coordinates $x_j^e, j \neq i$ whereas the ith coordinate is free, see Fig. 1 for examples.

We remark that every bipartite graph is representable in $\mathbb{R}^2$ because we can derive a grid-like representation as shown in Fig. 1a directly from the adjacency matrix of this graph. However, for $k \geq 3$ the class of $(k, 1)$-hypergraphs forms a non-trivial subclass of all k-hypergraphs.

3 Characterization of Representable Hypergraphs

We use the notation $[k] = \{1, \ldots, k\}$ for $k \in \mathbb{N}$. Let $G = (V = V_1 \cup \cdots \cup V_k, E)$ be a k-hypergraph.

Definition 1. *Let $s, t \in V$. An s–t path is a sequence of vertices $s = v_1, \ldots, v_r = t$ such that v_i and v_{i+1} are both contained in some hyperedge $e \in E$ for all $i \in [r-1]$. Similarly, if $e, e' \in E$ then an e–e' path is a v–v' path such that $v \in e$ and $v' \in e'$.*

The following two separability conditions based on vertex cuts are key for our characterization.

Definition 2 (Vertex separability). *For a given k-hypergraph G we say that two distinct vertices v and v' from the same part V_i, $i \in [k]$ are separable if there exists some $j \in [k]$ with $j \neq i$ such that every v–v' path contains a vertex in V_j. (Informally, removing V_j from the vertex set and from the edges separates v and v'.) A hypergraph is called vertex-separable if every two vertices from the same part are separable.*

Definition 3 (Edge separability). *For a given k-hypergraph G we say that two distinct hyperedges e and e' are separable if there exists some $j \in [k]$ such that every e–e' path contains a vertex in V_j. A hypergraph is called edge-separable if every two hyperedges are separable.*

Note that any pair of hyperedges sharing two or more vertices are not separable. Therefore, edge-separable hypergraphs do not contain such hyperedge pairs.

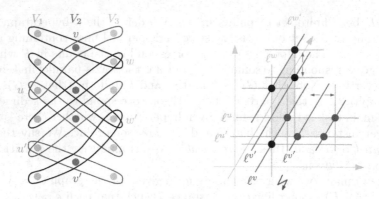

Fig. 2. A hypergraph G (on the left) that is edge-separable, but not vertex-separable (the vertices v and v' from V_2 are not separable). In a hypothetical representation of G (on the right), the line $\ell^{v'}$ must simultaneously intersect $\ell^{u'}$ and $\ell^{w'}$ and therefore must be equal to ℓ^v—a contradiction. (Color figure online)

Lemma 1. *For any given k-hypergraph, vertex separability implies edge separability.*

Proof. Assume that a given k-hypergraph G is not edge-separable. This means that there are two distinct hyperedges e and e' that are not separable. Then $\forall j \in [k]$ there is an e–e' path that does not contain a vertex from V_j. Because e and e' are distinct, there are distinct vertices v and v' with $v \in e$ and $v' \in e'$ from the same part V_i for some $i \in [k]$. Now, for each $j \in [k]$, $j \neq i$ there exists an e–e' path π_j that does not contain any vertex from V_j. But then v, π_j, v' forms a v–v' path not containing any vertex from V_j. This means that G is not vertex-separable. □

The converse is not true, see Fig. 2. In the instance depicted, the two red edges containing v,u and u', v' or the two black edges containing v, w and w', v', for example, are separated by removing the blue vertex part V_2 (which we can not do to show that vertices v and v' are separable).

Definition 4. *Let G be a k-hypergraph. For each $i \in [k]$ we construct a graph $G_i = (E, E_i)$ as follows: e and $e' \in E$ are adjacent if and only if e and e' have a common vertex in a part V_j with $j \neq i$.*

In the following theorem we state our characterization of k-hypergraphs representable by point line covers via vertex separability.

Theorem 1. *A k-hypergraph G is representable if and only if it is vertex-separable.*

Proof. We construct for each hyperedge e a point $p^e \in \mathbb{R}^k$ and for each vertex $v_i \in V_i$ with $i \in [k]$ a line $\ell^{v_i} \subseteq \mathbb{R}^k$ that is parallel to the x_i-axis. We do this as follows. For G we construct the graphs G_i, $i \in [k]$. For each graph G_i,

$i \in [k]$ we consider the connected components of G_i and assign to each connected component a unique (integer) value in $\{1, 2, \ldots, |V(G_i)|\}$.

Now, if p_i^e is the value of the connected component in G_i that contains e then we let the point $p^e = (p_1^e, \ldots, p_k^e)$ represent the hyperedge e, see Fig. 3 for an example.

Recall that any line parallel to the x_i-axis can be defined by fixing its x_j-coordinate for all $j \neq i$, while leaving x_i free. Now, if the hyperedge $e = \{v_1, \ldots, v_k\}$ is represented by $p^e = (p_1^e, \ldots, p_k^e)$ then for each $i \in [k]$, the line ℓ^{v_i} that represents the vertex v_i is defined by coordinates p_j^e, $j \neq i$ while leaving the x_i-coordinate free, see Fig. 3. It is important to note, that the representation ℓ^{v_i} is well-defined although v_i may be contained in multiple hyperedges in G. This follows from the fact that all the hyperedges containing v_i belong to the same connected component in G_j, $j \in [k]$, $j \neq i$ because each pair of them is joined by some edge in G_j corresponding to v_i and in particular these hyperedges form a clique. Therefore, there is no disagreement in the x_j-coordinate where $j \neq i$. Hence, we uniquely define the coordinates that determine a line.

($\Leftarrow$) Assume that G is vertex-separable. By the construction of the point line cover instance we have:

- every point p^e is in fact covered by the lines $\ell^{v_1}, \ldots, \ell^{v_k}$ where $e = \{v_1, \ldots, v_k\}$, because by construction every line ℓ^{v_i} and point p^e have the same x_j-coordinate with $j \neq i$.
- $\forall v \neq v' \in V$ it holds that $\ell^v \neq \ell^{v'}$. This is obviously true if vertices belong to different parts, because then the free coordinate of v is fixed for v' and vice versa. If $v, v' \in V_i$ for some $i \in [k]$ then, by vertex separability, there exists $j \neq i$ such that v and v' are not connected in graph G_j and get different x_j coordinates. So they represent distinct lines.
- $\forall e \neq e' \in E$ it holds that $p^e \neq p^{e'}$. Indeed, by Lemma 1, G is edge-separable and by definition of edge separability (see Def. 3) distinct hyperedges are not connected in at least one graph G_i and get different x_i-coordinates. So they represent distinct points.

By the above construction, for every incident vertex-hyperedge pair $v \in V, e \in E$, that is, $v \in e$, the corresponding geometric objects ℓ^v and p^e are incident as well. We claim that if v and e are not incident, that is, $v \notin e$ then ℓ^v and p^e are not incident as well. This is because every point p^e is already incident to precisely k lines ℓ^v by construction, because the lines ℓ^v are pairwise distinct, and because p^e cannot be incident on more than k axis-parallel lines. Thus we constructed a point line cover instance that represents the hypergraph G and this means that G is representable.

($\Rightarrow$) Assume that G is not vertex-separable but that it has a point line cover representation. This means that it contains at least two distinct vertices v and v' from the same part V_i that are not separable. Then for each part V_j with $j \neq i$, there exists a v–v' path $v = v_1, \ldots, v_r = v'$ such that $v_t \notin V_j$ for each $t \in [r]$. All lines ℓ^{v_t} with $t \in [r]$ that represent the vertices $v_1, \ldots, v_r$ lie on the same hyperplane H_j perpendicular to the x_j-axis. This is because successive line

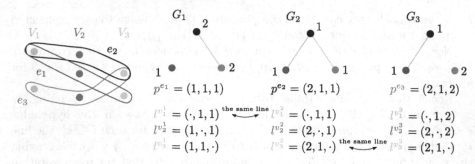

Fig. 3. The graphs G_1, G_2, G_3 and the coordinates of the points and lines corresponding to the hyperedges and vertices. The dots instead of coordinates mean that those coordinates are free.

pairs are joined by a common point (representing the hyperedge containing both) and since none of these lines is parallel to the x_j-axis and so the x_j-coordinate stays fixed. Since this holds for all $j \in [k]$, $j \neq i$, the lines ℓ^v and $\ell^{v'}$ lie in the intersection $\bigcap_{j \neq i} H_j$. But the intersection of such hyperplanes is a single line. This contradicts that v and v' correspond to the distinct lines.   □

Note that our characterization in Theorem 1 directly gives rise to an efficient recognition algorithm to check whether or not a given k-hypergraph is representable. We check for every vertex pair from the same part whether they are vertex-separable.

Below we give an algorithm to actually compute a representation for a given k-hypergraph (or output that a representation does not exist). The algorithm is an algorithmic implementation of the construction used in the proof Theorem 1 to recover a representation from a given (representable) k-hypergraph.

Computing a Representation. For each $i \in [k]$ we compute a graph G_i as in Definition 4. We use the construction from the proof of Theorem 1 to assign the coordinates for all points and lines (using connected components of G_i). In particular, we uniquely number the connected components of each graph G_i, $i \in [k]$. Then for any hyperedge e we construct the point p^e whose i-th coordinate is the number of the connected component containing e in G_i. Based on the representation of the hyperedges we can easily obtain the representation of the vertices as lines as well. To verify if the resulting candidate representation is in fact valid, we have to check if any two points $p^e, p^{e'}$ representing distinct hyperedges e, e' are in fact distinct, that is, differ in at least one coordinate. Similarly, we have to check distinctness of the lines. If this is the case then the hypergraph G is representable and we get the point line cover instance corresponding to G.

Constructing a point line cover instance for a given hypergraph can be done in $\mathcal{O}(k \cdot m^2)$, where m is the number of hyperedges. Checking distinctness of the vertex and edge representation takes $\mathcal{O}(k \cdot (m \log m + n \log n))$ using lexicographic sorting, where n is the number of vertices. In total the runtime of the algorithm is $\mathcal{O}(k \cdot (m^2 + n \log n))$.

4   Generalization to Point Subspace Cover

In this section, we generalize the previous result for $(d, 1)$-hypergraphs to the general case of (d, ℓ)-hypergraphs. That is, we want to characterize k-hypergraphs (where $k = \binom{d}{\ell}$) that are representable as sets of points in $\mathbb{R}^d$ covered by ℓ-*dimensional axis-aligned subspaces* where $1 \leq \ell \leq d - 1$ and $d \geq 2$. Notice that such an axis-aligned subspace is equal to the set of d-dimensional vectors for which $d - \ell$ coordinates are fixed and ℓ coordinates are *free*. Note that in the case $\ell = d - 1$ *every* k-hypergraph is representable analogously to the case of bipartite graphs each of which is representable in dimension two. This follows immediately from our characterization below but can also be seen in a direct way analogously to the two-dimensional setting.

For each $\ell \leq d - 1$ the class of (d, ℓ)-hypergraphs forms a subclass of k-hypergraphs (where $k = \binom{d}{\ell}$). Every part of the vertex partition in the hypergraph represents a set of ℓ-subspaces that have the same free coordinates. In what follows we assume that we know the *labeling* of the parts of the hypergraph. That is, we assume that for each part of the hypergraph we know to which ℓ-subset I of free coordinates this part corresponds; we write V_I to denote this correspondence. To this end, we guess the correct labeling in a brute force fashion by checking all $O(d^{\ell d^\ell})$ many labelings if it satisfies the characterization for fixed labeling as described below.

The below definitions and proofs are natural generalizations of the ones for $(d, 1)$-hypergraphs. In the case of $\ell = 1$ the idea of separability is related to cutting the part V_j of the graph corresponding to some coordinate $j \in [k]$. The key idea in the generalized setting is to cut instead all parts V_I that contain a specific coordinate j. Below, we give the full definitions and proofs, which are along the lines of $\ell = 1$ but rather use the generalized concept of cuts and separability.

Generalization of the Vertex Separability Property. Let $\binom{[d]}{\ell}$ denote the family of subsets of $[d]$ of cardinality ℓ. Note that there are exactly $\binom{d}{\ell}$ such sets. For a given $\binom{d}{\ell}$-hypergraph G and for all $I \in \binom{[d]}{\ell}$ we define the part V_I to be a set of subspaces, in which the coordinates with numbers from I are free. Now we generalize the vertex separability property using the definition of a path (Definition 1 from Sect. 3).

Definition 5 (Vertex separability). *For a given $\binom{d}{\ell}$-hypergraph G two distinct vertices v and v' from the same part V_I where $I \in \binom{[d]}{\ell}$ are separable if there exists a free coordinate $j \in [d]$, that is not free in V_I such that every v-v' path contains a vertex in one of the $\binom{d-1}{\ell-1}$ parts that share the free coordinate j. (Informally, removing all those $\binom{d-1}{\ell-1}$ parts from the vertex set and from the* *hyperedges separates v and v'.) A hypergraph is called* vertex-separable *if every two vertices from the same part are separable.*

In an analogous way we can generalize edge separability and Lemma 1 that vertex separability implies edge separability.

Definition 6 (Edge separability). *For a given $\binom{d}{\ell}$-hypergraph G two distinct hyperedges e and e' are* separable *if there exists a free coordinate $j \in [d]$ such that every e–e' path contains a vertex in one of the $\binom{d-1}{\ell-1}$ parts that share the free coordinate j. A hypergraph is called* edge-separable *if every two hyperedges are separable.*

Lemma 2. *For any given $\binom{d}{\ell}$-hypergraph where $\ell \leq d - 1$, vertex separability implies edge separability.*

Proof. Assume that a given $\binom{d}{\ell}$-hypergraph G is not edge-separable. This means that there are two distinct hyperedges e and e' that are not separable. Then $\forall j \in [d]$ there is an e–e' path that does not contain a vertex from any of the $\binom{d-1}{\ell-1}$ parts that share the free coordinate j. Because e and e' are distinct, there are distinct vertices v and v' with $v \in e$ and $v' \in e'$ from the same part V_I for some $I \in \binom{[d]}{\ell}$. Now, for each $j \in [d]$, that is not free in V_I, there exists an e–e' path π_j that does not contain a vertex from any of the $\binom{d-1}{\ell-1}$ parts that share the free coordinate j. But then v, π_j, v' forms a v–v' path not containing a vertex from any of the mentioned parts. This means that G is not vertex-separable.

As was already mentioned in the previous section, the converse is not true, see Fig. 2.

Definition 7. *For a $\binom{d}{\ell}$-hypergraph G we construct a graph $G_i = (E, E_i)$ for each $i \in [d]$ as follows: e and $e' \in E$ are adjacent iff e and e' have a common vertex in a part V_I where $I \in \binom{[d]}{\ell}$ and $i \notin I$.*

Theorem 2. *A $\binom{d}{\ell}$-hypergraph G is representable if and only if it is vertex-separable.*

Proof. The proof is a natural generalization of the proof for line case.

We construct for each hyperedge e a point $p^e \in \mathbb{R}^d$ and for each vertex $v_I \in V_I$ with $I \in \binom{[d]}{\ell}$ a subspace $\ell^{v_I} \in \mathbb{R}^d$ that is parallel to all x_i-axis where $i \in I$. We do this as follows: For G we construct the graphs G_i, $i \in [d]$. For each graph G_i we consider the connected components of the graph and assign to each of them a unique (integer) value.

Now, if p_i^e is the value of the connected component in G_i that contains e then we let the point $p^e = (p_1^e, \dots, p_d^e)$ represent the hyperedge e, see Fig. 3 for an example.

Recall that any subspace parallel to all x_i-axis where $i \in I$ can be defined by fixing its x_j-coordinate $j \neq i$, while leaving x_i coordinates free. Now, if the hyperedge $e = \{v_I\}$ for all $I \in \binom{[d]}{\ell}$ is represented by $p^e = (p_1^e, \dots, p_d^e)$ then for each $I \in \binom{[d]}{\ell}$, the subspace ℓ^{v_I} that represents the vertex v_I is defined by coordinates p_j^e, $j \notin I$ while leaving the x_i-coordinates free. It is important to note, that the representation ℓ^{v_I} is well-defined although v_I may be contained in multiple hyperedges in G. This follows from the fact that all the hyperedges containing v_I belong to the same connected component in G_j, $j \in [d]$ because each pair of them is joined by some edge in G_j corresponding to v_I and in particular these hyperedges form a clique. Therefore, there is no disagreement in the x_j-coordinate where $j \neq i$ for all $i \in I$. Hence, we uniquely define the coordinates that determine a subspace.

($\Leftarrow$) Assume that G is vertex-separable. By the construction of the point subspace cover instance we have:

1. every point p^e is in fact covered by the lines ℓ^{v_I} where $e = \{v_I\}$ for all $I \in \binom{[d]}{\ell}$. This is because by construction every line ℓ^{v_I} and point p^e have the same x_j-coordinate with $j \notin I$.
2. $\forall v \neq v' \in V$ it holds that $\ell^v \neq \ell^{v'}$. This is obviously true if vertices belong to different parts, because then there exist at least one free coordinate of v that is fixed for v' and vice versa. If $v, v' \in V_I$ for some $I \in \binom{[d]}{\ell}$ then, by vertex separability, there exists $j \notin I$ such that v and v' are not connected in graph G_j and get different x_j-coordinates. So they represent distinct subspaces.
3. $\forall e \neq e' \in E$ it holds that $p^e \neq p^{e'}$. Indeed, by Lemma 2, G is edge-separable and by the definition of edge separability (see Def. 6) distinct hyperedges are not connected in at least one graph G_i and get different x_i-coordinates. So they represent distinct points.

By the above construction, for every incident vertex-hyperedge pair $v \in V, e \in E$, that is, $v \in e$, the corresponding geometric objects ℓ^v and p^e are incident as well. We claim that if v and e are not incident, that is, $v \notin e$ then ℓ^v and p^e are not incident as well. This is because every point p^e is already incident to precisely $\binom{d}{\ell}$ subspaces ℓ^v by construction, because the subspaces ℓ^v are pairwise distinct, and because p^e cannot be incident on more than $\binom{d}{\ell}$ axis-parallel subspaces. Thus we constructed a point subspace cover instance that represents the hypergraph G and this means that G is representable.

($\Rightarrow$) Assume that G is not vertex-separable but that it has a point subspace cover representation. This means that it contains at least two distinct vertices v and v' from the same part V_I that are not separable. Then for each coordinate $j \notin I$, there exists a v-v' path such that none of the vertices on this path has j as a free coordinate. All subspaces that represent the vertices from the path lie in the same hyperplane H_j perpendicular to the x_j-axis, i.e. x_j-coordinate is fixed and all others are free. This is because successive subspace pairs are joined by a common point (representing the hyperedge containing both). Since none of these subspaces is parallel to the x_j-axis the x_j-coordinate stays fixed. Since this holds for all $j \in [d]$, such that $j \notin I$, the subspaces ℓ^v and $\ell^{v'}$ lie in the intersection $\bigcap_{j \notin I} H_j$. But the intersection of such hyperplanes is a single subspace that has all coordinates $i \in I$ free. This contradicts that v and v' correspond to the distinct subspaces. $\qquad\square$

Analogously to $(d,1)$-case the above proof leads directly to an algorithm computing a representation (or notifying about non-existence). A naive implementation of this algorithm gives a running time of $O(d^{\ell d^{\ell}+2}(m^2+n\log n))$ where the exponential factor in d comes from guessing the labeling of the parts. It is a very interesting question if the exponential dependence on d can be removed to get a polynomial algorithm also for non-constant d.

5   Conclusion and Outlook

There is a large body of literature in algorithms and graph theory on hypergraph problems. This motivates various future research directions, which we discuss in the full version [9].

1. Can the structure of (d, ℓ)-hypergraphs be leveraged in (optimization) problems for hypergraphs such as matching or vertex cover?
2. What is the relation to other classes of hypergraphs (for example geometrically representable hypergraphs)?

For example, maximum matching in k-hypergraphs is a very well-studied problem still exhibiting large gaps in our current understanding [7,13]. Another example is hypergraph vertex cover for which tight approximability results are known [11] on general k-hypergraphs. While the problem has been considered on $(k, 1)$-hypergraphs before [10] (from the geometric perspective of point-line covering) no improvement upon the general case is known. We hope that our structural characterization can help obtaining such improvements. In the full version [9], we state some first small results opening up these lines of research, namely we showed that matching on (3,1)-hypergraphs is NP-hard and considered the relation of $(d, 1)$-hypergraphs to point line cover representations on the plane where we drop the requirement of axis-alignment. We hope that our structural characterization helps make progress on these questions.

References

1. Aharoni, R., Holzman, R., Krivelevich, M.: On a theorem of lovász on covers in tau-partite hypergraphs. Combinatorica **16**(2), 149–174 (1996). https://doi.org/10.1007/BF01844843
2. Brönnimann, H., Goodrich, M.T.: Almost optimal set covers in finite VC-dimension. Discrete Comput. Geom. **14**(4), 463–479 (1995). https://doi.org/10.1007/BF02570718
3. Chan, T.M., Grant, E., Könemann, J., Sharpe, M.: Weighted capacitated, priority, and geometric set cover via improved quasi-uniform sampling. In: Rabani, Y. (ed.) Proceedings of Twenty-Third Annual ACM-SIAM Symposium on Discrete Algorithms (SODA'12), pp. 1576–1585. SIAM (2012). https://doi.org/10.1137/1.9781611973099.125
4. Chaplick, S., Fleszar, K., Lipp, F., Ravsky, A., Verbitsky, O., Wolff, A.: Drawing graphs on few lines and few planes. In: Hu, Y., Nöllenburg, M. (eds.) GD 2016. LNCS, vol. 9801, pp. 166–180. Springer, Cham (2016). https://doi.org/10.1007/978-3-319-50106-2_14

5. Chvátal, V.: A greedy heuristic for the set-covering problem. Math. Oper. Res. **4**(3), 233–235 (1979). https://doi.org/10.1287/moor.4.3.233
6. Coxeter, H.S.M.: Self-dual configurations and regular graphs. Bull. Amer. Math. Soc. **56**(5), 413–455 (1950)
7. Cygan, M.: Improved approximation for 3-dimensional matching via bounded pathwidth local search. In: Proceedings of 54th Annual IEEE Symposium on Foundations of Computer Science (FOCS'13), pp. 509–518. IEEE Computer Society (2013). https://doi.org/10.1109/FOCS.2013.61
8. Evans, W., Rzazewski, P., Saeedi, N., Shin, C.-S., Wolff, A.: Representing graphs and hypergraphs by touching polygons in 3D. In: Archambault, D., Tóth, C.D. (eds.) GD 2019. LNCS, vol. 11904, pp. 18–32. Springer, Cham (2019). https://doi.org/10.1007/978-3-030-35802-0_2
9. Firman, O., Spoerhase, J.: Hypergraph representation via axis-aligned point-subspace cover. Arxiv report. arXiv:2111.13555 (2021)
10. Gaur, D.R., Bhattacharya, B.: Covering points by axis parallel lines. In: Proceedings 23rd European Workshop on Computational Geometry (EuroCG'07), pp. 42–45. Citeseer (2007)
11. Guruswami, V., Saket, R.: On the inapproximability of vertex cover on k-partite k-uniform hypergraphs. In: Abramsky, S., Gavoille, C., Kirchner, C., Meyer auf der Heide, F., Spirakis, P.G. (eds.) ICALP 2010. LNCS, vol. 6198, pp. 360–371. Springer, Heidelberg (2010). https://doi.org/10.1007/978-3-642-14165-2_31
12. Hassin, R., Megiddo, N.: Approximation algorithms for hitting objects with straight lines. Discret. Appl. Math. **30**(1), 29–42 (1991). https://doi.org/10.1016/0166-218X(91)90011-K
13. Hazan, E., Safra, S., Schwartz, O.: On the complexity of approximating k-set packing. Comput. Complex. **15**(1), 20–39 (2006). https://doi.org/10.1007/s00037-006-0205-6
14. Ilie, L., Solis-Oba, R., Yu, S.: Reducing the size of NFAs by using equivalences and preorders. In: Apostolico, A., Crochemore, M., Park, K. (eds.) Combinatorial Pattern Matching, pp. 310–321. SV (2005). https://doi.org/10.1007/11496656_27
15. Kratsch, S., Philip, G., Ray, S.: Point line cover: the easy kernel is essentially tight. In: Chekuri, C. (ed.) Proceedings of 25th Annual ACM-SIAM Symposium on Discrete Algorithms (SODA'14), pp. 1596–1606. SIAM (2014). https://doi.org/10.1137/1.9781611973402.116
16. Kumar, V.S.A., Arya, S., Ramesh, H.: Hardness of set cover with intersection 1. In: Montanari, U., Rolim, J.D.P., Welzl, E. (eds.) ICALP 2000. LNCS, vol. 1853, pp. 624–635. Springer, Heidelberg (2000). https://doi.org/10.1007/3-540-45022-X_53
17. McKee, T.A., McMorris, F.R.: Topics in Intersection Graph Theory. Society for Industrial and Applied Mathematics (SIAM), Philadelphia (1999)
18. Pisanski, T., Randić, M.: Bridges between geometry and graph theory. In: Gorini, C.A. (ed.) Geometry at Work, MAA Notes 53, pp. 174–194. America (2000)
19. Varadarajan, K.R.: Weighted geometric set cover via quasi-uniform sampling. In: Schulman, L.J. (ed.) Proceedings of 42nd ACM Symposium on Theory of Computing (STOC'10), pp. 641–648. ACM (2010). https://doi.org/10.1145/1806689.1806777

Structural Parameterizations of Budgeted Graph Coloring

Susobhan Bandopadhyay[1(✉)], Suman Banerjee[3], Aritra Banik[1],
and Venkatesh Raman[2]

[1] National Institute of Science Education and Research, HBNI, Bhubaneswar, India
{susobhan.bandopadhyay,aritra}@niser.ac.in
[2] The Institute of Mathematical Sciences, HBNI, Chennai, India
vraman@imsc.res.in
[3] Indian Institute of Technology Jammu, Jammu, India
suman.banerjee@iitjammu.ac.in

Abstract. We introduce a variant of the graph coloring problem, which we denote as BUDGETED COLORING PROBLEM (BCP). Given a graph G, an integer c and an ordered list of integers $\{b_1, b_2, \ldots, b_c\}$, BCP asks whether there exists a proper coloring of G where the i-th color is used to color at most b_i many vertices. This problem generalizes two well-studied graph coloring problems, BOUNDED COLORING PROBLEM (BoCP) and EQUITABLE COLORING PROBLEM (ECP), and as in the case of other coloring problems, BCP is NP-hard even for constant values of c. So we study BCP under the paradigm of parameterized complexity, particularly with respect to (structural) parameters that specify how far (the deletion distance) the input graph is from a tractable graph class.

- We show that BCP is FPT (fixed-parameter tractable) parameterized by the vertex cover size. This generalizes a similar result for ECP and immediately extends to the BoCP, which was earlier not known.
- We show that BCP is polynomial time solvable for cluster graphs generalizing a similar result for ECP. However, we show that BCP is FPT, but unlikely to have polynomial kernel, when parameterized by the deletion distance to clique, contrasting the linear kernel for ECP for the same parameter.
- While the BoCP is known to be polynomial time solvable on split graphs, we show that BCP is NP-hard on split graphs. We also show that BCP is NP-hard on co-cluster graphs, contrasting the polynomial time algorithm for ECP and BoCP.

Finally we present an $\mathcal{O}^*(2^{|V(G)|})$ algorithm for the BCP, generalizing the known algorithm with a similar bound for the standard chromatic number.

Keywords: Graph Coloring · Exact Exponential algorithm · Parameterized complexity

© Springer Nature Switzerland AG 2022
P. Mutzel et al. (Eds.): WALCOM 2022, LNCS 13174, pp. 340–351, 2022.
https://doi.org/10.1007/978-3-030-96731-4_28

1   Introduction

A proper vertex coloring of a graph G is an assignment $\varphi : V(G) \rightarrow [c]$, of colors to its vertices such that, for any edge $(u, v) \in E(G)$, $\varphi(u) \neq \varphi(v)$, here $[c] = \{1, 2, 3, \cdots, c\}$. Minimum number of colors used by any proper coloring of G is the *chromatic number* of G and denoted by $\chi(G)$. For any proper coloring φ, we denote the set of vertices which gets color i by V_i, formally $V_i = \{v \in V(G) | \varphi(v) = i\}$.

We introduce a generalization of a well-studied variant called BOUNDED COLORING PROBLEM (BoCP) which asks for a proper coloring of G with each $|V_i|$ bounded by a given constant d. One motivation comes from the problem BIN PACKING PROBLEM WITH CONFLICTS (BPPC) [9]. Here we are given a set of n unit sized objects to be packed into at most c bins of size at most d each, except that some pairs of objects can not be placed in the same bin. This conflict information can be captured by a conflict graph and the problem is exactly an instance of BoCP. We consider a natural generalization where the bins have different, but given, sizes. The associated graph coloring problem which we call BUDGETED COLORING PROBLEM (BCP) is as follows.

BUDGETED COLORING PROBLEM (BCP)

Input: An undirected graph $G(V, E)$, an integer c and an ordered list $\mathcal{B} = \{b_1, b_2, \ldots, b_c\}$

Question: Does there exist a proper coloring $\varphi : V(G) \rightarrow [c]$ of G such that $|V_i| \leq b_i$ for all $1 \leq i \leq c$?

Given G and $\mathcal{B}$ we denote any proper coloring φ of G a PROPER BUDGETED COLORING if $|V_i| \leq b_i$ for all $1 \leq i \leq c$. BCP also generalizes another well-studied variant, the EQUITABLE COLORING PROBLEM (ECP). Given a graph G and an integer c, ECP asks whether G can be colored with c colors such that for each i, $|V_i| = \lfloor * \rfloor \frac{n}{c}$ or $|V_i| = \lceil * \rceil \frac{n}{c}$.

As coloring is hard even when c is 3 for general graphs [15], we can rule out any fixed parameter tractable (FPT) algorithm parameterized by the number of colors for BCP for general graphs. In this paper, we study the complexity of BCP for restricted graph classes and present FPT algorithms parameterized by some structural parameters.

Throughout the paper, we denote the number of colors by c. We follow the symbols and notations of graph theory and parameterized complexity theory as in the textbooks [11] and [10], respectively. A detailed description of notations used in the paper, definition of different graph classes, graph parameters and proof of lemmas marked with $\star$ can be found in the full arXiv version [2].

It follows from the definition of BoCP that for the graph classes for which BCP has a polynomial time or fixed-parameter tractable algorithms, BoCP also has a similar algorithm. On the other hand, if BoCP is NP-hard (W[i]-hard for a parameter) on a graph class then BCP is NP-hard (W[i]-hard for the same parameter) too. The following extension is not obvious though not difficult.

Lemma 1. [*1] *If there exists an algorithm to solve* BCP *for a graph class* $\mathcal{G}$ *with respect to some parameter* k *in time* $f(k, |V(G)|)$, *then we can solve* ECP *in time* $f(k, |V(G)|)$ *for any graph* $G \in \mathcal{G}$.

The following two corollaries are immediate from Lemma 1.

Corollary 1. *If for some graph class* $\mathcal{G}$ *(and a parameter* k), BCP *is polynomial time (FPT for the parameter* k) *then* BoCP *and* ECP *are also polynomial time (FPT for the same parameter respectively) for* $\mathcal{G}$.

Corollary 2. *If for some graph class* $\mathcal{G}$ *(and a parameter* k), BoCP *or* ECP *is NP-hard (W[i]-hard for the parameter* k), *then* BCP *is NP-hard (W[i]-hard for the same parameter respectively) for* $\mathcal{G}$.

It follows from Corollary 2 that BCP is NP-hard on interval graphs and co-graphs as ECP is NP-hard on these class of graphs [16]. Similarly as ECP is W[1]-hard on bounded treewidth graphs [12], BCP is also W[1]-hard parameterized by treewidth.

Table 1. Summary of Results in Different Graph Classes; Results marked with ⋆ are in this paper.

Graph Class	Chromatic Number	Equitable Coloring	Bounded Coloring	Budgeted Coloring
Bipartite	Polynomial Time	NP-hard even for three colors ⋆	NP-hard even for three colors [4]	NP-hard even for three colors
Cluster	Polynomial Time	Polynomial Time [16]	Polynomial Time	Polynomial Time⋆
Split	Polynomial Time	OPEN	Polynomial Time [6]	NP-hard ⋆
Co-Cluster	Polynomial Time	Polynomial Time	Polynomial Time	NP-hard ⋆

Our Results: In this paper, we first show NP-hard and polynomial time results for BCP in some graph classes. For the most part, we design fixed-parameter tractable algorithms parameterized by cluster vertex deletion set size (i.e. minimum number of vertices whose removal makes the graph a cluster graph – a collection of cliques) and vertex cover size. The results (including previously known results on ECP and BoCP, for contrast) are summarized in Table 1 and Table 2. We also give an $\mathcal{O}^*(2^{|V(G)|})^2$ exact algorithm for the problem generalizing the known algorithm with a similar bound for the standard chromatic number.

---

[1] Proofs of results marked ⋆ are in the full arXiv version [2].
[2] $\mathcal{O}^*$ hides the polynomial factor of the input size.

Table 2. Summary of Results in Parameterized Setting; Results marked ⋆ are in this paper.

Parameters	Chromatic Number	Equitable Coloring	Bounded Coloring	Budgeted Coloring
Cluster Vertex Deletion Number (CVD)	FPT	FPT [16]	OPEN	OPEN
CVD + Number of Colors	FPT	FPT [16]	FPT ⋆	FPT ⋆
CVD + Number of Clusters	FPT	FPT [16]	FPT ⋆	FPT ⋆
Vertex Cover	FPT [3]	FPT [13]	FPT ⋆	FPT ⋆, Polynomial kernel unlikely ⋆
Distance to Clique	FPT [17]	Linear Kernel [16]	FPT ⋆	FPT ⋆, Polynomial Kernel Unlikely ⋆

2 On Special Graphs

In this section we derive the complexity of BCP on cluster graphs (a collection of cliques) and co-cluster graphs (complete multipartite graphs), split graphs and bipartite graphs. The result on cluster graphs will be used in the next section when we generize to look at parameterization by cluster deletion set.

On Cluster and Co-Cluster Graphs. BCP is trivial on cliques as every vertex should get distinct color. Though BCP can be solved in polynomial time on cluster graphs by constructing a flow network [16], we provide a simpler and faster algorithm proving the following lemma.

Lemma 2. ⋆ *Let $\mathcal{I} = (G, c, \mathcal{B})$ be an instance of BCP where G is a cluster graph with a set $\{\mathcal{K}_1, \mathcal{K}_2, \cdots, \mathcal{K}_\ell\}$ of clusters sorted in non-increasing order of their sizes. If $\mathcal{I}$ is a YES instance, then there exists a proper coloring of G where the largest cluster $\mathcal{K}_1$ is colored with $|V(\mathcal{K}_1)|$ colors having the largest $|V(\mathcal{K}_1)|$ budgets.*

Lemma leads to a greedy algorithm (Algorithm 1 in the full arXiv version [2]) establishing the following theorem.

Theorem 1. BCP *on the class of Cluster Graphs can be solved in polynomial time.*

A co-cluster graph is a complement of a cluster graph. While ECP is known to be polynomial time solvable on co-cluster graphs, we are unable to extend our polynomial time algorithm for cluster graphs, and the following theorem explains why. We prove the following theorem by showing a reduction from 3-PARTITION PROBLEM.

Theorem 2. * BCP *on Co-Cluster Graphs is NP-hard.*

On Split Graphs. It is known that BoCP is polynomial-time solvable on the class of Split Graphs [7]. Surprisingly it turns out that BCP is NP-hard on Split Graphs. We give a reduction from the DOMINATING SET PROBLEM and prove the following theorem.

Theorem 3. *The* BCP *is NP-hard on Split Graphs.*

Proof. Let, $\mathcal{I} = (G, k)$ be an arbitrary instance of the DOMINATING SET PROBLEM, where $V(G) = \{w_1, w_2, \cdots, w_n\}$ and $k \leq n$. We construct a BCP instance $\mathcal{I}' = (G', n, \mathcal{B})$, where G' is a split graph as follows. $V(G') = C \cup I$, where $C = \{u_1, u_2, \ldots, u_n\}$ be the clique and $I = \{v_1, v_2, \ldots, v_n\}$ be the independent set. $E(G') = E_1 \cup E_2$, where $E_1 = \{(u_i, u_j) | \forall i \neq j\}$ and $E_2 = \{(u_i, v_j) | (w_i, w_j) \notin E(G)\}$. The number of colors c is n, with budgets $b_i = n + 1$ for $i \in [k]$ and the budget for the remaining colors is 1.

Next, we show that the graph G has a dominating set of size k if and only if there exists a proper budgeted coloring of G'. Assume that the graph G has a dominating set D of size k. Without loss of generality assume $D = \{w_1, w_2, \cdots, w_k\}$. Color u_i with color i where $i \in [k]$. As D is a dominating set in G, each vertex in I is adjacent to some vertex d of D in G, and hence is non-adjacent to that vertex in G', and so can be colored with the color of d. Thus we can color every vertex in I with the first k colors. Now, we have only $n - k$ uncolored vertices in C. Color them with the last $n - k$ colors. Hence we have a proper budgeted coloring of G'.

To prove the converse, note that in any proper budgeted coloring of G', all n colors must be used to color the vertices in C. Without loss of generality assume that u_i gets the color i, for $i = 1$ to n. Therefore the vertices in I are colored with the first k colors (as only they have budgets more than 1). Thus $\cup_{i \in [k]} \overline{N}(u_i) = I$. Hence $\{w_1, w_2, \ldots, w_k\}$ forms a dominating set in G. This completes the proof. □

On Bipartite Graphs. It was known that the bounded coloring problem is NP-hard on bipartite graphs when $c \geq 3$ [4]. We extend this result for ECP thus proving hardness for ECP as well as BCP (the later result follows from Corollary 2).

Theorem 4. * ECP *on Bipartite Graphs is NP-hard even when $c = 3$.*

3   Structural Parameterization of BCP

In this section, we address the parameterized complexity of BCP with respect to some structural parameters, parameters that measure the (deletion) distance to a tractable graph class. Formal definition of these parameters can be found in the full arXiv version [2].

3.1 BCP **Parameterized by Cluster Vertex Deletion Number (CVD)**

Recall from Sect. 2 that BCP is polynomial time solvable on cluster graphs. So, here we ask what if the input graph is k-vertices away from a cluster graph i.e., deleting k vertices from the graph makes the graphs a cluster graph? We show that BCP is FPT parameterized by k and an additional parameter which is the number of colors or the number of clusters in the resulting graph.

Consider any instance of $\mathcal{I} = (G, c, \mathcal{B})$ of BCP. Let S be the cluster vertex deletion set of G and $\mathcal{P} = \{P_1, P_2, \cdots, P_\ell\}$ be a partition of S. We remark that we do not need to assume that we are given the deletion set S, as we can find it in $\mathcal{O}^*(1.9102^k)$ time [5]. Let α be any proper coloring of S such that all vertices of P_i get the same color; i.e. for each part P_i and any two vertices $u, v \in P_i$, $\alpha(u) = \alpha(v)$. We define a new instance $\mathcal{I}' = (G, c, \mathcal{B}, \mathcal{P}, \alpha)$ of EXTENDED BUDGETED COLORING PROBLEM (EBCP) as follows. Given $\mathcal{I}'$, the EBCP asks whether there exists a proper budgeted coloring β of G such that $\forall v \in S$, $\beta(v) = \alpha(v)$. Without loss of generality assume that, in α for each i, P_i is colored with color i. Though the following lemma follows from Lemma 1 and Lemma 2 in [16], we give a slightly different proof in the full arXiv version [2] just for completeness.

Lemma 3. * *Given a partition $\mathcal{P}$ and its coloring α of cluster vertex deletion set S,* EBCP *instance $\mathcal{I}' = (G, c, \mathcal{B}, \mathcal{P}, \alpha)$ can be solved in polynomial time.*

Let Γ be the number of colorings of S. Observe that $|\Gamma| = c^k$. If $(G, c, \mathcal{B})$ is a YES instance of BCP then there must exist one coloring $\alpha \in \Gamma$ such that we have a solution to the EBCP with respect to α. Thus from Lemma 3 we have the following theorem.

Theorem 5. BCP *parameterized by cluster vertex deletion number and the number of colors can be solved in time $\mathcal{O}^*(c^k)$, where c is the number of colors and k is the cardinality of the cluster vertex deletion set.*

Next, we prove that BCP is FPT parameterization by CVD and the number of clusters in $G \setminus S$. Observe that the number of clusters can be significantly smaller than the number of colors, for example when $G \setminus S$ is a large clique.

Parameterization by CVD and the Number of Clusters. Let S be the cluster vertex deletion set and d be the number of clusters in $G - S$. Let $\mathcal{P} = \{P_1, P_2, \cdots, P_\ell\}$ be any partition of S into ℓ independent sets, and $D = \{d_1, d_2, \cdots, d_\ell\}$ be any ordered list of integers such that $0 \leq d_i \leq d$ for each $i \in [\ell]$. Without loss of generality assume $|P_1| \geq |P_2| \geq \cdots \geq |P_\ell|$. For any $\mathcal{P}$, D we define $\gamma_{\mathcal{P}D}$ as follows. $\gamma_{\mathcal{P}D}(P_1)$ be the least budgeted color with budget at least $|P_1| + d_1$. We inductively define $\gamma_{\mathcal{P}D}(P_i)$ to be the least budgeted color a with budget at least $|P_i| + d_i$ such that for all $j < i$, $\gamma_{\mathcal{P}D}(P_j) \neq a$. If there is no such a, we abandon $(\mathcal{P}, D)$. Observe that each color is assigned to at most one part in a partition. We modify $\mathcal{B}$ and create a new list of budgets $\mathcal{B}_{\mathcal{P}D} = \{b'_1, b'_2, \cdots, b'_\ell\}$ as follows. $b'_a = |P_i| + d_i$, if there exists a P_i

such that $\gamma_{\mathcal{P}D}(P_i) = a$ and $b'_a = b_a$ otherwise. Let $\mathcal{U}$ be the set of all pairs $(\mathcal{P}, D)$, where $\mathcal{P}$ be the all possible partition of S into independent sets and $D \in \{0, 1, 2, \cdots, d\}^{|\mathcal{P}|}$. Suppose we know the partition of S induced by a feasible coloring and for each part in the partition, we also know the number of vertices in $V \setminus S$ that get the same color. Given these two informations we can design a greedy algorithm to solve the problem in polynomial time using Lemma 3. Towards that we prove the following lemma.

Lemma 4. $\mathcal{I} = (G, c, \mathcal{B})$ *is a YES instance if and only if there exists at least one* $(\mathcal{P}^*, D^*) \in \mathcal{U}$ *such that* $\mathcal{I}_{\mathcal{P}^* D^*} = (G, c, \mathcal{B}_{\mathcal{P}^* D^*}, \mathcal{P}^*, \gamma_{\mathcal{P}^* D^*})$ *is a YES instance.*

Proof. Let us assume $\mathcal{I}$ is a YES instance of BCP i.e., there is a proper budgeted coloring β of G. Without loss of generality, assume that we have used the first ℓ colors to color the vertices of S. Recall, $V_i = \{v | \beta(v) = i\}$. We define $\mathcal{P}' = \{P'_1, P'_2, \cdots, P'_\ell\}$, where $P'_i = V_i \cap S$. Define $D' = \{d'_1, d'_2, d'_3, \cdots d'_\ell\}$ such that $d'_i = |V_i| - |P'_i|$ and $\mathcal{B}' = \{b'_1, b'_2, b'_3, \cdots, b'_c\}$, where $b'_i = |V_i|$, if $i \in [\ell]$ and $b'_i = b_i$, otherwise. Observe that, $(\mathcal{P}', D') \in \mathcal{U}$. As $\mathcal{I}$ is a YES instance, $\mathcal{I}' = (G, c, \mathcal{B}', \mathcal{P}', \beta)$ is also a YES instance of EBCP. Next we show that $\mathcal{I}_{\mathcal{P}'D'} = (G, c, \mathcal{B}', \mathcal{P}', \gamma_{\mathcal{P}'D'})$ is a YES instance.

Observe that if for all $1 \leq j \leq \ell$, $\gamma_{\mathcal{P}'D'}(P'_j) = \beta(P'_j)$ then nothing to prove. Let P'_j be the largest part in $\mathcal{P}'$ such that $\gamma_{\mathcal{P}'D'}(P_j) \neq \beta(P'_j)$. Let $\gamma_{\mathcal{P}'D'}(P'_j) = x$ and $\beta(P'_j) = y$. By construction $b_x \leq b_y$. We can exchange colors x and y to construct a new coloring β_1 from β which is also a feasible coloring for $\mathcal{I}'$. In β_1 one more part of S receives same color as $\gamma_{\mathcal{P}'D'}$. Thus applying the same step at most ℓ times we can create a feasible coloring β_ℓ such that for all $1 \leq j \leq \ell$, $\gamma_{\mathcal{P}'D'}(P'_j) = \beta_\ell(P'_j)$. Thus proving $\mathcal{I}_{\mathcal{P}'D'}$ is a YES instance.

For the converse, it is not hard to show that if $\mathcal{I}_{\mathcal{P},D}$ is a YES instance then $\mathcal{I}$ is also a YES instance.   □

As there can be k^k many partitions of S and d^k many choices of D, $|\mathcal{U}| = \mathcal{O}(d^k \cdot 2^{k \log k})$. For each pair of $(\mathcal{P}, D)$ using Lemma 3, we can solve the problem in polynomial time. Therefore we have the following theorem.

Theorem 6. BCP *parameterized by k, the cluster vertex deletion number and d, the number of clusters can be solved in time* $\mathcal{O}^*(d^k \cdot 2^{k \log k})$.

3.2   BCP Parameterized by the Distance to Clique

First, observe that when the number of clusters is one, the problem reduces to the BCP parameterized by the distance to a clique. Thus from Theorem 6 we get the following theorem as a corollary.

Theorem 7. BCP *parameterized by the distance to clique can be solved in time* $\mathcal{O}^*(2^{k \log k})$, *where k is the size of clique modulator.*

Next, we prove that there is no polynomial kernel for BCP parameterized by the distance to clique under standard complexity theoretic assumptions. We show a parameter preserving reduction from the clique problem (refer to the full arXiv version [2] for a formal definition) parameterized by vertex cover size. It is known that the clique problem parameterized by vertex cover does not admit a polynomial kernel unless $\mathsf{NP} \subseteq \mathsf{coNP/poly}$ [3].

Theorem 8. BCP *parameterized by the distance to clique does not admit a polynomial kernel unless* $\mathsf{NP} \subseteq \mathsf{coNP/poly}$.

Proof. Consider any instance of clique problem parameterized by vertex cover, $\mathcal{I} = (G, X, \ell)$. Here G is a graph, $X \subseteq V(G)$ is a vertex cover and we would like to find out whether there exists a clique of size ℓ in G. Parameter is $k = |X|$. We construct the following instance $\mathcal{I}' = (G^c, n - \ell + 1, \mathcal{B})$ of BCP as follows. Here G^c is the complement of G and we set the budgets as follows. First $n - \ell$ colors have budget one and the last color has budget ℓ.

Next, we prove that G has a clique of size ℓ if and only if $\mathcal{I}'$ admits a proper budgeted coloring. Assume that G has a clique of size ℓ. It is an independent set in G^c, and use the color with budget ℓ to color the independent set. There are $n - \ell$ colors with budget one and $n - \ell$ vertices left to color. Color them with all different colors. To prove the converse, assume $\mathcal{I}'$ is a YES instance. Observe that sum of the budgets of the colors is exactly n. So, the color with budget ℓ has been entirely used, which forms an independent set of size ℓ in G^c and hence a clique in G.

Observe that as X is a vertex cover in G, $V(G) \setminus X$ is an independent set in G. Thus $V(G) \setminus X$ is a clique in G^c and X is a clique modulator in G^c. This completes the proof.   □

3.3   Budgeted Graph Coloring Parameterized by the Vertex Cover Size

In this section, we study the BCP on the class of graphs that are k vertices away from an independent set, i.e. on graphs that have a k-sized vertex cover. Observe that if a graph has a k-sized vertex cover, then it also has a k-sized CVD because a vertex cover is also a CVD. Thus the FPT results of Sect. 3.1 follow for vertex cover size as well. However, in this section, we give a stronger result by showing that BCP is FPT parameterized just by vertex cover, independent of any other parameter.

Let, $(G, c, \mathcal{B})$ be an instance of the BCP where the graph G has a vertex cover S of size k. So, $I = V(G) \setminus S$ is an independent set. Note that we do not need to assume that we are given S, as we can find S in $\mathcal{O}^*(1.2738^k)$ time [8].

Our algorithm differs from the algorithm of Sect. 3.1 (Proof of Theorem 5) in only the first step of coloring the vertices of S. Instead of trying all possible colorings to color S in the first step, we apply a greedy method.

Let $\mathcal{P} = \{P_1, P_2, P_3, \cdots, P_\ell\}$ be the partition of S, where $\ell \leq k$, and each part P_i is independent. For each part P_i, we define the set of feasible colors for P_i by F_i where $F_i = \{j \in [c] | b_j \geq |P_i|\}$. Let L_i be the set of least budgeted ℓ colors in F_i; if $|F_i| \leq \ell$, we set $L_i = F_i$.

Next we show that if there exist a proper budgeted coloring for G then there exist a proper budgeted coloring where each P_i gets one of the colors from L_i. We denote a coloring φ, a RESTRICTED PROPER BUDGETED COLORING with respect to $\mathcal{P}$ if for all $P_i \in \mathcal{P}$ any two vertices $u, v \in P_i$, $\varphi(u) = \varphi(v)$.

Lemma 5. *If there exists a restricted proper budgeted coloring of G with respect to $\mathcal{P}$, then there exists a restricted proper budgeted coloring where the vertices in each partition P_i is colored with one of the colors in L_i.*

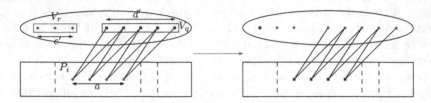

Fig. 1. Illustration of Lemma 5, color q and r are denoted by ■ and ● respectively (Color figure online)

Proof. Let φ^* be the coloring that maximizes the number of partitions P_i that are colored with one of the colors in L_i. Observe that if for a partition $|L_i| \leq \ell$ then $L_i = F_i$ thus must be colored with one of the colors in L_i in any restricted budgeted proper coloring. Let $P_i \in \mathcal{P}$ be the partition that is colored with a color $r \notin L_i$. As L_i contains at least ℓ colors there exist a color say $q \in L_i$ which is not used to color any of the partitions in $\mathcal{P}$. Observe that $b_q \leq b_r$ (see Fig. 1). Let $|P_i| = a$, $b_q = a + d$ and $b_r = a + e$ where $0 \leq d \leq e$. Next, we construct a coloring φ' from φ^* as follows. If $d = e$ then exchange the color r and q to construct φ' contradicting the maximality of φ^*. Thus assume that $d < e$. Let V_r be the set of vertices in $V \setminus S$ which are colored with color r. Observe that there are no edge between P_i and V_r. Let $|V_r| = e' \leq e$. Let V_q be the set of vertices in $V \setminus S$ which are colored with color q. Suppose $|V_q| = d' \leq a + d$.

We construct φ' from φ^* as follows. We color the vertices of P_i with the color q and the vertices of V_q with r (observe that $b_q < b_r$). We color V_r with rest of the colors of q and r. In order to prove that there are sufficient colors left to color V_r let us observe the following. $|V_r| = e' \leq e \leq e + a + d - d'$ (as $d' \leq a + d$). Thus $|V_r| \leq d + (e + a - d') = (b_q - a) + (b_r - d')$ We color the rest of the vertices with the same color as of φ^*. Observe that φ' is a restricted proper budgeted coloring and contradicts the maximality of φ^* and the claim holds.   □

For each part P_i in the partition, we first find a set of the least budgeted ℓ colors each of which is of size at least $|P_i|$ (if there aren't ℓ different colors satisfying the budget constraint, we pick all those that satisfy; if there aren't any color satisfying the budget constraint, we abandon this partition and move on). Then we try all possible colorings of S coloring each P_i with any of the colors we have found for P_i making sure that no pairs of P_i's get the same color. As soon as we fix the coloring of each P_i, we are left with the independent set vertices. Since each vertex in the independent set is itself a singleton cluster, we have to solve the EBCP. By Lemma 3, it is known that the problem can be solved in polynomial time.

As the number of partitions $\mathcal{P}$ of S is $\mathcal{O}(k^k)$, and as the number of colorings tried for each partition is $\mathcal{O}^*(k^k)$, we have the following theorem.

Theorem 9. BCP *parameterized by vertex cover can be solved in* $\mathcal{O}^*(2^{2k\log k})$ *time.*

It is known that ECP does not admit a polynomial kernel when parameterized by vertex cover and the number of colors unless NP $\subseteq$ coNP/poly [16]. Hence the same result is true for ECP when parameterized by vertex cover alone. As BCP is a generalization of ECP we have the following corollary.

Corollary 3. BCP *does not admit a polynomial kernel when parameterized by vertex cover unless* NP $\subseteq$ coNP/poly.

4   Exact Algorithm for Budgeted Coloring

In this section, we present an exact exponential time algorithm for the BCP based on dynamic programming. Let $G[X]$ denote the subgraph induced by X, for $X \subseteq V(G)$. A $\mathcal{O}^*(3^n 2^c)$ algorithm is easy by computing a table T of size $2^n \times 2^c$ whose entries are as follows. For a subset X of vertices, and a subset C of colors, the entry $T[X, C]$ will contain 1, if $G[X]$ can be colored with the colors in C along with their respective budgets and 0 otherwise. Now the recurrence relation for this problem is as follows: $T[S, C] = \bigvee\limits_{I \in G[S], c \in C} T[S \setminus I, C \setminus \{c\}]$ where I is an independent set in $G[S]$ of size at most b_c. Trivially, when both S and C are $\emptyset$, $T[S, C] = 0$. Our algorithm runs on all possible subsets S of $V(G)$, and for every subset S, it runs on all its independent sets, and all possible colors. For every S, we have table entries for all possible subset C of colors, whose number is bounded by 2^c. Hence, the total running time of this algorithm can be easily seen to be $\sum_{i=1}^{n} \binom{n}{i} 2^i \cdot 2^c \cdot c$ which is $= \mathcal{O}(3^n \cdot 2^c \cdot c)$.

In what follows we improve the runtime to $\mathcal{O}^*(2^n)$ using the principle of inclusion-exclusion, essentially generalizing the known algorithm for the proper c-coloring problem [14] to show the following.

Theorem 10. *Budgeted c-Coloring problem can be solved in* $\mathcal{O}^*(c2^n)$ *time.*

Proof. We give an algorithm with the claimed bound for the more general BUD-GETED SET COVER problem defined as follows.

Input: A ground set U of size n, and a family $\mathcal{F}$ of m subsets of U, and a set $b_1, b_2, \ldots, b_c$ of integers for $c \geq 1$.

Question: Are there subsets $F_1, F_2, \ldots, F_c$ in $\mathcal{F}$ such that $|F_i| \leq b_i$ and $\bigcup_{i=1}^{c} F_i = U$?

In this problem, we can even assume that the family $\mathcal{F}$ is given implicitly in the sense, given a subset S of U, one can test in time polynomial in $|S|$ whether or not S is in $\mathcal{F}$. We call a subfamily that witnesses a solution to our problem as a budgeted c-cover of $\mathcal{F}$.

Now by treating the family $\mathcal{F}$ as the family of independent sets of the input graph of the budgeted c-coloring, the claim in the theorem follows. To design an algorithm for BUDGETED SET COVER using inclusion-exclusion, we define our object of interest. The object of interest is simply a subfamily $\mathcal{F}' = \{F_1, F_2, \ldots, F_c\}$ of $\mathcal{F}$ such that $|F_i| \leq b_i$; observe that in our definition for $i \neq j$ F_i and F_j can be the same. The subfamily $\mathcal{F}'$ satisfies property $P(u)$ if $u \in \bigcup_{F \in \mathcal{F}'} F$.

Now it follows that the number of budgeted c-covers of $\mathcal{F}$ is simply the number of objects that satisfy $P(u)$ for every $u \in U$ (as our objects are ordered, the same budgeted c-cover will be counted a fixed number ($c!$) times, and hence the final number has to be divided by $c!$ which we ignore hereafter).

For a subset $W \subseteq U$, let $F(W, b_i)$ be the family of sets in $\mathcal{F}$ that avoids W (i.e. doesn't have any element of W) and are of size at most b_i, and let $f(W, b_i)$ be the number of such sets. Then the number of objects that do not satisfy property $P(u)$ for any element u of W is simply $\Pi_{i=1}^{c} f(W, b_i)$. This is simply because there are $f(W, b_i)$ choices for the i-th set of our object.

Hence, by the principle of inclusion-exclusion (see Theorem 4.7 of [14]), we have that the number of budgeted c-covers of $\mathcal{F}$ is given by

$$\sum_{W \subseteq U} (-1)^{|W|} \Pi_{i=1}^{c} f(W, b_i) \tag{1}$$

For a fixed W and b_i, $f(W, b_i)$ is the number of subsets of $U \setminus W$ of size at most b_i that are in $\mathcal{F}$ and hence can be computed in $\binom{n-|W|}{b_i}|b_i|^{\mathcal{O}(1)}$ which can result in an $\mathcal{O}^*(3^n)$ time for all W. In the following claim, we show that we can pre-compute and store $f(W, b_i)$ for each W and b_i using dynamic programming and compute them all in $\mathcal{O}^*(2^n)$ time.

Claim. * The quantity $f(W, b_i)$, for all W and $b_i, i = 1$ to c can be computed in $\mathcal{O}^*(2^n)$ time.

The theorem follows from the Eq. 1 and the Claim.   □

5   Conclusions and Open Problems

In this paper, we have introduced BCP and obtained hardness results, polynomial time algorithms, FPT algorithms and kernelization results. There are a number of open problems.

- What is the complexity of the BCP on trees when $c \geq 3$? The related ECP and BoCP are polynomial time solvable on trees [1,7].
- In Sect. 3.1, we showed that BCP when parameterized by the distance to a cluster graph and the number of colors or the number of clusters, is FPT. The parameterized complexity of the problem parameterized just by the distance to a cluster graph is an interesting open problem.

References

1. Baker, B.S., Coffman, E.G., Jr.: Mutual exclusion scheduling. Theoret. Comput. Sci. **162**(2), 225–243 (1996)
2. Bandopadhyay, S., Banerjee, S., Banik, A., Raman, V.: Structural parameterizations of budgeted graph coloring. arXiv preprint arXiv:2110.14498 (2021)
3. Bodlaender, H.L., Jansen, B.M.P., Kratsch, S.: Kernelization lower bounds by cross-composition. SIAM J. Discret. Math. **28**(1), 277–305 (2014)
4. Bodlaender, H.L., Jansen, K.: On the complexity of scheduling incompatible jobs with unit-times. In: Borzyszkowski, A.M., Sokołowski, S. (eds.) MFCS 1993. LNCS, vol. 711, pp. 291–300. Springer, Heidelberg (1993). https://doi.org/10.1007/3-540-57182-5_21
5. Boral, A., Cygan, M., Kociumaka, T., Pilipczuk, M.: A fast branching algorithm for cluster vertex deletion. Theory Comput. Syst. **58**(2), 357–376 (2016)
6. Chen, B.-L., Ko, M.-T., Lih, K.-W.: Equitable and m-bounded coloring of split graphs. In: Deza, M., Euler, R., Manoussakis, I. (eds.) CCS 1995. LNCS, vol. 1120, pp. 1–5. Springer, Heidelberg (1996). https://doi.org/10.1007/3-540-61576-8_67
7. Chen, B.-L., Lih, K.-W.: Equitable coloring of trees. J. Comb. Theory Ser. B **61**(1), 83–87 (1994)
8. Chen, J., Kanj, I., Xia, G.: Simplicity is beauty: improved upper bounds for vertex cover. Theor. Comput. Sci. **411**, 01 (2010)
9. Coffman, E.G., Csirik, J., Leung, J.: Variants of classical one-dimensional bin packing, pp. 33-1–33-14. CRC Press, January 2007
10. Cygan, M., et al.: Parameterized Algorithms. Springer, Cham (2015). https://doi.org/10.1007/978-3-319-21275-3
11. Diestel, R.: Graduate texts in mathematics (2012)
12. Fellows, M.R., et al.: On the complexity of some colorful problems parameterized by treewidth. Inf. Comput. **209**(2), 143–153 (2011)
13. Fiala, J., Golovach, P.A., Kratochvíl, J.: Parameterized complexity of coloring problems: treewidth versus vertex cover. Theoret. Comput. Sci. **412**(23), 2513–2523 (2011)
14. Fomin, F.V., Kratsch, D.: Exact Exponential Algorithms. TTCSAES, Springer, Heidelberg (2010). https://doi.org/10.1007/978-3-642-16533-7
15. Garey, M.R., Johnson, D.S.: Computers and Intractability; A Guide to the Theory of NP-Completeness. W.H. Freeman & Co., New York (1990)
16. Gomes, G., Guedes, M.R., dos Santos, V.F.: Structural parameterizations for equitable coloring. arXiv preprint arXiv:1911.03297 (2019)
17. Sæther, S., Telle, J.: Between treewidth and clique-width. Algorithmica **75**, 04 (2014)

Counting and Sampling Orientations
on Chordal Graphs

Ivona Bezáková[(✉)] and Wenbo Sun[(✉)]

Rochester Institute of Technology, Rochester, NY 14623, USA
ib@cs.rit.edu, ws3109@rit.edu

Abstract. We study counting problems for several types of orientations of chordal graphs: source-sink-free orientations, sink-free orientations, acyclic orientations, and bipolar orientations, and, for the latter two, we also present linear-time uniform samplers. Counting sink-free, acyclic, or bipolar orientations are known to be #P-complete for general graphs, motivating our study on a restricted, yet well-studied, graph class. Our main focus is source-sink-free orientations, a natural restricted version of sink-free orientations related to strong orientations, which we introduce in this work. These orientations are intriguing, since despite their similarity, currently known FPRAS and sampling techniques (such as Markov chains or sink-popping) that apply to sink-free orientations do not seem to apply to source-sink-free orientations. We present fast polynomial-time algorithms counting these orientations on chordal graphs. Our approach combines dynamic programming with inclusion-exclusion (going two levels deep for source-sink-free orientations and one level for sink-free orientations) throughout the computation. Dynamic programming counting algorithms can be typically used to produce a uniformly random sample. However, due to the negative terms of the inclusion-exclusion, the typical approach to obtain a polynomial-time sampling algorithm does not apply in our case. Obtaining such an almost uniform sampling algorithm for source-sink-free orientations in chordal graphs remains an open problem.

Little is known about counting or sampling of acyclic or bipolar orientations, even on restricted graph classes. We design efficient (linear-time) exact uniform sampling algorithms for these orientations on chordal graphs. These algorithms are a byproduct of our counting algorithms, but unlike in other works that provide dynamic-programming-based samplers, we produce a random orientation without computing the corresponding count, which leads to a faster running time than the counting algorithm (since it avoids manipulation of large integers).

1 Introduction

An orientation of an undirected graph is an assignment of a direction to each edge, converting the original graph to a directed graph. We initiate the study of counting *source-sink-free* orientations, where there are no sources, nor sinks (that

---

I. Bezáková—Partially supported by NSF awards 1819546 and 1821459.

P. Mutzel et al. (Eds.): WALCOM 2022, LNCS 13174, pp. 352–364, 2022.
https://doi.org/10.1007/978-3-030-96731-4_29

is, no vertices of indegree or outdegree 0). Our motivation is twofold: First, these orientations are related to the well-studied *sink-free* orientations, which can be counted (approximately) despite being #P-hard for general graphs. While clearly similar, source-sink-free orientations exhibit certain properties that prevent the application of the current techniques for counting or sampling of sink-free orientations. Therefore, new techniques are needed to understand this problem, and we are starting with a restricted but well-known graph class. Second, source-sink-free orientations can be thought of as a local (soft) version of strong orientations, another well-studied class of orientations, the counting of which is also #P-hard on general graphs [15] as well as on restricted graph classes such as planar and bipartite graphs [13]. Our study is a first step beyond sink-free and towards strong orientations.

Chordal graphs have attracted great attention in computer science theory as a natural graph class with real-world applications (for example, some inference techniques in probabilistic graphical models rely on sampling and counting of certain types of orientations on chordal graphs [9,16]), on which some problems that are NP-hard or #P-hard on general graphs can be solved in polynomial time; see, for example, [11].

Sink-free orientations are well understood. Bubley and Dyer [5] proved that counting these orientations is #P-complete on general graphs. They also provided a Markov Chain that samples sink-free orientations of an arbitrary input graph G approximately from the uniform distribution in time $O(|E(G)|^3 \log \epsilon^{-1})$, where ϵ is the degree of approximation. Additionally, they showed that the problem of counting sink-free orientations is self-reducible, yielding a fully polynomial randomized approximation scheme (FPRAS) for the counting problem, the running time of which is roughly $|E(G)|$ times the sampling running time. Huber [8] used the "coupling from the past" technique to obtain an exact sample in time $O(|E(G)|^4)$. Cohn, Pemantle, and Propp [6] proposed a "sink-popping" algorithm which can generate a sink-free orientation uniformly at random in $O(|V(G)||E(G)|)$ time. This algorithm fits the "partial rejection sampling through the Lovász Local Lemma" framework of Guo, Jerrum and Liu [7], yielding a uniformly random sink-free orientation in time $O(|V(G)|^2)$ time. Interestingly, none of these techniques appear to apply to the problems of counting and sampling of source-sink-free orientations.

Our main contribution is a polynomial (cubic in the worst case) exact counting algorithm for source-sink-free orientations in chordal graphs, combining dynamic programming with two-level inclusion-exclusion at every step of the dynamic programming computation. However, our combination with inclusion-exclusion prevents us from extending our algorithm to an exact uniform sampler, and we leave the sampling question as the main open problem of our work. We apply a similar approach, using one-level inclusion-exclusion, to count sink-free orientations of a given chordal graph in almost linear time, significantly improving the running time over the FPRAS for this graph class. Besides these two orientations, we also present almost linear time counting and linear time sampling algorithms for acyclic orientations and bipolar orientations, which are both

#P-complete on general graphs [10]. The problem of counting acyclic orienta-
tions has also attracted a lot of attention since it corresponds to a specific input
of the well-studied Tutte polynomial $T_G(x,y)$ [4], in particular to $T_G(2,0)$ [14]),
which plays an important role in graph theory and statistical physics. Acyclic
orientations can be counted efficiently on chordal graphs via the calculation of
the chromatic polynomial [1]. However, this result does not yield a(n almost)
uniform sampler since acyclic orientations are not known to be self-reducible.
In this work we present a simple linear-time exact uniform sampling algorithm
for acyclic orientations in chordal graphs. An interesting aspect of this sampling
algorithm is that it runs faster than its counting counterpart. This is atypical—
dynamic programming based samplers usually rely on a precomputation of the
corresponding counts and are efficient only after substantial preprocessing time.
We note that, with some extra work to maintain the desired (unique) source s and
sink t, our results extend to counting and sampling of bipolar (s,t)-orientations,
also known as st-numberings. Finally, we compare our work to the recent celebrated results of Wienöbst, Bannach, and Liśkiewicz [16], who count and sample another type of orientations, the so-called v-structure-free acyclic (or moral
acyclic) orientations in chordal graphs. The authors prove interesting structural
results for these orientations and employ dynamic programming over the clique
tree in order to count them. In their case, the dynamic programming consists
of additive quantities, which allows them to extend their counting approach to
sampling. In contrast, in our work we either do not need to compute the counts
in order to sample, or our dynamic programming does not appear to extend to
sampling due to the presence of negative terms in the computation.

The paper is organized as follows. Section 2 contains preliminaries on
chordal graphs and clique trees. Our main result, a fast counting algorithm for
source-sink-free orientations, combining dynamic programming with inclusion-
exclusion, is in Sect. 3. We summarize our other results in Sect. 4.

2   Preliminaries

For a graph G, we denote by $G[U]$ the graph induced in G on the vertex set
$U \subseteq V(G)$. An undirected graph is *chordal* if for every cycle of more than three
vertices there exists an edge, called a chord, not on this cycle connecting two
vertices on the cycle. Every chordal graph G can be represented by a tree T_G
where $V(T_G)$ is the set of maximal cliques of G, and the tree satisfies the *induced
subtree property*: For every vertex $v \in V(G)$, the induced subgraph $T_G[A_v]$ is
connected, where A_v is the set of maximal cliques of G containing v. Such a tree
T_G is called a *clique tree* of G, see, for example, [12]. Let T_{G,C_r} be the clique tree
T_G rooted at a maximal clique C_r. If G is clear from the context, we will simply
write T_{C_r}, or simply T if C_r is also clear. We denote by $T_{C_r,C}$ the subtree of T_{C_r}
containing C and its descendants; we write T_C if C_r is clear from the context.

Each clique C in T_{C_r} can be partitioned into a *separator set* $\mathrm{Sep}(C) =
C \cap \mathrm{Parent}(C)$ and a *residual set* $\mathrm{Res}(C) = C \setminus \mathrm{Sep}(C)$, where $\mathrm{Parent}(C)$ is
the parent clique of C in T_{C_r} (if $C = C_r$, then $\mathrm{Parent}(C) = \emptyset$). The following
properties hold, see, for example, [2,12].

- For each vertex v in G, there is a unique clique C_v that contains v in its residual set. This implies that $|V(T_G)| \leq |V(G)|$ and that C_v is the root of $T_{C_r}[A_v]$; we denote this rooted subtree by T_{C_v}. All other cliques in T_{C_v} that contain v have it in their separator set.
- For a clique C let $D(C)$ be the set of vertices in the descendant cliques of C in T_{C_r}, excluding the vertices in $\mathrm{Sep}(C)$, i.e., $D(C) := C' \in V(T_C)C' - \mathrm{Sep}(C)$. Let $A(C)$ be the vertices in the cliques not in T_C, excluding the vertices in $\mathrm{Sep}(C)$, i.e., $A(C) := \bigcup_{C' \in V(T_{C_r}) - V(T_C)} C' - \mathrm{Sep}(C)$. The separator $\mathrm{Sep}(C)$ separates $A(C)$ and $D(C)$ in G: there is no edge with one endpoint in $A(C)$ and the other endpoint in $D(C)$.
- Construction of a clique tree for a connected chordal graph can be done in time $O(|E(G)|)$.

We use $G[T_C]$ for the subgraph induced by the vertices that belong to cliques in T_C, i.e., $G[T_C] := G[\bigcup_{C' \in V(T_C)} C']$. We will often work with the following subgraph of $G[T_C]$: Let $\hat{G}[T_C]$ be $G[T_C]$ with the edges within the separator set $\mathrm{Sep}(C)$ removed, i.e., $\hat{G}[T_C] := G[T_C] - E(G[\mathrm{Sep}(C)])$.

The following lemma will be essential for our calculations.

Lemma 1. *Let C be a clique in the rooted clique tree T_{C_r} and let $C_1, C_2, \ldots, C_d$ be its child cliques. The edge sets of the graphs $\hat{G}[T_{C_i}]$, $i = 1, \ldots, d$, are mutually disjoint.*

Proof. By contradiction, suppose that there are $i \neq j \in \{1, \ldots, d\}$ such that $\hat{G}[T_{C_i}]$ and $\hat{G}[T_{C_j}]$ share an edge $e = (u, v)$. Since $\mathrm{Sep}(C_i)$ is a separator in G, separating vertices in $V(G[T_{C_i}]) - \mathrm{Sep}(C_i)$ from $V(G) - V(G[T_{C_i}])$, and since $V(G[T_{C_j}]) \subseteq V(G) - V(G[T_{C_i}])$, it follows that u and v must be in $\mathrm{Sep}(C_i)$. But then e is not in $\hat{G}[T_{C_i}]$, a contradiction. $\qquad\square$

In order to make the running times of our algorithms more readable, we assume that each arithmetic operation takes a constant time. This is, of course, a bit optimistic, since the ultimate number of orientations can be as high as 2^m for a graph with m edges, and, therefore, the true running time of each arithmetic operation adds a factor of about $m\,\mathrm{polylog}(m)$. We use $\tilde{O}()$ notation to indicate that this factor is omitted from our running time estimate.

Our sampling algorithms produce orientations uniformly at random: Each orientation is chosen with equal probability from the set of all desired orientations. We use $[d]$ to denote $\{1, 2, \ldots, d\}$.

3   Counting Source-Sink-Free Orientations

In this section we describe the main contribution of this paper. We show how to count source-sink-free orientations in chordal graphs using dynamic programming on the clique tree. While this approach is quite standard for algorithms on chordal graphs, the novel aspect of our work is to employ a two-level inclusion-exclusion principle as a subroutine of the dynamic programming. We prove the following theorem:

Theorem 1. *Let G be a chordal graph. The number of source-sink-free orientations of G can be computed in time $\tilde{O}(|C_{\max}||E(G)|) = \tilde{O}(|E(G)||V(G)|)$, where $C_{\max}$ is a maximum clique of G.*

Proof. We define the following quantities for each clique C in a rooted clique tree T of the given chordal graph G:

- SSFO(T_C): The number of orientations of the graph $\hat{G}[T_C]$ where every sink and every source is in Sep(C). Let $\mathcal{S}(T_C)$ be the set of all these orientations.
- SoO(T_C, v_1): The number of orientations in $\mathcal{S}(T_C)$, where $v_1 \in$ Sep(C) is a source.
- SiO(T_C, v_2): The number of orientations in $\mathcal{S}(T_C)$, where $v_2 \in$ Sep(C) is a sink.
- SoSiO(T_C, v_1, v_2): The number of orientations in $\mathcal{S}(T_C)$, where $v_1 \in$ Sep(C) is a source and $v_2 \in$ Sep(C) is a sink. Notice that since Res(C) $\neq \emptyset$, it follows that $v_1 \neq v_2$.

We will compute the quantities SSFO(T_C), SoO(T_C, v_1), SiO(T_C, v_2), and SoSiO(T_C, v_1, v_2) by dynamic programming on the rooted clique tree T_{C_r}. The quantity SSFO(T_{C_r}) represents the number of source-sink-free orientations of G.

To simplify our expressions, for a clique C we define quantities oa(C), os(C), ox(C), oss(C), and oxx(C) as follows. Let oa(C) be the number of all orientations of C, i.e., oa(C) $= 2^{\binom{|C|}{2}}$. Let os(C, v) be the number of orientations of C where the vertex $v \in C$ is a sink. All edges have to be oriented towards v, hence os(C, v) $= 2^{\binom{|C|-1}{2}}$. It also follows that v is the only sink in C. Moreover, since the quantity os(C, v) does not depend on the vertex v, we simplify the notation to just os(C). Let ox(C) be the number of orientations of C where no vertex is a sink. Since each orientation with a sink has a unique sink, we get that ox(C) $=$ oa(C) $- |C|$ os(C). Let oss(C, v_1, v_2) be the number of orientations of C where v_1 is a source and v_2 is a sink. It follows that oss(C, v_1, v_2) $= 2^{\binom{|C|-2}{2}}$. Since the value does not depend on v_1, v_2, we simplify the notation to just oss(C). Let oxx(C) be the number of orientations of C with no sources or sinks. An oriented clique can have at most one source and at most one sink. Therefore, from all orientations we can subtract those that have a sink and those that have a source; leading to "double penalization" of orientations with both a source and a sink. Therefore, oxx(C) $=$ oa(C) $- 2|C|$ os(C) $+ \binom{|C|}{2}$ oss(C).

Base case of the computation of SSFO(T_C), SoO(T_C, v_1), SiO(T_C, v_2), **and** SoSiO(T_C, v_1, v_2): Let C be a leaf of T. In SoO(T_C, v_1) and SoSiO(T_C, v_1, v_2) all edges incident to v_1 point away from v_1, and in SiO(T_C, v_2) and SoSiO(T_C, v_1, v_2) the edges incident to v_2 need to point towards v_2; the other edges can be oriented either way. We get:

$$\text{SoO}(T_C, v_1) = \text{SiO}(T_C, v_2) = \text{oa}(\text{Res}(C))2^{|\text{Res}(C)|(|\text{Sep}(C)|-1)},$$

$$\text{SoSiO}(T_C, v_1, v_2) = \text{oa}(\text{Res}(C))2^{|\text{Res}(C)|(|\text{Sep}(C)|-2)}.$$

For SSFO(T_C), we partition $\mathcal{S}(T_C)$ into these four mutually exclusive cases.

▶ The orientation restricted to $G[\text{Res}(C)]$ contains no sources or sinks. Then, the edges between $\text{Res}(C)$ and $\text{Sep}(C)$ can be oriented arbitrarily, leading to $\text{oxx}(\text{Res}(C))2^{|\text{Res}(C)||\text{Sep}(C)|}$ of such orientations.

▶ The orientation restricted to $G[\text{Res}(C)]$ contains a (single) source $u_1 \in \text{Res}(C)$ and no sinks. Then, at least one of the edges from $\text{Sep}(C)$ needs to be oriented towards u_1 to prevent it from remaining a source, and the other edges between $\text{Sep}(C)$ and $\text{Res}(C) - \{u_1\}$ can be oriented arbitrarily. The part of the orientation within $\text{Res}(C)$ has to have all edges outgoing from u_1, and the remaining edges must be oriented so that there is no sink within $\text{Res}(C) - \{u_1\}$. This corresponds to $\text{ox}(\text{Res}(C) - \{u_1\})(2^{|\text{Sep}(C)|} - 1)2^{|\text{Sep}(C)|(|\text{Res}(C)|-1)}$ of such orientations.

▶ The orientation restricted to $G[\text{Res}(C)]$ contains a (single) sink $u_2 \in \text{Res}(C)$ and no sources. The calculation is analogous to the previous case.

▶ The orientation restricted to $G[\text{Res}(C)]$ contains a (single) source u_1 and a (single) sink u_2. Then, u_1 needs to be "fixed" by at least one edge from $\text{Sep}(C)$, u_2 by at least one edge to $\text{Sep}(C)$, and the other edges between $\text{Sep}(C)$ and $\text{Res}(C)$ can be oriented arbitrarily. Likewise, the edges within $\text{Res}(C) - \{u_1, u_2\}$ can be oriented arbitrarily. We get $\text{oa}(\text{Res}(C) - \{u_1, u_2\})(2^{|\text{Sep}(C)|} - 1)^2 2^{|\text{Sep}(C)|(|\text{Res}(C)|-2)}$ of such orientations.

Therefore, summing across possible $u_1, u_2 \in \text{Res}(C)$, we get

$$\text{SSFO}(T_C) = \text{oxx}(\text{Res}(C))2^{|\text{Res}(C)||\text{Sep}(C)|} +$$
$$2|\text{Res}(C)|\,\text{ox}(\text{Res}(C)^{-1})(2^{|\text{Sep}(C)|} - 1)2^{|\text{Sep}(C)|(|\text{Res}(C)|-1)} +$$
$$\binom{|\text{Res}(C)|}{2}\text{oa}(\text{Res}(C)^{-2})(2^{|\text{Sep}(C)|} - 1)^2 2^{|\text{Sep}(C)|(|\text{Res}(C)|-2)},$$

where for a clique $\hat{C}$, the notation $\hat{C}^{-k}$ stands for removing k vertices from $\hat{C}$.

Inductive Case. Let C be a non-leaf of T, and let $C_1, C_2, \ldots, C_d$ be its child cliques in T. For $u \in \text{Res}(C)$ we denote by I_u the set of indices corresponding to the child cliques containing u, i.e., $I_u := \{i \in [d] \mid u \in C_i\}$.

To compute $\text{SoSiO}(T_C, v_1, v_2)$, the edges between v_1, respectively v_2, and $\text{Res}(C)$ are forced (away from v_1, towards v_2). This implies that no vertex in $\text{Res}(C)$ will be a source, or a sink, and hence the orientation of all other edges can be arbitrary. We get:

$$\text{SoSiO}(T_C, v_1, v_2) = \text{oa}(\text{Res}(C))2^{|\text{Sep}(C)|(|\text{Res}(C)|-2)} \prod_{i=1}^{d} \text{SSFO}(T_{C_i}).$$

For $\text{SoO}(T_C, v_1)$, the edges between v_1 and $\text{Res}(C)$ need to point away from v_1, and as such there will be no sources in $\text{Res}(C)$. We distinguish two cases:

▶ There is no sink in the orientation restricted to $G[\text{Res}(C)]$. There are $\alpha_1 :=$ $\text{ox}(\text{Res}(C))2^{|\text{Sep}(C)|(|\text{Res}(C)|-1)} \prod_{i=1}^{d} \text{SSFO}(T_{C_i})$ such orientations of $\hat{G}[T_C]$.

▶ The orientation restricted to $G[\text{Res}(C)]$ contains a (single) sink u_2. Then either there is an edge between u_2 and $\text{Sep}(C)$ pointing towards u_2, or all edges point away from u_2 and u_2 cannot be a sink at least one of the child subtrees. The number of orientations of $\hat{G}[T_C]$ corresponding to this case is

$$\alpha_2(u_2) := \mathrm{os}(\mathrm{Res}(C), u_2)(2^{|\,\mathrm{Sep}(C)|} - 1)2^{|\,\mathrm{Sep}(C)|(|\,\mathrm{Res}(C)|-1)} \prod_{i=1}^{d} \mathrm{SSFO}(T_{C_i}) +$$

$$\mathrm{os}(\mathrm{Res}(C), u_2)2^{|\,\mathrm{Sep}(C)|(|\,\mathrm{Res}(C)|-1)} \prod_{i\in[d]-I_{u_2}} \mathrm{SSFO}(T_{C_i}) \times$$

$$\left(\prod_{i\in I_{u_2}} \mathrm{SSFO}(T_{C_i}) - \prod_{i\in I_{u_2}} \mathrm{SiO}(T_{C_i}, u_2) \right).$$

Putting the two cases together, we get $\mathrm{SoO}(T_C, v_1) = \alpha_1 + \sum_{u_2 \in \mathrm{Res}(C)} \alpha_2(u_2)$. Note that $\mathrm{SiO}(T_C, v_1)$ can be computed analogously.

It remains to compute $\mathrm{SSFO}(T_C)$. We will split the possible orientations into these four mutually exclusive cases:

▶ The orientation restricted to $G[\mathrm{Res}(C)]$ contains no sources or sinks. Then, all the remaining edges within $\mathrm{Res}(C)$ can be oriented arbitrarily, and the child subtrees can be oriented recursively (provided, as always, that there are no sinks or sources outside their separator sets). Therefore, the number of these orientations is $\beta_1 := \mathrm{ox}(\mathrm{Res}(C))2^{|\,\mathrm{Sep}(C)||\,\mathrm{Res}(C)|} \prod_{i=1}^{d} \mathrm{SSFO}(T_{C_i})$.

▶ The orientation restricted to $G[\mathrm{Res}(C)]$ contains a (single) source u_1 and no sinks. Then, either there is an edge oriented from $\mathrm{Sep}(C)$ to u_1, or all edges are oriented from u_1 to $\mathrm{Sep}(C)$ and one of the child subtrees does not have u_1 as their source. Within $\mathrm{Res}(C)$, all edges point away from u_1 and the remainder of $\mathrm{Res}(C)$ needs to be sink-free. Thus, the number of orientations of $\hat{G}[T_C]$ corresponding to this case is:

$$\beta_2(u_1) := \mathrm{ox}(\mathrm{Res}(C)^{-1})(2^{|\,\mathrm{Sep}(C)|} - 1)2^{|\,\mathrm{Sep}(C)|(|\,\mathrm{Res}(C)|-1)} \prod_{i=1}^{d} \mathrm{SSFO}(T_{C_i}) +$$

$$\mathrm{ox}(\mathrm{Res}(C)^{-1})2^{|\,\mathrm{Sep}(C)|(|\,\mathrm{Res}(C)|-1)} \prod_{i\in[d]-I_{u_1}} \mathrm{SSFO}(T_{C_i}) \times$$

$$\left(\prod_{i\in I_{u_1}} \mathrm{SSFO}(T_{C_i}) - \prod_{i\in I_{u_1}} \mathrm{SoO}(T_{C_i}, u_1) \right).$$

▶ The orientation restricted to $G[\mathrm{Res}(C)]$ contains a (single) sink u_2 and no sources. The number of the corresponding orientations of $\hat{G}[T_C]$ can be computed analogously to the previous case; we refer to this quantity as $\beta_3(u_2)$.

▶ The orientation restricted to $G[\mathrm{Res}(C)]$ contains a (single) source u_1 and a (single) sink u_2. We will partition the corresponding orientations of $\hat{G}[T_C]$ into these subcases:

a) There is an edge from $\text{Sep}(C)$ to u_1 and from u_2 to $\text{Sep}(C)$ (i.e., both u_1 and u_2 are "fixed" by an edge from/to $\text{Sep}(C)$). The number of corresponding orientations is

$$\beta_{4a} := \text{oss}(\text{Res}(C))(2^{|\text{Sep}(C)|} - 1)^2 2^{|\text{Sep}(C)|(|\text{Res}(C)|-2)} \prod_{i=1}^{d} \text{SSFO}(T_{C_i}).$$

b) There is an edge from $\text{Sep}(C)$ to u_1 but no edge from u_2 to $\text{Sep}(C)$ (i.e., u_1 is "fixed" by $\text{Sep}(C)$ but u_2 is not). Then u_2 needs to be "fixed" by one of the child subtrees. The number of corresponding orientations is

$$\beta_{4b}(u_2) := \text{oss}(\text{Res}(C))(2^{|\text{Sep}(C)|} - 1)2^{|\text{Sep}(C)|(|\text{Res}(C)|-2)} \times$$

$$\prod_{i \in [d]-I_{u_2}} \text{SSFO}(T_{C_i}) \left(\prod_{i \in I_{u_2}} \text{SSFO}(T_{C_i}) - \prod_{i \in I_{u_2}} \text{SiO}(T_{C_i}, u_2) \right).$$

c) There is an edge from u_2 to $\text{Sep}(C)$ but no edge from $\text{Sep}(C)$ to u_1 (i.e., u_2 is "fixed" by $\text{Sep}(C)$ but u_1 is not). The number of corresponding orientations can be computed analogously to the previous subcase; we refer to this quantity as $\beta_{4c}(u_1)$.

d) There is no edge from $\text{Sep}(C)$ to u_1 and no edge from u_2 to $\text{Sep}(C)$ (i.e., neither u_1 nor u_2 is "fixed" by $\text{Sep}(C)$). Then both u_1 and u_2 need to be "fixed" by one of the child subtrees. Let X_{u_1} be the set of valid orientations of the subtrees where no subtree fixes u_1. (We call an orientation of the subtrees valid if sinks and sources are present only in the residual sets in the root cliques of each tree.) Then,

$$|X_{u_1}| = \prod_{i \in [d]-I_{u_1}} \text{SSFO}(T_{C_i}) \prod_{i \in I_{u_1}} \text{SoO}(T_{C_i}, u_1).$$

Let Y_{u_1} be the set of valid orientations of the subtrees where no subtree fixes u_2. Then,

$$|Y_{u_2}| = \prod_{i \in [d]-I_{u_2}} \text{SSFO}(T_{C_i}) \prod_{i \in I_{u_2}} \text{SiO}(T_{C_i}, u_2).$$

Let $Z_{u_1,u_2} = X_{u_1} \cap Y_{u_2}$. In particular, Z_{u_1,u_2} is the set of all valid orientations of the subtrees where no subtree fixes u_1 or u_2. In other words, the subtrees containing u_1 but not u_2 have u_1 as a source, the subtrees containing u_2 but not u_1 have u_2 as a sink, and the subtrees containing both u_1 and u_2 have u_1 as a source and u_2 as a sink. Therefore,

$$|Z_{u_1,u_2}| = \prod_{i \in [d]-I_{u_1}-I_{u_2}} \text{SSFO}(T_{C_i}) \prod_{i \in I_{u_1}-I_{u_2}} \text{SoO}(T_{C_i}, u_1) \times$$

$$\prod_{i \in I_{u_2}-I_{u_1}} \text{SiO}(T_{C_i}, u_2) \prod_{i \in I_{u_1} \cap I_{u_2}} \text{SoSiO}(T_{C_i}, u_1, u_2).$$

Then, when accounting for all orientations in case d, we use inclusion-exclusion as follows: We consider all valid orientations of the subtrees, then subtract those in X_{u_1} and Y_{u_2}, and then add those in Z_{u_1,u_2} to compensate for the double subtraction. Therefore, the number of orientations of $\hat{G}[T_C]$ corresponding to case d is

$$\beta_{4d}(u_1, u_2) := \text{oss}(\text{Res}(C)) 2^{|\operatorname{Sep}(C)|(|\operatorname{Res}(C)|-2)} \times$$

$$\left(\prod_{i \in [d]} \text{SSFO}(T_{C_i}) - |X_{u_1}| - |Y_{u_2}| + |Z_{u_1,u_2}| \right).$$

Putting it all together we get

$$\text{SSFO}(T_C) = \beta_1 + \sum_{u_1 \in \text{Res}(C)} \beta_2(u_1) + \sum_{u_2 \in \text{Res}(C)} \beta_3(u_2) +$$

$$\sum_{u_1, u_2 \in \text{Res}(C), u_1 \neq u_2} [\beta_{4a} + \beta_{4b}(u_2) + \beta_{4c}(u_1) + \beta_{4d}(u_1, u_2)].$$

Finally, we need to estimate the running time of the algorithm. After constructing the clique tree T (which is of size $O(|V(G)|)$), the algorithm performs a tree traversal of T. In the base case it performs $O(1)$ arithmetic operations per leaf clique.[1] To analyze the running time in the inductive case, we first pretend that we have access to the quantities $\text{SSFO}(T_{C_i})$. Notice that we do not need to store the quantities $\text{SoSiO}(T_C, v_1, v_2)$ for each v_1, v_2, since the computation is independent of v_1, v_2 and therefore we really need only one quantity $\text{SoSiO}(T_C)$ for each clique. The computation of $\text{SoSiO}(T_C)$ takes $O(d)$ arithmetic operations, assuming $\text{SSFO}(T_{C_i})$'s have been computed. Since $d = \text{outdeg}_T(C)$, the computation of all SoSiO's across the entire tree T takes $O(\sum_{C \in T} \text{outdeg}_T(C)) = O(|T|) = O(|V(G)|)$ arithmetic operations.

Next we consider the computations of $\text{SoO}(T_C, v_1)$; notice that the computations are independent of v_1. All α_1 quantities can be computed in $O(|V(G)|$ time following the same reasoning as for SoSiO. The same holds for the first term of the quantities $\alpha_2(u_2)$. We need to estimate the running time needed to compute the second (additive) term of the $\alpha_2(u_2)$'s. Notice that the size of the set I_{u_2} corresponding to the child cliques of C that contain u_2 is upper-bounded by $\deg_G(u_2)$. This is because for each child clique C_i for $i \in I_{u_2}$ we have a $w_i \in \text{Res}(C_i)$. All the w_i's are distinct since $\text{Sep}(C)$ separates them, yielding $|I_{u_2}| \leq \deg_G(u_2)$. We can rewrite the computation of the second term of $\alpha_2(u_2)$ as a product of $\text{os}(\text{Res}(C), u_2) 2^{|\operatorname{Sep}(C)|(|\operatorname{Res}(C)|-1)} \prod_{i \in [d]} \text{SSFO}(T_{C_i})$ and $(1 - \prod_{i \in I_{u_2}} \frac{\text{SiO}(T_{C_i}, u_2)}{\text{SSFO}(T_{C_i})})$. Computing the last term of this product takes $O(\sum_{u \in \text{Res}(C)} |I_u|)$ operations. Since $|I_u| \leq \deg_G(u)$, across the entire tree T we get $O(\sum_{C \in T} \sum_{u_2 \in \text{Res}(C)} \deg_G(u_2)) = O(\sum_{u_2 \in V(G)} \deg_G(u_2)) = O(|E(G)|)$ operations to compute all $\text{SoO}(T_C)$'s (and the same holds for the $\text{SiO}(T_C)$'s).

---

[1] Computation of the factorial of a k-bit number takes $O(k \operatorname{polylog} k)$, see [3], which will be subsumed by our $\tilde{O}()$ notation.

It remains to bound the running time needed to compute all the SSFO's. Using the same arguments as before we get that the computation of all β_1, $\beta_2(u_1)$, $\beta_3(u_2)$, β_{4a}, $\beta_{4b}(u_2)$, and $\beta_{4c}(u_1)$ takes $O(|E(G)|)$ arithmetic operations. The tricky part is to account for the computation of the $\beta_{4d}(u_1, u_2)$'s, due to the 2-level inclusion-exclusion depth. In particular, we want to efficiently compute

$$\sum_{u_1,u_2\in\mathrm{Res}(C),u_1\neq u_2} [\prod_{i\in[d]} \mathrm{SSFO}(T_{C_i}) - \prod_{i\in[d]-I_{u_1}} \mathrm{SSFO}(T_{C_i}) \prod_{i\in I_{u_1}} \mathrm{SoO}(T_{C_i}, u_1) -$$

$$\prod_{i\in[d]-I_{u_2}} \mathrm{SSFO}(T_{C_i}) \prod_{i\in I_{u_2}} \mathrm{SiO}(T_{C_i}, u_2) + \prod_{i\in[d]-I_{u_1}-I_{u_2}} \mathrm{SSFO}(T_{C_i})\times$$

$$\prod_{i\in I_{u_1}-I_{u_2}} \mathrm{SoO}(T_{C_i}, u_1) \prod_{i\in I_{u_2}-I_{u_1}} \mathrm{SiO}(T_{C_i}, u_2) \prod_{i\in I_{u_1}\cap I_{u_2}} \mathrm{SoSiO}(T_{C_i}, u_1, u_2)] =$$

$$\prod_{i\in[d]} \mathrm{SSFO}(T_{C_i}) \sum_{u_1,u_2\in\mathrm{Res}(C),u_1\neq u_2} [(1 - \prod_{i\in I_{u_1}} \frac{\mathrm{SoO}(T_{C_i}, u_1)}{\mathrm{SSFO}(T_{C_i})} - \prod_{i\in I_{u_2}} \frac{\mathrm{SiO}(T_{C_i}, u_2)}{\mathrm{SSFO}(T_{C_i})} +$$

$$\prod_{i\in I_{u_1}-I_{u_2}} \frac{\mathrm{SoO}(T_{C_i}, u_1)}{\mathrm{SSFO}(T_{C_i})} \prod_{i\in I_{u_2}-I_{u_1}} \frac{\mathrm{SiO}(T_{C_i}, u_2)}{\mathrm{SSFO}(T_{C_i})} \prod_{i\in I_{u_1}\cap I_{u_2}} \frac{\mathrm{SoSiO}(T_{C_i}, u_1, u_2)}{\mathrm{SSFO}(T_{C_i})}].$$

The first three products can be computed within the linear number of arithmetic operations discussed earlier. For the remaining part of the calculation, we get this bound on the number of arithmetic operations across all cliques in T:

$$O(\sum_{C\in T} \sum_{u_1,u_2\in\mathrm{Res}(C)} [\deg_G(u_1) + \deg_G(u_2)]) = O(\sum_{C\in T} \sum_{u_1\in\mathrm{Res}(C)} |C|\deg_G(u_1)),$$

which is $O(|C_{\max}||E(G)|)$. This concludes the proof of the theorem. $\square$

Counting algorithms based on dynamic programming can often be used to sample: If the algorithm is based on summing counts corresponding to disjoint subproblems, one first runs the counting algorithm, followed by the sampling which proceeds top-down, always choosing which subproblem to go into proportionally to its count. However, here we are subtracting quantities as part of our computations and, as such, a sampling algorithm does not seem to follow from the counting algorithm. For a single level inclusion-exclusion (a single subtraction), one could employ rejection sampling to reject the unfavorable (i.e. those that are subtracted) configurations. However, if almost all configurations are rejected, the probability of sampling success could be minuscule. For two-level inclusion-exclusion, as is the case for our algorithm, even this (potentially low-probability and hence large running time) approach is unclear. We leave the problem of efficient (almost) uniform sampling of source-sink-free orientations in chordal graphs open.

4 Results for the Other Types of Orientations

In this section we briefly sketch our results on counting and sampling of acyclic and bipolar orientations, and on counting sink-free orientations. We include the detailed proofs and discussion in the appendix.

An orientation is *acyclic* if it does not contain a directed cycle. A structural examination of the properties of acyclic orientations in chordal graphs leads to a simple relationship: $\mathrm{AO}(T_C) = \frac{|C|!}{|\operatorname{Sep}(C)|!} \prod_{i=1}^{d} \mathrm{AO}(T_{C_i})$, where $\mathrm{AO}(T_C)$ is the number of acyclic orientations of the clique subtree T_C and where $C_1, \ldots, C_d$ are the child cliques of C in the rooted T (and $d = 0$ if C is a leaf of T). This allows us to count all acyclic orientations in time $\tilde{O}(|V(G)| + |E(G)|)$.

To sample an acyclic orientation uniformly at random, we first construct a clique tree of the input graph and randomly pick a clique C_r as the root. We pick a uniformly random ordering of C_r. Then we process the remaining cliques in a depth-first manner. Let C be the current clique we are processing, we always pick an orientation on $G[C]$ that is consistent with $G[\operatorname{Sep}(C)]$. This can be done by choosing a random ordering π of C and replacing the relative order of $\operatorname{Sep}(C)$ in π by the given ordering. Once all cliques are processed, the resulting directed graph is just a uniformly generated random acyclic orientation. The running time is $O(\sum_{C \in T} |C|) = O(\sum_{v \in V(G)}(\deg_G(v) + 1)) = O(|E(G)|)$, assuming G is connected. The first equality follows from the fact that each v occurs in at most $\deg_G(v) + 1$ cliques. Both results are summarized in the following theorem, which also includes a more precise statement of the counting running time:

Theorem 2. *Let G be a connected chordal graph. The number of its acyclic orientations can be calculated in $\tilde{O}(|V(G)|) + O(|E(G)|)$ time, and a uniformly random acyclic orientation can be produced in time $O(|E(G)|)$.*

A *bipolar (s, t)-orientation* is an acyclic orientation with a unique source s and a unique sink t. We employ a similar strategy as we did for acyclic orientations. At the beginning, we construct a clique tree and randomly pick a clique C_s that contains the source s as the root clique. In order to maintain s and t as the unique source and sink, we differentiate between cliques in T that are or are not on the C_s-C_t path in T, and we recursively compute corresponding bipolar orientations of the subtrees (with some well-chosen restrictions to maintain the overall source and sink). While somewhat more complex than acyclic orientations, the structure still allows us to sample very efficiently analogously to acyclic orientations, as summarized in the following theorem:

Theorem 3. *Let G be a connected chordal graph and $s \neq t$ be two of its vertices. The number of bipolar (s, t)-orientations of G can be computed in $\tilde{O}(|V(G)|) + O(|E(G)|))$ time. A uniformly random bipolar (s, t)-orientation of G can be produced in time $O(|E(G)|)$.*

We conclude with sink-free orientations. Recall that for any graph there is an FPRAS counting these orientations [5] and an efficient exact uniform "sink-popping" sampler is also known [6]. Therefore, we focus on the counting problem, aiming to improve the running time compared to the FPRAS. In fact, our counting algorithm is deterministic, exact, and efficient, with (near) linear running time, as stated in this theorem:

Theorem 4. *Let G be a connected chordal graph. The number of sink-free orientations of G can be counted in $\tilde{O}(|V(G)|) + O(|E(G)|)$ time.*

The algorithm computes two separate quantities at every node of the clique tree: (i) NSFO(T_C), which counts orientations of $\hat{G}[T_C]$ where only sinks in Sep(C) are allowed, and (ii) ASFO(T_C, v), which counts orientations of $\hat{G}[T_C]$, where $v \in $ Sep(C) is a sink and there are no sinks in $V(\hat{G}[T_C]) - $ Sep(C). These quantities are reminiscent of the ones we used for the source-sink-free calculation, but they are significantly less involved. Due to the nature of these orientations, a single-level inclusion-exclusions is needed to compute these quantities, which allows us to run in the (near) linear running time, just as for acyclic and bipolar orientations. However, unlike for the other types of orientations, due to the negative term in the computation, this dynamic programming does not extend to a corresponding sampling algorithm. Understanding how to obtain a sampler from a dynamic programming approach combined with a single level inclusion-exclusion might help with solving our open problem related to sampling source-sink-free orientations in chordal graphs.

References

1. Agnarsson, G.: On chordal graphs and their chromatic polynomials. Math. Scand. **93**, 240–246 (2003)
2. Blair, J.R., Peyton, B.: An introduction to chordal graphs and clique trees. In: George, A., Gilbert, J.R., Liu, J.W.H. (eds.) Graph Theory and Sparse Matrix Computation, pp. 1–29. Springer, New York (1993). https://doi.org/10.1007/978-1-4613-8369-7_1
3. Borwein, P.B.: On the complexity of calculating factorials. J. Algorithms **6**(3), 376–380 (1985)
4. Brylawski, T., Oxley, J.: The tutte polynomial and its applications. Matroid Appl. **40**, 123–225 (1992)
5. Bubley, R., Dyer, M.: Graph orientations with no sink and an approximation for a hard case of #SAT. In: Proceedings of the Eighth Annual ACM-SIAM Symposium on Discrete Algorithms, pp. 248–257 (1997)
6. Cohn, H., Pemantle, R., Propp, J.: Generating a random sink-free orientation in quadratic time. arXiv preprint math/0103189 (2001)
7. Guo, H., Jerrum, M., Liu, J.: Uniform sampling through the Lovász local lemma. J. ACM (JACM) **66**(3), 1–31 (2019)
8. Huber, M.: Exact sampling and approximate counting techniques. In: Proceedings of the Thirtieth Annual ACM Symposium on Theory of Computing, pp. 31–40 (1998)
9. Koller, D., Friedman, N.: Probabilistic Graphical Models: Principles and Techniques. MIT Press, Cambridge (2009)
10. Linial, N.: Hard enumeration problems in geometry and combinatorics. SIAM J. Algebraic Discrete Methods **7**(2), 331–335 (1986)
11. Okamoto, Y., Uno, T., Uehara, R.: Counting the number of independent sets in chordal graphs. J. Discrete Algorithms **6**(2), 229–242 (2008)
12. Vandenberghe, L., Andersen, M.S.: Chordal graphs and semidefinite optimization. Found. Trends Optim. **1**(4), 241–433 (2015)
13. Vertigan, D.L., Welsh, D.J.: The computational complexity of the tutte plane: the bipartite case. Comb. Probab. Comput. **1**(2), 181–187 (1992)

14. Welsh, D.: The tutte polynomial. Random Struct. Algorithms **15**(3–4), 210–228 (1999)
15. Welsh, D.J.: Approximate counting. Surv. Comb. **241**, 287–324 (1997)
16. Wienöbst, M., Bannach, M., Liśkiewicz, M.: Polynomial-time algorithms for counting and sampling Markov equivalent dags. In: Proccedings of the 35th Conference on Artificial Intelligence, AAAI (2021)

Minimum t-Spanners on Subcubic Graphs

Renzo Gómez[1]([⊠]) [iD], Flávio Miyazawa[1] [iD], and Yoshiko Wakabayashi[2] [iD]

[1] Institute of Computing, University of Campinas, Campinas, Brazil
{rgomez,fkm}@ic.unicamp.br
[2] Institute of Mathematics and Statistics, University of São Paulo, São Paulo, Brazil
yw@ime.usp.br

Abstract. For a constant $t \geq 1$, a *t-spanner* of a connected graph G is a spanning subgraph of G in which the distance between any pair of vertices is at most t times its distance in G. We address two problems on spanners: the *minimum t-spanner problem* (MinS$_t$), and a minimization version of the *tree t-spanner problem* (TreeS$_t$). MinS$_t$ seeks a t-spanner with minimum number of edges. TreeS$_t$ is a decision problem concerning the existence of a t-spanner that is a tree. The concept of spanner was introduced by Peleg & Ullman in 1989, in a context regarding the construction of optimal synchronizers for the hypercube. MinS$_t$ is known to be NP-hard for every $t \geq 2$ even on some bounded-degree graphs. TreeS$_t$ is polynomially solvable for $t = 2$ and NP-complete for $t \geq 4$, but its complexity for $t = 3$ remains open.

We investigate both MinS$_3$ and TreeS$_2$ on the class of subcubic graphs. We prove that MinS$_3$ can be solved in polynomial time, using a similar technique as the one used by Cai & Keil (1994) for $t = 2$. This result also gives an alternative algorithm to solve TreeS$_3$ in polynomial time. Additionally, we study TreeS$_2$ from a polyhedral point-of-view and show a complete linear characterization of the associated polytope. This result, interesting on its own right, gives a polynomial-time algorithm to solve a natural minimization version of TreeS$_2$ on subcubic graphs with costs assigned to its edges.

Keywords: spanner · sparse spanner · tree spanner · subcubic graph · polyhedra

1 Introduction

For the problems considered here, the input graph is always connected (even if this is not stated explicitly). The *distance* between two vertices u and v in a graph G, $d_G(u, v)$, is the minimum length of a path between them. For a (rational) constant $t \geq 1$, a *t-spanner* of a graph $G = (V, E)$ is a spanning subgraph H of G in which the distance between any pair of vertices is at most t times its distance in G. That is,

$$d_H(u, v) \leq t \cdot d_G(u, v), \text{ for all } u, v \in V.$$

© Springer Nature Switzerland AG 2022
P. Mutzel et al. (Eds.): WALCOM 2022, LNCS 13174, pp. 365–380, 2022.
https://doi.org/10.1007/978-3-030-96731-4_30

This concept was introduced, in 1989, by Peleg & Ullman [20] when they presented a novel technique to construct an optimal synchronizer for the hypercube. This technique explored the close connection between synchronizers and a t-spanner on a network. Since then, spanners have raised much attention, both from theoretical and practical point-of-view. It has applications in many areas such as distributed systems and communication networks (synchronization, building succinct and efficient routing tables [21], distance oracles [3,22], roadmap planning [24]), computational geometry, robotics, etc.

The definition we have given refers to multiplicative t-spanner, in the sense that we require that the distances be preserved up to a multiplicative error; but one could require (solely or additionally) that they be preserved up to an additive error. We shall not mention here the different problems regarding spanners, but refer the readers to a recent survey by Ahmed et al. [1] that gives an overview of the rich body of literature on graph spanners.

We address here two problems on t-spanners, all of them on subcubic graphs (those with maximum degree at most three). The first one is the MINIMUM t-SPANNER PROBLEM (MINS$_t$), also known as the sparsest t-spanner problem, that asks for a t-spanner with minimum number of edges. The second one is an optimization version of the TREE t-SPANNER PROBLEM (TREES$_t$) which is a decision problem regarding the existence of a t-spanner that is a tree. In the latter problem we consider graphs with costs assigned to its edges.

Before mentioning some of the main results on these and related problems, we observe that when the graph is unweighted, case of MINS$_t$ and TREES$_t$, it suffices to study them only when t is an integer number. It is not difficult to see that results for this case carries over to the case t is a rational number. (This is not true when G is edge-weighted.)

In 1989, Peleg & Schäffer [19] proved that MINS$_2$ is NP-hard, and Cai [5] extended this result proving that MINS$_t$ is NP-hard for $t \geq 2$. Venkatesan et al. [23] showed that MINS$_t$ is NP-hard for $t \geq 2$ even if the graph is chordal. In an attempt to find classes of graphs that yield polynomial algorithms, another line of studies considered graphs of bounded degree. Let Δ denote the maximum degree of a graph. Cai & Keil [7] showed that MINS$_2$ can be solved in polynomial time if $\Delta \leq 4$. They also showed that MINS$_t$ is NP-hard when $t \geq 2$ and $\Delta \geq 9$. More recently, Kobayashi [14] improved this result showing that MINS$_t$ is NP-hard when $t = 2$ and $\Delta \geq 8$; and also when $3 \leq t \leq 4$ and $\Delta \geq 6$. We note that when $\Delta = 3$ (the input graph is subcubic), Cai & Keil [7] believed that, using an approach similar to the one they used for MINS$_2$ and $\Delta \leq 4$, the problem MINS$_3$ could be solved in polynomial time. Indeed, this approach worked for $t = 3$ and $\Delta = 3$, but turned out to be much more complicated. These results are shown here. In Table 1 we summarize the main results known in the literature regarding the computational complexity status of MINS$_t$.

In terms of approximation, the following is known. A first result in this context follows as a consequence of a greedy algorithm proposed by Althöfer et al. [2]. Given an n-vertex graph G and a real number $k > 0$, this algorithm finds a $(2k+1)$-spanner of G with $\mathcal{O}(n^{1+1/k})$ edges. Since every spanner contains at least

$n - 1$ edges, this algorithm can be seen as an $\mathcal{O}(n^{\frac{2}{t-1}})$-approximation for MINS_t, for $t \geq 4$. In 2007, Elkin & Peleg [11] showed that MINS_t, $t \geq 3$, is quasi-NP-hard to approximate by a factor of $\mathcal{O}(2^{\log^{1-\epsilon} n})$, for any $0 < \epsilon < 1$. It is interesting to note that there is a significant difference between the approximability for the cases $t = 2$ and $t \geq 3$. Indeed, Kortsarz & Peleg [16] obtained an $\mathcal{O}(\log n)$-approximation for MINS_2, shown to be best possible, unless $P = \mathrm{NP}$ [15].

Now, let us turn our attention to tree t-spanners. Clearly, a polynomial-time algorithm for MINS_t, for a fixed t, solves the corresponding TREES_t problem in polynomial time. Cai & Corneil [6] showed that TREES_t can be solved in polynomial time when $t \leq 2$, and that it is NP-complete when $t \geq 4$. On the other hand, when we consider the class of bounded-degree graphs, Fomin et al. [13] showed that TREES_t can be solved in polynomial time. (See additional comments on this result in Sect. 4.1). For $t = 3$, the complexity of TREES_t has not been settled. Looking for an answer to this question, it was shown that TREES_3 can be solved in polynomial time for several classes of graphs such as planar graphs [12], convex graphs [23], split graphs [23], line graphs [9], etc. We study a minimization version of TREES_t that considers an input graph with costs assigned to its edges, and seeks a tree t-spanner of minimum total cost. We called this problem the *minimum cost tree t-spanner problem* (MCTS_t). Our contribution on these problems are a polynomial-time algorithm for MINS_3 on subcubic graphs, and for MCTS_t we show a polynomial-time algorithm for $t = 2$ on subcubic graphs.

Table 1. Computational complexity of MINS_t on graphs with maximum degree at most Δ. The symbol (*) indicates a result presented here.

Δ	$t = 2$	$t = 3$	$t = 4$	$t \geq 5$
$= 3$	P [7]	P (*)	open	open
$= 4$	P [7]	open	open	open
$= 5$	open	open	open	open
$= 6$	open	NP-hard [14]	NP-hard [14]	open
$= 7$	open	NP-hard [14]	NP-hard [14]	open
$= 8$	NP-hard [14]	NP-hard [14]	NP-hard [14]	open
≥ 9	NP-hard [7]	NP-hard [7]	NP-hard [7]	NP-hard [7]

This work is organized as follows. In Sect. 2 we present some concepts that will be used throughout this text. In particular, we define a partition of the edges of a graph G, denoted by $\mathcal{C}_t(G)$, that helps us subdivide MINS_t (and TREES_t) into (possibly) smaller subproblems, each one on a (simpler) graph induced by the edges in a class of the partition. In Sect. 3, we study the polytope defined as the convex hull of the incidence vectors of the tree 2-spanners of a subcubic graph. We show a complete linear description of this polytope that yields a polynomial-time algorithm to solve MCTS_2 on subcubic graphs. To the best of our knowledge, this is a novel approach and result. In Sect. 4, we characterize the subgraphs that belong to $\mathcal{C}_3(G)$ when G is a subcubic graph. Using this result, we show how to solve MinS_3 in polynomial time for this class of graphs. As a byproduct, we obtain a simple algorithm to solve TreeS_3 on subcubic graphs. Owing to space limitation, most of the proofs are sketched and in some cases they are omitted.

2   Preliminaries

In this section, we present some results and define the main concepts that will be used throughout the text. The *length* of a path (resp. cycle) is its number of edges. We say that a path (resp. cycle) is a *k-path* (resp. *k-cycle*) if its length is k. The following result is very useful (and will be used henceforth) as it tells us that, to verify whether a spanning subgraph H is a t-spanner of a graph G, it suffices to test the distance condition only for pairs of adjacent vertices (see also Cai & Corneil [6]).

Proposition 1 (Peleg & Schäffer, 1989). *Let H be a spanning subgraph of a graph $G = (V, E)$. Then, H is a t-spanner of G if and only if for every edge $uv \in E$, $d_H(u, v) \leq t$.*

Let $G = (V, E)$ be a graph, and let H be a t-spanner of G. Observe that, if an edge $e \in E$ does not belong to H, there must exist in H a k-path, $k \leq t$, linking the ends of e. Moreover, if P is one such path, then $P + e$ is a $(k+1)$-cycle in G. Cai & Keil [7] studied MinS_2 and defined a partition of E based on the 3-cycles in G. Characterizing the subgraphs in this partition is the main idea behind their polynomial algorithm for MinS_2 on graphs of maximum degree four. To derive our result for MinS_3 on subcubic graphs, we also use this approach, and study a partition of E based on the 3-cycles and 4-cycles in G.

Now, we define formally this partition for any integer $t \geq 2$. Let $G = (V, E)$ be a graph, and let L be the graph associated with G defined as follows.

$$V(L) = \{v_e : e \in E\},$$
$$E(L) = \{v_e v_f : e, f \in E \text{ belong to a } k\text{-cycle in } G, \, k \leq t + 1\}.$$

We denote by $\mathcal{C}_t(G)$ the partition of E induced by L, defined as follows: two edges $e, f \in E$ belong to the same class (of $\mathcal{C}_t(G)$) if and only if v_e and v_f belong to the same connected component of L. Observe that, by the definition of $E(L)$, for every k-cycle in G, $k \leq t + 1$, all its edges belong to the same class in $\mathcal{C}_t(G)$. In Fig. 1, we show a graph G, its associated graph L, and the classes in $\mathcal{C}_2(G)$. Observe that, for the graph G in Fig. 1, $\mathcal{C}_3(G) = \mathcal{C}_2(G)$, and $\mathcal{C}_t(G) = \{E\}$, for $t \geq 4$. For simplicity, throughout this text we consider each class in $\mathcal{C}_t(G)$ as a subgraph of G.

Our first result shows that, if we are interested in finding a t-spanner of G, it suffices to find a t-spanner for each subgraph $H \in \mathcal{C}_t(G)$.

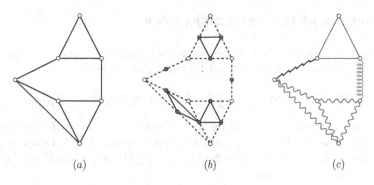

Fig. 1. (a) a graph G; (b) the graph L (its vertices are shaded, and its edges are depicted by full black edges); (c) the four classes in $\mathcal{C}_2(G)$ (represented by different types of edges).

Proposition 2. *A subgraph S of a graph G is a t-spanner if, and only if, $S \cap H$ is a t-spanner of H, for every $H \in \mathcal{C}_t(G)$.*

Proof. Let S be a subgraph of G. First, suppose that S is a t-spanner of G. Let $H \in \mathcal{C}_t(G)$. We will show that $S \cap H$ is a t-spanner of H. Let $e \in E(H)$. By the definition of $\mathcal{C}_t(G)$, H contains every path in G of length at most t between the ends of e. Since S is a t-spanner of G, $S \cap H$ must contain one such path. Thus, $S \cap H$ is a t-spanner of H.

Now, suppose that for every $H \in \mathcal{C}_t(G)$, $S \cap H$ is a t-spanner of H. We will show that S is a t-spanner of G. Since every edge $e \in E(G)$ belongs to a subgraph $H \in \mathcal{C}_t(G)$, and $S \cap H$ is a t-spanner of H, S must contain a k-path, $k \leq t$, between the ends of e. Therefore, S is a t-spanner of G. $\square$

The result above shows that, to find a t-spanner of a graph G, it suffices to find a t-spanner for each subgraph $H \in \mathcal{C}_t(G)$, and take the union of all these t-spanners. If we are interested in tree t-spanners, we can also base on this approach, but we may derive one of the conclusions: if the union of the tree t-spanners of each of the classes in $\mathcal{C}_t(G)$ is not a tree, we can conclude that G does not have a tree t-spanner. Otherwise, such a union gives a tree t-spanner of G.

We observe that, for fixed t, we can obtain $\mathcal{C}_t(G)$ in polynomial time. Now, we introduce some notation and terminology to be used in what follows. Let $G = (V, E)$ be a graph. We denote by $\mathbb{R}^E$ the set of real-valued vectors indexed by the elements in E. We denote by $\chi^F \in \mathbb{R}^E$ the *incidence vector* of the set $F \subseteq E$. That is, χ^F is the binary vector whose nonzero entries correspond to the elements in F. If H is a subgraph of G, we abbreviate $\chi^{E(H)}$ by χ^H.

3 Polytope of the TREE 2-SPANNERS

Throughout this section $G = (V, E)$ denotes a connected subcubic graph. Our aim is to study the polytope defined by the following set in $\mathbb{R}^E$.

$$\mathcal{T}_2(G) := \mathbf{conv}(\{\chi^T \in \mathbb{R}^E : T \text{ is a tree 2-spanner of } G\}),$$

where $\mathbf{conv}(X)$ denotes the convex hull of the vectors in X. In particular, we want to find a set of (linear) inequalities that describe $\mathcal{T}_2(G)$. Observe that, by Proposition 2, it suffices to describe the set $\mathcal{T}_2(H)$, for each $H \in \mathcal{C}_2(G)$, and write a restriction combining them.

Cai & Keil [7] characterized the subgraphs in $\mathcal{C}_2(F)$ when $\Delta(F) \le 4$. When specialized to subcubic graphs, their result gives the following result. For completeness, we present a simplified proof for this case.

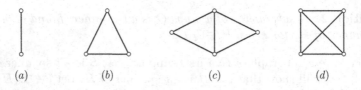

Fig. 2. (a) K_2;   (b) K_3;   (c) $K_4 - e$;   and (d) K_4.

Lemma 1 (Cai & Keil, 1994).   *Let G be a graph such that $\Delta(G) \le 3$. If $H \in \mathcal{C}_2(G)$, then H is isomorphic to K_2, K_3, $K_4 - e$, or K_4.*

Proof. Let $H \in \mathcal{C}_2(G)$, and let $uv \in E(H)$. If uv does not belong to any 3-cycle in G, then $H \cong K_2$. Now, suppose uv belong to a 3-cycle in G. We have that $N_G(u) \cap N_G(v) \ne \emptyset$. We distinguish two cases.

Case 1: $|N_G(u) \cap N_G(v)| = 2$
   Let $N_G(u) \cap N_G(v) = \{x, y\}$. If $xy \in E(G)$, then $H \cong K_4$. Otherwise, the set $\{u, v, x, y\}$ induces a $K_4 - e$ in H. Since $\Delta(H) \le 3$, we have that $\langle u, x, v \rangle$ is the unique 3-cycle that contains xu and xv. Similarly, $\langle u, y, v \rangle$ is the unique 3-cycle that contains yu and yv. Therefore, $H \cong K_4 - e$.

Case 2: $|N_G(u) \cap N_G(v)| = 1$
   Let $N_G(u) \cap N_G(v) = \{w\}$, and let $C = \langle u, v, w \rangle$. If $H \cong C$, the claim follows. So suppose that $H \ncong C$. Then, there must exist a 3-cycle $Q \ne C$ that contains either uw or vw. Without loss of generality, suppose that $uw \in E(Q)$. Since $\Delta(H) \le 3$, we conclude that C (resp. Q) is the only 3-cycle in H that contains the edges in $E(C) - uw$ (resp. $E(Q) - uw$). Therefore, $H \cong K_4 - e$.   □

In Fig. 2, we show the graphs that belong to $\mathcal{C}_2(G)$. In what follows, we show a linear formulation that describes $\mathcal{T}_2(G)$. The inequalities that compose this formulation are based on the following observations. First, if H is a complete graph, any tree 2-spanner of H must be a star. Thus, any such tree does not contain a matching of size two. Moreover, if C is 4-cycle in G, then any tree 2-spanner must contain at most two edges in C. This follows from the fact that, if a tree 2-spanner of G contains any three edges of C, these edges would induce a 3-path in the tree, violating the condition of 2-spanner. These observations explain the first two set of inequalities of the linear formulation below.

Now, consider the decision variables $x \in \mathbb{R}^E$ such that $x_e = 1$ if and only if e belongs to the solution. Let $P(G)$ be the polytope defined by the following set of inequalities.

$$(P(G)) \quad \begin{aligned} x(E(G)) &= |V(G)| - 1, \\ x(E(H)) &= |V(H)| - 1, \quad \forall H \in \mathcal{C}_2(G), \\ x(F) &\leq 1, \qquad\qquad \forall F \subseteq E(H),\ F \text{ matching},\ H \in \mathcal{C}_2(G),\ H \text{ clique}, \\ x(C) &\leq 2, \qquad\qquad \forall C \subseteq E(H),\ C \text{ is a 4-cycle},\ H \in \mathcal{C}_2(G), \\ x_e &\leq 1, \qquad\qquad \forall c \in E, \\ x_e &\geq 0, \qquad\qquad \forall e \in E. \end{aligned}$$

Next, we show the main result of this section.

Theorem 1. *Let $G = (V, E)$ be a connected graph such that $\Delta(G) \leq 3$. Then*

$$\mathcal{T}_2(G) = P(G).$$

Proof (sketch). By the previous arguments, we have that $\mathcal{T}_2(G) \subseteq P(G)$. Now, we show that $P(G) \subseteq \mathcal{T}_2(G)$. For this, it suffices to show that every vertex of $P(G)$ has integer coordinates. Let x^* be a vertex of $P(G)$. We say that an edge $e \in E$ is *fractional* if $0 < x_e^* < 1$. Moreover, we say that an edge e is *full* if $x_e^* = 1$. Let $e \in E$, and let $H \in \mathcal{C}_2(G)$ be the subgraph that contains e. To show that $x_e^* = 0$ or $x_e^* = 1$, we distinguish four cases.

Case 1: $H \cong K_2$
In this case, $E(H) = \{e\}$. Then, $x^*(E(H)) = x_e^* = 1$.

Case 2: $H \cong K_3$
In this case, we omit the proof, but the idea is to show that if x_e^* is fractional, then x^* is a convex combination of two vectors in $P(G)$.

Case 3: $H \cong K_4 - e$
In this case, we first prove the following claim. (We omit this proof owing to space limitation.)

Claim. Let C be the unique 4-cycle in H. If C contains a fractional edge, then $x^*(C) < 2$.

Since there is a unique edge in $E(H) \setminus E(C)$, the previous claim implies that C does not contain any fractional edge. Otherwise, we would have $x^*(E(H)) < 3$, a contradiction. Finally, since $x^*(E(H)) = 3$ and $x^*(C) \leq 2$, the unique edge in $E(H) \setminus E(C)$ must also be integral.

Case 4: $H \cong K_4$

Let f be an edge of H. We will denote by f' the unique edge in H such that $\{f, f'\}$ is a matching. In this case, the idea is to show first the following claim. We leave the proof to the reader.

Claim. If x_f^* is fractional, then $x_{f'}^* = 0$.

Consider that $E(H) = \{e, e', f, f', g, g'\}$. As $x^* \in P(G)$, we have that $x_e^* + x_{e'}^* \leq 1$, $x_f^* + x_{f'}^* \leq 1$, and $x_g^* + x_{g'}^* \leq 1$. Since $x^*(E(H)) = 3$, all the previous inequalities must be satisfied with equality. Therefore, the above claim implies that H has no fractional edge.   □

By Lemma 1, the graphs in $C_2(G)$ are all graphs on at most 4 vertices and at most 6 edges. Hence, the polytope $P(G)$ is defined by $\mathcal{O}(|E|)$ inequalities, and therefore, we can find an optimal solution of $P(G)$ in polynomial time on the size of the input graph G [10].

Now, suppose that there are costs $c_e \in \mathbb{R}$, $e \in E$, assigned to the edges of G, and consider the *minimum cost tree t-spanner problem* (MCTS_t): given a graph with costs assigned to its edges, find a tree t-spanner of minimum total cost. We observe that, the distance between two vertices is the minimum number of edges of a path between them. That is, when measuring the distance, we disregard the costs of the edges.

By Theorem 1, an optimal solution of $\min\{cx : x \in P(G)\}$ induces a tree 2-spanner of G which is an optimal solution for MCTS_2. Therefore, we obtain the following result.

Theorem 2. MCTS_2 *can be solved in polynomial time on subcubic graphs.*

4   MINIMUM 3-SPANNER on Subcubic Graphs

In this section, we show that MINS_3 can be solved in polynomial time if the input graph is subcubic. As in the previous section, we consider that $G = (V, E)$ is a connected subcubic graph. First, we study the structure of the subgraphs in $C_3(G)$. Our first result shows a constructive characterization of the subgraphs in $C_3(G)$. It says that for any graph H in $C_3(G)$, different from K_2, we can define a sequence of graphs $H_0, \ldots, H_n$ such that H_0 is a cycle of length at most four, $H_n \cong H$, and H_{i+1} is obtained from H_i by applying one of the four operations defined below. We now describe these operations. Let H_i be a subcubic graph. Let u and v be distinct vertices of degree two in H_i. The first three operations are the following:

a) Add to H_i an edge between uv if $uv \notin E(H_i)$ and $d_{H_i}(u, v) \leq 3$.
b) Add to H_i a 2-path between u and v if $d_{H_i}(u, v) \leq 2$.
c) Add to H_i a 3-path between u and v if $uv \in E(H_i)$.

Now, for the last operation, let uv and xy be edges in $E(H_i)$ such that u, v, x and y have degree two in H_i. We say that uv and xy *match* if $ux, vy \notin E(H_i)$ or if $uy, vx \notin E(H_i)$. Moreover, we call any of these pairs of edges a *matching* between uv and xy. Then, the fourth operation is the following.

d) Add to H_i a matching between the edges uv and xy.

Note that only operations b) and c) add vertices to H_i, and these new vertices have degree two in the resulting graph. Moreover, each operation increases the degree of the vertices where it is applied. Since H_0 is a 3-cycle or 4-cycle, we have that $\delta(H_i) \geq 2$, for $i = 0, \dots, n$. We show, in Fig. 3, the graphs obtained after applying one of these operations to a 4-cycle. The vertices and edges added to the graph are depicted by solid vertices and wavy edges, respectively.

In the next lemma we show that any graph in $\mathcal{C}_3(G)$, except for K_2, can be constructed using the previous operations. Its proof is omitted owing to space limitation.

Lemma 2. *Let $H \in \mathcal{C}_3(G)$ such that $H \ncong K_2$. Then, there exists a sequence of graphs $H_0, \dots, H_n$ such that H_0 is a cycle of length at most four, $H_n \cong H$, and H_{i+1} is obtained from H_i by applying one of the operations a), b), c) or d).*

Before we characterize the graphs in $\mathcal{C}_3(G)$, we define some classes of graphs related to it. The *k-ladder* graph, denoted by L_k, is the cartesian product of a k-path with K_2. We note that L_k has exactly four vertices of degree two. We denote these vertices by x_1, x_2, y_1 and y_2. Moreover, we consider that the edges x_1y_1, x_2y_2 belong to $E(L_k)$ (see Fig. 4 (a)). If we add the edges x_1y_2 and x_2y_1 (resp. x_1x_2 and y_1y_2) to L_k, for $k \geq 2$, we obtain the graph M_k (resp. N_k). From L_k we can also obtain two other graphs that are of our interest. If we add a 2-path between x_1 and y_1, we obtain the graph T_k^1. Moreover, if we add a 2-path between x_2 and y_2 in T_k^1, we obtain the graph T_k^2. We show, in Fig. 4, an example of those graphs for $k = 2$.

Let $M_k - e$ (resp. $N_k - e$) be the graph obtained from M_k (resp. N_k) by removing the edge x_1y_2 (resp. y_1y_2). In Fig. 5, we show an example of $M_2 - e$ and $N_2 - e$ along with the graphs G_1, G_2 and G_3 that will be important in our characterization.

Let $\mathcal{F}_3$ be the family composed of the following graphs:

a) K_2	e) $K_{2,3}$	i) G_1	m) $M_k, k \geq 2$
b) K_3	f) $M_2 - e$	j) G_2	n) $N_k, k \geq 2$
c) $K_4 - e$	g) $N_2 - e$	k) G_3	o) $T_k^1, k \geq 1$
d) K_4	h) $N_3 - e$	l) $L_k, k \geq 1$	p) $T_k^2, k \geq 1$

Next, we show the main result of this section.

Theorem 3. *Let G be a graph such that $\Delta(G) \leq 3$. If $H \in \mathcal{C}_3(G)$, then $H \in \mathcal{F}_3$.*

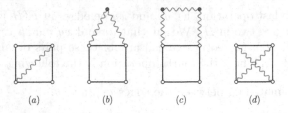

Fig. 3. Graphs obtained from a 4-cycle after applying operation a), b), c), or d)

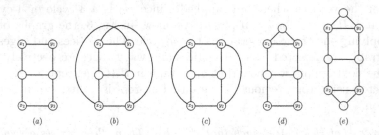

Fig. 4. (a) L_2; (b) M_2; (c) N_2; (d) T_2^1; (e) T_2^2.

Proof (sketch). Let $H \in \mathcal{C}_3(G)$. By the definition of $\mathcal{C}_3(G)$, if H is acyclic, then $H \cong K_2$. So, suppose that H is not acyclic. By Lemma 2, there is a sequence of graphs $H_0, H_1, \ldots, H_n$ such that

- $H_0 \cong K_3$ or $H_0 \cong L_1$,
- $H_n = H$,
- H_{i+1} is obtained from H_i by applying one of the operations a), b), c) or d).

The proof is by case analysis. We will show that $H_i \in \mathcal{F}_3$, for $i = 0, \ldots, n$. Moreover, if we can apply any operation to H_i, we will suppose that $H \neq H_i$, and consider each possibility for H_{i+1}.

By Lemma 2, $H_0 \cong K_3$ or $H_0 \cong L_1$. We will show the proof for the case $H_0 \cong K_3$. Suppose that $H_0 \cong K_3$. Observe that we can not apply operations a) or d) to H_0. That is, we may obtain H_1 as a result of operation b) or operation c). In the first case, $H_1 \cong K_4 - e$, and in the second case, $H_1 \cong T_1^1$. We distinguish these two cases.

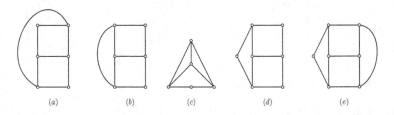

Fig. 5. (a) $M_2 - e$; (b) $N_2 - e$; (c) G_1; (d) G_2; (e) G_3.

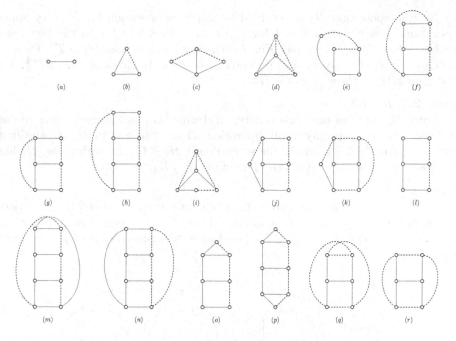

Fig. 6. Minimum 3-spanners for (representative) graphs in $\mathcal{F}_3$ (from (a) to (p)); and minimum 3-spanners for the special cases M_2 and N_2 (resp. (q) and (r)).

Case 1: $H_1 \cong K_4 - e$

In this case, H_1 contains only two vertices of degree two. Since these vertices are not adjacent, we can apply operations $a)$ or $b)$ to H_1. Thus, we have that either $H_2 \cong K_4$ or $H_2 \cong G_1$ (see Fig. 5). Observe that, in both cases, H_2 has at most one vertex of degree two. Therefore, $H = H_2$.

Case 2: $H_1 \cong T_1^1$

In this case, H_1 has three vertices of degree two. Let a, b and c be those vertices such that the neighbors of a have degree three (see Fig. 7 (a)). We may obtain H_2 by applying $a)$, $b)$ or $c)$. If we apply operation $a)$, H_2 is obtained by adding the edge ab or ac to H_1. In either case, we have $H_2 \cong G_1$ (see Fig. 7 (b)). Moreover, we must have $H = H_2$ since H_2 has only one vertex of degree two.

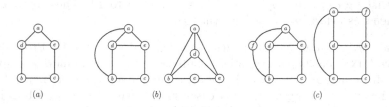

Fig. 7. (a) $H_1 \cong T_1^1$, (b) $H_2 \cong G_1$; and (c) $H_2 \cong N_2 - e$.

Now, suppose that H_2 is obtained by applying operation b). We may apply operation b) on vertices a and b (or c), or on vertices b and c. In the first case, we have that $H_2 \cong N_2 - e$ (see Fig. 7 (c)). In the other case, $H_2 \cong T_1^2$. Finally, we can only apply operation c) on vertices b and c. In this case, $H_2 \cong T_2^1$. We distinguish these three cases for H_2.

Case 2.1: $H_2 \cong N_2 - e$

Since H_2 contains only two vertices of degree two, and these vertices are at distance two, we can only apply operation a) or b) to obtain H_3. In the first case, we have $H_3 \cong N_2$, and in the second case, $H_3 \cong G_3$. In both cases, H_3 has at most one vertex of degree two. Therefore $H = H_3$.

Case 2.2: $H_2 \cong T_1^2$

In this case, H_2 has only two vertices of degree two. Moreover, these vertices are at distance three. Then, we can only obtain H_3 by applying operation a) to those vertices. Therefore, $H_3 \cong N_2$ (see Fig. 8 (a)). Since N_2 is cubic, we must have $H = H_3$.

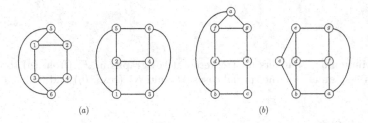

(a)   (b)

Fig. 8. (a) $H_3 \cong N_2$; and (b) $H_3 \cong G_3$.

Case 2.3: $H_2 \cong T_2^1$

Let a, b and c be the three vertices of degree two in T_2^1 such that b is adjacent to c. We may obtain H_3 by applying operations a), b) or c) to H_2. First, suppose that H_3 is obtained by applying a) to H_2. Note that, we can only apply this operation on vertices a and b (or c). In both cases, we have $H_3 \cong G_3$ (see Fig. 8 (b)). As G_3 has only one vertex of degree two, we have that $H = H_3$.

Now, suppose that we apply b) to obtain H_3. Observe that we can only apply this operation on vertices b and c. Thus, we have that $H_3 \cong T_2^2$. Since the vertices of degree two, in T_2^2, are at distance four, we must have $H = H_3$.

Finally, we can only apply c) to H_2 on vertices b and c. Then $H_3 \cong T_3^1$. Note that we can only apply operations b) and c) to T_3^1. Thus, by analogous arguments as before, we must have that $H \cong T_k^2$, $k \geq 3$. $\square$

The previous result characterizes the subgraphs in $\mathcal{C}_3(G)$. By Proposition 2, to solve MINS$_3$ on subcubic graphs, it suffices to find a minimum 3-spanner for each subgraph in $\mathcal{F}_3$. We show, in Fig. 6, a minimum 3-spanner (in solid edges) for each of these subgraphs. In most cases such 3-spanner is a tree which implies its minimality. We observe that, for K_2, K_3, $K_4 - e$ and K_4, such tree spanner is a star centered at a vertex of maximum degree. In a similar way, for $K_{2,3}$,

G_1, G_2 and G_3, we can obtain a tree 3-spanner, by case analysis. Finally, for the remaining graphs, a minimum 3-spanner is obtained from a tree 3-spanner of L_k.

In what follows, we describe a method to construct a minimum 3-spanner for L_k, T_k^1 and T_k^2, $k \geq 1$. Observe that, all the previous graphs contain L_k as a subgraph. Let x_1, y_1, x_2 and y_2 be the vertices of degree two, in L_k, such that x_i is adjacent to y_i, for $i = 1, 2$. A tree 3-spanner of L_k can be obtained in the following way:

i) Let P be the path between x_1 and x_2 of length k.
ii) Let S be the tree obtained from P by linking every vertex $v \in V(L_k) \setminus V(P)$ to its neighbor in P.

An example of the above construction is depicted in Fig. 6 (l). To show that S is indeed a tree 3-spanner of L_k, note that for each edge in $E(L_k) \setminus E(S)$, there is a 3-path between its ends. Now, we show how to add edges to S in order to obtain a minimum 3-spanner of T_k^1, T_k^2. Since L_k is a subgraph of T_k^1 and T_k^2, we will suppose that $V(L_k) \subset V(T_k^i)$, for $i = 1, 2$. Note that, by adding to S, the edges that link each vertex in $V(T_k^i) \setminus V(L_k)$ to P, we obtain a tree 3-spanner of T_k^i, for $i = 1, 2$. This construction is shown in Fig. 6 (o) and Fig. 6 (p). To conclude, we show how to construct a minimum 3-spanner of M_k and N_k, $k \geq 2$. We distinguish the cases $k = 2$ and $k \geq 3$. In Fig. 6 (q) and Fig. 6 (r), we show tree 3-spanners of M_2 and N_2, respectively, indicated with solid edges.

Since N_k is the cartesian product of a $(k+1)$-cycle with K_2, a result obtained by Lin & Lin [17] implies that N_k does not admit a tree 3-spanner, for $k \geq 3$. In what follows, we show that this also holds for M_k. For this, we first name some elements of a tree 3-spanner of L_k to state a property we need later. Let $B_1, B_2, \ldots, B_k$ be the 4-cycles in L_k such that $E(B_i) \cap E(B_{i+1}) \neq \emptyset$, for $i = 1, \ldots, k - 1$. Moreover, let e_0 (resp. e_k) be the edge that links x_1 and y_1 (resp. x_2 and y_2); and let e_i be the edge that belongs to B_i and B_{i+1}, for $i = 1, \ldots, k - 1$. In Fig. 9 we show an example for $k = 3$. We denote by f_i and g_i the edges in B_i that are different from e_{i-1} and e_i. The following result, regarding the edges e_i and the tree 3-spanners of L_k, is straightforward and is left to the reader.

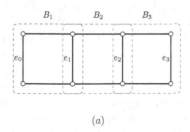

(a)

Fig. 9. (a) The 4-cycles B_1, B_2 and B_3 in L_3.

Lemma 3. *Let S be a tree 3-spanner of L_k. Then, $e_i \in E(S)$, for $i = 1, \ldots, k - 1$.*

The following result shows that a minimum 3-spanner of M_k has at least $|V(L_k)|$ edges.

Lemma 4. *The graph M_k does not admit a tree 3-spanner if $k \geq 3$.*

Proof (sketch). First, we consider that L_k is a subgraph of M_k. In particular, $V(L_k) = V(M_k)$. Let x_1, x_2, y_1 and y_2 be the vertices of degree two in L_k, such that $x_1 y_1$ and $x_2 y_2$ are edges in L_k. The proof is by contradiction. Suppose that M_k admits a tree 3-spanner, say S. First, we observe that $E(S) \cap \{x_1 y_2, y_1 x_2\} \neq \emptyset$. Otherwise, S is a subgraph of L_k. This implies that the distance between x_1 and y_2, in S, is at least $k + 1$ which is a contradiction for $k \geq 3$.

So, suppose that $E(S) \cap \{x_1 y_2, y_1 x_2\} \neq \emptyset$. By symmetry, we can suppose that $x_1 y_2 \in E(S)$. Let $S' = S - x_1 y_2$. The following claim holds. We leave its proof to the reader.

Claim. For every edge $uv \in E(M_k) \setminus \{x_1 y_2, y_1 x_2\}$, we have that u and v belong to the same component in S'.

Observe that, the previous claim implies that S' is connected, a contradiction. □

Then, any minimum 3-spanner of M_k or N_k, for $k \geq 3$, must have at least $|V(L_k)|$ edges. Let S be the tree 3-spanner of L_k constructed by the steps i) and ii). Note that, by adding the edge $x_1 y_2$ (resp. $x_1 x_2$) to S, we obtain a minimum 3-spanner of M_k (resp. N_k). An example of this construction is depicted in Fig. 6 (m) and Fig. 6 (n). Therefore, we have shown the following result.

Corollary 1. *If S^* is a minimum 3-spanner of M_k or N_k, then $|E(S^*)| = |V(L_k)|$, for $k \geq 3$.*

Theorem 4. MINS$_3$ *can be solved in polynomial time on subcubic graphs.*

Proof. First, we find the subgraphs in $\mathcal{C}_3(G)$. Let $H \in \mathcal{C}_3(G)$. We have already argued how to construct a minimum 3-spanner for the graphs in $\mathcal{F}_3$. Thus, we only need to show how to decide which graph, in $\mathcal{F}_3$, H is isomorphic to. First, if $|V(H)| \leq 8$, we do this by a brute-force algorithm. Second, if $|V(H)| = 9$, then $H \cong T_3^1$. Now, suppose that $|V(H)| \geq 10$. Note that the only candidates are L_k, T_k^1, T_k^2, M_k and N_k, for $k \geq 4$. Let s be the number of vertices of degree two in H. Observe that: (i) if $s = 4$, then $H \cong L_k$; (ii) if $s = 3$, then $H \cong T_k^1$; (iii) if $s = 2$, then $H \cong T_k^2$; and (iv) if $s = 0$, then $H \cong M_k$ or $H \cong N_k$.

For the case $s = 0$, we distinguish between M_k and N_k as follows. Let E' be the set of edges, in H, that belong to just one 4-cycle. Note that, we can obtain this information when we find $\mathcal{C}_3(G)$. Observe that E' induces either a Hamiltonian cycle of H, or two disjoint cycles of length $k + 1$. In the first case, we have $H \cong M_k$, and in the second case, we have $H \cong N_k$. □

4.1 Consequences for TREES₃

Fomin et al. [13] showed that if a graph G has a tree t-spanner then its treewidth is at most $\Delta(G)^t$. Using this result, they show that if G has bounded degree, say d, then TREES$_t$ can be solved in linear time, in the following way: (i) check whether G has treewidth at most d^t; and (ii) look for a tree t-spanner if (i) holds. Step (i) can be solved in linear time, using a well-known result proved by Bodlaender [4]. Step (ii) can be solved in linear time by Courcelle's theorem [8] since the property of admitting a tree t-spanner is expressible in monadic second order logic [13]. More recently, Papoutsakis [18] proposed a dynamic programming approach for TREES$_t$ on bounded-degree graphs that solves the problem in polynomial time (for fixed t and maximum degree).

We note that, by Proposition 2, our approach on MINS₃ gives an alternative algorithm for TREES₃ on subcubic graphs: for each $H \in \mathcal{C}_3(G)$, find a tree 3-spanner of H, say T_H. If $H \cong M_k$ or $H \cong N_k$, $k \geq 3$, then G does not admit a tree 3-spanner. Otherwise, let $\mathcal{T}$ be the union of the 3-spanners T_H. Then, G admits a 3-tree spanner if and only if $\mathcal{T}$ is a tree.

5 Concluding Remarks and Future Work

In this work, we studied MINS$_t$, and a minimization version of TREES$_t$, which we called MCTS$_t$. We showed that MINS₃ can be solved in polynomial time on subcubic graphs. This result answers partially an open question regarding the complexity of MINS$_t$ on bounded-degree graphs. As a byproduct, our approach yielded an alternative algorithm to solve TREES₃ on subcubic graphs. Currently, we are working on the remaining open questions regarding the computational complexity of MINS$_t$ on bounded-degree graphs.

We also studied the polytope defined as the convex hull of the incidence vectors of the tree 2-spanners of a subcubic graph. We showed a complete linear description of this polytope (of polynomial size). As a result, we obtained a polynomial-time algorithm for MCTS₂ on subcubic graphs. As far as we know, this is a novel result. The current (mixed) integer linear formulations for minimum t-spanner and its variants are able to solve only small instances in a reasonable amount of time. Possibly, finding strong and tight inequalities for the relaxed formulations may lead to approaches with better performance, but this seems to be a hard and challenging problem.

Acknowledgements. This research has been partially supported by FAPESP - São Paulo Research Foundation (Proc. 2015/11937-9). R. Gómez is supported by FAPESP (Proc. 2019/14471-1); F.K. Miyazawa is supported by CNPq (Proc. 314366/2018-0 and 425340/2016-3) and FAPESP (Proc. 2016/01860-1); Y. Wakabayashi is supported by CNPq (Proc. 306464/2016-0 and 423833/2018-9).

References

1. Ahmed, R., et al.: Graph spanners: a tutorial review. Comput. Sci. Rev. **37**, 100253 (2020)

2. Althöfer, I., Das, G., Dobkin, D., Joseph, D., Soares, J.: On sparse spanners of weighted graphs. Discrete Comput. Geom. **9**(1), 81–100 (1993). https://doi.org/10.1007/BF02189308
3. Baswana, S., Sen, S.: Approximate distance oracles for unweighted graphs in expected $O(n^2)$ time. ACM Trans. Algorithms **2**(4), 557–577 (2006)
4. Bodlaender, H.: A linear-time algorithm for finding tree-decompositions of small treewidth. SIAM J. Comput. **25**(6), 1305–1317 (1996)
5. Cai, L.: NP-completeness of minimum spanner problems. Discrete Appl. Math. **48**(2), 187–194 (1994)
6. Cai, L., Corneil, D.: Tree spanners. SIAM J. Discrete Math. **8**(3), 359–387 (1995)
7. Cai, L., Keil, M.: Spanners in graphs of bounded degree. Networks **24**(4), 233–249 (1994)
8. Courcelle, B., Engelfriet, J.: Graph Structure and Monadic Second-Order Logic, Encyclopedia of Mathematics and its Applications, vol. 138. Cambridge University Press, Cambridge (2012)
9. Couto, F., Cunha, L., Posner, D.: Edge tree spanners. In: Gentile, C., Stecca, G., Ventura, P. (eds.) Graphs and Combinatorial Optimization: from Theory to Applications. ASS, vol. 5, pp. 195–207. Springer, Cham (2021). https://doi.org/10.1007/978-3-030-63072-0_16
10. Dantzig, G., Thapa, M.: Linear Programming. 2: Theory and Extensions. Springer Series in Operations Research, Springer, New York (2003). https://doi.org/10.1007/b97283
11. Elkin, M., Peleg, D.: The hardness of approximating spanner problems. Theory Comput. Syst. **41**(4), 691–729 (2007)
12. Fekete, S., Kremer, J.: Tree spanners in planar graphs. Discrete Appl. Math. **108**(1–2), 85–103 (2001)
13. Fomin, F., Golovach, P., van Leeuwen, E.: Spanners of bounded degree graphs. Inf. Process. Lett. **111**(3), 142–144 (2011)
14. Kobayashi, Y.: NP-hardness and fixed-parameter tractability of the minimum spanner problem. Theoret. Comput. Sci. **746**, 88–97 (2018)
15. Kortsarz, G.: On the hardness of approximating spanners. Algorithmica **30**(3), 432–450 (2001)
16. Kortsarz, G., Peleg, D.: Generating sparse 2-spanners. J. Algorithms **17**(2), 222–236 (1994)
17. Lin, L., Lin, Y.: Optimality computation of the minimum stretch spanning tree problem. Appl. Math. Comput. **386**, 125502 (2020)
18. Papoutsakis, I.: Tree spanners of bounded degree graphs. Discrete Appl. Math. **236**, 395–407 (2018)
19. Peleg, D., Schäffer, A.: Graph spanners. J. Graph Theory **13**(1), 99–116 (1989)
20. Peleg, D., Ullman, J.: An optimal synchronizer for the hypercube. SIAM J. Comput. **18**(4), 740–747 (1989)
21. Peleg, D., Upfal, E.: A trade-off between space and efficiency for routing tables. J. ACM **36**(3), 510–530 (1989)
22. Thorup, M., Zwick, U.: Approximate distance oracles. J. ACM **52**(1), 1–24 (2005)
23. Venkatesan, G., Rotics, U., Madanlal, M., Makowsky, J., Pandu Rangan, C.: Restrictions of minimum spanner problems. Inf. Comput. **136**(2), 143–164 (1997)
24. Wang, W., Balkcom, D., Chakrabarti, A.: A fast online spanner for roadmap construction. Int. J. Rob. Res. **34**(11), 1418–1432 (2015)

Approximation Algorithms

Approximating the Bundled Crossing Number

Alan Arroyo[1] and Stefan Felsner[2]([✉])

[1] IST Austria, Klosterneuburg, Austria
[2] Institut für Mathematik, Technische Universität Berlin, Berlin, Germany
`felsner@math.tu-berlin.de`

Abstract. Bundling crossings is a strategy which can enhance the readability of graph drawings. In this paper we consider bundlings for families of pseudosegments, i.e., simple curves such that any two have share at most one point at which they cross. Our main result is that there is a polynomial-time algorithm to compute an 8-approximation of the bundled crossing number of such instances (up to adding a term depending on the facial structure). This 8-approximation also holds for bundlings of good drawings of graphs. In the special case of circular drawings the approximation factor is 8 (no extra term), this improves upon the 10-approximation of Fink et al. [6]. We also show how to compute a $\frac{9}{2}$-approximation when the intersection graph of the pseudosegments is bipartite.

1 Introduction

The study of bundled crossings is a promising topic in Graph Drawing due to its practical applications in Network Visualization and the rich connections with related areas such as Topological Graph Theory. One of the mantras motivating the study of crossing numbers is that "reducing crossings can improve the readability of a drawing, leading to better representation of graphs". The study of bundled crossings provide an alternative way to assess readability by allowing crossings of a drawing to be bundled into regular grid-patterns with the goal of minimizing the number of bundles instead of minimizing the number of individual crossings.

The *crossing number* of a graph G is the minimum integer $\mathrm{cr}(G)$ for which G has a drawing in the sphere with $\mathrm{cr}(G)$ crossings. Computing the crossing number of a graph is a notoriously hard problem. There are long standing conjectures

This work was initiated during the Workshop on Geometric Graphs in November 2019 in Strobl, Austria. We would like to thank Oswin Aichholzer, Fabian Klute, Man-Kwun Chiu, Martin Balko, Pavel Valtr for their avid discussions during the workshop. The first author has received funding from the European Union's Horizon 2020 research and innovation programme under the Marie Skłodowska-Curie grant agreement No 754411. The second author has been supported by the German Research Foundation DFG Project FE 340/12-1.

P. Mutzel et al. (Eds.): WALCOM 2022, LNCS 13174, pp. 383–395, 2022.
https://doi.org/10.1007/978-3-030-96731-4_31

regarding the crossing numbers of complete graphs [2,7] and complete bipartite graphs [10]. Another family whose crossing numbers have been intensely studied are Cartesian products of two cycles $C_m \square C_n$. It is conjectured that their crossing number is $\mathrm{cr}(C_m \square C_n) = (m-2)n$ for, $3 \leqslant m \leqslant n$ [8]. Figure 1(left) indicates how to draw products of cycles with few crossings, these drawings show $\mathrm{cr}(C_m \square C_n) \leqslant (m-2)n$.

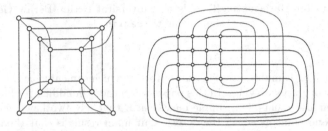

Fig. 1. The crossing number of $C_4 \square C_4$ is 8 (left) but there is a drawing with 16 crossings which can be viewed as a single bundled crossing (right).

To define bundled crossings, we consider the *planarization* of a drawing D in the sphere, this is the plane graph obtained by replacing crossings by degree-4 vertices (we always assume that a crossing point belongs to only two edges). A *bundled crossing* or *bundle* of D is a subgraph of the planarization of D isomorphic to an $n \times m$-grid graph $(n, m \geqslant 1)$, whose vertices are exclusively crossings. A drawing of $C_4 \square C_4$ where all crossings can be assigned to a single bundle is shown in Fig. 1(right).

A *bundling* of D is a partition of the crossings of D into disjoint bundles. The *bundled crossing number* $\mathrm{bc}(D)$ of a drawing D is the minimum number of bundles in any bundling of D, whereas the bundled crossing number $\mathrm{bc}(G)$ of a graph G is the minimum $\mathrm{bc}(D)$ taken over all drawings D of G. From the fact that $C_4 \square C_4$ is not planar and Fig. 1, it follows that $\mathrm{bc}(C_4 \square C_4) = 1$. Indeed, $\mathrm{bc}(C_m \square C_n) = 1$ for $n, m \geqslant 3$.

Previous Work. Schaefer in his survey on crossing number problems [9] suggests to consider bundlings of crossings. Alam et al. [1] were the first to study the problem from a graph drawing viewpoint. Later, Fink and coauthors considered the problem of computing $\mathrm{bc}(\cdot)$ both in the *free-drawing* variant, when a graph G is the input, and the goal is to compute $\mathrm{bc}(G)$, and in the *fixed-drawing* variant, where the drawing D is the input, and the goal is to find $\mathrm{bc}(D)$, i.e., to assign the crossings to as few bundles as possible. In this work we will focus on the fixed-drawing variant[1].

---

[1] In this work we are interested on bundling connected drawings. In this context, faces of a bundle bounded by "squares" are empty. In previous literature, the emptyness of squares was part of the definition of bundled crossings.

Fink et al. [6] showed that computing bc($\cdot$) is NP-hard in the fixed-drawing variant of the problem. The hardness of the free-drawing variant has been shown by Chaplick et al. [3]. An algorithm that computes a 10-approximation of bc(D) for circular drawings was presented by Fink et al. [6], here a *circular drawing* of a graph is one where vertices are drawn on a circle and edges are drawn inside the circle. Circular drawings are assumed to be simple. Simple drawings also known as *good drawings* obey: (0) there are no self-intersections of edges; (1) any two edges intersect at most once; (2) the intersection between any two edges is either a common vertex or a crossing; and (3) no three edges share a crossing. Our work was motivated by the question of whether bc(D) can be approximated on simple drawings or even more general classes of drawings of graphs.

Our Contribution. To prepare for the statement of our main results we first show how to reduce the problem of computing bc($\cdot$) for a graph drawing D to computing bc($\cdot$) for a *set of strings*. We then introduce a special configuration called a toothed-face which plays a special role in this work.

To a graph drawing D we associate a set $\mathcal{E}$ of strings obtained in two steps: first, delete all the uncrossed edges of D; second, for each edge e of D, remove a small bit of e at each endpoint to obtain a *string* (a closed arc). This results in a set $\mathcal{E}$ of strings which is the drawing of a matching, thus bc($\mathcal{E}$) is well-defined. Moreover, the bundlings of D are in one-to-one correspondence with the bundlings of $\mathcal{E}$, so often in this work we restrict ourselves to study bundlings of sets of strings. If the drawing D is a good drawing the associated family $\mathcal{E}$ is a *family of pseudosegments*, i.e., it is a set of strings with the property that no string self-intersects and no two strings cross more than once.

Any set of strings divides the plane into open regions called *faces*. A string ends in a face F if one of its endpoints is incident with the boundary of F. Before defining toothed-faces, keep in mind that the boundary of a face is not necessarily the same as the boundary of its closure. For instance, Fig. 2 shows two examples of faces where their boundaries include pieces of strings ending in the face while their closure is bounded by only four pieces of strings.

A *toothed-face* is a face F of a set of strings such that (1) at least one string ends in F; (2) the closure $\overline{F}$ is bounded by exactly four pieces of strings; and (3) all the crossings between the strings ending in F and the boundary of $\overline{F}$ occur in only two opposite string-pieces among the four string-pieces bounding $\overline{F}$. For a set $\mathcal{E}$ of strings we let $t(\mathcal{E})$ be the number of toothed-faces. The next is our first main result.

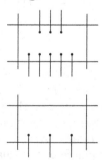

Theorem 1. *For a connected family of pseudosegments $\mathcal{E}$ there is a bundling with at most $8bc(\mathcal{E}) + t(\mathcal{E})$ bundles. This bundling can be computed in polynomial-time.*

Fig. 2. Examples of toothed-faces.

If $\mathcal{E}$ is associated to a drawing D, then define $t(D) := t(\mathcal{E})$. With this definition the approximation bound of $8bc(D) + t(D)$ applies to a class of drawings containing but not restricted to

simple drawings. In the absence of toothed-faces[2] we get a clean 8 approximation. Since circular drawings have no toothed-faces, Theorem 1 yields an
8-approximation for this class thus improving over the 10-approximation Fink
et al. [6].

In our second main result we consider bipartite families of pseudosegments,
i.e., families with a bipartite intersection graph, and obtain an improved approximation for such *bipartite instance*.

Theorem 2 (*[3]*). *For a connected bipartite family of pseudosegments $\mathcal{E}$ there is
a bundling with at most $\frac{9}{2}bc(\mathcal{E}) + \frac{1}{2}t(\mathcal{E})$ bundles. This bundling can be computed
in polynomial-time.*

1.1   An Easy Example: Bundling Laminar Families of Chords

In this section, in an informal approach, we consider a concrete example that
captures many of the concepts that will be used in the later parts of the paper.
The approximation algorithm of Fink et al. [6] is essentially based on the same
concepts.

A *laminar family of chords* is a drawing obtained from adding to a circle a
collection of vertical blue and horizontal red chords with ends on the circle.

Any such laminar family can be converted, by an appropriate crossing-
preserving transformation, into a family of blue vertical chords and red horizontal
chords drawn inside an *orthogonal polygon* P, i.e., a polygon whose edges are
parallel to the x- and y-axis (Fig. 3.1). Moreover, the polygon and the chords can
be chosen so that the chords are evenly spaced forming a regular grid inside P.

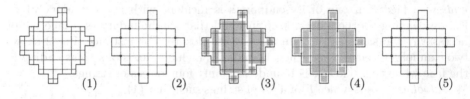

Fig. 3. Bundling a laminar family of chords. (Color figure online)

Subdivide the interior of P into squares, each of them containing a crossing
in its center. We will refer to the graph obtained from the union of P and the
perimeter of the squares as the *dual net* of the instance (Fig. 3.2). A bundled
crossing corresponds to a collection of squares in the dual net forming a *rectangle*; moreover, a *bundling* of the new instance corresponds to a partition of the
squares of the dual net into rectangles (Fig. 3.3). We refer to such a partition
as a *rectangulation* of P. A minimal bundling corresponds to a rectangulation
using a minimum number of rectangles.

---

[2] A toothed-face in a drawing D corresponds to a vertex v such that v is in a four face
of $D\backslash v$ and all edges incident to v leave the face through one or two opposite edges.

[3] Proofs of results marked by (∗) can be found in the full version: `arXiv:2109.14892`.

The problem of rectangulating an orthogonal polygon into the minimum number of rectangles was solved by at least three independent groups in the 80's, Eppstein [5] contains the relevant references. We next describe a polynomial-time algorithm for finding a minimum rectangulation.

The segments added to the polygon P to obtain a rectangulation are either horizontal or vertical (red bi-arrows in Fig. 3.4). In a rectangulation with R rectangles and S segments, the relation $R = S + 1$ holds. Therefore, the problem of finding a minimum rectangulation is equivalent to the problem of adding a set of segments of minimum cardinality that induces a rectangulation of P.

Each concave corner v of P (red vertices in Fig. 3.2) must be incident with at least one segment of any rectangulation. We call this the *requirement* at v (the requirement is related to what we later call the *exponent* of v). A *good segment* is a segment connecting two concave corners; it is called good because it satisfies the requirement of two corners. Figure 3.5 shows the given polygon together with all the good segments. The example shows that pairs of segments may cross or share an endpoint, we say that they are in *conflict*.

Ultimately the rectangulation problem for P reduces to study a conflict graph C whose vertices are the possible good segments. Two good segments in C are joined by an edge, when they are in conflict. The conflict graph is bipartite, with blue segments on one side and red segments on the other. A minimum family of segments that yields a rectangulation corresponds to a maximum independent set I of C plus a set of segments covering the corners which are not incident with elements in I. Computing a maximum independent set in a bipartite graph can be done in polynomial-time, the same holds true for computing a minimum family of segments rectangulating P.

Although we know how to find an exact solution for rectangulating P, we now describe how to find an approximate solution to illustrate the algorithm we will use to prove Theorem 1. Let S be a set of segments which is initially empty. Consider the concave corners of P one by one, if the current corner v is not incident to a segment S choose a direction d (horizontal or vertical) and *shoot a segment* in direction d from v, i.e., let s_v be the segment with one end at v extending into the interior of P and ending at the first point which belongs to a segment in S or the boundary of P, this segment s_v is added to S. From the discussion it should be clear that for laminar families of chords this process yields a rectangulation that uses at most twice as many the segments as an optimal one, i.e., it is a 2-approximation for the number of segments and also for the number of rectangles. We refer to this approach as the *greedy strategy*. Fink et al. [6] analyze this strategy in the setting where the input of the bundling problem is a circular drawing. They show that it yields a 10-approximation for the number of bundles. In Sects. 4, 5 and 6 we analyze the greedy strategy for the bundling problem for good drawings and show that this simple-minded strategy guarantees an 8-approximation (Theorem 1).

In Sect. 7 of the full version, we study bipartite collections of strings; this is somehow closer to the laminar family studied here. There we show that a slightly modified greedy strategy produces a solution which is a $\frac{3}{2}$-approximation for the number of segments needed to rectangulate P. The key is to first compute a set of segments of one color which maximizes a parameter called marginal gain. This set is extended using the greedy strategy until a rectangulation is obtained.

2   Strings and Nets

We think of a bundling instance as a set of strings, i.e., as a set of simple curves in the plane. Throughout this paper, unless otherwise stated, we assume that the endpoints of any two strings are different and the union of the strings forms a connected set. A set $\mathcal{E}$ of strings is *grounded* in $\mathcal{B}$ (see Fig. 4.2) if $\mathcal{B}$ is a set of pairwise disjoint blue simple closed-curves, we refer to them as *boundary curves*, such that (1) each string has its ends in the union of the boundary curves; (2) boundary curves only intersect the strings at their ends; (3) each boundary curve contains at least one end of a string; and (4) no two boundary curves are incident with the same cell of $\mathcal{E} \cup \mathcal{B}$.

One can always turn a set of strings into a grounded one by adding a single blue curve in each face where strings end (Fig. 4.1/4.2). Henceforth, unless otherwise stated, any set of strings we consider is grounded.

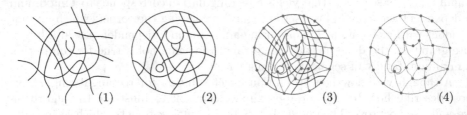

Fig. 4. A set of strings, its grounding, and the corresponding net

To a grounded set of strings $(\mathcal{E}, \mathcal{B})$ we associate a plane graph which is called the *dual net* or just the *net* (Fig. 4.3/4.4). The net $\mathcal{N}$ is obtained by placing a vertex in each cell which is not bounded by a single closed curve in $\mathcal{B}$ and by adding an edge between two vertices whenever their corresponding cells share a segment of a string connecting two consecutive intersection points. We will draw our dual net as in Fig. 4 so that each vertex corresponding to a cell incident to a boundary curve is drawn on the boundary curve, and edges connecting two consecutive vertices in the same boundary curve are also drawn along the boundary curve. Note that boundary holes containing one ore two ends of strings cause loops and double-edges in the net.

The faces of $\mathcal{N}$ come in two kinds: *boundary-holes*, defined as the faces bounded by the boundary blue curves; and *squares*, each of them enclosing exactly one crossing of strings.

A bundle of a set of strings corresponds to a rectangle in the dual net. To be more precise, a set of squares is a *rectangle* if their enclosed crossings induce a bundle, i.e., if the squares can be labeled with a vertex-set of an $n \times m$-grid, so that any two squares adjacent in the grid share an edge of their boundaries. A *rectangulation* is a partition of the squares into rectangles, and such rectangulation is *optimal* if it corresponds to a bundling with a minimum number of bundles.

3 Segments and Holes

In this section we consider a fixed set of grounded strings with its dual net $\mathcal{N} = (V, E)$. The *border* of $\mathcal{N}$ is the subgraph induced by vertices and edges drawn on boundary curves. Any vertex or edge of $\mathcal{N}$ not on the border is *interior*. A degree-4 interior vertex is called *regular* and any other interior vertex is a *vertex-hole*. A *hole* is a vertex or face of $\mathcal{N}$ that is either a vertex-hole or a boundary-hole.

A path of $\mathcal{N}$ is *straight* if all its inner vertices are regular and any two consecutive edges are opposite in the rotation at their common vertex. Since a single edge has no inner vertex it qualifies as straight path. A *cut-set* is a set $\mathcal{S}$ of edge-disjoint straight-paths where every end of a path is either on a hole or in the interior of another path of $\mathcal{S}$. We refer to the elements of a cut-set as *segments*.

Given a rectangulation $\mathcal{R}$ of $\mathcal{N}$, an edge of E is *separating* if it belongs to the boundary of two squares belonging to different rectangles. A cut-set *delimits* $\mathcal{R}$ if its segments only include separating edges and each separating edge is included in a segment.

Given $\mathcal{R}$ it is easy to iteratively build a delimiting cut-set: The first segment s_1 is obtained by considering any edge e_1 in the set $E' \subseteq E$ of separating edges, and by maximally extending e_1 into a straight path. In the i-th step, s_i is obtained by maximally extending an edge e_i of $E' \setminus E(s_1 \cup \cdots \cup s_{i-1})$ into a straight path that is edge-disjoint from $s_1, \ldots, s_{i-1}$. This is done until all the edges of E' are covered by segments.

If there is no regular vertex v whose four incident edges are separating, then the cut-set delimiting $\mathcal{R}$ is unique. Otherwise, for any such vertex v, one can choose the pair of opposite edges which belong to a common segment, the other two segments end at v. This binary choice at any such vertex v may give rise to exponentially many delimiting cut-sets. In practice, we will not be bothered by this technicality. We choose and fix an arbitrary cut-set of $\mathcal{R}$. With this in mind, we can now refer to the *segments of* $\mathcal{R}$ as the segments of the chosen cut-set delimiting $\mathcal{R}$.

Figure 5 shows a rectangulation with 10 segments and the corresponding bundling of the crossings with 6 bundles.

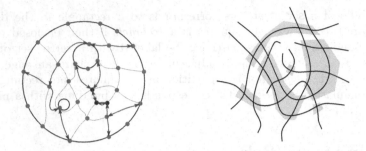

Fig. 5. A rectangulation with 10 segments (holes are blue) and the corresponding bundling. (Color figure online)

Next we relate the number of rectangles and the number of segments in a rectangulation. Let $H = H(\mathcal{N})$ denote the number of holes of $\mathcal{N}$.

Lemma 1. *In a net $\mathcal{N}$ with H holes the numbers R of rectangles and S of segments satisfy:*

$$R - S + H = 2. \tag{1}$$

To prove Lemma 1, to any rectangulation $\mathcal{R}$ of $\mathcal{N}$ we will associate a cubic plane graph $\Gamma = \Gamma(\mathcal{R})$. The construction will be used again in Sect. 5.

Construction of Γ: First, we consider the subgraph $\mathcal{H}$ of $\mathcal{N}$ obtained from the union of the segments in $\mathcal{R}$ and the border of $\mathcal{N}$. Color the edges of $\mathcal{H}$ included in segments such that for each segment all the edges on the segment have the same color and the colors used for different segments are distinct.

Next, we apply a local transformation at each vertex $v \in V(\mathcal{H})$: If v is a regular vertex, then we keep v unchanged unless $\deg_{\mathcal{H}}(v) = 4$. In this case we split v into two vertices v_1, v_2 joined by an edge as in Fig. 6, where each v_i has degree 3 and the color of $v_1 v_2$ is the color of the unique segment having v in its interior.

Fig. 6. Local transformation when v is regular and $\deg_{\mathcal{H}}(v) = 4$.

If v belongs to the border and $\deg_{\mathcal{H}}(v) \geqslant 4$ or if v is a vertex-hole of arbitrary degree we split v as follows. Let $\rho_{\mathcal{H}}(v) = (e_1, \ldots, e_d)$ be its rotation in $\mathcal{H}$, so that when v is in the border of $\mathcal{N}$, we assume that the edges e_1 and e_d are also in the border. Subdivide each vertex e_i by adding a degree-2 vertex in the middle of e_i. Remove v and all the half-edges incident with v. Next, add a path $(e_1, \ldots, e_d)$ or a cycle $(e_1, \ldots, e_d, e_1)$ depending whether v is in the border or not.

Finally, we suppress degree-2 vertices, i.e., if v is a vertex with $\deg_{\mathcal{H}}(v) = 2$ and incident edges e_1, e_2, then we delete v and make e_1, e_2 a single edge. This yields Γ. Figure 7 shows an example.

Proof of Lemma 1 Given a rectangulation $\mathcal{R}$, let $\Gamma = \Gamma(\mathcal{R})$ as above. Each segment of the rectangulation corresponds to a monochromatic path in Γ and each vertex of Γ is an end-vertex of exactly one of them. Thus $|V(\Gamma)| = 2S$. As Γ is cubic, $E(\Gamma) = \frac{3}{2}|V(\Gamma)| = 3S$. Finally, as each face of Γ corresponds to either a rectangle or a hole, Γ has $R + H$ faces. Equation 1 now follows from Euler's formula.

Fig. 7. The graph Γ corresponding to the rectangulation of Fig. 5. Vertices obtained by splitting vertex-holes or vertices on boundaries are shown in light blue. (Color figure online)

Before concluding this section, we make some remarks about Γ that will be used later. We let Γ^* denote the plane dual of Γ.

Remark 1. (i) Each face of Γ corresponds to a hole or to a rectangle.
(ii) Γ^* is a plane triangulation, i.e., the boundary of each face consists of three edges.
(iii) Γ and its planar dual Γ^* are simple graphs when every hole is incident with at least three segments of $\mathcal{R}$ (otherwise Γ has multi-edges and/or loops).
(iv) The vertices of Γ^* corresponding to holes form an independent set in Γ^*.

4 Approximating the Number of Segments

Let $\mathcal{N} = (V, E)$ be a net and $B \subset E$ be the set of edges in the border of $\mathcal{N}$. An edge-set $E_0 \subset E$ *saturates* a vertex $v \in V$ if each angle induced by the edges of $E_0 \cup B$ at v sees at most two squares, or, if v is regular and no edge of E_0 is incident with v. Moreover, E_0 *saturates* $\mathcal{N}$ if every vertex is saturated by E_0.

Naturally, the (edge-set of the) segments of a rectangulation saturate $\mathcal{N}$. The next lemma shows that saturation is also sufficient to induce a rectangulation.

Lemma 2 ($\star$). *A cut-set of a net $\mathcal{N}$ is saturating if and only if it delimits a rectangulation.*

Observe that this lemma does not extend to more general systems of curves. In Fig. 8 we depict more general systems of curves allowing closed curves and self-intersecting strings. Their corresponding dual nets are saturated by $\varnothing$ whereas $\varnothing$ does not induce a rectangulation in any of them.

Fig. 8. Two bad examples a square ring and a square loop.

The duals of the two nets shown in Fig. 8 contain special configuration of squares forbidden in nets of strings: A *square-ring* is a circular sequence $(s_0, s_1, \ldots, s_m, s_0)$ of squares, where s_i is glued to s_{i-1} and s_{i+1} by using opposite sides of s_i. A *square-loop* is similar to a square-ring, except that one square is glued by using two consecutive sides instead of opposite sides.

Definition 1 (Exponent). *Given a net $\mathcal{N} = (V, E)$ and $v \in V$, the exponent of v is the minimum number of edges in an edge-set saturating v. Hence $exp(v) = 0$ if v is a regular vertex and*

$$
exp(v) = \begin{cases} \left\lfloor \frac{deg_{\mathcal{N}}(v)}{2} \right\rfloor - 1 & v \text{ is in the border;} \\ \left\lceil \frac{deg_{\mathcal{N}}(v)}{2} \right\rceil & v \text{ is a vertex-hole.} \end{cases}
$$

We let $exp(\mathcal{N}) := \sum_{v \in V} exp(v)$.

A Greedy Strategy: This strategy consists on linearly ordering the vertices $v_1, \ldots, v_k$ of $\mathcal{N}$ with positive exponent and start adding segments at the vertices with increasing index. When it comes to v_i some incident edges may already be contained in segments belonging to earlier vertices. Select a minimal set of edges not covered by the segments such that shooting segments from these edges results in an edge set saturating v_i. Note that the number of segments introduced to saturate v_i is upper bounded by $exp(v_i)$.

Henceforth, we will denote the number of rectangles and segments in an optimal rectangulation of $\mathcal{N}$ by R_{opt} and S_{opt}, respectively. Likewise, we let R_{greed} and S_{greed} be the number of rectangles and segments, respectively, obtained after a run of the greedy strategy in $\mathcal{N}$ for some linear order of the vertices.

Observation 1 (∗). *The following hold true for a net $\mathcal{N}$:*

(i) $S_{greed} \leq exp(\mathcal{N})$;
(ii) $S_{opt} \geq \frac{1}{2} exp(\mathcal{N})$; and consequently
(iii) $S_{greed} \leq 2 \cdot S_{opt}$.

5   Rectangles and Holes

In this section we collect a few facts about rectangles and holes that will be used in our approximations. It is important to observe that a greedy rectangulation approximates the optimal when R_{greed} is bounded by a constant factor of R_{opt}. The next observation already gives a related bound by adding the holes.

Observation 2. $R_{greed} \leq 2R_{opt} + H - 2$.

Proof. Apply Lemma 1 to both sides of Observation 1.(iii)

Our task of approximating R_{opt} now reduces to understand under which circumstances $H = O(R_{opt})$. Let us start by deriving a bound for the odd holes. Let H_{odd} be the number of vertex-holes with odd degree in $\mathcal{N}$, i.e., H_{odd} is the number of vertices of odd degree $\geqslant 3$ in $\mathcal{N}$.

Observation 3 (∗). $H_{odd} \leqslant 4 \cdot R_{opt}$.

Our next observation about holes requires the following general observation about triangulations:

Observation 4 (∗). *In a simple plane triangulation with n vertices, an independent set has size at most $\frac{2}{3}n$.*

Definition 2. *Given a rectangulation $\mathcal{R}$ of $\mathcal{N}$, $\delta(\mathcal{R})$ denotes the minimum number of segments of $\mathcal{R}$ incident to any hole of $\mathcal{H}$ ($\delta(\mathcal{R})$ is the same as the minimum degree among the vertices of $\Gamma^*(\mathcal{R})$ that represent holes). We let $\delta(\mathcal{N})$ be the minimum integer k for which there is an optimal rectangulation $\mathcal{R}_0$ of $\mathcal{N}$ with $\delta(\mathcal{R}_0) = k$.*

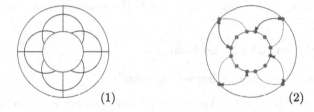

(1)   (2)

Fig. 9. A set of strings (1) and its dual net (2). The red segments induce an optimal rectangulation, indeed $\delta(\mathcal{N}) = 4$. The exponents only imply $\delta(\mathcal{N}) \geqslant 0$. (Color figure online)

Note that when h is a vertex-hole, then $\deg_{\Gamma^*(\mathcal{R})}(h) \geqslant \exp(h)$. For boundary-holes a corresponding lower bound is given by the sum of exponents of the incident vertices. Figure 9 shows that there are examples where $\delta(\mathcal{N})$ is much larger than given by these lower bounds.

Observation 5 (∗). *If $\delta(\mathcal{N}) \geqslant 3$, then $H \leqslant 2 \cdot R_{opt}$.*

5.1 Toothed-Holes and Toothed-Faces

Previously, we saw that if all holes are incident with at least three segments, then H is $O(R_{opt})$. Unfortunately, it is not true in general that H is $O(R_{opt})$: Fig. 10 shows that H can be arbitrarily large compared to R_{opt}. With the next lemma we prove that the unboundedness of H in terms of R_{opt} can always be attributed to the presence of toothed-faces.

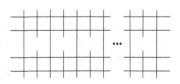

Fig. 10. An example where $R_{opt} = 2$ and H is arbitrarily large.

Lemma 3 (∗). $H \leqslant 6R_{opt} + t(\mathcal{E}) - 4$.

6   Approximations for the Number of Rectangles

The first proposition in this section states that the greedy strategy results in a 4-approximation when $\delta(\mathcal{N}) \geqslant 3$. Two families of strings whose elements have a dual net satisfying this condition are circular drawings with a bipartite sets of pseudosegments and triangle-free hyperbolic line arrangements. For references to triangle-free hyperbolic line arrangements we refer to Eppstein [5, Section 7] and his figure [4] in the Wikipedia article on circle graphs.

Proposition 1. *If $\delta(\mathcal{N}) \geqslant 3$, then $R_{greed} \leqslant 4 \cdot R_{opt} - 2$.*

Proof. Apply Observations 2 and 5.

Condition $\delta(\mathcal{N}) \geqslant 3$ is restrictive as it forbids in a set of strings the existence of a cell bounded by three pieces of strings. The next lemma handles very general sets of strings at the expense of a larger approximation factor.

Lemma 4. *If $\mathcal{N}$ is the dual net of a set of pseudosegments, then $R_{greed} \leqslant 8R_{opt} + t(\mathcal{E}) - 6$.*

Proof. Apply Observation 2 and Lemma 3.

Now our main result is an immediate corollary.

Proof (Proof of Theorem 1). Apply the greedy strategy and Lemma 3 to each connected component of the set of strings $\mathcal{E}$ associated to D to obtain the desired bundling.

7   Conclusion

In this paper we studied the bundled crossing number of connected good drawings and showed that the greedy strategy derived from the problem of rectangulating an orthogonal polygon leads to an 8-approximation (up to adding the number of toothed-faces in the drawing). In the full version we improved this strategy for bipartite instances by considering an initial good set of segments. We hope that the tools and the framework developed in this work will inspire more results about bundled crossings. We leave below some open questions.

1. Is there a constant c guaranteeing that, for any simple drawing D, the greedy algorithm produces bundling with at most $c \cdot bc(D)$ bundles? In other words, are toothed-faces relevant to approximate $bc(D)$?
2. What is the computational complexity of computing $bc(\mathcal{E})$ for bipartite instances?

References

1. Alam, M.J., Fink, M., Pupyrev, S.: The bundled crossing number. In: Hu, Y., Nöllenburg, M. (eds.) GD 2016. LNCS, vol. 9801, pp. 399–412. Springer, Cham (2016). https://doi.org/10.1007/978-3-319-50106-2_31
2. Beineke, L., Wilson, R.: The early history of the brick factory problem. Math. Intell. **32**(2), 41–48 (2010)
3. Chaplick, S., van Dijk, T.C., Kryven, M., Park, J., Ravsky, A., Wolff, A.: Bundled crossings revisited. In: Archambault, D., Tóth, C.D. (eds.) GD 2019. LNCS, vol. 11904, pp. 63–77. Springer, Cham (2019). https://doi.org/10.1007/978-3-030-35802-0_5
4. Eppstein, D.: https://en.wikipedia.org/wiki/File:Ageev5Xcirclegraph.svg
5. Eppstein, D.: Graph-theoretic solutions to computational geometry problems. In: Paul, C., Habib, M. (eds.) WG 2009. LNCS, vol. 5911, pp. 1–16. Springer, Heidelberg (2010). https://doi.org/10.1007/978-3-642-11409-0_1
6. Fink, M., Hershberger, J., Suri, S., Verbeek, K.: Bundled crossings in embedded graphs. In: Kranakis, E., Navarro, G., Chávez, E. (eds.) LATIN 2016. LNCS, vol. 9644, pp. 454–468. Springer, Heidelberg (2016). https://doi.org/10.1007/978-3-662-49529-2_34
7. Harary, F., Hill, A.: On the number of crossings in a complete graph. Proc. Edinburgh Math. Soc. **13**(4), 333–338 (1963)
8. Harary, F., Kainen, P.C., Schwenk, A.J.: Toroidal graphs with arbitrarily high crossing numbers. Nanta Math. **6**(1), 58–67 (1973)
9. Schaefer, M.: The graph crossing number and its variants: a survey. Electron. J. Comb., 90 (2013). Dynamic Survey 21
10. Turán, P.: A note of welcome. J. Graph Theory **1**(1), 7–9 (1977)

Path Cover Problems with Length Cost

Kenya Kobayashi[1], Guohui Lin[2], Eiji Miyano[1(✉)], Toshiki Saitoh[1],
Akira Suzuki[3], Tadatoshi Utashima[1], and Tsuyoshi Yagita[1]

[1] Kyushu Institute of Technology, Iizuka, Japan
{kobayashi.kenya472,utashima.tadatoshi965,
yagita.tsuyoshi307}@mail.kyutech.jp,
{miyano,toshikis}@ai.kyutech.ac.jp
[2] University of Alberta, Edmonton, Canada
guohui@ualberta.ca
[3] Tohoku University, Sendai, Japan
akira@tohoku.ac.jp

Abstract. For a graph $G = (V, E)$, a collection $\mathcal{P}$ of vertex-disjoint
(simple) paths is called a *path cover* of G if every vertex $v \in V$ is
contained in exactly one path of $\mathcal{P}$. The PATH COVER problem (PC
for short) is to find a minimum cardinality path cover of G. In this
paper, we introduce generalizations of PC, where each path is associated with a weight (cost or profit). Our problem, MINIMUM (MAXIMUM)
WEIGHTED PATH COVER (MinPC (MaxPC)), is defined as follows: Let
$U = \{0, 1, \ldots, n - 1\}$. Given a graph $G = (V, E)$ and a weight function $f : U \to \mathbb{R} \cup \{+\infty, -\infty\}$, which defines a weight for each path in its
length, MinPC (MaxPC) is to find a path cover $\mathcal{P}$ of G such that the total
weight of the paths in $\mathcal{P}$ is minimized (maximized). Let L be a subset
of U, and P^L be the set of paths such that each path is of length $\ell \in L$.
We especially consider $\text{Min}P^L\text{PC}$ with 0–1 cost, i.e., the cost function is
$f(\ell) = 1$ if $\ell \in L$; otherwise $f(\ell) = 0$. We also consider $\text{Max}P^L\text{PC}$ with
$f(\ell) = \ell + 1$, if $\ell \in L$; otherwise $f(\ell) = 0$. That is, $\text{Max}P^L\text{PC}$ is to maximize the number of vertices contained in the paths with length $\ell \in L$
in a path cover. In this paper, we first show that $\text{Min}P^{\{0,1,2\}}\text{PC}$ is NP-
hard for planar bipartite graphs of maximum degree three. This implies
that (i) for any constant $\sigma \geq 1$, there is no polynomial-time approximation algorithm with approximation ratio σ for $\text{Min}P^{\{0,1,2\}}\text{PC}$ unless
P=NP, and (ii) $\text{Max}P^{\{3,\ldots,n-1\}}\text{PC}$ is NP-hard for the same graph class.
Next, (iii) we present a polynomial-time algorithm for $\text{Min}P^{\{0,1,\ldots,k\}}\text{PC}$
on graphs with bounded treewidth for a fixed k. Lastly, (iv) we present a
4-approximation algorithm for $\text{Max}P^{\{3,\ldots,n-1\}}\text{PC}$, which becomes a 2.5-
approximation for subcubic graphs.

1 Introduction

Let $G = (V, E)$ be an unweighted and undirected graph, where V and E are
the sets of vertices and edges, respectively. Unless otherwise stated, we denote
$n = |V|$. A *path* is a sequence of vertices in which each vertex is connected by an

© Springer Nature Switzerland AG 2022
P. Mutzel et al. (Eds.): WALCOM 2022, LNCS 13174, pp. 396–408, 2022.
https://doi.org/10.1007/978-3-030-96731-4_32

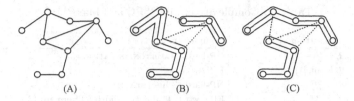

Fig. 1. (A) Graph G and (B)(C) two path covers of G; both of the path covers (B) and (C) are the optimal solutions of PC on G, while (C) is the optimal solution for MinP[2]PC under the current binary cost function. The path cover (B) contains one length-2 path and the cost of the path cover (B) is one. The path cover (C) only contains paths of length-3 or longer and the cost of the path cover (C) is zero.

edge to the next. The path length is the number of edges in the path. Given a graph G, a collection $\mathcal{P}$ of vertex-disjoint simple paths is called a *path cover* of G if every vertex $v \in V$ is contained in exactly one path of $\mathcal{P}$. The goal of the PATH COVER problem (PC for short) is to find a minimum cardinality path cover of G. PC is a fundamental problem and contains the well-known HAMILTONIAN PATH problem, which is to seek a single path visiting every vertex exactly once. In other words, PC generalizes the HAMILTONIAN PATH problem. Therefore, if the HAMILTONIAN PATH problem is NP-hard for a certain graph class $\mathcal{G}$, PC is also NP-hard for $\mathcal{G}$. In this paper, we introduce generalizations of PC, where each path is associated with a weight (which represents cost or profit). Our problems are defined as follows:

Minimum (Maximum) Weighted Path Cover (MinPC (MaxPC))

Let $U = \{0, 1, \ldots, n-1\}$ denote a set of path lengths (where the length of an isolated vertex is defined as 0). Given a graph $G = (V, E)$ and a cost (profit) function $f : U \to \mathbb{R} \cup \{+\infty, -\infty\}$, which defines a cost (profit) for each path in its length, find a path cover $\mathcal{P}$ of G such that the total cost (profit) of the paths in $\mathcal{P}$ is minimized (maximized).

Let L be a subset of U. We denote the set consisting of all the paths whose length is in L as P^L. We, especially, consider MinPC whose cost function is $f(\ell) = 1$ if $\ell \in L$; otherwise $f(\ell) = 0$. The problem is denoted by $\mathrm{Min}P^L\mathrm{PC}$. Under the current setting, namely, $\mathrm{Min}P^L\mathrm{PC}$ is a problem to find a path cover with the minimum number of paths of length $\ell \in L$ since we consider a binary cost function. We also consider the problem $\mathrm{Max}P^L\mathrm{PC}$ with $f(\ell) = \ell + 1$, if $\ell \in L$, and $f(\ell) = 0$, otherwise. We can see that the objective function of $\mathrm{Max}P^L\mathrm{PC}$ is to maximize the number of vertices contained in the paths with length $\ell \in L$ in a path cover. For ease of expression, let $[k] = \{0, 1, \ldots, k\}$ and let $\overline{[k]} = U \setminus [k] = \{k+1, k+2, \ldots, n-1\}$, i.e., $\mathrm{Min}P^{[k]}\mathrm{PC}$ means $\mathrm{Min}P^{\{0,\ldots,k\}}\mathrm{PC}$. Consider the graph G in Fig. 1(A). Since G has no Hamiltonian paths, both of the path covers (B) and (C) are the optimal solutions for PC containing two paths. For $\mathrm{Min}P^{[2]}\mathrm{PC}$, however, the path cover (B) has cost one for $\mathrm{Min}P^{[2]}\mathrm{PC}$ since it contains one length-2 path, and the path cover (C) achieves cost zero for $\mathrm{Min}P^{[2]}\mathrm{PC}$ since the two paths are of length three and four, respectively.

Table 1. Complexity of $\mathrm{Min}P^L\mathrm{PC}$ and $\mathrm{Max}P^L\mathrm{PC}$

Problem	The setting of L	Complexity
$\mathrm{Min}P^L\mathrm{PC}$	$L = [n-2]$	NP-hard (HAMILTONIAN PATH)
	$L = [0], [1]$	P ([2])
	$L = [2]$	NP-hard (**Theorem** 2)
	$L = [k]$	FPT w.r.t. k plus treewidth (**Theorem** 3)
$\mathrm{Max}P^L\mathrm{PC}$	$L = \{0\} \cup L'$ for any L'	P (the optimal solution is trivially V)
	$L = \{1\}$	P (MAXIMUM MATCHING)
	$L = \{2\}$	NP-hard ([4])
	$L = \overline{[2]}$	NP-hard (**Corollary** 2) 4-approximation algorithm (**Theorem** 4)

Related Work. Note that several classical problems can be seen as special cases of MinPC or MaxPC. For example, the HAMILTONIAN PATH problem is a decision version with cost zero of $\mathrm{Min}P^{[n-2]}\mathrm{PC}$ and the MAXIMUM MATCHING problem is equivalent to $\mathrm{Max}P^{\{1\}}\mathrm{PC}$. Also, the $\mathrm{Max}P_k\mathrm{PACKING}$ problem (defined later) is a special case of MaxPC. In Table 1, we summarize the computational complexity results on $\mathrm{Min}P^L\mathrm{PC}$ and $\mathrm{Max}P^L\mathrm{PC}$.

One of the most natural variants of PC is to have some bounds of lengths for paths, for example, the number of vertices of each path, in the path cover. In fact, such a problem has been well-researched for several years [3,5,8,11]. Chen et al. [3] state that, if the number of vertices of each path in the path cover is restricted to *exactly* some constant k, then the problem is called P_k-PARTITIONING. Garey and Johnson [4] show that P_3-PARTITIONING is NP-complete. This implies that $\mathrm{Max}P^{\{2\}}\mathrm{PC}$ is NP-hard. Similarly, if the number of vertices of each path in the path cover is *at most* some constant k, then the problem is called k-PATH PARTITION (kPP), which is to find a minimum collection of vertex-disjoint paths. We remark that kPP is equivalent to MinPC with cost function f such that the cost of each path of length at most $k - 1$ is 1, or $+\infty$ otherwise. kPP is a nice relaxation of P_k-PARTITIONING, since it always ensures the existence of a feasible solution on any given graph, while P_k-PARTITIONING not.

The maximization variant of P_k-PARTITIONING is also studied, which is often called $\mathrm{Max}P_k\mathrm{PACKING}$. Given a graph G, $\mathrm{Max}P_k\mathrm{PACKING}$ aims to find the maximum number of vertex-disjoint paths of exactly k vertices. For example, Monnot and Toulouse [11] discuss its (in)tractabilities.

$\mathrm{Max}P_k\mathrm{PACKING}$ is equivalent to MaxPC with cost function $f(k-1) = 1$ and $f(\ell) = -\infty$ for $\ell \in \{0,\ldots,n-1\} \setminus \{k-1\}$.

On the other hand, for the variant of PC with a *lower bound* of the length of each path, there are only a few algorithmic results so far. Almost all of the known results for this variant are obtained from the graph-theoretical point of view, such as the *existence* of such path covers and its necessary and sufficient conditions. For example, we can find a related recent result in [12]. A $P_{\geq 3}$-factor F of a graph G is a spanning subgraph of G such that every component of F is a path of length at least two [7]. A graph G is called a $P_{\geq 3}$-factor covered graph if G has a $P_{\geq 3}$-factor including all edges of G. Zhou et al. [12] show some sufficient conditions for graphs to be $P_{\geq 3}$-factor covered graphs. Very recently, Cai et al. [2] and Gómez and Wakabayashi [6] have independently studied this variant from the algorithmic point of view. In the former paper, they propose algorithms to find a path cover with the minimum number of length-0 paths and with the minimum total number of length-0 and length-1 paths. The latter paper studies a path cover problem without using length-0 paths. Given a graph, their algorithm returns such a path cover if it exists, or answers no together with a certificate. The maximization variant is also introduced. In the previous papers [2, 6], it is known that $\mathsf{Min}P^{[0]}\mathsf{PC}$ and $\mathsf{Min}P^{[1]}\mathsf{PC}$ with the same cost function as ours can be solved in polynomial time. This also shows that $\mathsf{Max}P^{[0]}\mathsf{PC}$ and $\mathsf{Max}P^{[1]}\mathsf{PC}$ are both polynomial-time solvable.

Our Results. In this paper, we show that $\mathsf{Min}P^{[2]}\mathsf{PC}$ is NP-hard even on planar bipartite graphs with maximum degree three. Our proof also implies that $\mathsf{Max}P^{[2]}\mathsf{PC}$ is NP-hard on the same graph class. Our result can also be evaluated with the following perspective. When one tries to seek the complexity of PC on some graph class, usually it depends on the hardness of HAMILTONIAN PATH problem on the same class. In other words, it has never been argued which length of path affects the hardness of PC. Thus, our result can be recognized as the first step of revealing the hidden structures in PC. Our main goal is to reveal the computational complexities of this path cover variant with respect to path lengths, restricting the graph structures. Our contribution is as follows.

- First, we show that $\mathsf{Min}P^{[2]}\mathsf{PC}$ is NP-hard on planar bipartite subcubic graphs, by giving a polynomial-time reduction from PLANAR 3-SAT. Accordingly, we obtain the NP-hardness of $\mathsf{Max}P^{[2]}\mathsf{PC}$ on the same graph class. Furthermore, our reduction shows that, for any constant $\sigma \geq 1$, there is no polynomial-time approximation algorithm with approximation ratio σ for $\mathsf{Min}P^{[2]}\mathsf{PC}$ unless P = NP.
- Fortunately, however, we can solve $\mathsf{Min}P^{[k]}\mathsf{PC}$ on graphs with bounded treewidth in polynomial time. In other words, we show that $\mathsf{Min}P^{[k]}\mathsf{PC}$ is fixed-parameter tractable when parameterized by treewidth plus k. Assuming we are given an n-vertex graph of treewidth at most W together with its tree decomposition, the running time of our algorithm is $O(4^{2W} \cdot W^{2W+2} \cdot (k+2)^{2W+2} \cdot n)$.
- Finally, we focus on the approximability of the maximization variant; we design a polynomial-time approximation algorithm for $\mathsf{Max}P^{[2]}\mathsf{PC}$. We prove that its approximation ratio is 4 in general, and can be improved to 2.5 if the maximum degree of the input graph is three.

Due to the space constraint, some results and proofs are omitted from this extended abstract.

2 Preliminaries

Definitions. All graphs considered in this paper are unweighted, undirected, and simple. That is, graphs are containing no loops or multiple edges. For a graph $G = (V, E)$, $V(G)$ and $E(G)$ denote the vertex set and the edge set of G, respectively. A subgraph of G induced by a vertex set $S \subseteq V$ is denotes by $G[S]$. The open neighborhood of a vertex v is denoted by $N(v)$. The *degree* of a vertex v on G is denoted by $\deg_G(v)$, which is defined as $\deg_G(v) = |N(v)|$.

A (simple) *path* p of k vertices from a vertex v_1 to v_k is represented by a sequence $p = \langle v_1, \ldots, v_k \rangle$ of distinct vertices. If a path has no vertices, then it is called an *empty path*. The vertex set of p is referred to as $V(p)$ and v_1 and v_k are called *endpoints* of p. A *length* of path p is defined by the number of edges contained in p, denoted by $\ell(p)$. If the length of a path p is ℓ, then we say that p is a length-ℓ path. We often use P^ℓ to denote a set of all length-ℓ paths. To simplify the expression, let the set of lengths $\ell_1, \ell_2, \ldots, \ell_c$ be denoted by "length-$\{\ell_1, \ell_2, \ldots, \ell_c\}$." Also, let the set of paths of length ℓ_1, ℓ_2, $\ldots$, or ℓ_c be denoted by $P^{\{\ell_1, \ell_2, \ldots, \ell_c\}}$. Consider two vertex-disjoint paths $p_1 = \langle v_{1,0}, \ldots, v_{1,k_1} \rangle$ and $p_2 = \langle v_{2,0}, \ldots, v_{2,k_2} \rangle$ such that v_{1,k_1} is adjacent to $v_{2,0}$. A *concatenation* of p_1 and p_2 is a path $p = \langle v_{1,0}, \ldots, v_{1,k_1}, v_{2,0}, \ldots, v_{2,k_2} \rangle$ and denoted by $p = p_1 \oplus p_2$. Since the graph is simple, this concatenation is defined uniquely if we are given two paths. For a graph G, a *path cover* $\mathcal{P}$ of G is a set of vertex-disjoint paths such that every vertex in G belongs to exactly one path of $\mathcal{P}$.

Given a graph G, a *tree decomposition* of G is a pair $\mathcal{T} = (T, \{X_t\}_{t \in V(T)})$, where T is a tree and X_t is a subset of $V(G)$ called a *bag*, such that (i) every vertex of G is contained in at least one bag, (ii) for every edge $uv \in E(G)$, there exists a node $t \in V(T)$ such that bag X_t contains both u and v, and (iii) for every vertex $v \in V(G)$, the set $T_v = \{t \in V(T) : v \in X_t\}$, i.e., the set of nodes whose corresponding bags contain v, induces a connected subtree of T. The *width* of a tree decomposition is $\max_{t \in V(T)} |X_t| - 1$, and the *treewidth* of the graph G is the minimum possible width of a tree decomposition of G. Note that, given a graph, it is NP-hard to obtain the optimal treewidth of the graph [1]. If the treewidth of a graph G is bounded in some constant W, then G is called a bounded-treewidth graph, or a graph with bounded treewidth.

An algorithm ALG is called a σ-approximation algorithm and ALG's approximation ratio is σ if $ALG(x)/OPT(x) \leq \sigma$ ($OPT(x)/ALG(x) \leq \sigma$, resp.) holds for every input x of a minimization (maximization, resp.) problem, where $ALG(x)$ and $OPT(x)$ are the total costs (profits, resp.) of solutions obtained by ALG and an optimal algorithm, respectively.

Hardness and Tractability Results. Before proceeding to our results, we show some basic results. We define the decision version of $\mathsf{Min}P^L\mathsf{PC}$, named $\mathsf{Min}P^L\mathsf{PC}(t)$, as follows: Given a graph $G = (V, E)$ and an integer t, $\mathsf{Min}P^L\mathsf{PC}(t)$ determines whether there is a path cover $\mathcal{P}$ of G such that the total cost of the paths in $\mathcal{P}$ is at most t or not. Recall that the cost function in $\mathsf{Min}\mathsf{PC}$ is $f(\ell) = 1$ if $\ell \in L$; otherwise $f(\ell) = 0$. Therefore, the decision problem is equivalent to determine whether there is a path cover $\mathcal{P}$ of G such that the number of length-L paths is at most t. Similarly, we define the decision version of $\mathsf{Max}P^L\mathsf{PC}$. Recall that this problem is MaxPC with $f(\ell) = \ell + 1$, if $\ell \in L$, and $f(\ell) = 0$ otherwise. $\mathsf{Max}P^L\mathsf{PC}(t)$ is a problem to determine whether there exists a path cover such that the number of vertices contained in the paths with length $\ell \in L$ in the path cover is at least t. First, we can show the following proposition:

Proposition 1. *$\mathsf{Min}P^{[k]}\mathsf{PC}(0)$ is equivalent to $\mathsf{Max}P^{\overline{[k]}}\mathsf{PC}(n)$ for any integer k and n-vertex graphs.*

Suppose that we have a path cover $\mathcal{P}$ without length-0, length-1, ..., and length-k paths. Then it is obvious that such a $\mathcal{P}$ contains only length-$(k+1)$, ..., and length-$(n-1)$ paths. Additionally, we should emphasize that, for example, $\mathsf{Min}P^{\overline{[n-2]}}\mathsf{PC}(0)$ coincides with HAMILTONIAN PATH problem. Summing up the results shown in [2,6] and the NP-completeness of HAMILTONIAN PATH problem, we obtain the following (in)tractabilities:

Theorem 1 [2,6]. *$\mathsf{Min}P^{[0]}\mathsf{PC}$ and $\mathsf{Min}P^{[1]}\mathsf{PC}$ are in P, while $\mathsf{Min}P^{[n-2]}\mathsf{PC}$ is NP-hard.*

Furthermore, we can show the following hardness of $\mathsf{Min}P^{[2]}\mathsf{PC}(0)$:

Theorem 2. *$\mathsf{Min}P^{[2]}\mathsf{PC}(0)$ is NP-complete for planar bipartite subcubic graphs.*

The proof is by a polynomial-time reduction to $\mathsf{Min}P^{[2]}\mathsf{PC}(0)$ from PLANAR 3-SAT for 3-CNF-formulas such that each variable occurs in exactly three clauses, exactly once negatively and twice positively [9,10]. Since Theorem 2 implies that it is NP-complete to determine whether the cost is zero or not, we can derive the following hardness of approximation:

Corollary 1. *For any constant $\sigma \geq 1$, there is no polynomial-time approximation algorithm with approximation ratio σ for $\mathsf{Min}P^{[2]}\mathsf{PC}$ unless $P = NP$.*

Fortunately, however, $\mathsf{Min}P^{[k]}\mathsf{PC}$ is solvable in polynomial time if the input is restricted to graphs with bounded treewidth:

Theorem 3. *Let G be an n-vertex graph together with its tree decomposition of treewidth at most W. Then, for a fixed constant k, $\mathsf{Min}P^{[k]}\mathsf{PC}$ can be solved in time $O(4^{2W} \cdot W^{2W+2} \cdot (k+2)^{2W+2} \cdot n)$.*

3 Approximation Algorithm for $\mathsf{Max}P^{\overline{[2]}}\mathsf{PC}$

In this section, we turn our attention to the maximization variant $\mathsf{Max}P^{\overline{[2]}}\mathsf{PC}$. Recall that the goal of $\mathsf{Max}P^L\mathsf{PC}$ is to maximize the total profit, i.e., the number of vertices contained in the paths with length $\ell \in L$ in a path cover. First, by Proposition 1 and Theorem 2, one can verify that $\mathsf{Max}P^{\overline{[2]}}\mathsf{PC}$ is also NP-hard for planar bipartite subcubic graph.

Algorithm 1: ALG

Input: A graph $G = (V, E)$
Output: A path cover P'
1 $P := \emptyset$, $P' := \emptyset$, $\pi = $ null, $i := 1$, $j := 1$;
2 **while** G contains at least one length-3 path **do**
3 | Find a length-3 path p in G;
4 | $P := P \cup \{p\}$, $G := G[V \setminus V(p)]$, $\pi(i) := p$, $i := i + 1$;
   // Next, along π, try the extension procedure
5 **while** $j \neq i$ **do**
6 | Extend $\pi(j)$ as long as possible and rename it p';
7 | $P' := P' \cup \{p'\}$, $G := G[V \setminus V(p')]$, $j := j + 1$;
8 $P' := P' \cup \{\langle u \rangle \mid$ a vertex u is not included in any paths in $P'\}$;
9 **return** P';

Corollary 2. *MaxP$^{\overline{[2]}}$PC is NP-hard for planar bipartite n-vertex subcubic graphs.*

In the following we consider the approximability of MaxP$^{\overline{[2]}}$PC. As shown before in Sect. 2, the minimization variant MinP$^{[2]}$PC is hard from the viewpoint of the approximability; Corollary 1 shows that the minimization variant MinP$^{[2]}$PC is NP-hard to approximate even if the input graph is very restricted. On the other hand, for the maximization variant MaxP$^{[2]}$PC, we show that there is a simple 4-approximation algorithm. Furthermore, we sharpen the approximation algorithm when restricted to subcubic graphs as input, and prove the approximation ratio can be improved to 2.5.

The proposed algorithm ALG is based on the following strategies: (1) First, obtain a *maximal* set of vertex-disjoint length-3 paths. Then, (2) pick an arbitrary path, say, p up, and extend its length by adding a vertex (i.e., length-0 path), a length-1 path, or a length-2 path to the head and/or the tail of p if the extended path does not intersect with any other paths obtained in (1). If several paths can be added, then the longest path is always added, i.e., p is extended to the longest possible direction. In the following, we denote a pick-up ordering of paths by π, where $\pi(i)$ denotes the i-th path in the ordering. See the detailed descriptions of our algorithm ALG. The above (1) and (2) are implemented in the first **while**-loop on the second through fourth lines and in the "Extend" operation on the sixth line, respectively.

Theorem 4. *Algorithm ALG is a 4-approximation algorithm for MaxP$^{\overline{[2]}}$PC.*

Proof. The running time of our algorithm ALG is clearly polynomial. We show the approximation ratio of ALG by estimating an upper bound of the total profit of an optimal solution. Note that the profit of every path of length at most two is zero. Therefore, let ALG and OPT be sets of paths of length at least three obtained by ALG and an optimal algorithm, respectively. Let $V(OPT) = \bigcup_{p \in OPT} V(p)$ and also let $V(ALG) = \bigcup_{p \in ALG} V(p)$. That is, $|V(OPT)|$ and $|V(ALG)|$ are the optimal profit and the profit obtained by ALG, respectively. Suppose that ALG and OPT have α vertex-disjoint paths, p_1 through p_α, and β vertex-disjoint paths, q_1 through q_β, respectively. Now, to "locally" bound the optimal profit of OPT from above, we divide every path in OPT into subpaths such that each subpath intersects with exactly one path in ALG as follows: Consider a path $q_i \in OPT$, assuming that q_i intersects with k paths $p_{i_1}, p_{i_2}, \ldots, p_{i_k}$ in ALG in this order, where $\{i_1, i_2, \ldots, i_k\} \subseteq \{1, 2, \ldots, \alpha\}$. Note that q_i might intersect with one path twice or more, and hence some two paths p_{i_j} and $p_{i_{j'}}$ may be identical for $i_j \neq i_{j'}$ ($1 \leq j \leq k$ and $1 \leq j' \leq k$). Suppose that q_i and p_{i_j} ($1 \leq j \leq k$) share a path $r_{i,j} = \langle v_{i,j,0} \rangle \oplus r'_{i,j} \oplus \langle v_{i,j,\ell_j} \rangle$ of length ℓ_j, where if $\ell_j = 0$, then $r_{i,j}$ is the single vertex $v_{i,j,0}$, and if $\ell_j = 1$, then $r_{i,j}$ is the edge $\langle v_{i,j,0}, v_{i,j,1} \rangle$ and $r'_{i,j}$ is an empty path. One sees that we can represent the path q_i by the concatenation $q_i = q_{i,0} \oplus r_{i,1} \oplus q_{i,1} \oplus r_{i,2} \oplus q_{i,2} \oplus \ldots, \oplus q_{i,k-1} \oplus r_{i,k} \oplus q_{i,k}$, where the subpath $q_{i,j''}$ might be an empty one for $0 \leq j'' \leq k$. Then, we construct the following k subpaths from q_i: $q_{i,0} \oplus r_{i,1} \oplus q_{i,1}$, $r_{i,2} \oplus q_{i,2}$, $r_{i,3} \oplus q_{i,3}$, $\ldots$, $r_{i,k} \oplus q_{i,k}$. It is important to note here that the length of every $q_{i,j}$ is at most two because of the maximality of the set of vertex-disjoint length-3 paths obtained in the first **while**-loop of ALG. After dividing all the paths in OPT into subpaths as mentioned above, we define OPT^* by the set of such subpaths, i.e., $OPT^* = \bigcup_{q_i \in OPT} \{q_{i,0} \oplus r_{i,1} \oplus q_{i,1}, r_{i,2} \oplus q_{i,2}, r_{i,3} \oplus q_{i,3}, \ldots, r_{i,k_i} \oplus q_{i,k_i}\}$, supposing that q_i is divided into k_i subpaths. For $1 \leq i \leq \alpha$, let $OPT_{p_i} = \{q \mid q \in OPT^*, V(q) \cap V(p_i) \neq \emptyset\}$, i.e., a subset of subpaths in OPT^* which share at least one vertex in the path $p_i \in ALG$.

For ease of understanding, see an example graph illustrated in Fig. 2(A). Here, suppose that ALG finds two paths $\langle v_1, v_2, v_3, v_4 \rangle$ and $\langle v_5, v_6, v_7, v_8 \rangle$ of length three in the first **while**-loop. Also suppose that ALG selects (1) the former path $\langle v_1, v_2, v_3, v_4 \rangle$ and obtains the extended path $p_1 = \langle w_3, w_2, w_1, v_1, v_2, v_3, v_4, w_4, w_5 \rangle$ by concatenating two paths $\langle w_1, w_2, w_3 \rangle$ and $\langle w_4, w_5 \rangle$, and then selects (2) the latter path $\langle v_5, v_6, v_7, v_8 \rangle$ and obtains the extended path $p_2 = \langle w_6, v_5, v_6, v_7, v_8 \rangle$ by concatenating a path $\langle w_6 \rangle$ in the second **while**-loop. That is, ALG outputs $ALG = \{p_1, p_2\}$. On the other hand, suppose that OPT has three paths, $q_1 = \langle w_3, w_2, w_1, v_1, u_1, u_2, v_2 \rangle$, $q_2 = \langle u_3, w_4, v_4, v_3, u_4, u_5, u_6, v_7, u_7, u_8, u_9 \rangle$ and $q_3 = \langle u_{10}, u_{11}, u_{12}, v_6, u_{13}, u_{14}, u_{15} \rangle$. Here, q_1 intersects with p_1 twice. The path q_2 intersects with both p_1 and p_2. The path q_3 intersects with p_2. As mentioned above, we now define OPT_{p_i} for $1 \leq i \leq \alpha$ ($\alpha = 2$ in this example). See Fig. 2(B). (1) First, take a look at q_1, and set $p_{1_1} = p_1$ and $p_{1_2} = p_1$. One sees that $q_{1,0}$ is an empty path, $r_{1,1} = \langle w_3, w_2, w_1, v_1 \rangle$, $q_{1,1} = \langle u_1, u_2 \rangle$, $r_{1,2} = \langle v_2 \rangle$, and $q_{1,2}$ is an empty path. That is, we can represent q_1 by the concatenation $\langle w_3, w_2, w_1, v_1 \rangle \oplus \langle u_1, u_2 \rangle \oplus \langle v_2 \rangle$. Hence, we divide q_1 into two subpaths $\langle w_3, w_2, w_1, v_1, u_1, u_2 \rangle$, and $\langle v_2 \rangle$. Next, (2) consider q_2, $p_{2_1} = p_1$, and $p_{2_2} = p_2$. One sees that $r_{2,1} = \langle w_4, v_4, v_3 \rangle$, $r_{2,2} = \langle v_7 \rangle$, and thus $q_2 = \langle u_3 \rangle \oplus r_{2,1} \oplus \langle u_4, u_5, u_6 \rangle \oplus r_{2,2} \oplus \langle u_7, u_8, u_9 \rangle$. Thus, q_2 is divided into two subpaths $\langle u_3, w_4, v_4, v_3, u_4, u_5, u_6 \rangle$ and $\langle v_7, u_7, u_8, u_9 \rangle$. On the other hand, (3) since q_3 intersects with only one path in ALG, q_3 is not divided.

In summary, we define OPT^* as follows:

$$OPT^* = \{\langle w_3, w_2, w_1, v_1, u_1, u_2 \rangle, \langle v_2 \rangle, \langle u_3, w_4, v_4, v_3, u_4, u_5, u_6 \rangle,$$
$$\langle v_7, u_7, u_8, u_9 \rangle, \langle u_{10}, u_{11}, u_{12}, v_6, u_{13}, u_{14}, u_{15} \rangle\}.$$

Then, we obtain OPT_{p_1} and OPT_{p_2}:

$$OPT_{p_1} = \{\langle w_3, w_2, w_1, v_1, u_1, u_2 \rangle, \langle v_2 \rangle, \langle u_3, w_4, v_4, v_3, u_4, u_5, u_6 \rangle\},$$
$$OPT_{p_2} = \{\langle v_7, u_7, u_8, u_9 \rangle, \langle u_{10}, u_{11}, u_{12}, v_6, u_{13}, u_{14}, u_{15} \rangle\}.$$

Roughly, we compare the number $|V(p_i)|$ of vertices in p_i with the upper bound of the total number $|V(OPT_{p_i})|$ of vertices in OPT_{p_i}.

The following simple observations play an important role to prove the approximation ratio of ALG:

Observation 1. *(1) $V(OPT) = V(OPT^*)$, i.e., $|V(OPT^*)|$ is equal to the optimal profit $|V(OPT)|$. (2) Every path in OPT^* must intersect exactly one path in ALG. (3) $OPT^* = \bigcup_{p_i \in ALG} OPT_{p_i}$, and $OPT_{p_i} \cap OPT_{p_j} = \emptyset$ for different two paths $p_i, p_j \in ALG$. (4) The graph consisting of $V(p_i) \cup V(OPT_{p_i})$ is a connected component (i.e., subgraph) for every $p_i \in ALG$.*

One sees that the number $|V(p_i) \cup V(OPT_{p_i})|$ of vertices in the connected subgraph is a (trivial) upper bound of $|V(OPT_{p_i})|$. It follows that the maximum ratio of the number $|V(p_i) \cup V(OPT_{p_i})|$ of vertices in the graph consisting of p_i and all the paths in OPT_{p_i} to the number $|V(p_i)|$ of vertices in p_i over i is an upper bound of the ratio of $|V(OPT)|$ to $|V(ALG)|$, i.e., the approximation ratio can be bounded above by $\max_{p_i \in ALG} \left\{ |V(p_i) \cup V(OPT_{p_i})| / |V(p_i)| \right\}$.

In the following, (1) we first suppose that $\langle v_1, v_2, v_3, v_4 \rangle$ is the path obtained in the first **while**-loop of ALG. Next, (2) we see the possibilities on the extended paths in the second **while**-loop of ALG. Then, (3) for each extended path, represented by $p = w \oplus \langle v_1, v_2, v_3, v_4 \rangle \oplus w'$ where w and w' are respectively concatenated to v_1 and v_4, we consider the "largest" set OPT_p of paths which intersect with p. (4) Let $H = (V(p) \cup V(OPT_p), E(p) \cup E(OPT_p))$ be a connected graph consisting of the path p and all the paths in OPT_p. Then, we count the number $|V(H)|$ of vertices in H and the number $|V(p)|$ of vertices in p.

Let ℓ and ℓ' be the numbers of vertices in two concatenated paths w and w', respectively. Let $|OPT(H)|$ and $|ALG(H)|$ be the solution sizes obtained by an optimal algorithm and ALG for the connected subgraph H, respectively. We consider ten possibilities on the numbers of vertices in w and w', $(\ell, \ell') = (3,3), (2,3), (1,3), (0,3), (2,2), (1,2), (0,2), (1,1), (0,1), (0,0)$ since $(\ell, \ell') = (i,j)$ is essentially the same as $(\ell, \ell') = (j,i)$ for $0 \le i, j \le 3$.

Case 1: $(\ell, \ell') = (3,3)$. Assume that $p = \langle w_1, w_2, w_3, v_1, v_2, v_3, v_4, w_4, w_5, w_6 \rangle$ be the path obtained by ALG. Then, the connected subgraph H_1 which includes the largest set OPT_p of paths intersecting with p is shown in Fig. 3. For example, if w_1 has an additional edge, say, $\langle u_{27}, w_1 \rangle$, then after the first **while**-loop

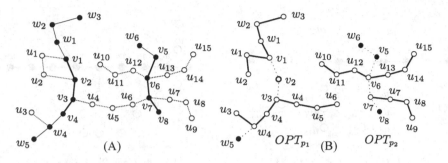

Fig. 2. How to divide paths into several subpaths: For example, if the optimal path cover contains a path $\langle u_3, w_4, v_4, v_3, u_4, u_5, u_6, v_7, u_7, u_8, u_9 \rangle$ as (A), then we divide the path into two paths $\langle u_3, w_4, v_4, v_3, u_4, u_5, u_6 \rangle$ and $\langle v_7, u_7, u_8, u_9 \rangle$ as (B).

of ALG, one vertex-disjoint length-3 path $\langle u_{27}, w_1, w_2, w_3 \rangle$ is not selected into ALG, which contradicts to the maximality of the set of vertex-disjoint length-3 paths. As another example, if w_3 has an additional edge, say, $\langle u_{28}, w_3 \rangle$, then there exists a path $\langle w_1, w_2, w_3, u_{28} \rangle$ of length 3, which is again a contradiction. Therefore, in this case, $|OPT(H_1)| \leq |V(H_1)| = 38$ holds. Since $|ALG(H_1)| = 10$ (= the number of black vertices in Fig. 3), the approximation ratio $|OPT(H_1)|/|ALG(H_1)|$ is bounded above by $|V(H_1)|/|ALG(H_1)| = 3.8$.

Case 2: $(\ell, \ell') = (2, 3)$. Assume that $p = \langle w_1, w_2, v_1, v_2, v_3, v_4, w_4, w_5, w_6 \rangle$ is the path obtained by ALG. In this case, the connected subgraph H_2 which includes the largest set OPT_p of paths intersecting with p is illustrated in Fig. 4. If u_1 has a neighbor vertex, say, u_0, then the longer path $\langle u_0, u_1, u_2 \rangle$ must be concatenated to v_1 in the second **while**-loop of ALG, which is a contradiction to the assumption that $\langle w_1, w_2 \rangle$ is concatenated to v_1. By the same reason, u_5 (and u_6) has no additional neighbor vertex. Therefore, in this case, $|OPT(H_2)| \leq |V(H_2)| = 35$ holds. Since $|ALG(H_2)| = 9$ (= the number of black vertices in Fig. 4), the approximation ratio $|OPT(H_2)|/|ALG(H_2)|$ is bounded above by $|V(H_2)|/|ALG(H_2)| = 35/9 < 3.9$.

Case 3: $(\ell, \ell') = (1, 3)$. Assume that $p = \langle w_1, v_1, v_2, v_3, v_4, w_2, w_3, w_4 \rangle$ is the path obtained by ALG. In this case, the connected subgraph H_3 which includes the largest set OPT_p of paths intersecting with p is illustrated in Fig. 5. Note that u_1 dose not have any additional neighbor vertex since the path $\langle w_1 \rangle$ of length zero is selected in the second **while**-loop of ALG. One sees that $|V(H_3)| = 30$ and $|ALG(H_3)| = 8$. Therefore, the approximation ratio is at most 3.75.

Case 4: $(\ell, \ell') = (0, 3)$. Assume that $p = \langle v_1, v_2, v_3, v_4, w_1, w_2, w_3 \rangle$ is the path obtained by ALG. In this case, it is sufficient to consider the subgraph H_4 shown in Fig. 6. Since v_1 cannot be extended, all the neighboring vertices of v_1 are included in some other paths in ALG. One sees that $|OPT(H_4)| \leq 27$ and $|ALG(H_4)| = 7$. Hence, the approximation ratio is at most 3.86.

One can observe that vertices connecting to v_2 and v_3 in the "middle v-part" are the same in every case. Moreover, for example, in the case $(\ell, \ell') = (2, 2)$, all

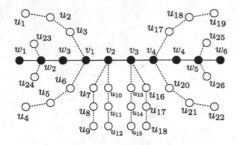

Fig. 3. Case 1: $(\ell, \ell') = (3, 3)$

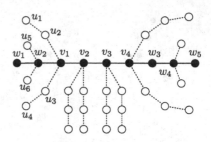

Fig. 4. Case 2: $(\ell, \ell') = (2, 3)$

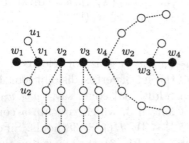

Fig. 5. Case 3: $(\ell, \ell') = (1, 3)$

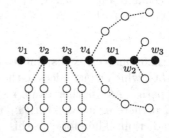

Fig. 6. Case 4: $(\ell, \ell') = (0, 3)$

we have to do is check the subgraph consisting of vertices connecting to w_1, w_2 and v_1 in Fig. 4 (i.e., the case where $\ell = 2$) as the "left w-part" subgraph or the "right w'-part" subgraph of the case $(\ell, \ell') = (2, 2)$. In the case $(\ell, \ell') = (1, 2)$, we should observe the subgraph including the left-part ($\ell = 1$) in Fig. 5 and the left-part ($\ell = 2$) in Fig. 4. Although details are omitted here, we can show that the approximation ratios are **(Case 5)** at most 4 for $(\ell, \ell') = (2, 2)$, **(Case 6)** at most 3.86 for $(\ell, \ell') = (1, 2)$, **(Case 7)** at most 4 for $(\ell, \ell') = (0, 2)$, **(Case 8)** at most 4 for $(\ell, \ell') = (1, 1)$, **(Case 9)** at most 3.8 for $(\ell, \ell') = (0, 1)$, and **(Case 10)** at most 4 for $(\ell, \ell') = (0, 0)$. As a result, the approximation ratio of ALG is bounded above by 4. □

If the maximum degree of the input graph is three, then the approximation ratio of ALG can be slightly improved as follows:

Corollary 3. *Algorithm ALG is a 2.5-approximation algorithm for MaxP[2]PC on subcubic graphs.*

Remark 1. The analysis on the approximation ratio of Corollary 3 is tight in the following sense: There exists a bad example such that the optimal solution has value $10k$ but the solution by ALG has value $4k$ for any integer $k \geq 1$.

See a graph G illustrated in Fig. 7. The graph has $4k$ vertices $v_{1,1}$, $v_{1,2}$, $v_{1,3}$, $v_{1,4}$, $v_{2,1}$, $\ldots$, $v_{k,1}$, $v_{k,2}$, $v_{k,3}$, $v_{k,4}$ and $6k$ vertices $u_{1,1}$ through $u_{k,6}$. The

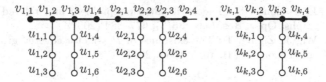

Fig. 7. Bad example for MaxP$^{[2]}$PC on subcubic graphs

former $4k$ vertices form a simple path. For $1 \leq i \leq k$, $\langle v_{i,2}, u_{i,1}, u_{i,2}, u_{i,3} \rangle$ and $\langle v_{i,3}, u_{i,4}, u_{i,5}, u_{i,6} \rangle$ are paths of length three. If the graph is given as input, ALG first might obtain a maximal set of k vertex-disjoint length-3 paths, $\langle v_{1,1}, v_{1,2}, v_{1,3}, v_{1,4} \rangle$ through $\langle v_{k,1}, v_{k,2}, v_{k,3}, v_{k,4} \rangle$. One sees that at this moment, ALG cannot do anything, i.e., $|ALG(G)| = 4k$. On the other hand, an optimal solution consists of $2k$ length-4 paths, $\langle v_{i,1}, v_{i,2}, u_{i,1}, u_{i,2}, u_{i,3} \rangle$ and $\langle v_{i,4}, v_{i,3}, u_{i,4}, u_{i,5}, u_{i,6} \rangle$ for each $1 \leq i \leq k$. That is, the maximum number of vertices is $10k$.

Acknowledgments. This work is partially supported by NSERC Canada, JSPS KAKENHI Grant Numbers JP17K00016, JP18H04091, JP19K12098, JP20H05794, JP20K11666 and JP21K11755, and JST CREST JPMJCR1402.

References

1. Arnborg, S., Corneil, D.G., Proskurowski, A.: Complexity of finding embeddings in a k-tree. SIAM J. Algeb. Discrete Meth. **8**(2), 277 284 (1987)
2. Chen, Y., et al.: Path cover with minimum nontrivial paths and its application in two-machine flow-shop scheduling with a conflict graph. J. Combin. Optim. (online) (2021). https://doi.org/10.1007/s10878-021-00793-3
3. Chen, Y., et al.: A local search 4/3-approximation algorithm for the minimum 3-path partition problem. In: Chen, Y., Deng, X., Lu, M. (eds.) FAW 2019. LNCS, vol. 11458, pp. 14–25. Springer, Cham (2019). https://doi.org/10.1007/978-3-030-18126-0_2
4. Garey, M.R., Johnson, D.S.: Computers and Intractability: A Guide to the Theory of NP-Completeness. W.H. Freeman and Company, New York (1990)
5. George, S.: On the k-path partition of graphs. Theoret. Comput. Sci. **290**(3), 2147–2155 (2003)
6. Gómez, R., Wakabayashi, Y.: Nontrivial path covers of graphs: existence, minimization and maximization. J. Comb. Optim. **39**(2), 437–456 (2019). https://doi.org/10.1007/s10878-019-00488-w
7. Kaneko, A.: A necessary and sufficient condition for the existence of a path factor every component of which is a path of length at least two. J. Combin. Theory Ser. B **88**(2), 195–218 (2003)
8. Kirkpatrick, D.G., Hell, P.: On the complexity of general graph factor problems. SIAM J. Comput. **12**(3), 601–609 (1983)
9. Lichtenstein, D.: Planar formulae and their uses. SIAM J. Comput. **11**(2), 329–343 (1982)
10. Manuch, J., Gaur, D.R.: Fitting protein chains to cubic lattice is NP-complete. J. Bioinform. Comput. Biol. **6**(1), 93–106 (2008)

11. Monnot, J., Toulouse, S.: The path partition problem and related problems in bipartite graphs. Oper. Res. Lett. **35**(5), 677–684 (2007)
12. Zhou, S., Wu, J., Zhang, T.: The existence of $P_{\geq 3}$-factor covered graphs. Discuss. Math. Graph Theory **37**(4), 1055–1065 (2017)

On Approximating Shortest Paths
in Weighted Triangular Tessellations

Prosenjit Bose[1], Guillermo Esteban[1,2(✉)], David Orden[2],
and Rodrigo I. Silveira[3]

[1] School of Computer Science, Carleton University, Ottawa, Canada
`jit@scs.carleton.ca`
[2] Departamento de Física y Matemáticas, Universidad de Alcalá,
Alcalá de Henares, Spain
`{g.esteban,david.orden}@uah.es`
[3] Departament de Matemàtiques, Universitat Politècnica de Catalunya,
Barcelona, Spain
`rodrigo.silveira@upc.edu`

Abstract. We study the quality of weighted shortest paths when a continuous 2-dimensional space is discretized by a weighted triangular tessellation. In order to evaluate how well the tessellation approximates the 2-dimensional space, we study three types of shortest paths: a weighted shortest path $SP_w(s,t)$, which is a shortest path from s to t in the space; a weighted shortest vertex path $SVP_w(s,t)$, which is a shortest path where the vertices of the path are vertices of the tessellation; and a weighted shortest grid path $SGP_w(s,t)$, which is a shortest path whose edges are edges of the tessellation. The ratios $\frac{\|SGP_w(s,t)\|}{\|SP_w(s,t)\|}$, $\frac{\|SVP_w(s,t)\|}{\|SP_w(s,t)\|}$, $\frac{\|SGP_w(s,t)\|}{\|SVP_w(s,t)\|}$ provide estimates on the quality of the approximation.

Given any arbitrary weight assignment to the faces of a triangular tessellation, we prove upper and lower bounds on the estimates that are independent of the weight assignment. Our main result is that $\frac{\|SGP_w(s,t)\|}{\|SP_w(s,t)\|} = \frac{2}{\sqrt{3}} \approx 1.15$ in the worst case, and this is tight.

Keywords: Shortest Path · Tessellation · Weighted Region Problem

1 Introduction

Geometric shortest path problems, where the goal is to find an optimal path in a geometric setting, are fundamental problems in computational geometry. In contrast to the classical shortest path problem in graphs, where the space of possible paths is discrete, in geometric settings the space is continuous: the source and target points can be anywhere within a certain geometric domain (e.g., a

Partially supported by NSERC, project PID2019-104129GB-I00/MCIN/AEI/ 10.13039/501100011033 of the Spanish Ministry of Science and Innovation, and H2020-MSCA-RISE project 734922 - CONNECT.

P. Mutzel et al. (Eds.): WALCOM 2022, LNCS 13174, pp. 409–421, 2022.
https://doi.org/10.1007/978-3-030-96731-4_33

polygon, the plane, a surface), and the set of possible paths to consider has infinite size. Many variations of geometric shortest path problems exist, depending on the geometric domain, the objective function (e.g., Euclidean metric, link-distance, geodesic distance), or specific domain constraints (e.g., obstacles in the plane, or holes in polygons). Applications of geometric shortest path problems are ubiquitous, appearing in diverse areas such as robotics, video game design, or geographic information science. We refer to Mitchell [15] for a complete survey on geometric shortest path problems.

One of the most general settings for geometric shortest path problems arises when the cost of traversing the domain varies depending on the region. That is, the domain consists of a planar subdivision, that without loss of generality can be assumed to be triangulated. Each region i of the subdivision has a weight w_i, that represents the cost per unit of distance of traveling in that region. Thus, the cost of traversing a region is typically given by the Euclidean distance traversed in the region, multiplied by the corresponding weight. The resulting metric is often called the *weighted region metric*, and the problem of computing a shortest path between two points under this metric is known as the *weighted region problem* (WRP) [13,14]. The WRP is very general, since it allows to model many well-known variants of geometric shortest path problems. Indeed, having that all weights are equal makes the metric equivalent to the Euclidean metric (up to scaling), while using two different weight values, such as 1 and ∞, allows to model paths amidst obstacles.

Perhaps not surprisingly, the WRP turns out to be a challenging problem. The first algorithm for WRP was a $(1 + \varepsilon)$-approximation proposed by Mitchell and Papadimitriou [14], which runs in time $O(n^8 \log\left(\frac{nNW}{w\varepsilon}\right))$, where N is the maximum integer coordinate of any vertex of the subdivision, W and w are, respectively, the maximum finite and the minimum nonzero integer weight assigned to the regions. Substantial research has been devoted to studying faster approximation algorithms and different variants of the problem [1–3]. Approximation schemes for WRP are sophisticated methods that usually are based on variants of continuous Dijkstra, subdividing triangle edges in parts for which crossing shortest paths have the same combinatorial structure (e.g., [14]), or work by computing a discretization of the domain by carefully placing Steiner points (e.g., see [9] for the currently best method of this type). The lack of exact algorithms for WRP is probably justified by algebraic reasons: WRP was recently shown to be impossible to solve in the Algebraic Computation Model over the Rational Numbers [10]. This is a model of computation where one can compute exactly any number that can be obtained from rational numbers by a finite number of basic operations. Efficient algorithms for WRP only exist for a few special cases, e.g., rectilinear subdivisions with L_1 metric [8], or weights restricted to $\{0, 1, \infty\}$ [11].

In applications where the WRP arises, like robotics, gaming or simulation, which usually require efficient and practical algorithms, the problem is simplified in two ways. First, the domain is approximated by using a (weighted) plane subdivision with a simpler structure. Secondly, optimal shortest paths in that simpler subdivision are approximated. The typical way to represent a 2D (or 3D) environment where shortest paths need to be computed is by using *navigation meshes* [18]. These are polygonal subdivisions together with a graph that models the adjacency between the regions. Path planning is then done first on the graph to obtain a sequence of regions to be traversed, and then within each region, for which a shortest geometric path is extracted. Triangles, convex polygons, disks or squares—of different sizes—are among the most frequently used region shapes [18]. Navigational meshes allow efficient path planning in large environments as long as the region weights are limited to $\{1, \infty\}$ (i.e., obstacles only). In case general weights are needed, the complexity of computing the shortest path inside each region requires the use of the simplest possible navigational mesh: *regular grids*. In 2D, the only three types of regular polygons that can be used to tessellate continuous environments are triangles, squares and hexagons. The drawback with a grid is that it imposes a fixed resolution, requiring in general a large number of cells or regions. Still, grids are often used as navigation meshes (even for weights $\{1, \infty\}$), since they are easy to implement, are a natural choice for environments that are grid-based by design (e.g., many game designs), and popular shortest path algorithms such as A^* can be optimized for grids [16].

Even when a regular grid is used as a navigation mesh, in practice exact weighted shortest paths are not computed: instead, an approximation is obtained by computing a shortest path on a weighted graph associated to the grid. To this end, two different graphs have been considered in the literature [5], *corner-vertex graph* and *k-corner grid graph*, defined next. The baseline to analyze the quality of any approximate path is the weighted shortest path that takes into account the full geometry of each region, as in the WRP. A weighted shortest path will be denoted $SP_w(s, t)$. As already mentioned, exact weighted shortest paths for regions with weights in $\{0, 1, \infty\}$ were studied in [13, 14].

In a *corner-vertex graph* G_{corner}, the vertex set is the set of corners of the tessellation and every pair of vertices is connected by an edge. These graphs can be seen as the complete graphs over the set of vertices. Figure 1a depicts some of the neighbors of a vertex v in the corner-vertex graph. Note that in this graph some edges overlap. A path in this graph is called a *vertex path*; a shortest vertex path between s and t will be denoted $SVP_w(s, t)$, where w makes explicit that this path depends also on a particular weight assignment w.

In a *k-corner grid graph* $G_{k\text{corner}}$, which is a subgraph of a corner-vertex graph, the vertex set is the set of corners of the tessellation, and each vertex is connected by an edge to a predefined set of k neighboring vertices, depending on the tessellation and other design decisions. See Fig. 1b for the 6-corner grid graph in a triangular tessellation. (Analogous k-corner grid graphs can be defined for square and hexagonal tessellations.) A path in this graph is called a *grid path*; a shortest grid path between s and t will be denoted $SGP_w(s, t)$.

Shortest vertex paths and shortest grid paths for the case of weights of the cells being 1 or ∞ have been previously studied in [17] and [4], respectively. In all cases, the weight of each graph edge is defined based on the cost of the associated line segment, depending on the weights of the regions that it goes through. More

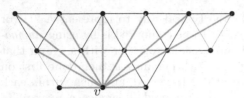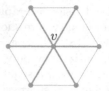

(a) Some neighbors of a vertex v in the corner-vertex graph.

(b) All neighbors of a vertex v in the 6-corner grid graph.

Fig. 1. Vertex v is connected to its neighbors in a triangular tessellation.

formally, let T_i be a region in a subdivision with weight $\omega_i \in \mathbb{R}_{>0}$. The cost of a segment π_i in the interior of a cell T_i is given by $\omega_i \|\pi_i\|$, where $\|\cdot\|$ is the Euclidean norm. In the case where a segment π lies in the boundary of two cells T_j and T_k, the cost is $\min\{\omega_j, \omega_k\} \|\pi\|$.

Figure 2 shows an example, illustrating the three paths considered: the shortest path $SP_w(s,t)$ (blue), the shortest vertex path $SVP_w(s,t)$ (green), and the shortest grid path $SGP_w(s,t)$ (red) in a 6-corner grid graph. Note that in all figures in this work, cells that are not depicted are considered to have infinite weight.

1.1 Quality Bounds for Approximation Paths

The goal of this work is to understand the relation between $SGP_w(s,t)$, $SVP_w(s,t)$, and the baseline $SP_w(s,t)$. Since $SVP_w(s,t)$ and $SGP_w(s,t)$ are approximations of $SP_w(s,t)$, a fundamental question is: what is the worst-case approximation factor that they can give?

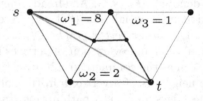

In this paper we focus on weighted tessellations where every face is an equilateral triangle (analog ideas can be used for square and hexagonal grids). In particular, we are interested in upper-bounding the ratios $\frac{\|SGP_w(s,t)\|}{\|SP_w(s,t)\|}$ and $\frac{\|SVP_w(s,t)\|}{\|SP_w(s,t)\|}$, since they

Fig. 2. $SP_w(s,t)$ (blue), $SVP_w(s,t)$ (green), and a $SGP_w(s,t)$ (red) between two corners s and t in $G_{6corner}$. The cost of each path is 16.75, 17.32 and 18, respectively.

indicate the approximation factor of the shortest grid and vertex path, respectively. The ratio $\frac{\|SGP_w(s,t)\|}{\|SVP_w(s,t)\|}$ is also studied, to see which approximation is better.

Almost all previous bounds on the ratio $\frac{\|SGP_w(s,t)\|}{\|SP_w(s,t)\|}$ consider a limited set of weights for the cells. Nash [17] considered only weights in the set $\{1, \infty\}$ and proved that the weight of $SGP_w(s,t)$ in hexagonal $G_{6corner}$, and $G_{12corner}$, square $G_{4corner}$, and $G_{8corner}$, triangle $G_{6corner}$, and $G_{3corner}$ can be, respectively, up to

Table 1. Bounds on the quality of approximations of shortest paths in weighted triangular tessellations for $G_{6corner}$. The upper bound for the ratio $\frac{\|SGP_w(s,t)\|}{\|SVP_w(s,t)\|}$, and the bounds for the ratio $\frac{\|SVP_w(s,t)\|}{\|SP_w(s,t)\|}$ are shown in the full version [6].

$\frac{\|SGP_w(s,t)\|}{\|SVP_w(s,t)\|}$		$\frac{\|SGP_w(s,t)\|}{\|SP_w(s,t)\|}$		$\frac{\|SVP_w(s,t)\|}{\|SP_w(s,t)\|}$	
Lower bound	Upper bound	Lower bound	Upper bound	Lower bound	Upper bound
$\frac{2}{\sqrt{3}} \approx 1.15$ [17]	$\frac{2}{\sqrt{3}} \approx 1.15$ [6]	$\frac{2}{\sqrt{3}} \approx 1.15$ [17]	$\frac{2}{\sqrt{3}} \approx 1.15$ (Thm. 1)	$\frac{2\sqrt{7\sqrt{3}-12}}{(7-4\sqrt{3})(6\sqrt{2}+\sqrt{7\sqrt{3}-12})} \approx 1.11$ [6]	$\frac{2}{\sqrt{3}} \approx 1.15$ [6]

≈ 1.15, ≈ 1.04, ≈ 1.41, ≈ 1.08, ≈ 1.15, and 2 times the weight of $SP_w(s,t)$. When the weights of the cells are allowed to be in $\mathbb{R}_{>0}$, the only result that we are aware of is for square tessellations and another type of shortest path, with vertices at the center of the cells, for which Jaklin [12] showed that $\frac{\|SGP_w(s,t)\|}{\|SP_w(s,t)\|} \leq 2\sqrt{2}$.

The main contribution of this paper is the analysis of the quality of the three types of shortest paths for a triangular grid for $G_{6corner}$, which is the most natural graph defined on a triangular grid. In contrast to previous work, we allow the weights ω_i to take any value in $\mathbb{R}_{>0}$, so the main challenge here is to obtain tight upper bounds that hold for *any* assignment of region weights. Surprisingly, we show that this is possible: the ratios are upper bounded by constants that are independent of the weights assigned to the regions in the tessellation. Our main result is that $\frac{\|SGP_w(s,t)\|}{\|SP_w(s,t)\|} = \frac{2}{\sqrt{3}}$ in the worst case, for any (positive) weight assignment. This implies bounds for the other two ratios considered. Moreover, our upper bound for $\frac{\|SGP_w(s,t)\|}{\|SP_w(s,t)\|}$ is tight, since it matches the lower bound claimed by Nash [17]. Table 1 summarizes our results, together with the previously known lower bounds.

2 $\frac{\|SGP_w(s,t)\|}{\|SP_w(s,t)\|}$ Ratio in $G_{6corner}$ for Triangular Cells

This section is devoted to obtaining, for two vertices s and t, an upper bound on the ratio $\frac{\|SGP_w(s,t)\|}{\|SP_w(s,t)\|}$ in $G_{6corner}$ in a triangular tessellation $\mathcal{T}$. We suppose, without loss of generality, that the length of each edge of the triangular cells is 2, in order to have a non-fractional length for the cell height.

Let $(s = u_1, u_2, \ldots, u_n = t)$ be the ordered sequence of consecutive points where $GP_w(s,t)$ and $SP_w(s,t)$ coincide; in case $GP_w(s,t)$ and $SP_w(s,t)$ share one or more segments, we define the corresponding points as the endpoints of each of these segments, see Fig. 3 for an illustration. Observation 1 below is a special case of the mediant inequality.

Observation 1. *Let $GP_w(s,t)$ and $SP_w(s,t)$ be, respectively, a weighted grid path, and a weighted shortest path, from s to t. Let u_i and u_{i+1} be two consecutive points where $GP_w(s,t)$ and $SP_w(s,t)$ coincide. Then, the ratio $\frac{\|GP_w(s,t)\|}{\|SP_w(s,t)\|}$ is at most the maximum of all ratios $\frac{\|GP_w(u_i,u_{i+1})\|}{\|SP_w(u_i,u_{i+1})\|}$.*

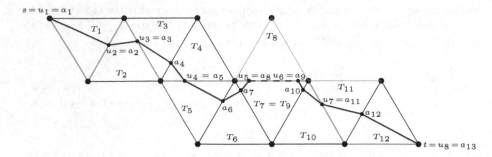

Fig. 3. Weighted shortest path $SP_w(s,t)$ (blue) and the crossing path $X(s,t)$ (orange) from s to t in a triangular tessellation. (Color figure online)

2.1   Crossing Paths and Weakly Simple Polygons

In the weighted version of the problem, conversely to the unweighted version, we need to take into account all the different weights of the regions intersected by $SP_w(s,t)$. In addition, we do not know the shape of the shortest paths $SP_w(s,t)$ and $SGP_w(s,t)$. To solve all these inconveniences, for each $SP_w(s,t)$ we will define a particular grid path called *crossing path* $X(s,t)$, whose behavior will be easier to control. See orange path in Fig. 3. Then, the key idea to prove the upper bound on the ratio $\frac{\|SGP_w(s,t)\|}{\|SP_w(s,t)\|}$ will be to upper-bound it by the ratio $\frac{\|X(s,t)\|}{\|SP_w(s,t)\|}$. To do so, we will analyze the components resulting from the intersection between $SP_w(s,t)$ and $X(s,t)$. Each component will be a weakly simple polygon, which will be the basic unit that we will analyze to obtain our main result. Also, a relation between the weights of some cells intersected by $SP_w(s,t)$ and $X(s,t)$ will be obtained.

Let $(T_1, \ldots, T_n)$ be the ordered sequence of consecutive cells intersected by $SP_w(s,t)$ in the tessellation $\mathcal{T}$. Let v_1^i, v_2^i, v_3^i be the three consecutive corners of the boundary of T_i, $1 \le i \le n$. Let $(s = a_1, a_2, \ldots, a_{n+1} = t)$ be the sequence of consecutive points where $SP_w(s,t)$ changes cell in $\mathcal{T}$. In particular, let a_i and a_{i+1} be, respectively, the points where $SP_w(s,t)$ enters and leaves T_i. In a triangular tessellation, the crossing path $X(s,t)$ from a vertex s to a vertex t is defined as follows:

Definition 1. *The crossing path $X(s,t)$ between two vertices s and t in a triangular tessellation $\mathcal{T}$ is defined by the sequence $(X_1, \ldots, X_n)$, where X_i is a sequence of vertices determined by the pair (a_i, a_{i+1}), $1 \le i \le n$, as follows. Let $e_1^i \in T_i$ be an edge containing a_i, then:*

- *If $a_{i+1} \in e_1^i$, let $[v, w]$ be the endpoints of e_1^i, where a_i is encountered before a_{i+1} when traversing e_1^i from v to w. Then $X_i = (v, w)$, see Fig. 4a.*
- *If a_i is an endpoint of e_1^i, let p be the midpoint of the edge $e_2^i \in T_i$ not containing a_i. If $a_{i+1} \in e_2^i$ is to the left of $\overrightarrow{a_i p}$, X_i is a_i and the endpoint of e_2^i to the right of $\overrightarrow{a_i p}$, see Fig. 4b. Otherwise, X_i is a_i and the endpoint of e_2^i to the left of $\overrightarrow{a_i p}$.*

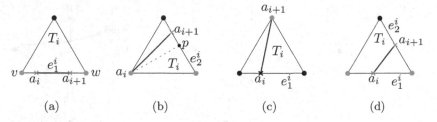

Fig. 4. Some of the positions of the intersection points between $SP_w(s,t)$ (blue) and a cell. The vertices of the crossing path $X(s,t)$ in a triangular cell are depicted in orange. (Color figure online)

- If a_i is in the interior of e_1^i and a_{i+1} is a corner, $X_i = (a_{i+1})$, see Fig. 4c.
- If a_i and a_{i+1} belong to the interior of two different edges e_1^i and e_2^i, X_i is the common endpoint of e_1^i and e_2^i, see Fig. 4d.

Let $(s = u_1, u_2, \ldots, u_\ell = t)$ be the sequence of consecutive points where $X(s,t)$ and $SP_w(s,t)$ coincide. The union of $SP_w(s,t)$ and $X(s,t)$ between two consecutive points u_j and u_{j+1}, for $1 \leq j < \ell$, induces a weakly simple polygon (see [7] for a formal definition). We distinguish six different types of weakly simple polygons, denoted $P_1, \ldots, P_6$, depending on the number of edges intersected by $SP_w(u_j, u_{j+1})$, see Fig. 5. Observe that, by definition of $X(s,t)$, these are the only weakly simple polygons that can arise.

The weakly simple polygons will be an important tool in our proof, since it will be enough to upper bound $\frac{\|X(s,t)\|}{\|SP_w(s,t)\|}$ for each of $P_1, \ldots, P_6$.

Definition 2. *Let u_j and u_{j+1} be two consecutive points in a triangular tessellation, where $X(s,t)$ and $SP_w(s,t)$ coincide. Let p be a common endpoint of the edges of the tessellation that contain u_j and u_{j+1}. A weakly simple polygon induced by u_j and u_{j+1} is of type P_k, for $1 \leq k \leq 6$, if the subpath $SP_w(u_j, u_{j+1})$ intersects k consecutive edges around p.*

2.2 Bounding the Ratio for Weakly Simple Polygons

We are now ready to upper bound the ratio $\frac{\|X(u_j, u_{j+1})\|}{\|SP_w(u_j, u_{j+1})\|}$ for each of the six types of weakly simple polygons in $G_{6\text{corner}}$.

First we make a geometric observation that will be needed later. Let p and q be two points that are in the interior of two different edges on the boundary of the same triangular cell. Then, the length of the subpath of the weighted shortest path between p and q is given in Observation 2, which can be proved using the law of cosines.

Observation 2. *Let T_i be a triangular cell, and let (u, v, w) be the three vertices of T_i, in clockwise order. Let $p \in [u,v]$ and $q \in [v,w]$ be two points on the boundary of T_i. Then, $|pq| = \sqrt{|pv|^2 + |vq|^2 - |pv||vq|}$, see Fig. 6a.*

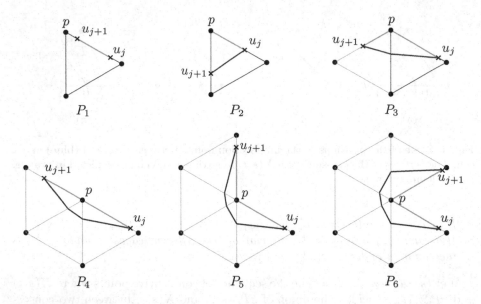

Fig. 5. Some weakly simple polygons P_k, and the subpath of the crossing path $X(s,t)$ (orange) from u_j to u_{j+1} intersecting consecutive triangular cells. (Color figure online)

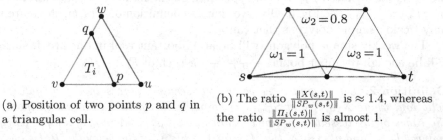

(a) Position of two points p and q in a triangular cell.

(b) The ratio $\frac{\|X(s,t)\|}{\|SP_w(s,t)\|}$ is ≈ 1.4, whereas the ratio $\frac{\|\Pi_i(s,t)\|}{\|SP_w(s,t)\|}$ is almost 1.

Fig. 6. Weighted shortest path $SP_w(s,t)$ (blue), crossing path $X(s,t)$ (orange), and shortcut path $\Pi_i(s,t)$ (purple) intersecting a weakly simple polygon P_2. (Color figure online)

We observe that, by definition, for P_1 we have $\frac{\|X(u_j,u_{j+1})\|}{\|SP_w(u_j,u_{j+1})\|} = 1$. Therefore the focus will be on bounding $P_2, \ldots, P_6$. We will begin from the simpler case of $P_3, \ldots, P_6$, and later we will consider P_2, which is substantially more involved. The proof of the lemma below can be found in the full version [6].

Lemma 1. *Let u_j, $u_{j+1} \in P_k$, for $3 \le k \le 6$, be two consecutive points where a shortest path $SP_w(s,t)$ and the crossing path $X(s,t)$ coincide in a triangular tessellation $\mathcal{T}$. An upper bound on the ratio $\frac{\|X(u_j,u_{j+1})\|}{\|SP_w(u_j,u_{j+1})\|}$ in $G_{6corner}$ is $\frac{2}{\sqrt{3}}$.*

The grid paths in $G_{6corner}$ are paths whose edges are edges of the triangular cells. Thus, the ratio $\frac{\|X(u_j,u_{j+1})\|}{\|SP_w(u_j,u_{j+1})\|}$ between two consecutive crossing points u_j

and u_{j+1} depends on the weights of these regions. So, the next difficulty that we encountered, related to the crossing path, was that it is possible to find an instance where $SP_w(s,t)$ intersects a weakly simple polygon P_2 such that the ratio $\frac{\|X(s,t)\|}{\|SP_w(s,t)\|}$ is much larger than $\frac{\|SGP_w(s,t)\|}{\|SP_w(s,t)\|}$, see Fig. 6b. However, between s and t there are other grid paths shorter than $X(s,t)$ that intersect a P_2. So, in order to obtain an upper bound when $SP_w(s,t)$ intersects a P_2, we will need a finer analysis.

Since the ratio $\frac{\|X(u_j,u_{j+1})\|}{\|SP_w(u_j,u_{j+1})\|}$ is at most $\frac{2}{\sqrt{3}}$ for weakly simple polygons P_k, $k \neq 2$, we will assume from now on, that the ratio is maximized when all weakly simple polygons are of type P_2. Otherwise, we are done.

Definition 3 determines another class of grid paths called *shortcut paths* that gives a tighter upper bound on the ratio $\frac{\|SGP_w(s,t)\|}{\|SP_w(s,t)\|}$ when a weakly simple polygon P_2 is intersected by $SP_w(s,t)$.

Let $\{v_1, \dots, v_n\}$ be a sequence of corners of a triangular tessellation. Then, the grid path $\Pi(s, [v_1, \dots, v_n], t)$ is defined as the path $X(s, v_1) \cup \pi(v_1, \dots, v_n) \cup X(v_n, t)$, where $\pi(v_1, \dots, v_n)$ is the grid path through the vertices $v_1, \dots, v_n$ in that order. We now define shortcut paths.

Definition 3. *Let (u, v, w) be the sequence of vertices of a cell $T_i \in T$ in clockwise order. If $X(s,t)$ contains the subpath (u, v, w), the shortcut path $\Pi_i(s,t)$ is defined as the grid path $\Pi(s, [u, w], t)$, see purple path in Fig. 6a.*

Now, we have all the tools needed to obtain an upper bound on the ratio $\frac{\|X(u_j,u_{j+1})\|}{\|SP_w(u_j,u_{j+1})\|}$ for P_2. By using the shortcut path $\Pi_i(s,t)$, we will be able to obtain a relation between the weights of the cells adjacent to $T_i \in T$ intersected by the crossing path $X(s,t)$. This relation is given in the next lemma.

Lemma 2. *Let $(T_k, \dots, T_m)$ be the sequence of consecutive cells for which there exists a shortcut path $\Pi_i(s,t)$, $k \leq i \leq m$, for a given assignment of weights w to the cells of the triangular tessellation. The ratio $\frac{\|SGP_w(s,t)\|}{\|SP_w(s,t)\|}$ is maximized when $\|X(s,t)\| = \|\Pi_i(s,t)\|$ in $G_{6corner}$.*

Proof. Consider an instance for which the ratio $\frac{\|SGP_w(s,t)\|}{\|SP_w(s,t)\|}$ is maximized. Recall this instance just contains weakly simple polygons of type P_2. We will argue that if there is a grid path $GP_w(s,t)$ among $X(s,t)$, $\Pi_i(s,t)$, $k \leq i \leq m$, that is strictly shorter than the other grid paths, then this instance cannot maximize $\frac{\|SGP_w(s,t)\|}{\|SP_w(s,t)\|}$.

Suppose that there is one grid path $GP_w(s,t)$ among $X(s,t)$, $\Pi_i(s,t)$, $k \leq i \leq m$, that is strictly shorter than the other grid paths in the set. Since $GP_w(s,t)$ is a grid path, the ratio $\frac{\|SGP_w(s,t)\|}{\|SP_w(s,t)\|}$ is upper bounded by $\frac{\|GP_w(s,t)\|}{\|SP_w(s,t)\|}$. The objective of the proof is to find another assignment of weights w' for the cells, such that $GP_{w'}(s,t)$ is still a shortest grid path among the grid paths in the set, and $\frac{\|GP_{w'}(s,t)\|}{\|SP_{w'}(s,t)\|} > \frac{\|GP_w(s,t)\|}{\|SP_w(s,t)\|}$.

Let u_j, u_{j+1} be two consecutive points where $GP_w(s,t)$ and $SP_w(s,t)$ coincide. Let T_ℓ be the cell that shares the edge of $\Pi_i(s,t)$ with T_i, see Fig. 7. We first set to infinity the weight of all the cells that are not traversed by $SP_w(s,t)$. This way, we ensure that when modifying the weights of some cells, the combinatorial structure of the shortest path is preserved. The weight of the crossing path $X(s,t)$ along the edges of T_i is $2\min\{\omega_{i-1},\omega_i\} + 2\min\{\omega_i,\omega_{i+1}\}$, and the weight of the shortcut path $\Pi_i(s,t)$ along the edges of T_i is $2\min\{\omega_i,\omega_\ell\} = 2\omega_i$ (because $\omega_\ell = \infty$). Let $[p,q]$, and $[p',q']$ be, respectively, the edges containing u_j and u_{j+1}, where $p, p' \in T_\ell$.

– If $GP_w(s,t) = X(s,t)$ then $\|X(s,t)\| < \|\Pi_i(s,t)\|$, and we have that

$$\min\{\omega_{i-1},\omega_i\} + \min\{\omega_i,\omega_{i+1}\} < \omega_i. \tag{1}$$

- Suppose $\omega_i \le \omega_{i-1}$, then $\omega_i + \min\{\omega_i,\omega_{i+1}\} < \omega_i$, which is not possible since $\min\{\omega_i,\omega_{i+1}\} > 0$. Hence, $\omega_i > \omega_{i-1}$.
- Suppose $\omega_i \le \omega_{i+1}$, then $\min\{\omega_{i-1},\omega_i\}+\omega_{i+1} < \omega_i$, which is not possible since $\min\{\omega_{i-1},\omega_i\} > 0$. Hence, $\omega_i > \omega_{i+1}$.

These two facts together with Eq. 1 imply that $\omega_{i-1} + \omega_{i+1} < \omega_i$. We also have that

$$\frac{\|X(s,t)\|}{\|SP_w(s,t)\|} = \frac{\|X(s,p)\| + 2(\omega_{i-1}+\omega_{i+1}) + \|X(p',t)\|}{\|SP_w(s,u_j)\| + |u_ju_{j+1}|\omega_i + \|SP_w(u_{j+1},t)\|},$$

being $|u_ju_{j+1}| > 0$, so if we decrease the weight ω_i until $\omega_{i-1} + \omega_{i+1} = \omega_i$, the denominator $\|SP_w(s,t)\|$ will decrease, and the numerator $\|X(s,t)\|$ will remain. Hence, the ratio $\frac{\|X(s,t)\|}{\|SP_w(s,t)\|}$ will increase, so we found another weight assignment w' such that $\frac{\|X(s,t)\|}{\|SP_{w'}(s,t)\|} > \frac{\|X(s,t)\|}{\|SP_w(s,t)\|}$ and $\|X(s,t)\| = \|\Pi_i(s,t)\|$.

– Otherwise, if $GP_w(s,t) = \Pi_i(s,t)$ then $\|\Pi_i(s,t)\| < \|X(s,t)\|$, and we have that $\omega_i < \min\{\omega_{i-1},\omega_i\} + \min\{\omega_i,\omega_{i+1}\}$. We also have that

$$\frac{\|\Pi_i(s,t)\|}{\|SP_w(s,t)\|} = \frac{\|\Pi_i(s,p)\| + 2\omega_i + \|\Pi_i(p',t)\|}{\|SP_w(s,u_j)\| + |u_ju_{j+1}|\omega_i + \|SP_w(u_{j+1},t)\|},$$

given $|u_ju_{j+1}| < 2$. If we increase the weight ω_i until $\omega_i = \omega_{i-1} + \omega_{i+1}$, the numerator $\|\Pi_i(s,t)\|$ will increase faster than the denominator $\|SP_w(s,t)\|$. Hence, the ratio $\frac{\|\Pi_i(s,t)\|}{\|SP_w(s,t)\|}$ will increase, so we found another weight assignment w' such that $\frac{\|\Pi_i(s,t)\|}{\|SP_{w'}(s,t)\|} > \frac{\|\Pi_i(s,t)\|}{\|SP_w(s,t)\|}$ and $\|X(s,t)\| = \|\Pi_i(s,t)\|$.

We are now ready to prove the upper bound on the ratio $\frac{\|X(u_j,u_{j+1})\|}{\|SP_w(u_j,u_{j+1})\|}$ in a P_2. Lemma 3 presents an upper bound on the ratio $\frac{\|X(u_j,u_{j+1})\|}{\|SP_w(u_j,u_{j+1})\|}$, where $u_j, u_{j+1} \in T_i$ are two consecutive points where $X(s,t)$ and $\Pi_i(s,t)$ coincide. Lemma 2 implies that the ratio $\frac{\|X(s,t)\|}{\|SP_w(s,t)\|}$ in $G_{6\text{corner}}$ is maximized when $\|X(s,t)\| = \|\Pi_i(s,t)\|$ for each i such that the shortcut path $\Pi_i(s,t)$ exists. Thus, the ratio is obtained in a weakly simple polygon P_2 when $\|X(s,t)\| = \|\Pi_i(s,t)\|$. Since the exact shape of $SP_w(s,t)$ is unknown, when calculating the ratio in the following Lemma 3, we will maximize the ratio for any position of the points u_j and u_{j+1} where $SP_w(s,t)$ and $X(s,t)$ coincide. The prove of the result is given in the full version [6].

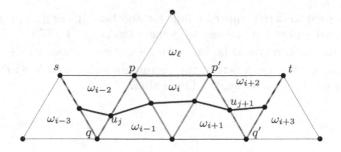

Fig. 7. $SP_w(s,t)$ through a P_2, where $\|\Pi_i(s,t)\| < \|X(s,t)\|$.

Lemma 3. *Let u_j, $u_{j+1} \in P_2$ be two consecutive points in a triangular tessellation $\mathcal{T}$, where a shortest path $SP_w(s,t)$ and the crossing path $X(s,t)$ coincide. Let u_j, $u_{j+1} \in T_i$ and $\|X(s,t)\| = \|\Pi_i(s,t)\|$, then an upper bound on the ratio $\frac{\|X(u_j,u_{j+1})\|}{\|SP_w(u_j,u_{j+1})\|}$ in $G_{6corner}$ is $\frac{2}{\sqrt{3}}$.*

Finally, we have all the pieces to prove our main result.

Theorem 1. *In $G_{6corner}$, an upper bound on the ratio $\frac{\|SGP_w(s,t)\|}{\|SP_w(s,t)\|}$ is $\frac{2}{\sqrt{3}}$.*

Figure 8 provides an illustration of the lower bound $\frac{2}{\sqrt{3}}$ on the ratio between the weighted shortest grid path $SGP_w(s,t)$ (red) and the weighted shortest path $SP_w(s,t)$ (blue) claimed by Nash [17]. Hence, the upper bound in Theorem 1 is tight for $G_{6corner}$.

3   Discussion and Future Work

We presented upper bounds on the ratio between the lengths of three types of weighted shortest paths in a triangular tessellation. The fact that a compact grid graph such as $G_{6corner}$ guarantees an error bound of $\approx 15\%$, regardless of weights used, justifies its widespread use in applications in areas such as gaming and simulation, where performance is a priority over accuracy.

Our analysis techniques, presented here for triangular grids, can also be

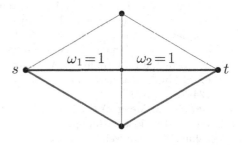

Fig. 8. The ratio $\frac{\|SGP_w(s,t)\|}{\|SP_w(s,t)\|}$ is $\frac{2}{\sqrt{3}}$.

applied to obtain upper bounds for the same ratios in the other two types of regular tessellations, square and hexagonal. The main differences lie in the exact definition of the crossing paths and the weakly simple polygons. Our techniques can also be used to derive upper bounds for another type of grid graphs, where the vertices are cell centers instead of corners (see, e.g., [12,17]).

For future work, it would be interesting to close the gap for $\frac{\|SVP_w(s,t)\|}{\|SP_w(s,t)\|}$. It is an intriguing question whether the seemingly richer graph $SVP_w(s,t)$ can actually guarantee a better quality factor than $G_{6corner}$.

References

1. Aleksandrov, L., Lanthier, M., Maheshwari, A., Sack, J.-R.: An ε — approximation algorithm for weighted shortest paths on polyhedral surfaces. In: Arnborg, S., Ivansson, L. (eds.) SWAT 1998. LNCS, vol. 1432, pp. 11–22. Springer, Heidelberg (1998). https://doi.org/10.1007/BFb0054351

2. Aleksandrov, L., Maheshwari, A., Sack, J.-R.: Approximation algorithms for geometric shortest path problems. In: Proceedings of the Thirty-Second Annual ACM Symposium on Theory of Computing, pp. 286–295 (2000)

3. Aleksandrov, L., Maheshwari, A., Sack, J.-R.: Determining approximate shortest paths on weighted polyhedral surfaces. J. ACM **52**(1), 25–53 (2005)

4. Ammar, A., Bennaceur, H., Châari, I., Koubâa, A., Alajlan, M.: Relaxed Dijkstra and A^* with linear complexity for robot path planning problems in large-scale grid environments. Soft Comput. **20**(10), 4149–4171 (2015). https://doi.org/10.1007/s00500-015-1750-1

5. Bailey, J., Tovey, C., Uras, T., Koenig, S., Nash, A.: Path planning on grids: the effect of vertex placement on path length. In: Proceedings of the AAAI Conference on Artificial Intelligence and Interactive Digital Entertainment, vol. 11 (2015)

6. Bose, P., Esteban, G., Orden, D., Silveira, R.I.: On approximating shortest paths in weighted triangular tessellations. arXiv preprint arXiv:2111.13912 (2021)

7. Chang, H.C., Erickson, J., Xu, C.: Detecting weakly simple polygons. In: Proceedings of the Twenty-Sixth Annual ACM-SIAM Symposium on Discrete Algorithms, pp. 1655–1670. SIAM (2014)

8. Chen, D.Z., Klenk, K.S., Tu, H.: Shortest path queries among weighted obstacles in the rectilinear plane. SIAM J. Comput. **29**(4), 1223–1246 (2000)

9. Cheng, S., Jin, J., Vigneron, A.: Triangulation refinement and approximate shortest paths in weighted regions. In: Indyk, P. (ed.) Proceedings of the Twenty-Sixth Annual ACM-SIAM Symposium on Discrete Algorithms, SODA 2015, San Diego, CA, USA, 4–6 January 2015, pp. 1626–1640. SIAM (2015)

10. de Carufel, J.L., Grimm, C., Maheshwari, A., Owen, M., Smid, M.: A note on the unsolvability of the weighted region shortest path problem. Comput. Geom. **47**(7), 724–727 (2014)

11. Gewali, L., Meng, A.C., Mitchell, J.S.B., Ntafos, S.C.: Path planning in 0/1/∞ weighted regions with applications. INFORMS J. Comput. **2**(3), 253–272 (1990)

12. Jaklin, N.S.: On weighted regions and social crowds: autonomous-agent navigation in virtual worlds. Ph.D. thesis, Utrecht University (2016)

13. Mitchell, J.: Shortest paths among obstacles, zero-cost regions, and roads. Technical report, Cornell University Operations Research and Industrial Engineering (1987)

14. Mitchell, J., Papadimitrou, C.: The weighted region problem: finding shortest paths through a weighted planar subdivision. J. ACM **38**(1), 18–73 (1991)

15. Mitchell, J.S.B.: Shortest paths and networks. In: Goodman, J.E., O'Rourke, J., Toth, C.D. (eds.) Handbook of Discrete and Computational Geometry, 2nd edn., pp. 811–848. Chapman and Hall/CRC (2017)
16. Nagy, B.N.: Shortest paths in triangular grids with neighbourhood sequences. J. Comput. Inf. Technol. **11**(2), 111–122 (2003)
17. Nash, A.: Any-angle path planning. Ph.D. thesis, University of Southern California (2012)
18. Van Toll, W., et al.: A comparative study of navigation meshes. In: Proceedings of the 9th International Conference on Motion in Games, pp. 91–100 (2016)

Author Index

Printed in the United States
by Baker & Taylor Publisher Services